The Changing World Religion Map

Stanley D. Brunn
Editor

The Changing World Religion Map

Sacred Places, Identities, Practices and Politics

Volume II

Donna A. Gilbreath
Assistant Editor

Editor
Stanley D. Brunn
Department of Geography
University of Kentucky
Lexington, KY, USA

Assistant Editor
Donna A. Gilbreath
UK Markey Cancer Center
Research Communications Office
Lexington, KY, USA

ISBN 978-94-017-9375-9 ISBN 978-94-017-9376-6 (eBook)
DOI 10.1007/978-94-017-9376-6

Library of Congress Control Number: 2014960060

Springer Dordrecht Heidelberg New York London

Printed on acid-free paper

Springer Science+Business Media B.V. Dordrecht is part of Springer Science+Business Media (www.springer.com)

Preface: A Continuing Journey

Religion has always been a part of my life. I am a Presbyterian PK (preacher's kid). From my father I inherited not only an interest in the histories and geographies of religions, not just Christianity, but also a strong sense of social justice, a thread that has been part of my personal and professional (teaching, research, service) life. My mother was raised as a Quaker and from her I also learned much about social justice, peace and reconciliation and being a part of an effective voice calling for ends to war, social discrimination of various types, and other injustices that seem to be a continual part of daily life on the planet. My father had churches mostly in the rural Upper Middle West. These were open country and small town congregations in Illinois, Wisconsin, Minnesota, South Dakota, Nebraska, and Missouri. The members of these congregations were Germans, Czech, Scandinavians (Norwegians, and Swedes), and English. Perhaps or probably because of these experiences, I had friendships with many young people who comprised the mosaic of the rural Middle West. Our family moved frequently when I was living at home, primarily because my father's views on social issues were often not popular with the rural farming communities. (He lost his church in northwest Missouri in 1953 because he supported the Supreme Court's decision on desegregation of schools. By the time I graduated from high school in a small town in southeastern Illinois, I had attended schools in a half-dozen states; these include one-room school house experiences as well as those in small towns.

During my childhood days my interests in religion were, of course, important in the views I had about many subjects about those of different faiths and many places on the planet. I was born in a Catholic hospital, which I always attribute to the beginning of my ecumenical experiences. The schools I attended mixes of Catholics and Protestants; I had few experiences with Native Americans, Jews and African, and Asian Americans before entering college. But that background changed, as I will explain below. My father was always interested in missionaries and foreign missions and once I considered training for a missionary work. What fascinated me most about missionaries were that they were living in distant lands, places that I just longed to know about; an atlas was always my favorite childhood book, next to a dictionary. I was always glad when missionaries visited our churches and stayed in our homes. The fascination extended to my corresponding with missionaries in Africa, Asia, and Latin America. I was curious what kind of work they did. I also found them a source for stamps, a hobby that I have pursued since primary school. Also I collected the call letters of radio stations, some which were missionary stations, especially in Latin America. (Some of these radio stations are still broadcasting.)

When I enrolled as an undergraduate student at Eastern Illinois University, a small regional university in east central Illinois, I immediately requested roommates from different countries. I very much wanted to make friends with students from outside the United States and learn about their culture. During my 3 years at EIU, I had roommates from Jordan, Samoa, Costa Rica, Ethiopia, and South Korea; these were very formative years in helping me understand cross-cultural, and especially, religious diversity. On reflection, I think that most of the Sunday services I attended were mostly Presbyterian and Methodist, not Catholic, Lutheran, or Baptist. When I entered the University of Wisconsin, Madison, for the M.A. degree, I was again exposed to some different views about religion. The Madison church that fascinated me the most was the Unitarian church, a building designed by Frank Lloyd Wright. I remember how different the services and sermons were from Protestant churches, but intellectually I felt at home. My father was not exactly pleased I found the Unitarian church a good worship experience. The UW-Madison experience also introduced me to the study of geography and religion. This was brought home especially in conversations with my longtime and good friend, Dan Gade, but also a cultural geography course I audited with Fred Simoons, whose new book on religion and food prejudices just appeared and I found fascinating. Also I had conversations with John Alexander, who eventually left the department to continue in his own ministry with the Inter Varsity Christian Fellowship. A seminar on Cultural Plant Geography co-taught with Fred Simoons, Jonathan Sauer, and Clarence Olmstead provided some opportunities to explore cultural and historical dimensions of religion, which were the major fields where geographers could study religion. The geographers I knew who were writing about religion were Pierre Deffontaines, Eric Issac, and Xavier de Phanol. That narrow focus, has, of course, changed in the past several decades, as I will discuss below.

The move to Ohio State University for my doctoral work did not have the strong religious threads that had emerged before. I attended a variety of Protestant churches, especially Presbyterian, Congregational, and Methodist. I took no formal courses in geography that dealt with religion, although I was very interested when Wilbur Zelinsky's lengthy article on church membership patterns appeared in the *Annals of the Association of American Geographers* in 1961. I felt then that this was, and would be, a landmark study in American human geography, as the many maps of denominational membership patterns plus extensive references would form the basis for future scholars interested in religion questions, apart from historical and cultural foci which were the norm at that time. My first article on religion was on religious town names; I wrote it when I was at Ohio State with another longtime friend, Jim Wheeler, who had little interest in religion. I can still remember using my knowledge of biblical place names and going through a Rand McNally atlas with Jim identifying these town names. This study appeared in *Names*, which cultural geographers acknowledge is one of the premier journals concerned with names and naming processes. Even though my dissertation on changes in the central place functions in small towns in northwest Ohio and southeast Ohio (Appalachia) did not look specifically at churches, I did tabulate the number and variety during extensive fieldwork in both areas.

My first teaching job was at the University of Florida in fall 1966. I decided once I graduated from OSU that I wanted to live in a different part of the United States where I could learn about different regional cultures and politics. I was discouraged by some former teachers about teaching in Florida, especially about the region's segregation history, recent civil rights struggles in the South and also the John Birch Society (which was also active in Columbus when I lived there). The 3 years (1966–1969) in Gainesville were also very rewarding years. These were also very formative years in developing my interests in the social geography, a new field that was just beginning to be studied in the mid-1960s. Included in the forefront of this emerging field of social geographers were Anne Buttimer, Paul Claval, Yi-Fu Tuan, Dick Morrill, Richard Peet, Bill Bunge, Wilbur Zelinsky, David Harvey, and David Smith, all who were challenging geographers to study the social geographies of race, employment, school and housing discrimination, but also poverty, environmental injustice, inequities in federal and state programs promoting human welfare, the privileges of whiteness and the minorities' participation in the voting/political process. Living in northern Florida in the late 1960s or "Wallace years" could not help but alert one to the role that religion was playing in rural and urban areas in the South. Gainesville had distinct racial landscapes. I was definitely a "northerner" and carpetbagger who was an outcast in many ways in southern culture. One vivid memory is attending a University of Florida football game (a good example of regional pride and nationalism) and being about the only person seated while the band played "Dixie." I joined a Congregational/United Church of Christ church which was attended by a small number of "northern faculty" who were supportive of initiatives to end discriminatory practices at local, university, and broader levels. At this time I also was learning about the role of the Southern Baptist Church, a bastion of segregation that was very slow to accommodate to the wishes of those seeking ends to all kinds of overt and subtle discrimination (gender, race, class) practices. The term Bible Belt was also a label that rang true; it represented, as it still does, those who adhere to a literal interpretation of the Bible, a theological position I have never felt comfortable. I soon realized that if one really wanted to make a difference in the lives of those living with discrimination, poverty, and ending racial disenfranchisement in voting, religion was a good arena to express one's feelings and work with others on coordinated efforts. Published research that emerged from my Florida experiences included studies on poverty in the United States (with Jim Wheeler), the geographies of federal outlays to states, an open housing referendum in Flint, Michigan (with Wayne Hoffman), and school levies in cities that illustrated social inequities (with Wayne Hoffman and Gerald Romsa). My Florida years also provided me the first opportunity to travel in the developing world; that was made possible with a summer grant where I visited nearly 15 different Caribbean capitals where I witnessed housing, social, and infrastructure gaps. This experience provided my first experiences with the developing world and led to a Cities of the World class I taught at Michigan State University and also co-edited several editions with the same title of a book with Jack Williams.

The 11 years at Michigan State University did not result in any major research initiatives related to religion, although it did broaden my horizons about faiths other than Christianity. I began to learn about Islam, especially from graduate students in the department from Saudi Arabia, Libya, Kuwait, and Iran. Many of these I advised on religion topics about their own cultures, especially those dealing with pilgrimages and sacred sites. Probably the main gain from living in Michigan was support for and interest in an emerging secular society. The religious "flavors" of Michigan's religious landscape ran the full gamut from those who were very traditional and conservative to those who were globally ecumenical, interfaith, and even agnostic. I continued to be active in Presbyterian and United Churches of Christ, both which were intellectually and spiritually challenging places for adult classes and singing in a choir.

When I moved to the University of Kentucky in 1980, I knew that living in the Bluegrass State would be different from Michigan in at least two respects. One is that Kentucky was considered a moderate to progressive state with many strong traditional and conservative churches, especially the Southern Baptist denomination. Zelinsky's map accurately portrayed this region as having a dominance of conservative and evangelical Protestantism. Second, I realized that for anyone interested in advancing social issues related to race and gender equality or environmental quality (especially strip mining in eastern Kentucky), there would likely be some conflicts. I also understood before coming to Kentucky that alcoholic beverage consumption was a big issue in some countries; that fact was evident in an innovative regional map Fraser Hart prepared in a small book about the South. And then there was the issue about science and religion in school curricula. With this foreknowledge, I was looking forward to living in a region where the cross-currents of religion and politics meshed, not only experiencing some of these social issues or schisms firsthand, but also having an opportunity to study them, as I did.

I realized when I moved to Lexington, it was in many ways and still is a slowly progressing socially conscious city. Southern Baptist churches, Christian churches, and Churches of Christ were dominant in the landscape and in their influences on social issues. One could not purchase alcoholic beverages on Sunday in restaurants until a couple referenda were passed in the mid-1980s that permitted sales. I think 90 of the state's 120 counties were officially dry, although everyone living in a dry county knew where to purchase liquor. One could not see the then-controversial "The Last Temptation of Christ" movie when it appeared unless one would drive three hours to Dayton. "Get Right with God" signs were prominent along rural highways. The University of Kentucky chimes in Memorial Hall on campus played religious hymns until this practice stopped sometime in the middle of the decade; I am not exactly sure why. Public schools had prayers before athletic events; some still do. Teachers in some public schools could lose their jobs if they taught evolution. Creationism was (and still is) alive and well. I was informed by university advisors that the five most "dangerous" subjects to new UK students were biology, anthropology, astronomy, geology, and physical geography. Students not used to other than literal biblical interpretations were confused and confounded by evolutional science. Betting on horses was legal, even though gambling was frowned on by some religious leaders. Cock fighting and snake handling still existed (and still do) in pockets in rural eastern Kentucky. In many ways living in Kentucky was like living "on the dark side of gray." Lexington in many ways was and still is an island or outlier. Desegregation was a slow moving process in a city with a strong southern white traditional heritage. Athletic programs were also rather slow to integrate, especially UK basketball. In short, how could one not study religion in such an atmosphere. Living in Kentucky is sort of the antipode to living in agnostic-thriving New England and Pacific Northwest. I would expect that within 100 miles of Lexington one would discover one of the most diverse religious denominational and faith belief landscapes in the United States. There are the old regular mainline denominations, new faiths that have come into the Bluegrass and also many one-of-a-kind churches, especially in rural eastern, southern, and southeastern Kentucky.

I have undertaken a number of studies related to religion in Kentucky and the South in the past three decades. Some of these have been single-authored projects, others with students and faculty at UK and elsewhere. Some were presentations at professional meetings; some resulted in publications. The topics that fascinated me were ones that I learned from my geography colleagues and those in other disciplines that were understudied. These include the history and current patterns of wet/dry counties in Kentucky, a topic that appears in local and statewide media with communities deciding whether to approve the sale of alcoholic beverages. This study I conducted with historian Tom Appleton. With regularity there were clergy of some fundamentalist denominations who decried the sale of such drinks; opposing these clergy and their supporters were often those interested in promoting tourism and attracting out-of-state traffic on interstates. Also I looked into legislation that focused on science/education interfaces in the public schools and on the types of religious books (or avoidance of such, such as dealing with Marx, Darwin, and interfaith relations) in county libraries. Craig Campbell and I published an article in *Political Geography* on Cristo Redentor (Christ of the Andes statue) as an example of differential locational harmony. At the regional level I investigated with Esther Long the mission statements of seminaries in the South, a study that led to some interesting variations not only in their statements, but course offerings and visual materials on websites. I published with Holly Barcus two articles in *Great Plains Research* about denominational changes in the Great Plains. Missionaries have also been relatively neglected in geography, so I embarked on a study with Elizabeth Leppman that looked at the contents of a leading Quaker journal in the early part of the past century. Religions magazines, as we acknowledged in our study, were (and probably still are) a very important medium for educating the public about places and cultures, especially those where most Americans would have limited first-hand knowledge. The music/religion interface has long fascinated me, not only as a regular choir member, but for the words used to convey messages about spirituality, human welfare and justice, religious traditions and promises of peace and hope.

After 11 September 2001, I collected information from a number of churches in eastern Kentucky about how that somber event was celebrated and also what hymns they sung on the tenth anniversary. As expected, some were very somber and dignified, others had words about hope, healing, and reaching across traditional religious boundaries that separate us. I also co-authored an article (mostly photos) in *Focus* on the Shankill-Falls divided between Catholic and Protestant areas of Belfast with three students in my geography of religion class at the National University of Ireland in Maynooth. The visualization theme was integral to a paper published in *Geographica Slovenica* on ecumenical spaces and the web pages of the World Council of Churches and papers I delivered how cartoonists depicted the controversial construction of a mosque at Ground Zero. How cartoonists depicted God-Nature themes (the 2011 Haitian earthquake and Icelandic volcanic eruption) were the focus of an article in *Mitteilungen der Ősterreichsten Geographischen Gesellschaft*. I published in *Geographical Review* an article how the renaissance of religion in Russia is depicted on stamp issues since 1991. A major change in my thinking about the subject of religion in the South was the study that I worked on with Jerry Webster and Clark Archer, a study that appeared in the *Southeastern Geographer* in late 2011. We looked at the definition and concept of the Bible Belt as first discussed by Charles Heatwole (who was in my classes when I taught at Michigan State University) in 1978 in the *Journal of Geography*. We wanted to update his study and learn what has happened to the Bible Belt (or Belts) since this pioneering effort. What we learned using the Glenmary Research Center's county data on adherents for the past several decades was that the "buckle" has relocated. As our maps illustrated, the decline in those counties with denominations adhering to a literal interpretation of the Bible in western North Carolina and eastern Tennessee and a shift to the high concentration of Bible Belt counties in western Oklahoma and panhandle Texas. In this study using Glenmary data for 2000, we also looked at the demographic and political/voting characteristics of these counties. (In this volume we look at the same phenomenon using 2010 data and also discuss some of the visual features of the Bible Belt landscapes.)

What also was instrumental in my thinking about religion and geography interfaces were activities outside my own research agenda. As someone who has long standing interests in working with others at community levels on peace and justice issues, I worked with three other similarly committed adults in Lexington to organize the Central Kentucky Council for Peace and Justice. CKCPJ emerged in 1983 as an interfaith and interdenominational group committed to working on peace and justice issues within Lexington, in Central Kentucky especially, but also with national and global interests. The other three who were active in this initiative were Betsy Neale (from the Friends), Marylynne Flowers (active in a local Presbyterian church) and Ernie Yanarella (political scientist, Episcopalian, longtime friend, and also contributor to a very thoughtful essay on Weber in this volume). This organization is a key agent in peace/justice issues in the Bluegrass; it hosts meetings, fairs, conferences, and other events for people of all ages, plans annual marches on Martin Luther King Jr. holiday, and is an active voice on issues related to capital punishment, gun control, gay/lesbian issues, fair trade and employment, environmental responsibility and stewardship, and the rights of women, children, and minorities.

I also led adult classes at Maxwell Street Presbyterian Church where we discussed major theologians and religious writers, including William Spong, Marcus Borg, Joseph Campbell, Philip Jenkins, Diane Eck, Kathleen Norris, Diana Butler Bass, Francis Collins, Sam Harris, Paul Alan Laughlin, James Kugel, Dorothy Bass, and Garry Wills. We discuss issues about science, secularism, death and dying, interfaith dialogue, Christianity in the twenty-first century, images of God, missions and missionaries, and more. I also benefitted from attending church services in the many countries I have traveled, lived, and taught classes in the past three decades. These include services in elaborate, formal, and distinguished cathedrals in Europe, Russian Orthodox services in Central Asia, and services in a black township and white and interracial mainline churches in Cape Town. Often I would attend services where I understood nothing or little, but that did not diminish the opportunity to worship with youth and elders (many more) on Sunday mornings and listen to choirs sing in multiple languages. These personal experiences also became part of my religion pilgrimage.

While religion has been an important part of my personal life, it was less important as part of my teaching program. Teaching classes on the geography of religion are few and far between in the United States; I think the subject was accepted much more in the instructional and research arenas among geographers in Europe. I think that part of my reluctance to pursue a major book project on religion was that for a long time I considered the subject too narrowly focused, especially on cultural and historical geography. From my reading of the geography and religion literature, there were actually few studies done before 1970s. (See the bibliography at the end of Chap. 1). I took some renewed interest in the subject in the mid-1980s when a number of geographers began to examine religion/nature/environment issues. The pioneering works of Yi-Fu Tuan and Anne Buttimer were instrumental in steering the study of values, ethics, spirituality, and religion into some new and productive directions. These studies paved the way for a number of other studies by social geographers (a field that was not among the major fields until the 1970s and early 1980s). The steady stream of studies on geography and religion continued with the emergence of GORABS (Geography of Religion and Belief Systems) as a Specialty Group of the Association of American Geographers. The publication of more articles and special journal issues devoted to the geography of religion continued into the last decade of the twentieth century and first decade of this century. The synthetic works of Lily Kong that have regularly appeared in *Progress in Human Geography* further supported those who wanted to look at religion from human/environmental perspectives. These reviews not only introduced the study of religion within geography, but also to those in related scholarly disciplines.

As more and more research appeared in professional journals and more conferences included presentations on religion from different fields and subfields, it became increasing apparent that the time was propitious for a volume that looked at religion/geography interfaces from a number of different perspectives. From my own vantage point, the study of religion was one that could, should, might, and would benefit from those who have theoretical and conceptual training in many of the discipline's major subfields. The same applied to those who were regional specialists; there were topics meriting study from those who looked a political/religion issues in Southeast Asia or Central America as well as cultural/historical themes in southern Africa and continental Europe and symbolic/architectural features and built environments of religions landscapes in California, southeast Australia, and southwest Asia. Studying religions topics would not have to be limited to those in human geography, but could be seen as opportunities for those studying religion/natural disaster issues in East Asia and southeast United States as well as the spiritual roots of early and contemporary religious thinking in Central Asia, East Asia, Russia, and indigenous groups in South America. For those engaged in the study of gender, law, multicultural education, and media disciplines, there were also opportunities to contribute to the study of this emerging field. In short, there were literally "gold mines" of potential research topics in rural and urban areas everywhere on the continent.

About 7 years ago I decided to offer a class on the geography of religion in the Department of Geography at the University of Kentucky. The numbers were never larger (less than 15), but these were always enlightening and interesting, because of the views expressed by students. Their views about religion ran the gamut from very conservative to very liberal and also agnostic and atheist, which made, as one would expect, some very interesting exchanges. Students were strongly encouraged (not required, as I could not do this in a public university) to attend a half dozen different worship services during the semester. This did not mean attending First Baptist, Second Baptist, Third Baptist, etc., but different kinds of experiences. For some this course component was the first some had ever attended a Jewish synagogue, Catholic mass, Baptist service, an African American church, Unitarian church, or visited a mosque. Some students used this opportunity to attend Wiccan services, or visit a Buddhist and Hindu temple. Their write-ups about these experiences and the ensuing discussion were one of the high spots of the weekly class. In addition, the classes discussed chapters in various books and articles from the geography literature about the state of studying religion. And we always discussed current news items, using materials from the RNS (Religion News Service) website.

Another ingredient that stimulated my decision to edit a book on the geography of religion emerged from geography of religion conferences held in Europe in recent years. These were organized by my good friends Ceri Peach (Oxford), Reinhard Henkel (Heidelberg), and also Martin Baumann and Andreas Tunger-Zanetti (University of Lucerne). These miniconferences, held in Oxford, Lucerne, and Gottingen, usually attracted 20–40 junior and senior geographers and other religious scholars, and were a rich source of ideas for topics that might be studied. The opportunities for small group discussions, the field trips, and special events were conducive to learning about historical and contemporary changes in the religious landscapes of the European continent and beyond. A number of authors contributing to this volume presented papers at one or more of these conferences. Additional names came from those attending sessions at annual meetings of the Association of American Geographers.

Some of my initial thoughts and inspiration about a book came from the course I taught, conversations with friends who studied and did not study religion, and also the book I edited on megaengineering projects. This three-volume, 126-chapter book, *Engineering Earth: The Impacts of Megaengineering Projects*, was published in 2011 by Springer. There were only a few chapters in this book that had a religious content, one on megachurches, another on liberation theologians fighting megadevelopment projects in the Philippines and Guatemala. When I approached Evelien Bakker and Bernadette Deelen-Mans, my first geography editors at Springer, about a religion book, they were excited and supportive, as they have been since day one. They gave me the encouragement, certainly the latitude (and probably the longitude) to pursue the idea, knowing that I would identify significant cutting-edge topics about religion and culture and society in all major world regions. The prospectus I developed was for an innovative book that would include the contributions of scholars from the social sciences and humanities, those from different counties and those from different faiths. For their confidence and support, I am very grateful. The reviews they obtained of the prospectus were encouraging and acknowledging that there was a definite need for a major international, interdisciplinary, and interfaith volume. Springer also saw this book as an opportunity to emphasize its new directions in the social sciences and humanities. I also want to thank Stefan Einarson who came on board late in the project and shepherded the project to its completion with the usual Springer traits of professionalism, kindness, and commitment to the project's publication. And I wish to thank Chitra Sundarajan and her staff for helpful professionalism in preparing the final manuscript for publication.

The organization of the book, which is discussed in Chapter One, basically reflects the way I look at religion from a geographical perspective. I look at the subject as more than simply investigations into human geography's fields and subfields, including cultural and historical, but also economic, social, and political geography, but also human/environmental geography (dealing with human values, ethics, behavior, disasters, etc.). I also look at the study of religious topics and phenomena with respects to major concepts we use in geography; these include landscapes, networks, hierarchies, scales, regions, organization of space, the delivery of services, and virtual religion. I started contacting potential authors in September 2010. Since then I have sent or received over 15,000 emails related the volume.

I am deeply indebted to many friends for providing names of potential authors. I relied on my global network of geography colleagues in colleges and universities around the world, who not only recommended specific individuals, but also topics they deemed worthy of inclusion. Some were geographers, but many were not; some taught in universities, others in divinity schools and departments of religion around the world. Those I specifically want to acknowledge include: Barbara Ambrose, Martin Checa Artasu, Martin Baumann, John Benson, Gary Bouma, John Benson, Dwight Billings, Marion Bowman, John D. Brewer, David Brunn, David Butler, Ron Byars, Heidi Campbell, Caroline Creamer, Janel Curry, David Eicher, Elizabeth Ferris, Richard Gale, Don Gross, Wayne Gnatuk, Martin Haigh, Dan Hofrenning, Wil Holden, Hannah Holtschneider, Monica Ingalls, Nicole Karapanagiotis, Aharon Kellerman, Judith Kenny, Jean Kilde, Ted Levin, James

Munder, Alec Murphy, Tad Mutersbuagh, Garth Myers, Lionel Obidah, Sam Otterstrom, Francis Owusu, Maria Paradiso, Ron Pen, Ivan Petrella, Adam Possamai, Leonard Primiano, Craig Revels, Heinz Scheifinger, Anna Secor, Ira Sheskin, Doug Slaymaker, Patricia Solis, Anita Stasulne, Jill Stevenson, Robert Strauss, Tristan Sturm, Greg Stump, Karen Till, Andreas Tunger-Zenetti, Gary Vachicouras, Viera Vlčkova, Herman van der Wusten, Stanley Waterman, Mike Whine, Don Zeigler, Shangyi Zhou, and Matt Zook.

And I want to thank John Kostelnick who provided the GORABS Working Bibliography; most of the entries, except dissertations and theses, are included in Chap. 1 bibliography. Others who helped him prepare this valuable bibliography also need to be acknowledged: John Bauer, Ed Davis, Michael Ferber, Julian Holloway, Lily Kong, Elizabeth Leppman, Carolyn Prorock, Simon Potter, Thomas Rumney, Rana P.B. Singh, and Robert Stoddard. These are scholars who devoted their lifetimes to advancing research on geography and religion.

Finally I want to thank Donna Gilbreath for another splendid effort preparing all the chapters for Springer. She formatted the chapters and prepared all the tables and illustrations per the publisher's guidelines. Donna is an invaluable and skilled professional who deserves much credit for working with multiple authors and the publisher to ensure that all text materials were correct and in order. Also I am indebted to her husband, Richard Gilbreath, for helping prepare some of the maps and graphics for authors without cartographic services and making changes on others. As Director of the Gyula Pauer Center for Cartography and GIS, Dick's work is always first class. And, finally, thanks are much in order to Natalya Tyutenkova for her interest, support, patience, and endurance in the past several years working on this megaproject, thinking and believing it would never end.

The journey continues.

February 2014 Stanley D. Brunn

Contents of Volume II

Contributors

Jamaine Abidogun Department of History, Missouri State University, Springfield, MO, USA

Afe Adogame School of Divinity, University of Edinburgh, Edinburgh, Scotland, UK

Christopher A. Airriess Department of Geography, Ball State University, Muncie, IN, USA

Kaarina Aitamurto Aleksanteri Institute, University of Helsinki, Helsinki, Finland

Mikael Aktor Institute of History, Study of Religions, University of Southern Denmark, Odense, Denmark

Elizabeth Allison Department of Philosophy and Religion, California Institute of Integral Studies, San Francisco, CA, USA

Johan Andersson Department of Geography, King's College London, London, UK

Stephen W. Angell Earlham College, School of Religion, Richmond, IN, USA

J. Clark Archer Department of Geography, School of Natural Resources, University of Nebraska-Lincoln, Lincoln, NE, USA

Ian Astley Asian Studies, University of Edinburgh, Edinburgh, UK

Steven M. Avella Professor of History, Marquette University, Milwaukee, WI, USA

Yulier Avello COPEXTEL S.A., Ministry of Informatics and Communications, Havana, Cuba

Erica Baffelli School of Arts, Languages and Cultures, University of Manchester, Manchester, UK

Bakama BakamaNume Division of Social Work, Behavioral and Political Science, Prairie View A&M University, Prairie View, TX, USA

Josiah R. Baker Methodist University, Fayetteville, NC, USA

Economics and Geography, Methodist University, Fayetteville, USA

Holly R. Barcus Department of Geography, Macalester College, St. Paul, MN, USA

David Bassens Department of Geography, Free University Brussels, Brussels, Belgium

Ramon Bauer Wittgenstein Centre for Demography and Global Human Capital (IIASA, VID/ŐAW, WU), Vienna Institute of Demography/Austrian Academy of Sciences, Vienna, Austria

Whitney A. Bauman Department of Religious Studies, Florida International University, Miami, FL, USA

Gwilym Beckerlegge Department of Religious Studies, The Open University, Milton Keynes, UK

Michael Bégin Department of Global Studies, Pusan National University, Pusan, Republic of Korea

Demyan Belyaev Collegium de Lyon/Institute of Advanced Studies, Lyon, France

Alexandre Benod Research Division, Department of Japanese Studies, Université de Lyon, Lyon, France

John Benson School of Teaching and Learning, Minnesota State University, Moorhead, MN, USA

Sigurd Bergmann Department of Philosophy and Religious Studies, Norwegian University of Science and Technology, Trondheim, Norway

Rachel Berndtson Department of Geographical Sciences, University of Maryland, College Park, MD, USA

Martha Bettis Gee Compassion, Peace and Justice, Peace and Justice Ministries, Presbyterian Mission Agency, Presbyterian Church (USA), Louisville, KY, USA

Warren Bird Research Division, Leadership Network, Dallas, TX, USA

Andrew Boulton Department of Geography, University of Kentucky, Lexington, KY, USA

Humana, Inc., Louisville, KY, USA

Kathleen Braden Department of Political Science and Geography, Seattle Pacific University, Seattle, WA, USA

Namara Brede Department of Geography, Macalester College, St. Paul, MN, USA

John D. Brewer Institute for the Study of Conflict Transformation and Social Justice, Queen's University Belfast, Belfast, UK

Laurie Brinklow School of Geography and Environmental Studies, University of Tasmania, Hobart, Australia

Interim Co-ordinator, Master of Arts in Island Studies Program, University of Prince Edward Island, Charlottetown, PE Canada

Dave Brunn Language and Linguistics Department, New Tribes Missionary Training Center, Camdenton, MO, USA

Stanley D. Brunn Department of Geography, University of Kentucky, Lexington, KY, USA

David J. Butler Department of Geography, University of Ireland, Cork, Ireland

Anne Buttimer Department of Geography, University College Dublin, Dublin, Ireland

Éric Caron Malenfant Demography Division, Statistics Canada, Ottawa, Canada

Lori Carter-Edwards Gillings School of Global Public Health, Public Health Leadership Program, University of North Carolina, Chapel Hill, NC, USA

Clemens Cavallin Department of Literature, History of Ideas and Religion, University of Gothenburg, Göteborg, Sweden

Martin M. Checa-Artasu Department of Sociology, Universidad Autónoma Metropolitana, Unidad Iztapalapa, Mexico, DF, Mexico

Richard Cimino Department of Anthropology and Sociology, University of Richmond, Richmond, VA, USA

Paul Claval Department of Geography, University of Paris-Sorbonne, Paris, France

Paul Cloke Department of Geography, Exeter University, Exeter, UK

Kevin Coe Department of Communication, University of Utah, Salt Lake City, UT, USA

Noga Collins-Kreiner Department of Geography and Environmental Studies, Centre for Tourism, Pilgrimage and Recreation, University of Haifa, Haifa, Israel

Louise Connelly Institute for Academic Development, University of Edinburgh, Edinburgh, Scotland, UK

Thia Cooper Department of Religion, Gustavus Adolphus College, St. Peter, MN, USA

Catherine Cottrell Department of Geography and Earth Sciences, Aberystwyth University, Aberystwyth, UK

Thomas W. Crawford Department of Geography, East Carolina University, Greenville, NC, USA

Janel Curry Provost, Gordon College, Wenham, MA, USA

Seif Da'Na Sociology and Anthropology Department, University of Wisconsin-Parkside, Kenosha, WI, USA

Erik Davis Department of Religious Studies, Rice University, Houston, TX, USA

Jenny L. Davis Department of American Indian Studies, University of Illinois, Urbana-Champaign, Urbana, USA

Kiku Day Department of Ethnomusicology, Aarhus University, Aarhus, Denmark

Renée de la Torre Castellanos Centro de Investigaciones y Estudios Superiores en Antropologia Social-Occidente, Guadalajara, Jalisco, Mexico

Frédéric Dejean Institut de recherche sur l'intégration professionnelle des immigrants, Collège de Maisonneuve, Montréal (Québec), Canada

Veronica della Dora Department of Geography, Royal Holloway University of London, UK

Sergio DellaPergola The Avraham Harman Institute of Contemporary Jewry, The Hebrew University of Jerusalem, Mt. Scopus, Jerusalem, Israel

Antoinette E. DeNapoli Religious Studies Department, University of Wyoming, Laramie, WY, USA

Matthew A. Derrick Department of Geography, Humboldt State University, Arcata, CA, USA

C. Nathan DeWall Department of Psychology, University of Kentucky, Lexington, KY, USA

Jualynne Dodson Department of Sociology, American and African Studies Program, Michigan State University, East Lansing, MI, USA

David Domke Department of Communication, University of Washington, Seattle, WA, USA

Katherine Donohue M.A. Diplomacy and International Commerce, Patterson School of Diplomacy and International Commerce, University of Kentucky, Lexington, KY, USA

Lizanne Dowds Northern Ireland Life and Times Survey, University of Ulster, Belfast, UK

Kevin M. Dunn School of Social Sciences and Psychology, University of Western Sydney, Penrith, NSW, Australia

Claire Dwyer Department of Geography, University College London, London, UK

Patricia Ehrkamp Department of Geography, University of Kentucky, Lexington, KY, USA

Paul Emerson Teusner School of Media and Communication, RMIT University, Melbourne, VIC, Australia

Chad F. Emmett Department of Geography, Brigham Young University, Provo, UT, USA

Ghazi-Walid Falah Department of Public Administration and Urban Studies, University of Akron, Akron, OH, USA

Yasser Farrés Department of Philosophy, University of Zaragoza, Pedro Cerbuna, Zaragoza, Spain

Timothy Joseph Fargo Department of City Planning, City of Los Angeles, Los Angeles, CA, USA

Michael P. Ferber Department of Geography, The King's University College, Edmonton, AB, Canada

Tatiana V. Filosofova Department of World Languages, Literatures, and Cultures, University of North Texas, Denton, TX, USA

John T. Fitzgerald Department of Theology, University of Notre Dame, Notre Dame, IN, USA

Colin Flint Department of Political Science, Utah State University, Logan, UT, USA

Daniel W. Gade Department of Geography, University of Vermont, Burlington, VT, USA

Armando Garcia Chiang Department of Sociology, Universidad Autónoma Metropolitana Iztapalapa, Iztapalapa, Mexico

Jeff Garmany King's Brazil Institute, King's College London, London, UK

Martha Geores Department of Geographical Sciences, University of Maryland, College Park, MD, USA

Hannes Gerhardt Department of Geosciences, University of West Georgia, Carrolton, GA, USA

Christina Ghanbarpour History Department, Saddleback College, Mission Viejo, CA, USA

Danilo Giambra Department of Theology and Religion, University of Otago-Te Whare Wänanga o Otägo, Dunedin, New Zealand/Aotearoa

Banu Gökarıksel Department of Geography, University of North Carolina, Chapel Hill, NC, USA

Margaret M. Gold London Guildhall Faculty of Business and Law, London Metropolitan University, London, UK

Anton Gosar Faculty of Tourism Studies, University of Primorska, Portorož, Slovenia

Anne Goujon Wittgenstein Centre for Demography and Global Human Capital (IIASA, VID/ŐAW, WU), International Institute for Applied Systems Analysis (IIASA), Laxenburg, Austria

Vienna Institute of Demography/Austrian Academy of Sciences, Vienna, Austria

Alyson L. Greiner Department of Geography, Oklahoma State University, Stillwater, OK, USA

Daniel Jay Grimminger Faith Lutheran Church, Kent State University, Millersburg, OH, USA

School of Music, Kent State University, Kent, OH, USA

Zeynep B. Gürtin Department of Sociology, University of Cambridge, Cambridge, UK

Cristina Gutiérrez Zúñiga Centro Universitario de Ciencias Sociales y Humanidades, El Colegio de Jalisco, Zapopan, Jalisco, Mexico

Martin J. Haigh Department of Social Sciences, Oxford Brookes University, Oxford, UK

Anna Halafoff Centre for Citizenship and Globalisation, Deakin University, Burwood, VIC, Australia

Airen Hall Department of Theology, Georgetown University, Washington, DC, USA

Randolph Haluza-DeLay Department of Sociology, The Kings University, Edmonton, AB, Canada

Tomáš Havlíček Faculty of Science, Department of Social Geography and Regional Development, Charles University, Prague 2, Czechia

C. Michael Hawn Sacred Music Program, Perkins School of Theology, Southern Methodist University, Dallas, TX, USA

Bernadette C. Hayes Department of Sociology, University of Aberdeen, Aberdeen, Scotland, UK

Peter J. Hemming School of Social Sciences, Cardiff University, Cardiff, Wales, UK

William Holden Department of Geography, University of Calgary, Calgary, AB, Canada

Edward C. Holland Havighurst Center for Russian and Post-Soviet Studies, Miami University, Oxford, OH, USA

Beverly A. Howard School of Music, California Baptist University, Riverside, CA, USA

Martina Hupková Faculty of Science, Department of Social Geography and Regional Development, Charles University, Prague 2, Czechia

Tim Hutchings Post Doc, St. John's College, Durham University, Durham, UK

Ronald Inglehart Institute of Social Research, University of Michigan, Ann Arbor, MI, USA

World Values Survey Association, Madrid, Spain

Marcia C. Inhorn Anthropology and International Affairs, Yale University, New Haven, CT, USA

Adrian Ivakhiv Environmental Program, University of Vermont, Burlington, VT, USA

Maria Cristina Ivaldi Dipartimento di Scienze Politiche "Jean Monnet", Seconda Università degli Studi di Napoli, Caserta, Italy

Thomas Jablonsky Professor of History, Marquette University, Milwaukee, WI, USA

Maria Jaschok International Gender Studies Centre, Lady Margaret Hall, Oxford University, Norham Gardens, UK

Philip Jenkins Institute for the Study of Religion, Baylor University, Waco, TX, USA

Wesley Jetton Student, University of Kentucky, Lexington, KY, USA

Shui Jingjun Henan Academy of Social Sciences, Zhengzhou, Henan Province, China

Mark D. Johns Department of Communication, Luther College, Decorah, IA, USA

James H. Johnson Jr. Kenan-Flagler Business School and Urban Investment Strategies Center, University of North Carolina, Chapel Hill, NC, USA

Lucas F. Johnston Department of Religion and Environmental Studies, Wake Forest University, Winston-Salem, NC, USA

Peter Jordan Austrian Academy of Sciences, Institute of Urban and Regional Research, Wien, Austria

Yakubu Joseph Geographisches Institut, University of Tübingen, Tübingen, Germany

Deborah Justice Yale Institute of Sacred Music, Yale University, New Haven, CT, USA

Akel Ismail Kahera College of Architecture, Art and Humanities, Clemson University, Clemson, SC, USA

P.P. Karan Department of Geography, University of Kentucky, Lexington, KY, USA

Sya Buryn Kedzior Department of Geography and Environmental Planning, Towson State University, Towson, MD, USA

Kevin D. Kehrberg Department of Music, Warren Wilson College, Asheville, NC, USA

Laura J. Khoury Department of Sociology, Birzeit University, West Bank, Palestine

Hans Knippenberg Department of Geography, Planning and International Development Studies, University of Amsterdam, Velserbroek, The Netherlands

Katherine Knutson Department of Political Science, Gustavus Adolphus College, St. Peter, MN, USA

Miha Koderman Science and Research Centre of Koper, University of Primorska, Koper-Capodistria, Slovenia

Lily Kong Department of Geography, National University of Singapore, Singapore, Singapore

Igor Kotin Museum of Anthropology and Ethnography, Russian Academy of Sciences, St. Petersburg, Russia

Katharina Kunter Faculty of Theology, University of Bochum, Bochum, Germany

Lisa La George International Studies, The Master's College, Santa Clarita, CA, USA

Shirley Lal Wijesinghe Faculty of Humanities, University of Kelaniya, Kelaniya, Sri Lanka

Ibrahim Badamasi Lambu Department of Geography, Faculty of Earth and Environmental Sciences, Bayero University Kano, Kano, Nigeria

Michelle Gezentsvey Lamy Comparative Education Research Unit, Ministry of Education, Wellington, New Zealand

Justin Lawson Health, Nature and Sustainability Research Group, School of Health and Social Development, Deakin University, Burwood, VIC, Australia

Deborah Lee Department of Geography, National University of Singapore, Singapore, Singapore

Karsten Lehmann Senior Lecturer, Science des Religions, Bayreuth University, Fribourg, Switzerland

Reina Lewis London College of Fashion, University of the Arts, London, UK

Micah Liben Judaic Studies, Kellman Brown Academy, Voorhees, NJ, USA

Edmund B. Lingan Department of Theater, University of Toledo, Toledo, OH, USA

Rubén C. Lois-González Departamento de Xeografía, Universidade de Santiago de Compostela, Galiza, Spain

Naomi Ludeman Smith Learning and Women's Initiatives, St. Paul, MN, USA

Katrín Anna Lund Department of Geography and Tourism, Faculty of Life and Environmental Sciences, University of Iceland, Reykjavik, Iceland

Avril Maddrell Department of Geography and Environmental Sciences, University of West England, Bristol, UK

Juraj Majo Department of Human Geography and Demography, Faculty of Sciences, Comenius University in Bratislava, Bratislava, Slovak Republic

Virginie Mamadouh Department of Geography, Planning and International Development Studies, University of Amsterdam, Amsterdam, The Netherlands

Mariana Mastagar Department of Theology, Trinity College, University of Toronto, Toronto, Canada

Alberto Matarán Department of Urban and Spatial Planning, University of Granada, Granada, Spain

René Matlovič Department of Geography and Applied Geoinformatics, Faculty of Humanities and Natural Sciences, University of Prešov, Prešov, Slovakia

Kvetoslava Matlovičová Department of Geography and Applied Geoinformatics, Faculty of Humanities and Natural Sciences, University of Prešov, Prešov, Slovakia

Hannah Mayne Department of Anthropology, University of Florida, Gainesville, FL, USA

Shampa Mazumdar Department of Sociology, University of California, Irvine, CA, USA

Sanjoy Mazumdar Department of Planning, Policy and Design, University of California, Irvine, CA, USA

Andrew M. McCoy Center for Ministry Studies, Hope College, Holland, MI, USA

Daniel McGowin Department of Geology and Geography, Auburn University, Auburn, AL, USA

James F. McGrath Department of Philosophy and Religion, Butler University, Indianapolis, IN, USA

Nick Megoran Department of Geography, University of Newcastle-upon-Tyne, Newcastle, UK

Amy Messer Department of Sociology, University of Kentucky, Lexington, KY, USA

Sarah Ann Deardorff Miller Researcher, Refugee Studies Centre, Oxford, UK

Kelly Miller Centre for Integrative Ecology, School of Life and Environmental Sciences, Deakin University, Burwood, VIC, Australia

Nathan A. Mosurinjohn Center for Social Research, Calvin College, Grand Rapids, MI, USA

Sven Müller Institute for Transport Economics, University of Hamburg, Hamburg, Germany

Erik Munder Institut für Vergleichende Kulturforschung - Kultur- u. Sozialanthropologie und Religionswissenschaft, Universität Marburg, Marburg, Germany

David W. Music School of Music, Baylor University, Waco, TX, USA

Kathleen Nadeau Department of Anthropology, California State University, San Bernadino, CA, USA

Caroline Nagel Department of Geography, University of South Carolina, Columbia, SC, USA

Pippa Norris John F. Kennedy School of Government, Harvard University, Cambridge, MA, USA

Government and International Relations, University of Sydney, Sydney, Australia

Orville Nyblade Makumira University College, Usa River, Tanzania

Lionel Obadia Department of Anthropology, Université de Lyon, Lyon, France

Daniel H. Olsen Department of Geography, Brigham Young University, Provo, UT, USA

Samuel M. Otterstrom Department of Geography, Brigham Young University, Provo, UT, USA

Barbara Palmquist Department of Geography, University of Kentucky, Lexington, KY, USA

Grigorios D. Papathomas Faculty of Theology, University of Athens, Athens, Greece

Nikos Pappas Musicology, University of Alabama, Tuscaloosa, AL, USA

Mohammad Aslam Parvaiz Islamic Foundation for Science and Environment (IFSE), New Delhi, India

Valerià Paül Departamento de Xeografía, Universidade de Santiago de Compostela, Galiza, Spain

Miguel Pazos-Otón Departamento de Xeografía, Universidade de Santiago de Compostela, Galiza, Spain

David Pereyra Toronto School of Theology, University of Toronto, Toronto, Canada

Bruce Phillips Loucheim School of Judaic Studies at the University of Southern California, Hebrew Union College-Jewish Institute of Religion, Los Angeles, CA, USA

Awais Piracha School of Social Sciences and Psychology, University of Western Sydney, Penrith, NSW, Australia

Linda Pittman Department of Geography, Richard Bland College of the College of William and Mary, Petersburg, VA, USA

Richard S. Pond Department of Psychology, University of North Carolina, Wilmington, NC, USA

Carolyn V. Prorok Independent Scholar, Slippery Rock, PA, USA

Steven M. Radil Department of Geography, University of Idaho, Moscow, ID, USA

Esther Long Ratajeski Independent Scholar, Lexington, KY, USA

Daniel Reeves Faculty of Science, Department of Social Geography and Regional Development, Charles University, Prague 2, Czechia

Arthur Remillard Department of Religious Studies, St. Francis University, Loretto, PA, USA

Claire M. Renzetti Department of Sociology, University of Kentucky, Lexington, KY, USA

Friedlind Riedel Department of Musicology, Georg-August-University of Göttingen, Göttingen, Germany

Sandra Milena Rios Oyola Department of Sociology and the Compromise after Conflict Research Programme, University of Aberdeen, Aberdeen, Scotland, UK

C.K. Robertson Presiding Bishop, The Episcopal Church, New York, NY, USA

Arsenio Rodrigues School of Architecture, Prairie View A&M University, Prairie View, TX, USA

Andrea Rota Institute for the Study of Religion, University of Bern, Bern, Switzerland

Rainer Rothfuss Geographisches Institut, University of Tübingen, Tübingen, Germany

Jeanmarie Rouhier-Willoughby Department of Modern and Classical Languages, Literatures and Cultures, University of Kentucky, Lexington, KY, USA

Rex J. Rowley Department of Geography-Geology, Illinois State University, Normal, IL, USA

Bradley C. Rundquist Department of Geography, University of North Dakota, Grand Forks, ND, USA

Simon Runkel Department of Geography, University of Bonn, Bonn, Germany

Joanna Sadgrove Research Staff, United Society, London, UK

Michael Samers Department of Geography, University of Kentucky, Lexington, KY, USA

Åke Sander Department of Literature, History of Ideas and Religion, University of Gothenburg, Göteborg, Sweden

Xosé M. Santos Departamento de Xeografía, Universidade de Santiago de Compostela, Galiza, Spain

Alessandro Scafi Medieval and Renaissance Cultural History, The Warburg Institute, University of London, London, UK

Anthony Schmidt Department of Communication Studies, Edmonds Community College, Edmonds, WA, USA

Mallory Schneuwly Purdie Institut de sciences sociales des religions contemporaines, Observatoire des religions en Suisse, Université de Lausanne – Anthropole, Lausanne, Switzerland

Anna J. Secor Department of Geography, University of Kentucky, Lexington, KY, USA

Hafid Setiadi Department of Geography, University of Indonesia, Depok, West Java, Indonesia

Fred M. Shelley Department of Geography and Environmental Sustainability, University of Oklahoma, Norman, OK, USA

Ira M. Sheskin Department of Geography and Regional Studies, University of Miami, Coral Gables, FL, USA

Lia Dong Shimada Conflict Mediator, Methodist Church in Britain, London, UK

Caleb Kwang-Eun Shin ABD, Korea Baptist Theological Seminary, Daejeon, Republic of Korea

J. Matthew Shumway Department of Geography, Brigham Young University, Provo, UT, USA

Dmitrii Sidorov Department of Geography, California State University, Long Beach, Long Beach, CA, USA

Caleb Simmons Religious Studies Program, University of Arizona, Tucson, AZ, USA

Devinder Singh Department of Geography, University of Jammu, Jammu, Jammu and Kashmir, India

Rana P.B. Singh Department of Geography, Banaras Hindu University, Varanasi, UP, India

Nkosinathi Sithole Department of English, University of Zululand, KwaZulu-Natal, South Africa

Vegard Skirbekk Wittgenstein Centre for Demography and Global Human Capital (IIASA, VID/ŐAW, WU), International Institute for Applied Systems Analysis, Laxenburg, Austria

Alexander Thomas T. Smith Department of Sociology, University of Warwick, Coventry, UK

Christopher Smith Independent Scholar, Tecumseh, OK, USA

Ryan D. Smith Compassion, Peace and Justice Ministries, Presbyterian Ministry at the U.N., Presbyterian Mission Agency, Presbyterian Church (USA), New York, NY, USA

Sara Smith Department of Geography, University of North Carolina, Chapel Hill, NC, USA

Leslie E. Sponsel Department of Anthropology, University of Hawaii, Honolulu, HI, USA

Chloë Starr Asian Christianity and Theology, Yale Divinity School, New Haven, CT, USA

Jeffrey Steller Public History, Northern Kentucky University, Highland Heights, KY, USA

Christopher Stephens Southlands College, University of Roehampton, London, UK

Jill Stevenson Department of Theater Arts, Marymount Manhattan College, New York, NY, USA

Anna Rose Stewart Department of Religious Studies, University of Kent, Canterbury, UK

Nancy Palmer Stockwell Senior Contract Administrator, Enerfin Resources, Houston, TX, USA

Robert Strauss President and CEO, Worldview Resource Group, Colorado Springs, CO, USA

Tristan Sturm School of Geography, Archaeology and Palaeoecology, Queen's University Belfast, Belfast, UK

Edward Swenson Department of Anthropology, University of Toronto, Toronto, ON, Canada

Anna Swynford Duke Divinity School, Duke University, Durham, NC, USA

Jonathan Taylor Department of Geography, California State University, Fullerton, CA, USA

Francis Teeney Institute for the Study of Conflict Transformation and Social Justice, Queen's University Belfast, Belfast, UK

Mary C. Tehan Stirling College, University of Divinity, Melbourne, Australia

Andrew R.H. Thompson The School of Theology, The University of the South, Sewanee, TN, USA

Scott L. Thumma Professor, Department of Sociology, Hartford Seminary, Hartford, CT, USA

Meagan Todd Department of Geography, University of Colorado, Boulder, CO, USA

Soraya Tremayne Fertility and Reproduction Studies Group, Institute of Social and Cultural Anthropology, University of Oxford, Oxford, UK

Gill Valentine Faculty of Social Sciences, University of Sheffield, Sheffield, UK

Inge van der Welle Department of Geography, Planning and International Development Studies, University of Amsterdam, Amsterdam, The Netherlands

Herman van der Wusten Department of Geography, Planning and International Development Studies, University of Amsterdam, Amsterdam, The Netherlands

Robert M. Vanderbeck Department of Geography, University of Leeds, West Yorkshire, UK

Jason E. VanHorn Department of Geography, Calvin College, Grand Rapids, MI, USA

Viera Vlčková Department of Public Administration and Regional Development, Faculty of National Economy, University of Economics in Bratislava, Bratislava, Slovakia

Geoffrey Wall Department of Geography and Environmental Management, University of Waterloo, Waterloo, ON, Canada

Robert H. Wall Counsel, Spilman Thomas & Battle, Winston-Salem, NC, USA

Kevin Ward School of Theology and Religious Studies, University of Leeds, West Yorkshire, UK

Barney Warf Department of Geography, University of Kansas, Lawrence, KS, USA

Stanley Waterman Department of Geography and Environmental Studies, University of Haifa, Haifa, Israel

Robert H. Watrel Department of Geography, South Dakota State University, Brookings, SD, USA

Gerald R. Webster Department of Geography, University of Wyoming, Laramie, WY, USA

Paul G. Weller Research, Innovation and Academic Enterprise, University of Derby, Derby, UK

Oxford Centre for Christianity and Culture, University of Oxford, Oxford, UK

Cynthia Werner Department of Anthropology, Texas A&M University, College Station, TX, USA

Geoff Wescott Centre for Integrative Ecology, School of Life and Environmental Sciences, Deakin University, Burwood, VIC, Australia

Carroll West Center for Historic Preservation, Middle Tennessee State University, Murfreesboro, TN, USA

Gerald West School of Religion, Philosophy and Classics, University of KwaZulu-Natal, Scottsville, South Africa

Mark Whitaker Department of Anthropology, University of Kentucky, Lexington, KY, USA

Thomas A. Wikle Department of Geography, Oklahoma State University, Stillwater, OK, USA

Justin Wilford Department of Geography, University of California, Los Angeles, CA, USA

Joseph Witt Department of Philosophy and Religion, Mississippi State University, Mississippi State, MS, USA

John D. Witvliet Calvin Institute of Christian Worship, Calvin College and Calvin Theological Seminary, Grand Rapids, MI, USA

Teresa Wright Department of Political Science, California State University, Long Beach, CA, USA

Ernest J. Yanarella Department of Political Science, University of Kentucky, Lexington, KY, USA

Yukio Yotsumoto College of Asia Pacific Studies, Ritsumeikan Asia Pacific University, Beppu, Oita, Japan

Samuel Zalanga Department of Anthropology, Sociology and Reconciliation Studies, Bethel University, St. Paul, MN, USA

Donald J. Zeigler Department of Geography, Old Dominion University, Virginia Beach, VA, USA

Shangyi Zhou School of Geography, Beijing Normal University, Beijing, China

Teresa Zimmerman-Liu Departments of Asian/Asian-American Studies and Sociology, California State University, Long Beach, CA, USA

Part IV
Pilgrimage Landscapes and Tourism

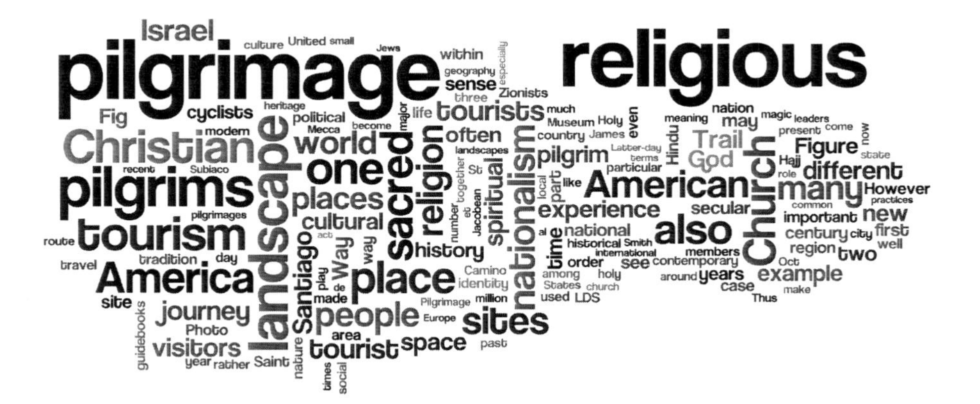

Chapter 34
Tourism and Religion: Spiritual Journeys and Their Consequences

Noga Collins-Kreiner and Geoffrey Wall

34.1 Introduction

Many years ago, while visiting south-central Nigeria, the second author was taken to visit the Oba's (King's) palace, which was a large compound with mud walls that were adorned with impressive designs. He was then taken to a sacred grove where a natural spring emerged amidst the trees in what was otherwise a dry place. On completing the visits, the guide stressed how different this experience was from that available in the western world. Superficially, this was the case for the palace structure and sacred grove were distinctive symbols of the local culture. However, at another level, there were remarkable similarities with many visits by tourists made elsewhere to the centers of secular and spiritual power, which are sometimes separate and elsewhere the same place. Thus, in London one might visit Buckingham Palace, the Houses of Parliament, Westminster Abby and St. Paul's Cathedral.

Also, as an undergraduate student in England, he learned about Lourdes, the catholic shrine in France (Wall 2010). The evolution of the pilgrimage site and the steady accumulation of hotels and other facilities for visitors were described. Notes from the lecture state that one million pilgrims visited Lourdes in 1931 and 2.75 million in 1962, roughly evenly divided by train and road travel. There were 392 hotels in the town with 25,000 beds and 563 special trains were arranged to bring people to the shrine. Over half of these trains came from elsewhere in France, but many came from other countries. For someone familiar with British seaside resorts,

N. Collins-Kreiner (✉)
Department of Geography and Environmental Studies, Centre for Tourism, Pilgrimage and Recreation, University of Haifa, Haifa 31095, Israel
e-mail: nogack@geo.haifa.ac.il

G. Wall
Department of Geography and Environmental Management, University of Waterloo, Waterloo, ON, Canada
e-mail: gwall@uwaterloo.ca

S.D. Brunn (ed.), *The Changing World Religion Map*,
DOI 10.1007/978-94-017-9376-6_34

it was easy to see parallels in the process that resulted in the development and evolution of distinctive resort morphology in coastal England. On eventually visiting Lourdes within the last decade, the grotto, shrine and infrastructure for visitors are still prominent features in the landscape, although most people now arrive by road rather than rail.

Centers of spiritual power are places that are visited by seekers of enlightenment, devotees and the curious. As such, they are tourist attractions, although visitor motivations may be diverse and not all visitors to such places regard themselves as tourists, which is sometimes used as a derogatory term, in comparison with other terms such as traveler or pilgrim. Nevertheless, large numbers of people congregate at sacred sites. Their motivations and behaviors may not fit together easily, for example the quiet desired for prayer and contemplation being at odds with the laughter of pleasure-seekers, and the monologue of the tour guide. Thus, as at tourism sites of other types, there may be tensions between locals and visitors, and among visitors with different expectations, with implications for the site and the community in which it is located, and the need for management. Sometimes visitors will be welcomed as sources of income and potential converts. At other times and places they may be tolerated, perhaps with instructions concerning appropriate behavior and dress. Elsewhere, tourists may be discouraged or banned: for example, some aboriginal sites in Australia are deemed to be so sacred that they are off-limits to tourists who are denied access (Altman 1989; Altman and Finlayson 1993). Furthermore, the situation may change over time. Thus, for example, formerly tourists were merely discouraged from climbing Uluru (Ayer's Rock) by a simple sign at the main access point, but now climbing is forbidden.

Thus, relationships between religion and tourism have a long history, are extremely varied and have a many implications, for the sites themselves and the people who visit and are visited. This chapter provides a comprehensive assessment of the primary issues and concepts related to the intersection of tourism and religion, while providing a balanced discussion of both subjects from theoretical and applied perspectives. It also explores substantial empirical cases from Christianity, Hinduism, Islam, Judaism and Baha'ism.

34.2 Theoretical Context

The scholarly research has barely touched upon the interrelationship between religion and tourism, although there is growing interest as revealed in the increase in publications on aspects of the topic (Cohen 1998; Collins-Kreiner 2010; Kong 2001). Our aim here is to address the relationship of these complex and global phenomena by presenting and analyzing the religion-tourism nexus from a number of different perspectives. Thus, we will delineate the category of religious tourists and other types of religiously-motivated tourists, identify holy sites and draw attention to numerous other contemporary issues requiring additional research. Finally, we address both sides of the religion-tourism connection: *supply*, stemming from the large number of major tourist destinations that have been developed over the years due largely to their association with sacred people, places, and events; and *demand*, fuelled by visitors, who embody the intersection of spirituality, religiosity, and tourism, although at times they may be unaware of their true motives and of the role that this intersection plays in the tourism system.

34.3 Tourism

In order to collect information on the number of tourists and tourism businesses, it is necessary to have precise definitions. Many definitions of a tourist exist, depending upon the purpose at hand and the scale of the destination of interest. Some definitions of a tourist have received wide acceptance, such as that of the United Nations World Tourism Organization, which is very detailed, as is necessary for the measurement of the magnitude of the phenomenon in a way that permits comparison from place to place, in this case, usually between countries. Most such definitions involve spatial, temporal and motivational components. Thus, spatially, it is usually necessary to travel a certain distance or cross a specified border in order to be considered a tourist. Temporally, it is usually necessary to stay a minimum length of time, often at least one night in order to exclude individuals who commute or are participating in forms of local recreation but, at the upper bound, if travelers stay for more than a year they are usually regarded as migrants. Often the reasons for travel are also considered, for example, making a distinction between business and pleasure travelers, but motivations are complex and often mixed and difficult to ascertain. Thus, as will be discussed later, it is difficult to distinguish between tourists and pilgrims on the basis of motivations.

Tourism involves much more than tourists. In order to encompass tourism as a global phenomenon, it is appropriate to adopt a much broader definition. Thus, Wall and Mathieson (2006, 1) defined tourism as follows:

> Tourism is the temporary movement of people to destinations outside their normal places of work and residence, the activities undertaken during their stay in those destinations, and the facilities created and services provided to cater to their needs. The study of tourism is the study of people away from their usual habitat, of the establishments which respond to the requirements of travellers, and of the impacts that they have on the economic, environmental and social well-being of their hosts. It involves the motivations and experiences of the tourists, the expectations of and adjustments made by residents of reception areas, and the roles played by the numerous agencies and institutions which intercede between them.

34.4 Religion and Tourism

Today, religion and tourism are inextricably bound together. Religion is still among the most common motivations for travel, and religiously motivated pilgrimage, which remains one of the world's oldest and most basic forms of population mobility, is emerging as a major tourism phenomenon in the twenty-first century (Collins-Kreiner 2010). In this context, religious sites are becoming main tourist attractions visited by religious visitors and tourists alike.

Although modern tourism is regarded as a relatively new phenomenon, it is clear that its origins are rooted in the age-old phenomenon of pilgrimage. The study of the relationship among religion, pilgrimage, and tourism often focuses separately on either religion or tourism alone, and pays little attention to the actual interaction between these two phenomena, or to a comparison between the two. This is surprising, as the development of tourism is difficult to understand without a thorough understanding of religion and the practice of pilgrimage in ancient times (Timothy and Olsen 2006; Vukonic 2002).

34.5 Demand: Pilgrims, Religious Tourists and Tourists

Until the 1970s, tourism studies barely existed as an academic field (Nash 2005), and studies of the relationship between religion and tourism frequently approached religion and tourism as two separate fields warranting little interrelated or comparative exploration.

Initial dedifferentiation between tourism and pilgrimage began to emerge in the 1970s, based on MacCannell's (1973) assertion that the tourist was a pilgrim in search of something different: authenticity. Later in the decade, Graburn (1977) characterized tourism as a kind of ritual, suggesting the existence of parallel processes, in both formal pilgrimage and tourism, that could be interpreted as "sacred journeys." These journeys, he contended, are about self-transformation and the gaining of knowledge and status through contact with the extraordinary or sacred.

Since then, theories have concentrated on different typologies of tourists and pilgrims as part of the differentiation between visit-related experiences and *everyday* life (Cohen 1979, 1992a, b; Smith 1992, 1989; MacCannell 1973). Over the past two decades, a new focus on pilgrimage has emerged via researchers interested in the field of tourism who have explored interesting political, cultural, behavioral, economic and geographical subjects of research (Timothy and Olsen 2006).

A number of researchers have recognized that the ties between tourism and pilgrimage are unclear, blurred, and poorly classified. This relationship is the subject of Eade's (1992) article, which describes the interaction between pilgrims and tourists at Lourdes; of Bowman's (1991) work on the place of Jerusalem in Christianity; and of Rinschede's (1992) description of the touristic uses of pilgrimage sites.

Smith (1992) claimed that, according to current usage, the term "pilgrimage" connotes a religious journey, one of a pilgrim, particularly to a shrine or another type of sacred place. However, its derivation from the Latin *peregrinus* allows broader interpretations, including foreigner, wanderer, exile, traveler, newcomer or stranger. The term "tourist" – one that makes a tour for pleasure or culture – also originally evolved from Latin, namely from the term *tornus*: one who makes a circuitous journey, usually for pleasure, and returns to the starting point.

The contemporary use of terminology that identifies religious travelers as "pilgrims" and vacationers as "tourists" is a culturally constructed polarity that obfuscates the true range of traveler motivations (Smith 1992).

Cohen (1992a) also maintains that pilgrimage and tourism differ with regard to the direction of the journey. The "pilgrim" and the "pilgrim-tourist" peregrinate toward their socio-cultural center, while the "traveler" and the "traveler-tourist" move in the opposite direction. This distinction applies particularly to journeys where the destination is a formal pilgrimage center. However, journeys to popular pilgrimage centers, which are typically "centers out there," are often characterized by a combination of features typical of both pilgrimage and tourism.

Pilgrims and tourists are distinct actors situated at opposite ends of Smith's continuum of travel, which first appeared in 1992. The poles of the pilgrimage-tourism axis are labeled *sacred* and *secular* respectively. Between these two poles exists an almost endless range of possible sacred-secular combinations, with a central area (c) which has come to be referred to generally as "religious tourism." These combinations reflect the multiple and changing motivations of travelers, whose interests and activities may change – consciously or subconsciously – from tourism to pilgrimage and vice-versa. Jackowski and Smith (1992) used the term "knowledge-based tourism" synonymously with "religious tourism." Most researchers associate "religious tourism" with an individual's quest for shrines and locales where, in lieu of piety, they seek sites of historical and cultural meaning (Nolan and Nolan 1989). Smith (1992) understood the difference as stemming from individual beliefs and worldviews.

According to Gatrell and Reid (2002), tourism, like pilgrimage, is embedded within a complex of socio-spatial processes that are historically, culturally and locally dependent. Both phenomena are complex systems comprising perceptions, expectations and experiences (Gatrell and Reid 2002; McCann 2002; Petric and Mrnjavac 2003).

In 2004, it was Badone and Roseman who first claimed that: "Rigid dichotomies between pilgrimage and tourism or pilgrims and tourists no longer seem tenable in the shifting world of postmodern travel" (2004, 2). On this basis, their book *Intersecting Journeys: The Anthropology of Pilgrimage and Tourism* seeks to highlight the similarities between these two categories of travel, which have frequently been understood as conceptual opposites of one another. Although modern tourism is often regarded as one of the newer phenomena of the modern world, we are reminded that the origins of tourism are rooted in pilgrimage. Indeed, this dedifferentiation has been one of the main subjects of current research in the field of religion and tourism.

Gaps still remain in understanding of the differences between pilgrimage and tourism from the perspective of the pilgrims themselves, the tourism industry, and scholars (Timothy and Olsen 2006). The distinction between pilgrims and tourists should be examined on two levels: on the level of religious organizations and on the level of the travelers themselves. From the view of the travelers themselves, pilgrims do not usually consider themselves to be tourists, or at least regard themselves as being distinct from tourists. This view suggests that pilgrims are not tourists because they travel for spiritual reasons, while tourists travel or visit sites for reasons that are more secular in nature, such as curiosity or pleasure. From the viewpoint of the tourism industry, however, pilgrims and tourists are very similar and should be treated as such, for they require transportation, food and beverages, and accommodation, and buy souvenirs. Such an approach has great relevance for the development of related economic activities and the development of infrastructure and services, such as the establishment and operation of hotels, restaurants, shops, hospices, and religious centers.

We conclude this section by pointing out that most new studies reflect a tendency toward dedifferentiation, and that some researchers have argued that the differences between tourism, pilgrimage, and even secular pilgrimage continue to narrow (Bilu 1998; Kong 2001). This distinction is increasingly being seen as misplaced, as the religious and secular spheres of tourism are quickly converging, and as religious tourism assumes a more prominent market niche in international tourism.

34.6 Supply: Holy Sites as Tourist Attractions

In many parts of the world, pilgrimage shapes economic activity and public space. One of the most interesting case studies on this topic is the Camino de Santiago in northern Spain. The last two decades have witnessed an extraordinary revival of interest in the pilgrimage to Santiago de Compostela. The route known as the El Camino Francés was declared the first European Cultural Route by the Council of Europe in October 1987, and was registered as a UNESCO World Heritage Site in 1993. In the holy year of 2010, approximately 270,000 people made their way along the ancient routes, on foot or bicycle, and in some cases on horseback. Today (2012), the Camino and Santiago are undoubtedly a tourist attraction, and not just a religious route. In this way, an attraction that was once religious at its core now serves both secular tourists and spiritual visitors, in addition to religious ones.

Sacred space is contested space, just as the sacred is a "contested category" (Kong 2001). Ivakhiv (2006) showed how several scholars of religion have argued in recent years that there is no stable and historically valid definition of religion and the sacred. Chidester and Linenthal (1995, 15) argued that a sacred space "is not merely discovered, or founded, or constructed; it is claimed, owned, and operated by people advancing specific interests." Hecht (1994, 222) described sacred space and time as "situational or relational categories," as matters of "setting boundaries and negotiating relationships."

Given the simultaneous status of pilgrimage as center, periphery, other, and liminal, the process and places it embodies occupy a unique space in the imagination of religious and secular tourism alike – what political geographer Edward Soja (1980) referred to as a "third space." By following this lead and perceiving religious sites as a third space existing beyond and between the lived and *anticipated worlds*, researchers should be able to unlock and deconstruct the social practices of religious and secular tourists at religious sites. The concept of "third space" enables them to avoid using simplified notions such as "religious traveler" and "vacationer" to denote pilgrims and tourists respectively, insofar as these two groups are linked in a shared space. Indeed, a revised religious tourism paradigm based in part on the notion of 'third space' acknowledges – in both implicit and explicit terms – the interdependent nature of the two actors and the social construction of sites as simultaneously sacred and secular (Gatrell and Collins-Kreiner 2006).

Alderman (2002) employed the term "pilgrimage landscape" to highlight relationships between people and place. Indeed, no place is intrinsically sacred. Rather, pilgrimages and their attendant landscapes are "social constructions." They do not simply emerge but undergo what Seaton (1999, 2002) called "sacralisation," a sequential process by which tourism attractions are marked as meaningful, quasi-religious shrines.

Today, scholars have started thinking about other forms of pilgrimage, such as those embarked on by spiritual tourists, the latest romantics, or the hippies who began frequenting India and the Himalayas in the 1960s. There is also a growing market in "new age" spiritual travel for pilgrimage, personal growth and non-traditional spiritual practices (Attix 2002), and an increasing amount of research is being done on modern secular pilgrimage where the search for the miraculous is a trait shared by religious and secular pilgrims alike.

Growth in faith-based tourism is a worldwide phenomenon, with expansion not just in numbers of tourists but in sites and the emergence of new kinds of activities. According to the U.S. National Tour Association (NTA), faith-based travel has a "centuries-old legacy" and "remains one of the most enduring activities in tourism today." "Among our 700 tour operator members," explains NTA President Lisa Simon, "40 % report some level of involvement in faith-based tourism" (Simon 2011). Simon also noted that although the roots of faith-based tourism can be found in the phenomenon of pilgrimage, it now encompasses a multiplicity of other activities, such as cruises, conferences, volunteer vacations, and adventure trips. "Many people think of religious travel as a trip to the Holy Land, but it involves many aspects of the tourism industry right here in North America" (Simon 2011).

In addition, the word "pilgrimage" itself is becoming widely used in a broad range of secular contexts, such as visits to war graves or the graves and residences of deceased celebrities, and visits to churchyards and to funerary sites. A good example is Elvis Presley's tomb and mansion in Memphis, Tennessee (Alderman 2002; Reader and Walter 1993). Today, this kind of tourism is also known as "thana-tourism" or as an aspect of "dark tourism" (Stone 2006; Seaton 1999, 2002).

34.7 Empirical Studies

This section is based on the analysis of studies addressing different religious contexts in which the tourism-religion nexus discussed in the first part of this chapter is visibly manifested.[1]

[1] For more details on the studies, please see the authors' papers in the references section.

34.7.1 Christian Visitors to the Holy Land

Between one-quarter and one-third of all international tourists who visit Israel are Christians visiting for religious reasons (Ministry of Tourism 2011). The research exploring this phenomenon assesses the Christian visitor experience through the eyes of the visitors themselves, based on before-and-after interviews (Collins-Kreiner et al. 2006). According to the findings of this study, trips taken by most religious Christian visitors to the Holy Land are motivated primarily by the profound religious beliefs they espoused even before embarking on the sacred journey, which surpass all economic, family and health considerations. Apparently, it is the pilgrims' strong religious faith that motivates them and shapes their expectations of their trip. The Holy Land's unique world of images and perceptions is also imbued with religio-spiritual content, and typical itineraries enable pilgrims to "walk in the footsteps of Jesus." The spiritual contents of their religious wishes also reflect the different dogmas of Catholicism and various Protestant churches (Fig. 34.1).

There are many indications of the minor role played by touristic components in these pilgrims' sacred journeys. Their touristic motivations and expectations are marginal, although, according to their own testimony, they are nonetheless well fulfilled. Still, a number of tourism-related aspects are reflected in the comments and responses of some of these visitors. Like all tourists in Israel, religious Christian visitors frequent well-known Israeli sites and spend time shopping. As tourists, they enjoy the fact that the sacred sites are not crowded, but they complain about the tight

Fig. 34.1 Christian pilgrims in the Jordan River (Photo by Noga Collins-Kreiner)

schedule and lack of free time. Yet, as stated before, all of these points are minor compared to what, for them, is the most important aspect of the tour – the religious aspect.

The prototype Christian pilgrim to the Holy Land is a person who is focused entirely on his or her experiential sense of fulfillment from completing their religious mission. The world of religious Christian tourism in Israel leaves little room for mundane touristic experiences, and the pilgrims themselves felt no need to expand on the issue (Collins-Kreiner and Kliot 2000).

These findings highlight a gap in the perception of the pilgrimage-tourism connection advanced by Timothy and Olsen (2006). For pilgrims, the religious perspective is dominant, while the tourism industry and tourism researchers perceive such pilgrimages as an economic activity, choosing instead to focus on the economic benefits and the touristic development of the country's Christian holy sites.

34.7.2 The Balinese Landscape as an Expression of Tourism and Religion

Bali, Indonesia's most visited tourism destination is often called "a Hindu island in a Moslem sea." While Indonesia is home to the world's largest Moslem population, the residents of Bali are predominantly Hindu. Bali possesses a tropical environment, sandy beaches and coral reefs, but these are fairly ordinary when compared with other islands across the world. Rather, it is Balinese culture and its brand of cultural tourism, infused with the distinctive Balinese Hindu religion that makes Bali special (Picard 1992). It is a religion that operates at all scales from that of the whole island to the individual family compound (Budihardjo 1986), with colorful and impressive ceremonies, and a gentle atmosphere that charms all but the hardest visitor, as described at length in Gilbert's (2006) autobiography, *Eat, Pray, Love*. Gunung (Mount) Agung, is an active volcano in the centre of Bali. It is the abode of the Gods and on its flanks is Besakih, the most important temple in the island (Fig. 34.2).

In Balinese one uses the terms *kelod* and *kajur*, towards Agung and away from Agung, rather than the points of the compass (Wall 1998). Each village (*desa*), and there are more than 600 of them, is entered via an ornamental 'split gate' and contains three major temples: the *pura puseh* which is located nearest to the mountain, the *pura desa* in the centre of the village where village meetings and fertility rites are held; and the *pura dalem*, the farthest from Mount Agung where the dark forces are worshipped and near which burials occur. In addition, each *banjar* (neighborhood unit) has a temple. All of these temples have their own calendar of events and festivals. The large number of temples and the events that focus upon them provide an authentic cultural display that captivates visitors. Balinese religious events have become tourist attractions and even cremations have become part of the tourism itinerary (as is also the case elsewhere in Indonesia in Tana Toraja where

Fig. 34.2 Mount Agung and the Besakih Temple in Bali (Photo by Geoffrey Wall)

funerals celebrations and effigies of the dead have become the major tourism attractions (Adams 1984, 2006) (Fig. 34.3).

Extended families live in family compounds or walled enclosures which contain a series of largely open structures (*bales*) that are laid out in accordance with Balinese cosmology. Thus, for example, the family temple will be in the corner closest to the mountain and the waste will be collected in the corner closest to the sea. This has worked well for centuries in village settings with low population densities. However, in high-density resorts like Kuta in the south of the island, where ground water tables have declined and salt-water intrusion is occurring, and one family's well may be in close proximity to another's waste, there is a danger of water contamination. The numerous small structures in the family compound can be readily converted into simple tourist accommodation or even small hotels (Fig. 34.4). However, in Bali it was rare to have two-story buildings and when these are built, it is necessary to ensure that they are linked to the ground by running a line to maintain contact with the earth, much like a lightning rod.

Balinese families make offerings to the Gods three times each day and this is a woman's task. The offering is made at the family shrine and usually consists of a little rice or meat, with flowers of six different colors, in a small tray made from a banana leaf. Traditionally, the offerings have been made from materials available in the compound but, with the growth of tourism and a money economy, some families now purchase their offerings, thereby providing jobs for women who make them to sell (Wall 1995).

Most of the villages are located on ridges and, especially in the south central part of the island, irrigated rice fields cascade down the hillsides, in a landscape that

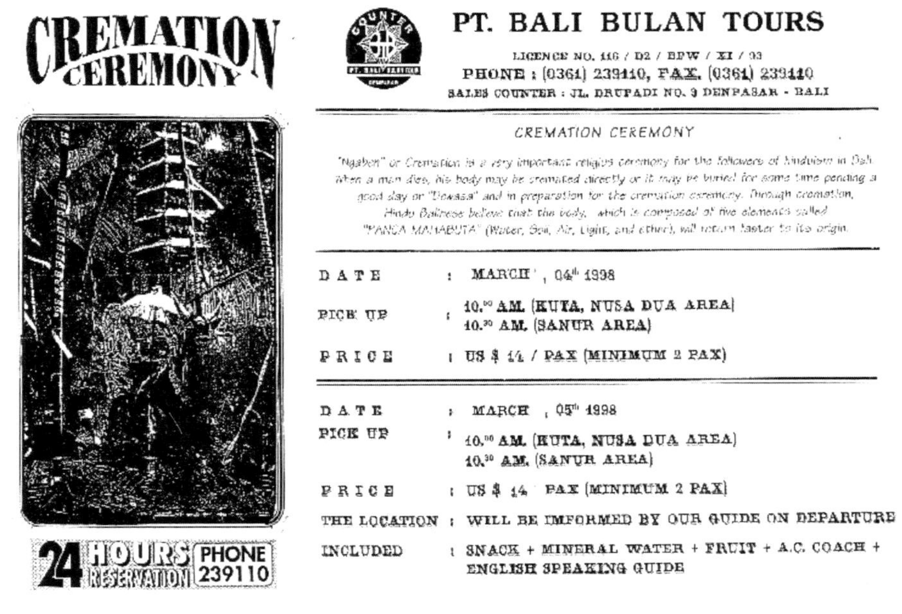

Fig. 34.3 Advertisement for a cremation tour in Bali (Photo by Geoffrey Wall)

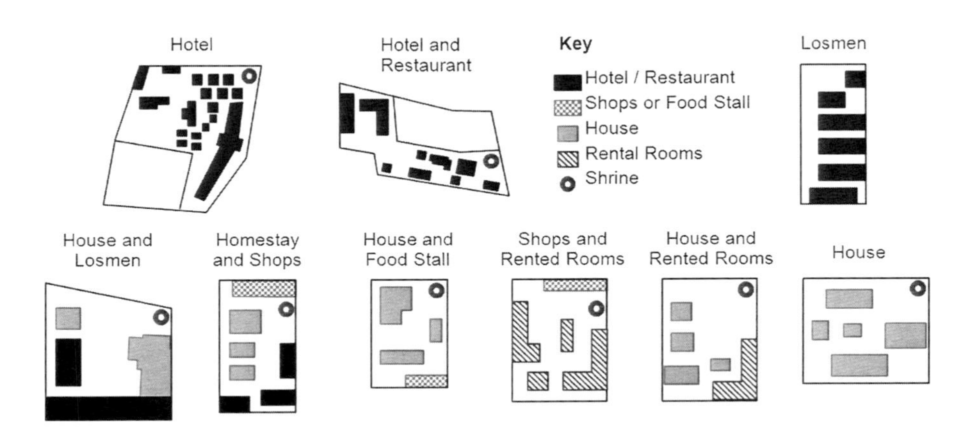

Fig. 34.4 Balinese compounds converted into tourism facilities (Modified from Rahmi 1992; Sulistyawati 1989, 44; Budihardjo 1986)

most tourists will visit. This agricultural landscape is a tourism resource with religious meaning or, alternatively, a cultural landscape that is photographed by secular tourists on a brief visit from their hotels in the coastal resorts (Lansing 1991) (Fig. 34.5).

Fig. 34.5 Rice terraces in Bali (Photo by Geoffrey Wall)

The village *subak* (irrigation authority) determines the distribution of water and the time for planting, arranges festivals to celebrate the agricultural round (and there can be three crops of rice each year) and small temples have been constructed where the main irrigation channels diverge. However, in some places, hotels have been built in sites with magnificent views but on steep slopes, disrupting the irrigation systems, and golf courses have been constructed, increasing the demand for water that once sustained the rice fields. In protest, Balinese people asked the following question: "How many grains of rice does it take to make a golf ball?" Elsewhere, as at Tanah Lot, which is one of the most important temples on the south coast, hotels and condominiums have been built in close proximity to the temple, overriding the recommendations of the religious authorities and detracting from it both as a religious and touristic place (Cohen 1994).

Thus, Bali's magnificent landscape is imbued with religious symbolism. Tourists are attracted by the beauty of a landscape which, at first sight, seems to represent a harmonious relationship between human activities and natural forces. In the 1970s, the tourism plan identified designated tourists routes through the rice fields to the volcanoes to enable visitors to access the most spectacular routes. Now, these routes are lined with structures that sell things to tourists and service their needs, forming a barrier between the traveler and the special landscapes that lie behind. Most visitors, however, still have a good time, though few fully understand the meaning of the landscape or the pressures that it is under.

34.7.3 The Hajj to Mecca

A third illustrative example of the religion-tourism nexus can be seen in the case of the Hajj – the annual Muslim pilgrimage to Mecca. More than two million people are estimated to make this pilgrimage on a specific date each year (Fig. 34.6).

In the past pilgrims and organizers alike saw the Hajj purely as a phenomenon of pilgrimage. However, recent years have witnessed a shift in this attitude. The Hajj is still defined as pure religious travel. However, when we look more carefully at site-infrastructure in terms of accommodations, food and beverage, transportation, and at the character of the trip as a whole, it becomes clear that the Hajj is becoming part of the global phenomenon of tourism. An excerpt from a *New York Times* article regarding the 2010 Hajj is reflective of this change: "Rashed Abdullah displays Oriental perfumes on a glass table to late-night shoppers in his small shop in Mecca ready for what he hopes will be a sales bonanza during this month's hajj pilgrimage" (*NY Times*, Nov. 3, 2010).

The annual Hajj is one of the world's largest religious gatherings, bringing together millions of pilgrims in Saudi Arabia, the birthplace of Islam. A 2010 syndicated news report effectively captures the current economic importance of the event:

> According to John Sfakianakis, chief economist at Banque Saudi Fransi, foreigners were forecast to spend 8 billion riyals ($2.1 billion) during the Hajj, up from 7.2 billion last year, while locals would spend 3.6 billion riyals versus 3.3 billion riyals. "The multiplier effect from the hajj season for the entire economy could surpass 35 billion riyals this year alone," he said, adding last year it was 31 billion riyals. Tourism GDP would surpass 7.2 % of

Fig. 34.6 Mecca: pilgrims on the hajj filling the Great Mosque, Mecca, Saudi Arabia (Photo by Dr Razaq Raj, Leeds Metropolitan University, UK, used with permission)

> non-oil GDP this year, up from 6.8 % in 2009, and 2.8 % of overall GDP, up from 2.6 % last year, he estimated. Although revenues from pilgrims are small compared to the massive wealth of the kingdom because of its oil and gas resources, it helps a sector the government is developing. Faced with a fast-rising population of 18 million Saudis, the government needs to create jobs and developing tourism is one area to do that… So far pilgrim visas are limited to visit the two holy cities of Mecca and Medina but officials have said they are working on plans to allow visits to tourist sites as well. (Reuters 2010)

Plans are also in place to develop a variety of other potential tourist sites in and around the Mecca region as part of the touristification of the old religious stronghold (see Fig. 34.6).

34.7.4 Jewish Pilgrims

The main finding of studies examining current domestic Jewish pilgrimage to the sites of holy graves in Israel is the existence of a continuum, like that proposed by Smith (1992), along which different groups of visitors may be represented. At one end of the continuum are pure pilgrims (19 %) and religious Jews (27 %) whose visits are motivated by religious belief. The main activity of these groups is prayer. At the opposite end of the continuum are so-called spiritual or heritage tourists (15 %), whose motivations are curiosity, cultural interest, and the quest for a new meaning in life. Between these extremes are traditional visitors (36 %), who believe in the power of the buried holy personage and his advice. It should be noted, however, that this belief on the part of traditional visitors is anchored not in religious faith but in personal outlook, as reflected in their tendency to make specific requests rather than to engage in formal prayer (Fig. 34.7).

Differences between these groups may also be observed in the customs and behavior of their members. Whereas pilgrims pray and secular visitors just visit, the mid-continuum group takes part in many local folklore activities such as the lighting of candles, the placing of supplications and notes, and the purchase of souvenirs such as holy water, pictures of the holy personage in question (*tsaddik*), candles, cards, pamphlets, and amulets against the 'evil eye.' The differences noted in the visitors' attitudes to the sites were found to depend mainly on religious affiliation, and not on age, ethnic origin, socio-economic status, self-perception, gender, or other factors. Each pilgrim's location on the scale is personal and subjective, with a virtually infinite number of sacred-secular combinations existing between the extremes. The survey results highlight the increasing convergence of traditional pilgrimage and current tourism, which appear to have a great deal in common.

34.7.5 Baha'i Pilgrimage Sites

Another study examines the practices that transform the Baha'i´ Gardens in Haifa, Israel, into a secular shared community asset. The contemporary status of the garden as both a holy site and a tourist attraction makes the case of the Baha'i´

Fig. 34.7 Traditional Jewish visitors to a holy grave in Israel (Photo by Noga Collins-Kreiner)

Gardens and its cultural and economic context more distinct but also somewhat ambiguous, due to the unclear boundaries between the two (Collins-Kreiner and Gatrell 2006).

Visitors to the Baha'i gardens can be classified according to a basic typology of religious Baha'i visitors and non-Baha'i visitors. The religious visitors are in "existential mode" (Cohen 1979, 190), fully committed to an elective spiritual center. They are Baha'i´ who view their travel as a pilgrimage and a once-in-a-lifetime experience. Their visits are not dominated by recreational or diversionary experiences, but rather are completely occupied by the restorative effects of the inherent spirituality of their trip. The secular tourists examined in this study correspond predominantly to Cohen's "recreational mode" (1979, 190) of tourism. Their trip is a form of entertainment not unlike cinema, theater, or television. This kind of visitor, who is typically a domestic Israeli tourist, enjoys his/her trip because it restores physical and mental powers and provides a general sense of well-being. In addition to "existential mode" and "recreational mode," a number of visitors may be classified as "experiential tourists," as the gardens and the Baha'i´ World Centre provide them with an authentic "other" experience distinct from their everyday life and normal social reality. Indeed, the beauty of the gardens and of the site as a whole, and the site's connection to the emerging Baha'i faith system, make the experience one that, for many visitors, goes well beyond their everyday life experiences.

The case of the Baha'i´ Gardens in Haifa demonstrates how differing visitor motives can shape different visitor activities and the activity-space they define. The separation of the tourist and pilgrim experiences at the Baha'i´ Gardens in Haifa is unique, as the site has been specially designed to prevent any potential conflict with

Fig. 34.8 The Baha'i Holy Shrine and Gardens in Haifa, Israel (Photo by Noga Collins-Kreiner)

local residents which, in turn, enables the municipality to emphasize the secular and aesthetic benefits of the gardens to the local community.

The outcome of this place-based strategy of avoidance and mitigation toward religion-tourism conflict has been the creation of a layered collection of spatial practices that preserve the sacred nature of the Baha'i´ complex while simultaneously enabling the non-Baha'i community to enjoy a variety of secular benefits. This strategy has a great deal to offer planners and practitioners in other places in the world, who would be wise to consider its adoption (Fig. 34.8).

34.8 Summary

The time has come for the expansion of the contemporary terminology that identifies religious travelers as "pilgrims" and vacationers as "tourists" to accommodate broader interpretations in accordance with the Latin and Greek origins of the terms themselves. Understanding tourism to religion-related sites according to the continuum discussed in the second part of this chapter reinforces the emerging connection between the two motilities of tourism and religion. Today, any real distinction between modern day "pilgrimage" and modern day tourism is difficult to discern, and pilgrims simply cannot be differentiated from tourists. Both kinds of travelers can be motivated to embark upon journeys and seek out experiences in order to add meaning to their lives.

Geographers have much to contribute to the contemporary debates about religion, pilgrimage, tourism, space, and experience currently emerging across a range of disciplines. By reviewing these central themes, which highlight the importance of a new direction in the geography of religion, this chapter has not only contextualized recent geographical work on the subject, but has also shown how this field of research is advancing, drawing upon specific influential theoretical developments.

Pilgrimage and religious tourism has been relatively ignored in mobilities research (Urry 2007), despite its scale, frequency and variety across time and space. As far as today, there is weak conceptualization of the connection between mobility and pilgrimage, and especially within the new mobility paradigm, which has contributed to relative neglect of the relationships between these two. Researchers should develop an explanatory framework for locating religious movements within the conceptual domain of the mobilities paradigm.

We all need to understand that studying the meaning of the connection between religion and tourism transcends geographical and sociological aspects. It involves an interpretative approach to seeking an alternative meaning, one hitherto neglected. Present studies assume that religious travels are products of the culture in which they were created; hence they tell us 'stories' from political, religious, cultural and social standpoints. These mobilities are products of the norms and values of social tradition and order and, at the same time, are the creators of such culture and tradition.

References

Adams, K. M. (1984). Come to Tana Toraja, "Land of the Heavenly Kings:" Travel agents as brokers in ethnicity. *Annals of Tourism Research, 11*(3), 469–485.

Adams, K. M. (2006). *Art as politics: Re-crafting identities, tourism and power in Tana Toraja, Indonesia*. Honolulu: University of Hawaii Press.

Alderman, D. H. (2002). Writing on the Graceland wall: On the importance of authorship in pilgrimage landscapes. *Tourism Recreation Research, 27*(2), 27–33.

Altman, J. (1989). Tourism dilemmas for aboriginal Australians. *Annals of Tourism Research, 16*(4), 456–476.

Altman, J., & Finlayson, J. (1993). Aborigines, tourism and sustainable development. *Journal of Tourism Studies, 4*(1), 38–50.

Attix, S. A. (2002). New age-oriented special interest travel: An exploratory study. *Tourism Recreation Research, 27*(2), 51–58.

Badone, E., & Roseman, S. (Eds.). (2004). *Intersecting journeys: The anthropology of pilgrimage and tourism*. Champaign: University of Illinois Press.

Bilu, Y. (1998). Divine worship and pilgrimage to holy sites as universal phenomena. In R. Gonen (Ed.), *To the holy graves: Pilgrimage to the holy graves and Hillulot in Israel* (pp. 11–26). Jerusalem: The Israel Museum.

Bowman, G. (1991). Christian ideology and the image of a holy land: The place of Jerusalem in the various Christianities. In M. J. Sallnow & J. Eade (Eds.), *Contesting the sacred: The anthropology of Christian pilgrimage* (pp. 98–121). London: Routledge.

Budihardjo, E. (1986). *Architectural conservation in Bali*. Yogyakarta: Gadjah Mada University Press.

Chidester, D., & Linenthal, E. (1995). *American sacred space*. Bloomington: Indiana University Press.
Cohen, E. (1979). A phenomenology of tourist experiences. *Sociology, 132*, 179–201.
Cohen, E. (1992a). Pilgrimage centres: Concentric and excentric. *Annals of Tourism Research, 19*(1), 33–50.
Cohen, E. (1992b). Pilgrimage and tourism: Convergence and divergence. In A. Morinis (Ed.), *Sacred journeys: The anthropology of pilgrimage* (pp. 47–61). New York: Greenwood Press.
Cohen, M. (1994, May 26). God and mammon: Luxury resort triggers outcry over Bali's futures. *Far Eastern Economic Review*, pp. 28–33.
Cohen, E. (1998). Tourism and religion: A comparative perspective. *Pacific Tourism Review, 2*, 1–10.
Collins-Kreiner, N. (2010). Researching pilgrimage: Continuity and transformations. *Annals of Tourism Research, 37*(2), 440–456.
Collins-Kreiner, N., & Gatrell, J. D. (2006). Tourism, heritage and pilgrimage: The case of Haifa's Baha'i gardens. *Journal of Heritage Tourism, 1*(1), 32–50.
Collins-Kreiner, N., & Kliot, N. (2000). Pilgrimage tourism in the holy land: The behavioural characteristics of Christian pilgrims. *GeoJournal, 501*, 55–67.
Collins-Kreiner, N., Kliot, N., & Sagie, K. (2006). *Christian tourism to the holy land: Pilgrimage during security crisis*. Hampshire: Ashgate Publications.
Eade, J. (1992). Pilgrimage and tourism at Lourdes, France. *Annals of Tourism Research, 19*, 18–32.
Gatrell, J. D., & Collins-Kreiner, N. (2006). Negotiated space: Tourists, pilgrims and the Baha´´ı´ terraced gardens in Haifa. *Geoforum, 37*, 765–778.
Gatrell, J., & Reid, N. (2002). The cultural politics of local economic development: The case of Toledo jeep. *Tijdschrift voor Economische en Sociale Geografie, 93*, 397–411.
Gilbert, E. (2006). *Eat, prey, love*. New York: Penguin.
Graburn, N. H. H. (1977). Tourism: The sacred journey. In V. L. Smith (Ed.), *Hosts and guests: The anthropology of tourism* (pp. 17–31). Philadelphia: University of Pennsylvania Press.
Hecht, R. D. (1994). The construction and management of sacred time and space: Sabta Nur in the Church of the Holy Sepulchre. In R. Friedland & D. Boden (Eds.), *Now here: Space, time and modernity* (pp. 181–235). Berkeley: University of California Press.
Israel, Ministry of Tourism. (2011). *Incoming tourists – 2010* (Final report). Jerusalem.
Ivakhiv, A. (2006). Toward a geography of "religion:" Mapping the distribution of an unstable signifier. *Annals of the Association of American Geographers, 96*(1), 169–175.
Jackowski, A., & Smith, V. L. (1992). Polish pilgrim-tourists. *Annals of Tourism Research, 19*, 92–106.
Kong, L. (2001). Mapping 'new' geographies of religion: Politics and poetics in modernity. *Progress in Human Geography, 25*, 211–233.
Lansing, J. S. (1991). *Priests and programmers: Technologies of power in the engineered landscape of Bali*. Princeton: Princeton University Press.
MacCannell, D. (1973). Staged authenticity: Arrangements of social space in tourist settings. *American Journal of Sociology, 793*, 589–603.
McCann, E. (2002). The cultural politics of local economic development: Meaning making, place-making, and the urban policy process. *Geoforum, 33*, 385–398.
Nash, D. (2005). *The study of tourism: Anthropological and sociological beginning*. London: Elsevier Science.
Nolan, M. L., & Nolan, S. (1989). *Christian pilgrimage in modern Western Europe*. Chapel Hill: University of North Carolina Press.
Petric, L., & Mrnjavac, Z. (2003). Tourism destinations as a locally embedded system. *Tourism, 51*, 403–415.
Picard, M. (1992). *Bali: Tourisme Culturel et culture touristique*. Paris: L'Harmattan.
Rahmi, D. (1992). Integrated development for spatial, water supply and sanitation systems in the tourism area of Kuta, Bali, Indonesia. MA thesis, Department of Geography, University of Waterloo.

Reader, I., & Walter, T. (Eds.). (1993). *Pilgrimage in popular culture*. London: The Macmillan Press.
Reuters, Ulf Laessing. (2010). http://in.mobile.reuters.com/article/worldNews/idINIndia-52651820101103. Accessed 6 July 2011.
Rinschede, G. (1992). Forms of religious tourism. *Annals of Tourism Research, 19*, 51–67.
Seaton, A. V. (1999). War and thanatourism: Waterloo 1815–1914. *Annals of Tourism Research, 261*, 130–158.
Seaton, A. V. (2002). Thanatourism's final frontiers? Visits to cemeteries, churchyards and funerary sites as sacred and secular pilgrimage. *Tourism Recreation Research, 27*(2), 27–33.
Simon, L. (2011). N.T.A. News Releases. www.ntaonline.com/media-center/news-releases. Accessed July 2012.
Smith, V. L. (1989). *Hosts and guests – The anthropology of tourism*. Philadelphia: University of Pennsylvania Press.
Smith, V. L. (1992). Introduction: The quest in guest. *Annals of Tourism Research, 19*, 1–17.
Soja, E. (1980). Socio-spatial dialectic. *Annals of the Association of American Geographers, 70*, 207–225.
Stone, P. (2006). A dark tourism spectrum: Towards a typology of death and macabre related tourist sites, attractions and exhibitions. *Tourism: An Interdisciplinary International Journal, 52*(2), 145–160.
Sulistyawati. (1989). *The Balinese home: Factors that influence change in its architecture*. Unpublished M.A. thesis, University of Indonesia.
Timothy, D. J., & Olsen, D. H. (Eds.). (2006). *Tourism, religion and spiritual journeys*. London/New York: Routledge.
Urry, J. (2007). *Mobilities*. Cambridge: Polity.
Vukonic, B. (2002). Religion, tourism and economics: A convenient symbiosis. *Tourism Recreation Research, 27*(2), 59–64.
Wall, G. (1995). Forces for change: Tourism. In S. Martopo & B. Mitchell (Eds.), *Bali: Balancing environment, economy and culture* (Department of Geography publication series 44, pp. 57–74). Waterloo: University of Waterloo.
Wall, G. (1998). Landscape resources, tourism and landscape change in Bali. In G. Ringer (Ed.), *Destinations: Cultural landscapes of tourism* (pp. 51–62). London: Routledge.
Wall, G. (2010). With a little help from my friends. In S. Smith (Ed.), *The discovery of tourism* (pp. 153–162). Bingley: Emerald.
Wall, G., & Mathieson, A. (2006). *Tourism: Change, impacts and opportunities*. Harlow: Pearson.

Chapter 35
The Way of Saint James: A Contemporary Geographical Analysis

Rubén C. Lois-González, Valerià Paül, Miguel Pazos-Otón, and Xosé M. Santos

35.1 Introduction

The contemporary *Camino de Santiago*[1] has become a global pilgrimage route in all respects in recent years. To make this conversion possible, a series of profound changes in the Jacobean[2] peregrination has had to happen. Thus, the strict Catholic Christian significance it had in the early twentieth century has become what can be considered today as mass tourism as the motivations that drive people to engage in it vary greatly. Indeed, it is not uncommon to find companies in countries as far away as Australia promoting the Way as a package tour and, curiously, it is promoted free of religious attributes, referring to "walkers" not "pilgrims," and even confusing the Apostle who allegedly is buried in Santiago.[3] This fact clearly shows how the religious significance has been diluted and how the secular and the cultural dimension in general has emerged. Set in this context, in this chapter we propose a geographical analysis of the Way in several venues. Its contemporary dimension is favored, providing the basic keys to understanding the changes

[1] The Way of Saint James is internationally known by its Spanish name, *Camino*. We will use both terms interchangeably.

[2] "Jacobean" is the adjective for the Apostle James. In Galician the word *Xacobeo* is widely used, as in Spanish *Jacobeo*, to refer to all the culture dimensions related with the Apostle James, the City of Santiago and the Way of Saint James. "Jacobean" will be used in this sense in this contribution.

[3] An advertisement appeared in the March 2012 issue of the journal *WEA Course Guide*, published in Adelaide (South Australia). The advert says that the *Camino de Santiago* leads to "the final resting place of Saint John the Elder" and defines the experience as a "cultural exchange."

R.C. Lois-González • V. Paül (✉) • M. Pazos-Otón • X.M. Santos
Departamento de Xeografía, Universidade de Santiago de Compostela,
15703 Santiago de Compostela, Galiza, Spain
e-mail: rubencamilo.lois@usc.es; v.paul.carril@usc.es;
miguel.pazos.oton@usc.es; xosemanuel.santos@usc.es

S.D. Brunn (ed.), *The Changing World Religion Map*,
DOI 10.1007/978-94-017-9376-6_35

mentioned, but certainly a focus is on the past. The considerations build on the recent published research by two of the authors.[4]

The article begins with a theoretical review of the peregrination issue from a geographic perspective. Subsequently, we propose a brief explanation of the rise of the Jacobean phenomenon from a particular geopolitical point of view, so that the reader can discover in this text some of the latest claims. Then we proceed to explain the contemporary revival of the Way, conveyed through various geographical scales that have discursively used it. Beyond the narrative, the next section deals with the contemporary spatial production of the Way. The final section is devoted to the interaction between landscape (as understood within current cultural geography) and the *Camino*. The concluding remarks systematize the binomial tourism / pilgrimage nexus in the current Jacobean phenomenon, including reviewing arguments presented throughout this work.

35.2 Geographical Theorization on Pilgrimages

It is now understood that the *pilgrimage* is primarily a movement of people in an open space, whose fate is often associated with God or the divine in many ways. This is the preeminent meaning of pilgrimage and would explain mass movements of people of faith to the Mecca, Jerusalem, Rome or Lourdes. According to García Cantero (2010), *pilgrimage* is derived from Latin *peragrare* which refers to one who passes through the fields *(per agrum)*. Originally the concept was of legal and administrative value, but it later gained a generic meaning of movement or to take the road and travel to somewhere, a sense in which the religious connotation was introduced. In this sense, pilgrimage is constitutive of the geography of religions to the extent that it implies necessary space and also that the hegemonic reason (at least in origin) of the displacement is of a religious nature. Nevertheless, it is clear that pilgrimage in present secularized societies has mutated and some of its religious and spiritual sense has been lost (Ivakhiv 2003). Similarly, just as there is a territorial religion, there are also secular spaces and, indeed, the sacred and secular often interact in complex and even bellicose fashion (Yorgason and della Dora 2009; Stump 2008; Kong 2004).

Pilgrimages in a secular context may respond well to tourist logic (Collins-Kreiner 2010). In fact, in the past tourism and pilgrimage basically meant the same, that is, the movement from the usual place of existence or origin to another. For Collins-Kreiner (2010) or Coleman and Eade (2004), traditional pilgrimage is but the origin of modern tourism, with a passage located in the *grand tour* which was the discovery of the German and British elite. Medieval pilgrimages also maintained an elitist tinge (see Lopez and Lois 2011) in places, shrines, culture or civilization, many of

[4]Among other important works are Santos and Lois (2011), Lois and Santos (1999), Lois (2013, 2011) and Santos (2006, 2002, 1999).

them from ancient Christian centers. Today, people walk to a place in search of peace and spiritual renewal, tranquility, culture, landscape stimuli and contact with others. Although a pilgrimage often passes through medieval sacred spaces, today the phenomenon is very different, despite certainly remaining a metaphor for the beliefs and feelings of each.

In regards to understanding the current pilgrimage, there are a number of concepts or recurrent themes in its study and that fall within cultural geography. On the one hand, as Maddrell (2009) has clearly shown, pilgrimage involves an experience of liminality, that is to say, transit from one stage of life to another (Turner 1974, 1969). This translation is manifested in the very notion of routes, always shown ranging from beginning to end. It also relates to the notion of *communitas* developed by Turner himself, understood as a meeting with others, that is, building a group that gets to know each other while walking and which will maintain its identity. Also, pilgrimage can be associated at present with slow mobility (Urry 2007) and a general emotional experience (Bondi et al. 2005). Be as it may, calm contemplation of the landscape can be carried out by on foot movement in an idealized landscape with a series of well-rooted stereotypes (Cosgrove and Daniels 1988).

However, an analysis of pilgrimages must not be limited to a geographical perspective to the immaterial, so it must delve into the construction of space in the line of research initiated by Lefebvre (1974). Kong (2004) referred to the sacred spaces imprinted within the territory, with business logic and a constant reply or even conflict. A perspective particularly suggestive of this direction is the *post-secular* city and those areas in which religions reinvigorate and rebuild the area (Beaumont and Baker 2011). Under similar premises urban restoration should be emphasized as an ideology aimed at beautifying the city (Rossi 1966), an evident trend throughout the twentieth century (Hall 1988) with which it maintains obvious post-secular parallels. Nevertheless, Maddrell (2009) has drawn attention to the need to understand the construction of sacred places not only from collective perspectives, but also by covering individual interventions.

The emphasis on the material perspective in the geography of religion is especially significant in the case of pilgrimages. It is known that in medieval times it was not uncommon for pilgrims to stray from the beaten track and engage in some erratic practices from a spatial point of view (Soria y Puig 1991); this could infer certain degree of spatial dispersion. Without detracting from peregrination, the building capacity during the Middle Ages in the form of bridges, temples and roads is no less true that the contemporary conception of peregrination space, which usually occurs in a (re)invention of tradition in accord with Hobsbawm and Ranger's interpretation (1983). Associated with this material dimension, there is a derived immaterial interest which also seems appropriate to mention here. While information available to the pilgrimage in the medieval world was often limited to little more than knowledge of the destination and some very vague references on how to get to there, there is now a wide range of literature produced on routes, which includes tourism, the landscape, literary, etc. This output again connects to an idea of landscape as immaterial construction linked to sites (Roger 1997; Cosgrove and Daniels 1988).

35.3 Origin and Formation of the Pilgrimage to Santiago from a Geopolitical Point of View

As stated above, Kong (2004) attaches critical importance to the political explanation of the geography of religions. In this section we intend to briefly interpret the emergence and development of the peregrination to Santiago from a geopolitical perspective. As it is known, it is assumed that sometime in the first third of the ninth century, the *inventio*[5] of the remains of Saint James was produced at the site now occupied by the City of Santiago de Compostela. It is impossible to prove the truth of the remains and there is some consensus that it is more of a myth than a plausible reality (Rey 2006). What is considered proven is that in the year 834 Alfonso II[6] gave up a court to preserve the original church built there, which demonstrates the genuine interest in protecting the brand new temple (López Ferreiro 1898–1909). From these early events, a number of pilgrim routes to Santiago appeared during the Middle Ages with the worship of Saint James urging pilgrims to travel and thus asserting the site itself as the most popular pilgrimage in medieval Europe (Barreiro 1999; Moralejo 1993), more popular than those to Jerusalem and Rome due to the ongoing instability which these latter places found themselves involved in for centuries. Only this account studied here supports a geopolitical reading which will be unraveled below. Indeed, without attempting to amend the religious and spiritual reasons of the Ways to Santiago being maintained during the Middle Ages, it is clear that the Ways corresponded to reasons that go far beyond the strict reasons of being strictly religious in nature.

The *inventio* was, first and foremost, an instrument for the assertion of political power in the Kingdom of Galicia in the eighth and ninth centuries and it allied with ecclesiastical power, which quickly moves *de facto* from Iria Flavia, the capital of the nearest Bishopric, to Santiago (Rey 2006; López Carreira 2005).[7] The infant political structure could also be situated in the orbit of the Islamic caliphate (the majority rule in the Iberian Peninsula) or the Frankish kingdom. Geopolitically, the *inventio* of Santiago constitutes a strategic orientation of Galicia and the entire north of the peninsula towards France. In this sense it is not surprising that Charlemagne was consistently associated with Santiago. Even medieval literature proposed a direct link of Charlemagne to the *inventio*.[8] The distance to the caliphate should be understood not only as political or religious opposition to Islam, but also as an internal struggle of Christianity, erecting an Apostolic Santiago challenged the ecclesiastical dominion of

---

[5] Historiography prefers the Latin term *inventio*, which literally means "discovery." However, there is a current agreement that the term means a real "invention" in this particular case.

[6] In contemporary Spanish historiography, Alfonso II is considered "King," but possibly he was a modest feudal Lord. See López Carreira (2005) for a critical review of de Alfonso II's "reign."

[7] The transfer has also to do with a practical circumstance: the risks and attacks coming from the sea, due to the coastal setting of Iria Flavia.

[8] The *Codex Calixtinus* (popular name for *Liber Sancti Iacobi*, dated from the mid twelfth century) and medieval Galician texts such as *Miragres de Santiago* (dated from the fifteenth century). The *Codex Calixtinus* is a fundamental text in peregrination to Santiago history and the foresight of a canon on the peregrination tradition, not only in the Middle Ages, but also at present.

Toledo and delegitimized its religious ideology. Secondly, the northwestern Iberian Peninsula did not conform to a uniformed kingdom as one might infer from the traditional Spanish historiography;[9] rather, it was a diverse conglomerate of competing lordships. In this scenario, for the Lord who aspires to control the northwestern Iberian Peninsula, it is strategic to erect a temple under his protection and privileges granted to it, that is, the connection between the church and the monarchy in this consolidation process is essential. Thus, the *inventio* constitutes an act of external geopolitics, and at the same time, internal.

The figure of Xelmírez (Bishop of Santiago from 1101 to 1139 and Archbishop by papal privilege from 1120) stands out here. Xelmírez appealed to the Pope to strengthen and enforce the status of Santiago. The *Codex Calixtinus* reflected the rise Santiago experienced under Xelmírez, describing a sacred map of four cardinal points: Rome, Jerusalem, Santiago and Cluny (Rey 2006). Santiago under Xelmírez was able to challenge the title of metropolis from Braga (restored in 1070) and the Peninsula seat from Toledo. In 1120 a papal bull obtained by Xelmírez confirmed the veracity of the apostolic remains, which saw the beginning of the golden age of medieval pilgrimages. Thus, Santiago was consolidated in the complex internal geopolitics of the Peninsula, where the balance between bishoprics and kingdoms, and between them and Islam, was very unstable. Santiago also succeeded in definitively linking up with Europe and on certain equal terms with Rome. A map of the Iberian Peninsula in the early eleventh century can be interpreted within this geopolitical logic (Fig. 35.1). The Way of Saint James allowed the connection beyond the Pyrenees with obvious geopolitical, economical and demographical implications. In this regard, a fledgling network that functioned as integrated urban development axis was substantial (López Alsina 1993). Overall, we can say that the Jacobean routes developed sacred symbols and ways (churches, bridges, fountains and so on), which allowed the configuration of a *sacred space* in the sense given its theoretical background. In turn, the medieval splendor of the City of Santiago de Compostela was reflected in the current importance of the heritage that originated from that period, mainly Romanesque, that is, a dominant medieval style on the Way and the city itself (Moralejo 1993).

Since the sixteenth century this scenario was broken with the triumph of Protestantism in much of Europe, with the schism with the Anglican Church and with the wars of the Spanish monarchy with other European monarchies (France, England, etc.). Sight should not be lost on the condemnation of pilgrimages and the veneration of relics by the Protestants, Lutherans or Reformists; a variant of this Christianity prevailed in much of Europe. Broadly speaking, the new situation brought about a long crisis for the Way and for the City of Santiago which "disconnected" itself from the Way, disassociating itself from its once powerful international

[9] We can cite, for example, a book that reflects in its very own title the deep ideological content: *The Kingdom of Asturias. Origins of the Spanish Nation* (Sánchez Albornoz 1979), which contains "the dominant thesis" in Spanish historiography that the northwest peninsular quickly constituted a uniform kingdom that was able to repel the Muslim power, formed around Asturias-León, which, centuries later, would become Spain. It must be said that this simple literal interpretation continues to be quite preached still today in Spain.

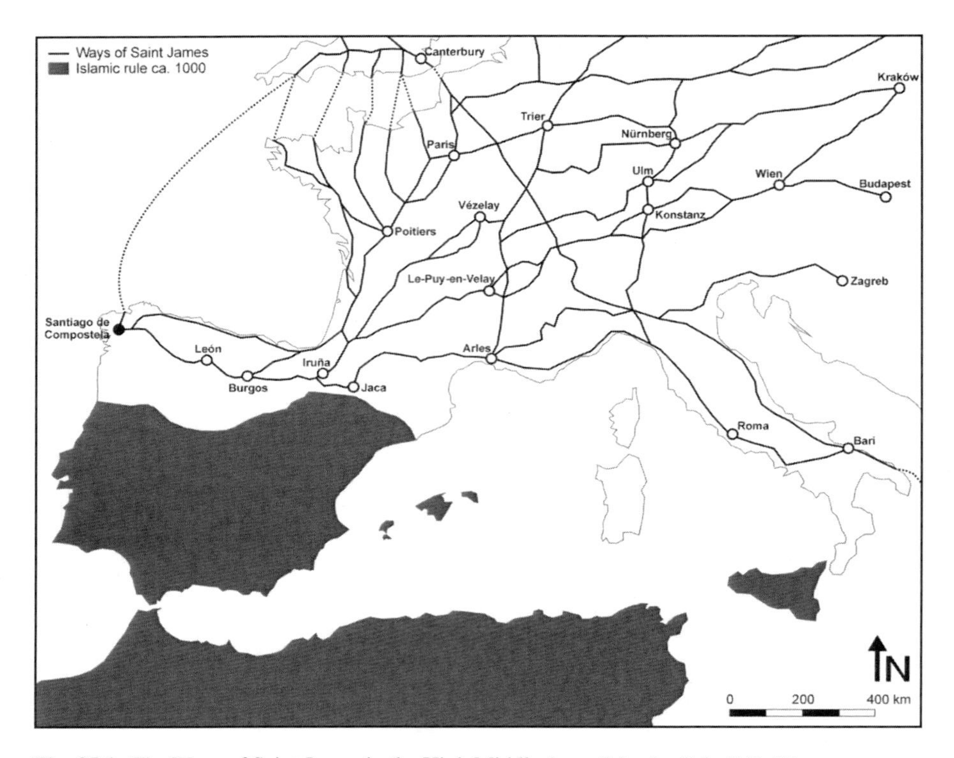

Fig. 35.1 The Ways of Saint James in the High Middle Ages (Map by Valerià Paül)

connections, except for the Italian territories (Lopez and Lois 2011), which, together with the Hispanic kingdoms, are the only Catholics who remained under the Counter-Reformation that culminated in the Council of Trent. In fact, we see an abandonment of the appeal of the figure of Saint James the pilgrim, very common in medieval iconographic representations, and the consolidation and dissemination of Saint James the Moor-slayer (*Santiago Matamoros*) or Warrior, who in the medieval period was not widespread (Rey 2006). In logic correspondence with this symbolic reconfiguration of the figure, Saint James the Warrior will be the figure transferred to Hispanic America, where we see lavish Jacobean worship during the Modern Era with hundreds of places being named Santiago and where many temples are built and dedicated in his honor. Ironically, recent years have seen the construction of "Ways of Saint James" in Latin America[10] in order to introduce a pilgrim meaning in a land where previously there was little or no interest. The Jacobean resurgence and is now the subject of next section.

[10] In the state of Bahia (Brazil) one finds the *Caminho de Santiago do Iguape* and the *Caminho de Santiago do Piripiri*, the first from Cachoeira and the second from Vitória da Conquista. These ways are used by Brazilian pilgrims before going to the *Camino de Santiago* in Europe or, after doing it, by veteran pilgrims. They are also increasingly being used by people who have no connection with the European *Camino*.

35.4 Rebirth of the Way in a Game of Multiple Scales

From the late nineteenth century onwards signs can be traced indicating the mobilization of the Way of Saint James again, but, above all in the early twenty-first century the pilgrimage acquired a new relevance in a complex multifaceted process. Consistent with the geographical perspective practiced, this section is structured by means of scales, that is, where the Way was appropriated within the framework developed in classical political geography, "the act of adopting something by its agent" (Sànchez Pérez 1992: 66). It should be recalled that the scales are constructed, both politically and socially (Marston 2000), "not simply an external awaiting discovery but a way of framing conceptions of reality" (Delaney and Leitner 1997: 94–95). Thus, we distinguish five scales which cannot be understood in isolation, but interact between each other (Delaney and Leitner 1997):

- Spain as a nation-state construction.
- Galicia as national construction.
- The devolved regions and nationalities that attain autonomy within the framework of a decentralized political system in Spain since the late 1970s. Since the 1978 Spanish Constitution, they are euphemistically so-called "communities."
- Local governments draw up their own urban strategies.
- The European Union, which is regarded internationally as one of the clearest cases for the emergence of a new world regional scale.

35.4.1 Spain as a Nation-State Construction

According to Álvarez Junco (2001), the idea of Spain as a nation-construction took place mainly in the nineteenth century by means of a dominant discourse that defends the identification of the Spanish national entity with Catholicism. The key ideology in this respect has been National-Catholicism, preponderant in state elites from the mid-nineteenth century and certainly official during the Franco dictatorship (1936/39–1975/78). Within this ideological national construction the prominent figure of Saint James took pride of place, as Álvarez Junco (2001: 556) himself mentions. The figure of Cardinal Miguel Payá (Archbishop of Santiago between 1875 and 1886) is crucial in this respect as he "reinvents" the Apostle, which, as mentioned above, had lost his appeal for centuries. In 1878 Payá promotes an excavation to find the tomb of Saint James, allegedly hidden since the sixteenth century, with prospecting carried out by Canon López Ferreiro. In 1884 the Pope, through a papal bull, recognizes the authenticity of the tomb (Villares 2003).[11]

[11] In historiography this is considered the second *inventio*. Villares (2003) examines the archaeological excavations carried out at those times and points out that the remains have never been scientifically analyzed.

The Jacobean years[12] and pilgrimages were quickly reinstated, first from different localities of the bishopric of Santiago and then from the 1920s from more distant points in Spain. The association between Santiago and the Spanish nation is born and in this way it was emphasized from 1936 onwards, when Franco carried out a coup d'état and took power in Spain through a civil war that lasted until 1939. In fact, the 1937 Holy Year is extended to 1938 (the only time in history), because of the civil war. This activity was clearly due to the Church's backing Franco's cause, associating his struggle as a "crusade" in which the image of Saint James the Warrior was frequently present.

As Castro Fernández (2010) has written, Franco developed the National-Catholic narrative, denouncing Spain as having lost its "eternal" spiritual and religious values and proposing that Saint James would help rediscover them and unify the nation. Thus, the Holy Years of 1948 and 1954 were widely used by the Franco regime for massive historicist rehabilitation which affected the Way and the City of Santiago, as will be seen in the next section. However, the importance of the National-Catholic state machine granted to Saint James should be emphasized here, in direct collaboration with Cardinal Quiroga Palacios (Archbishop from 1949 to 1971) and spread the worship of Saint James. As Castro Fernández (2010) suggests, from the Holy Year of 1965 onwards, the Franco Government discursively redirected the Way to tourism interests, without in any way neglecting the values of faith and nation (Fig. 35.2). Political attention on the Jacobean increases and in this regard was paid to the establishment of a National Foundation for Santiago and a Central Board for the Holy Year of 1965. The then Minister of Information and Tourism, Manuel Fraga, stated in 1965 that the Ways will acquire "apart from its religious mission" another "nature, very close to tourist and information" (quoted in Castro Fernández 2010: 88).

35.4.2 Galicia as a Nation Construction

Since the nineteenth century, Galician nation building does not include the Jacobean. However, according to Gaspar (1996), the Galician nationalist movement before the civil war featured a certain image of the Apostle as a pilgrim and understood the Way as a key mechanism to explain the Europeanness of the Galician nation, for the connection of a *land's end* with the rest of the continent. For Otero Pedrayo, one of the key nationalist thinkers, the City of Santiago assumed on a fundamental value for the nation as the "eternal capital" that symbolizes both the sublimation of national values and the connection to Europe. Consequently, it is not surprising that Galician nationalism chooses the 25th of July as Galician National Day or better known to nationalists as the Day of the Galician Fatherland. It was first held in 1920. During the Franco dictatorship the date in the Galician nationalist sense was

[12] In the Jacobean tradition, Holy Years (the Jubilees or years of remission of sins and plenary indulgence) are declared when Saint James Day (25th of July) falls on a Sunday. Holy Years fall every 6, 5, 6, and 11 years.

Fig. 35.2 Promotional poster of the Holy Year of 1965, by Francesc Català-Roca (Source: Santos 2005: 132, used with permission)

forbidden, but in the 1960s banned nationalist demonstrations on this day cause clashes in the City of Santiago. Since the founding of the Galician autonomy in 1981 (Statute of Autonomy), the date has become a bank holiday in Galicia. It is interesting that in the 1980s the Spanish Government converted the 25th of July into a working day while Our Lady of the Pillar (12th of October) is maintained as a non-working day (and Spanish National Day). The 12th of October combined a traditional religious meaning (now Our Lady of the Pillar seems to be "more Spanish" than Saint James) and connections with Hispanic peoples (the American countries formerly ruled by the Spanish Empire), as soon as America was discovered on October 12th, 1492.

35.4.3 The Autonomous Communities, with Special Reference to Galician Autonomy

The development of a consolidated and decentralized system in Spain in the 1980s involved the creation of 17 devolved governments that have, among its constitutionally conferred powers, economic policy development and tourism planning and management. In the case of Galicia in the early 1990s, it is decided that the Way should become the leading tourism product (Santos 2006). Although the reasons for this decision vary, the role played by Manuel Fraga, who had been minister under the dictator Franco and was president of Galicia from 1990 until 2005, was critical. While in office, his agenda included the touristification of the Way. Touristification of the *Camino* had been limited during the Franco regime. That the Galician Government's marked 1993 as a Holy Year was seen a true leap forward in this respect. The political developments of the 1990s can be summarized as follows:

- The improvement of the Way's paths, including new sections, decisions to relocate the old tracks, minor repairs and public works, etc. An Act of Parliament in 1996 sought to define precisely the routes of the Ways of Saint James.[13]
- The creation of a network of public hostels along the Ways, viz., the French Way. This was considered the backbone of the tourist development. Although most private rural tourism houses were also implemented, the preferred accommodation by pilgrims is, even today, the public infrastructure of hostels, as they had been free of charge since their origins in 1993 until the mid-2000s (today they are still cheap).
- The creation of public companies dependent on the Galician Government that dealt with developing, marketing and branding:
  - *Turgalicia* (Company for the Image and Promotion of Tourism in Galicia) is dedicated to tourist promotion and management.
  - *Xacobeo* is a company dedicated to the promotion of the Jacobean culture.
- Promotional campaigns, including the sponsorship of sports teams, singers, etc., in which an association of important companies in Spain intensively uses Jacobean brands set up by the Government in its advertisements.
- Investments in the city of Santiago: public works, especially city access infrastructures. In this respect, the role of the Consortium of Santiago, with the participation of the Galician Government, the Municipality of Santiago and the Spanish Government, has to be highlighted.[14]

The effects of these aggressive policy developments were immediate. In the 1980s there were less than 1,000 pilgrims each year, according to the Archbishop of Santiago. In 1992 there were 9,764 pilgrims recorded, yet 1993 saw an impressive ten-fold increase of almost 100,000 pilgrims.[15] Although in 1994 there was a decline, a decision was made that each Holy Year would be the main aim of tourist promotion, campaigning, marketing and branding. In this respect, it is decided that the tourism policy in Galicia would be dependent on the Holy Years and that tourist promotion would always target the next Holy Year as the tourist catalyst.

[13] Act 3/1996 for the Protection of the Ways of Saint James.

[14] Created in 1992, the Consortium is public administration that includes participation of the Spanish Government (60 %), the Galician Government (35 %) and the Municipality of Santiago de Compostela (5 %). The Consortium proposes, promotes and safeguards, financially and technically as parts of its goals various initiatives and projects for the City of Santiago de Compostela.

[15] Official figures for the number of pilgrims as declared by the Catholic Church. See www.peregrinossantiago.es/ (Accessed 30 Apr 2012) and footnote 25.

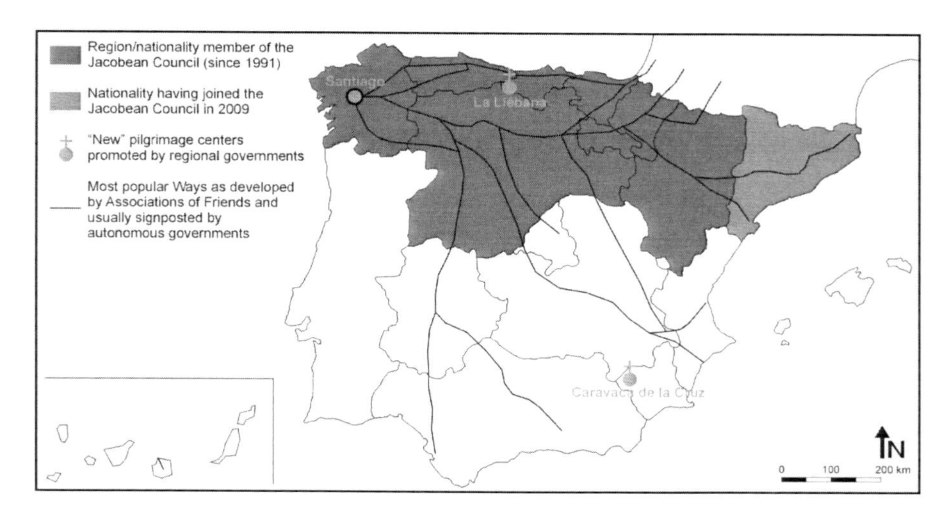

Fig. 35.3 The autonomous regions and nationalities of Spain and the pilgrimages promoted through tourism policies (Map by Valerià Paül)

It is important to note that other autonomous governments have followed the Galician tourism policy development and use the *Camino de Santiago* as a significant element. This especially applies to governments with sections along the French Way (Castile and León, La Rioja and Navarre), which have not only put considerable effort into placing signposts similar to those in Galicia, but also other areas such as Catalonia and the Basque Country (Fig. 35.3). In 1991 the Spanish Government created the Jacobean Council, a body for cooperation and collaboration which assumes a cooperative role consistent with the decentralized nature of Spain.[16] The "Jacobean success" has no doubt inspired some Spanish autonomous regions to seek and develop tourism of their own using ancient pilgrimage centers of more limited significance from an historical point of view than that of Santiago. Now they are trying to "reactivate" these centers by using an association with Saint James, for example, Caravaca de la Cruz (Murcia) and La Liébana (Cantabria) (see Fig. 35.3).

35.4.4 The Cities on the Way with Special Reference to Santiago de Compostela

In a sense similar to tourist developments that are reported for regional governments in Spain, several cities have discussed the Way as a strategic opportunity for them to promote an urban project since the advent of democracy.[17] This is the case of Astorga

[16] The Jacobean Council essentially is responsible for carrying out programs and performances related to the *Camino de Santiago*, especially Holy Year celebrations.

[17] In Spain, local authorities did not have any real independent decision-making power until the Constitution of 1978. The first democratic elections were in 1979.

Fig. 35.4 "Santiago Is Big" brochure by the Municipality of Santiago, 2008, used with permission of © Turismo de Santiago de Compostela – http://www.santiagoturismo.com/

and León (in Castile and León), and especially Santiago (Lois and Somoza 2003). The latter city, the Way's termination point, deserves special attention. Since the 1980s the city has initiated a strategic project to construct a symbolic image of Santiago as a globally relevant city. Its first great milestone was its declaration as a World Heritage site in 1985, the *Camino de Santiago* was also later declared a World Heritage site in 1993 in Spain (in 1998 in France). It is significant that one of the most discussed issues in Galicia at the start of devolution was the designation of a capital.[18] Santiago was chosen not only as a Solomonic decision (Vigo and A Coruña are major metropolitan areas in Galicia, with over half million inhabitants each, while Santiago is a medium-sized city that sits between the two), but also because of its symbolic value. The City of Santiago has a strategically active renovation planning policy that promotes historical parts of the city, something which will be discussed in the next section, and is accompanied by the association of consistently recurring institutional images of the city with the *Camino de Santiago* and Galicia (Fig. 35.4).

35.4.5 The European Union

There are cultural references of enormous cultural significance within Europe, such as Goethe or Dante, both who spoke of the Way's importance to the continent. It is certainly a leading cultural heritage that supports the European dimension of the

[18] The establishment of the capital was the first Act passed by the Galician Parliament (Act 1/1982 establishing the institutions' seat). However, it cannot be considered a closed case, in the way that there always remains public and political debate on this question. In fact, the capital statute for Santiago was agreed two decades after the city was selected as the capital (Act 4/2002 of the statute of the capital of the City of Santiago de Compostela). These developments show the continuing debate of this theme.

Jacobean phenomenon. However, what has occurred in recent years is a clear Europeanization of the Way led by the European institutions and driven largely by the progressive Spanish (left-wing) government in the 1980s, during which time Europe overcame the perceived traditional backwardness of Spain and took the peregrination, free of its religious and nationalist meaning, as a discursive mechanism in this direction. It is important in this regard to indicate that the Socialist Party won the 1982 Spanish elections, with a strong pro-European discourse, and that since 1983 the Santiago Mayor's office has been in the hands of the Socialist Party as well (until 2011). Both policy-makers worked in this direction in the 1980s. Above all, it highlights the fact that the Way of Saint James was declared the first European Cultural Route in 1987, a collaborative decision that marked Spain's entry into the European Communities (later, the European Union) in 1986. Moreover, symbolically, the selected colors for the Way of Saint James signs are blue and yellow (the same as the European flag) since the 1950s.[19] In addition, the designation of Santiago as the European Capital of Culture in 2000, together with eight other cities, represented a kind of culmination of this association process of the city and the Jacobean phenomenon with Europe as a whole.

35.5 Contemporary Space Production of the Way and Santiago

In previous sections we have privileged geopolitical views, including key appropriations of essentially immaterial spaces. What was theorized in the initial theoretical background, we wish to continue by looking at the physical changes of the Way and the City of Santiago as a route and a place of peregrination respectively. During the last century, sacred spaces inherited from the Middle Ages (which we have discussed above) have been renovated, reconfigured, or their space value reinterpreted, and there have even been new additions that have little or nothing to do with the past. In the contemporary context, we start by referring to the intense restoration work related to Way's construction led by Pons-Sorolla between the mid-1940s and early 1970s, that is, during most of the dictatorial period. As demonstrated by Castro Fernández (2010), the peaks of increased activity occurred prior to the Holy Years (especially 1965), along with the vast majority of interventions occurred in religious architecture (particularly churches and monasteries) and especially those with a Romanesque art style. Among the interventions, two are important. One is Cebreiro, the legendary place of arrival in Galicia, on the French route. This mountain village was restored to accommodate pilgrims, thus converting Cebreiro into a museum or exhibition piece that displaced the original inhabitants. Second, there is Portomarín, where the construction of a reservoir flooded the original nucleus and only churches and other medieval elements were transferred to the new settlement. This decision

[19] It is noteworthy that Jacobean routes are signposted using these same colors in countries such as Brazil.

showed that the dictatorial regime corresponded to the National-Catholicism and aimed to place religion in a prominent position through intentional and partial use of the past (especially medieval) as the ideological support of this decision to pursue a traditionalist model for Spain.

Other interventions highlighted during the Franco period with a strong ideological connotation also occurred in the City of Santiago de Compostela itself. In regards to restoration, it should be noted the renovation of the Royal Hospital as a luxury hotel was included in the public network of hotels in Spain, so-called *Paradores*. This renovation promoted the creation of a tourist infrastructure in the city by picking up on the peregrination tradition (Castro Fernández and Lois 2006). New installations, including the construction and inauguration of the Burgo das Nacións in the late 1960s should be emphasized; this was originally a hostel designed to receive Peregrines in the Holy Year that has subsequently was transformed into a university residence hall for students.

Since the advent of democracy, the production of space linked to the *Camino* has largely abandoned religious rationales and, on a local level, there was need for cities to compete in the global context with a defined individualized brand image. Thus, comprehensive and strategic plans emerged in many cities on the Way of Saint James (Lois and Somoza 2003) in which there was a desire for both urban beautification (in line with Rossi 1966) and international positioning. The Comprehensive Plan for the City of Santiago in 1989 and the Special Plan for the historic area, which was finally adopted in 1997 (but with progress since 1990), should be stressed in this respect. These are only possible because the council was democratic and, therefore, able to develop its own strategies. The case of the old town, for example, favored a reflection on the historic urban fabric within the city itself, which goes beyond the monuments and the desire to keep the historic area alive (Dalda 2007). Also, the 1989 Comprehensive Plan determined how to provide meaning to the urban path of the French Way, as it serves as an axis for the pedestrian citizen (Dalda 2007). Specific urban treatment for the Porta do Camiño (the area that gave pilgrims entrance to the historic city through a walled gate), was where there is now a museum and a new park. It was also within this area that gave new attention to the Jacobean urban project within the democratic City of Compostela. The Porta do Camiño is a sort of "flagship" *avant la lettre* (see Vila Vázquez 2011 or Bianchini et al. 1992 for the concept of "flagship," that is, a marketing initiative to (re)develop an urban area to (re)launch the city beyond its area of influence and to the widest possible scale).

The production of space in the democratic period also included the enormous work undertaken by the Galician Government as a result of activities of a tourism policy linked to the *Caminos* since the early 1990s. Beyond what has been noted above, the implementation of the hostels as a clear mechanism for creating a new Jacobean geography has to be analyzed. There are currently about 52 public hostels according to the Galician Government (data from 2011) on all the Ways, including ex novo buildings and restorations of old constructions. As Cebreiro in the 1960s, these restorations sometimes affect several sets of buildings and make partial use of architectural historicist quaintness, but, overall, there is a certain commitment to the introduction of modern elements combined with traditional architecture.

Fig. 35.5 Sculpture on the top of Monte de Gozo Hill (Photo by Valerià Paül, 2009)

Within the democratic period, a special case that deserves mention is Monte do Gozo.[20] On this hill Pope John Paul II held a crowded Mass in 1989 for World Youth Day; a modern religious sculpture has been erected to commemorate the event (Fig. 35.5). Subsequently, the management of the Monte do Gozo hill for such masses, and afterwards for tourist purposes and major events, has required major landscape construction. The project has been criticized by the Church because of the religious and spiritual significance of the mount. In particular, a "city of pilgrims" was built for the 1993 Holy Year with 3,000 beds available that are now being used for tourism. Be it as it may, the striking sculpture crowns the late 1980s as a whole and largely symbolizes the strength of the religious power in the Jacobean tale.

One last point to the production of space has a particular nuance: the "City of Culture of Galicia;" it is located on the top of a hill and has been controversial and disputed due to the huge investment it has required and the (still now) unclear use of the complex. Since 1999 the project has been under development and construction

[20] This hill is renowned for being the place where pilgrims on the Way of Saint James get their first views of their destination: the Cathedral of Santiago. Gozo means "enjoyment" in Galician.

Fig. 35.6 The City of the Culture of Galicia, on the top of Gaiás Hill, in Santiago (Photo by Valerià Paül, 2011)

and in financial terms has been assumed solely by the Galician Government, a truly "flagship initiative" in the sense given by Vila Vázquez (2011) or Bianchini et al. 1992. In terms of design, the City is designed by the American architect Peter Eisenman as a huge new urban landscape combining Jacobean, Santiago City and Galician motifs (Fig. 35.6). For instance, the dimensions of the flagship precinct are identical to the dimensions of the Old Quarter of Santiago and the external morphology of the area is like an almond, inspired by the shape of the Old City. The internal distribution of the precinct reminds one of the urban street map of the Old Quarter. In addition, the site is supposed to form a shell shape (the pilgrim symbol) that is orientated towards the city center, like the Ways of Saint James are. This work is a hideously monumental structure and possibly can be considered post-secular due to the contemporary relevance of the Jacobean motifs (in part, religious) used therein.

35.6 The Emergence of the Way's Landscape

The Western idea of landscape is a modern construction, developed mainly between the eighteenth and nineteenth centuries, straddling the Enlightenment and Romanticism periods (López Silvestre 2004).[21] Thus, it is untimely to consider there is a landscape dimension in a medieval pilgrimage, so that pilgrims could not

[21] See also the "seven criteria" for a landscape to be considered, according to Berque (2008).

experience or enjoy (or suffer) the Way in the scenic sense. Indeed, the quintessential medieval Jacobean guide, the *Codex Calixtinus* contains no type of landscape arguments and that has a character between religious and utilitarian. As reported in the theoretical framework above, the landscape can be understood as a cultural reality, halfway between thought, perception or representation on the one hand, and a certain piece of respected land area on the other, that is, as a transition point between the immaterial and the material (Berque 2008; Roger 1997; Cosgrove and Daniels 1988). In this sense, the revised spatial productions discussed in the previous section can be considered landscape as far as they constitute ideological imprints in several areas, joining both dimensions of the landscape. Beyond these dimensions, in this section we will focus on the landscape of the *Camino* as current cultural discourse.

Since the early twentieth century, from the revival of the Jacobean theme discussed above, the peregrination literature has been produced about Santiago de Compostela telling of the delights of the City and the Way. In this literature both the devout and the religious, as well as the touristic, can be traced (Villares 2003). This is where the first discussion talk of landscape on the *Camino* can be found. However, not until the development of the Way of Saint James Friends Associations in the 1960s and the effort of Elías Valiña, Cebreiro priest from 1959 to 1989, can we properly speak of the generation of a landscape discourse. Valiña, who worked with Pons-Sorolla in rebuilding Cebreiro, devoted much of his life to spreading the Jacobean peregrination tradition and published several works (guides, catalogues, etc.). Likewise, he carried out a complete mapping of the Ways. All these narratives embody a vision of the Way with significant emphasis on landscape fundamentally rooted in historical elements. So, it is modeled on medieval sacred spaces and generated a coherently structured discourse that emphasized the importance of tradition. Thus, bridges, chapels, fountains and so on[22] emerged as a "landscape of the Way" and integrated within it its own logic, which can be traced back to cartographically accurate maps that are placing the different elements while the pilgrim advanced his/her paces and the landscape can be seen and perceived. This form of telling and enjoying the scenery can be considered consolidated when taking into account the importance of history and tradition in current Way guidebooks, both Spanish and international ones. In fact, this direction has been instrumental in the success of the Jacobean in those "new" countries where the elite tend to "mystify history" (for example, Australia, the United States or Brazil). Also, the preference for the historical stereotype in landscape construction is understandable in an uprooted society that is, therefore, eager to discover its past and its roots. The *Camino* is in this sense a strategy used to get in touch with previous generations through a given landscape.

Nevertheless, the landscape of the Way of Saint James is now a vector that is not limited to the past, but incorporates a number of relevant connotations without an historical basis. Thus, the association between *Camino* and the building of spirituality and

[22]Among other contributions, see Soria y Puig (1991) and Moralejo (1993) as works that extensively cataloged these items.

personality in a common place is widespread, but recent. It is an intangible impression in which the experience of walking allows a personal exercise of reflection, and even emotion, so that the landscape becomes the most confidant of silent introspection. We must also emphasize the importance of mutual understanding and coexistence among the pilgrims as they walk. Both aspects are directly related to the notions of liminality and *communitas* by Turner (1974, 1969) as mentioned in the theoretical background, with the nuance that they occur on the Way while also enjoying the landscape. This spiritual landscape, both individual and collective, has been conveyed largely through literature that has enjoyed wide circulation. Thus, it has come to create a literary landscape that, far beyond allowing the Way to be promoted around the world, gives a prejudical perspective to walkers/pilgrims before they experience the Way by themselves. Thus, they expect to find what is presented in books about the *Camino*. This is, ultimately, a mechanism of prime importance for the production and reproduction of a landscape as a cultural discourse.

At this point it should be stressed *O diário de um mago*, by Brazilian novelist Paulo Coelho (1987), has been translated into over twenty languages (in English, *The Pilgrimage*). Coelho reports the Way of Saint James as a mystical and magical experience. Nevertheless, there are so many books that have been written in a similar vein; for example, *Ich bin dann mal weg* (by the German actor Hape Kerkeling 2006; *I'm off then* in English) has been very popular; in it spiritual and personal arguments related to the landscape also emerge with the idea that doing the *Camino* offers the chance to meet very interesting people from around the world with whom one weaves multiple affinities. The *Camino* acts as a meeting point in the film *The Way* (2010, directed by Emilio Estevez); in this the landscapes occupy a prominent place on a par with the personal meanings mentioned (Fig. 35.7). Recently in Australia *A Food Lover's Pilgrimage to Santiago de Compostela*, by Dee Nolan (2010), has been highly successful under the guise of a cookbook. It offers a full reading of Way through a generous photographic apparatus that correlates landscape, food and introspection. Its main idea is in the following quote, which in turn summarizes what has been said in this point: "Why has this medieval Christian pilgrimage made such a comeback? This was the question everyone asked me when I returned home. […] I think I know the answer. I had been surprised how calmed and soothed I felt when I walked day after day. I relished experiencing nature at human speed and rediscovered the luxury of conversations without the interruptions that are usual in our daily lives."[23]

In fact, this quote from Dee Nolan can refer to a point related to the formation of a landscape of the contemporary Way and links to trends seen in other areas. This is interpreting all that is seen while doing the *Camino* as natural, as an untouched paradise or of little human presence, where it is possible to be alone (or with very few other people), relaxed and in harmony with nature. However, the fact is that the

[23] Edited abstract of the book published in a Qantas passenger magazine in December 2010, pp. 200–206.

Fig. 35.7 *The Way* (2010) film posters (Source: Internet Movie Poster Awards – http://www.impawards.com/2010/way.html)

Way crosses intensively modified areas and many of them are densely populated and even with prominent environmental and notable landscape impacts that the pilgrim/walker may not want to see them. It is about a deep-routed landscape discourse and, in some way, distant from reality and responds to a chlorophyll or pro-environmentalist perception dominant in Western urbanized society (Ojeda 1999; Roger 1997). In fact, just as past interventions were performed on elements constructed with a clear ideological purpose, at present there is a growing desire to "make up" certain environmental points on the Way in order to hide shocking elements or locations such as landfills, quarries or wind turbines (Fig. 35.8). It is no more than just a certain way of understanding the landscape, that is, chasing the stereotypical idealized landscape of the Way (as historical, spiritual and green, as we have here traced) and not to find contradictions with reality. But surely this idealized green scenery acts as a subterfuge that hides a much more complex reality in which, for example, the Galician Government has not yet been able to pass the Special Plan for Protection and Promotion of the Way which it was required to do in 1996.[24] Without this plan, current management of the *Camino* area is inconsistent and results in disjointed and uncoordinated decisions.

[24] Additional Provision of Act 3/1996 (cited in footnote 13) required the Galician Government in 2 years to pass this Special Plan. It has been 14 years since the legal deadline.

Fig. 35.8 Deteriorated landscape in the Way of Saint James, near Santiago (Photo by Miguel Pazos, 2011)

35.7 Concluding Remarks

As we have been seen, the *Camino de Santiago* has an ambivalent nature: between tangible and intangible, between past and present, between secular and religious. Indeed, religion still has an important role, for example in the latest rationale for landscape discourse of historical character or in the process of the appropriation of Saint James for Spain as a nation state construction. However, it seems clear that what strongly emerges is secular and that this novelty is ostensibly expressed in space, notably in the growing touristification of the Way and, associated with it, its progressive loss of religious meaning. In parallel, we have observed a spiritual consolidation in everything referring to the *Camino*, either individually or in groups, a trend that is consistent with the emotional geographies of Bondi et al. (2005). The fact that today most of the travelers on the Way profess no religious motivations[25] proves how secularization is underway. The current demand for the *Camino* is not only composed of Catholics, but also people from other religious backgrounds and people who declare themselves agnostic or atheist.

[25] In this sense, it should be noted that data supplied by CETUR (2007–2010) are appropriate. Data provided by the Catholic Church, which are often used in Jacobean research (in this work this has been done, see footnote 15), cannot be considered strictly statistics as they only count people who have received the religious certification (so-called *Compostella*). This certification represents an underestimation of the flow, because the "pilgrims" must compulsory declare doing the Way for religion reasons in order to receive certification, which many times people only casually fill out.

Be as it may, this chapter has argued that there is evidence of resacralization or, as advocated in the theoretical framework, post-secularism (Beaumont and Baker 2011). While it is true that the secular has taken over the Way, the fact remains that the process by which it was recovered from the late nineteenth century responds to clearly religious facts and materialized in the space: rehabilitation of churches and monasteries, historic-artistic embellishment of stone crosses, etc., that is, new sacred spaces resituate into the contemporary reality of centuries ago do not always resurrect the past but, as we have seen, come with notable innovations. Also, in recent years, there is the reiterated institutional use of the Jacobean, with constant repositions of religious elements in the space, albeit mostly for tourist purposes. One case in point is Monte do Gozo, although essentially it is devoted to being a tourist complex because it contains visible religious landmarks.

It should be added that from the strictly tourist point of view, we have developed a critical literature to this regard (among other works, see Santos 2006, 1999; or Santos and Lois 2011). Indeed, we doubt that the enormous economic resources spent by the public sector on the promotion and development of tourism on the Jacobean product have real positive effects. It is true that there has been a steady increase in the flow of pilgrims that moves along the Way (Fig. 35.9), but the fact remains that the average expenditure is very low and that this flow is a high debtor to public investment (for example, lodging) and has low multiplier effects.

Finally, this discussion has shown the enormous explanatory power of geography to address the *Camino de Santiago* in its many contemporary facets. We have not

Fig. 35.9 Bike pilgrims in the French Way, in O Pino municipality, near Santiago (Photo by Valerià Paül, 2009)

only used arguments from the geography of religions in the strictest sense (for example, sacred spaces and the debate on secularization), but also issues of cultural geography (landscape narratives, emotional turn, etc.) political geography (political-administrative scales, spatial appropriation, etc.), urban geography (heritage restoration, flagship initiatives, etc.), economic geography (tourism, development, etc.) and regional and urban planning, all which present a way to look globally at a different *Camino*, one which complements previous approaches exclusively focused on specific aspects or one only within historical or anthropological interpretations. Our main conclusion is that the Way offers us one of the most comprehensive geography lessons in the early beginnings of the twenty-first century.

References

Álvarez Junco, J. (2001). *Mater dolorosa. La idea de España en el siglo XIX*. Madrid: Taurus.

Barreiro, X. L. (1999). *The construction of political space: Symbolic and cosmological elements* [Jerusalem and Santiago in western history]. Santiago de Compostela: The Araguaney Foundation.

Beaumont, J., & Baker, C. (Eds.). (2011). *Postsecular cities: Space, theory and practice*. London: Continuum.

Berque, A. (2008). *La pensée paysagère*. Paris: Archibooks.

Bianchini, F., Dawson, J., & Evans, R. (1992). Flagship projects in urban regeneration. In P. Healey, S. Davoudi, M. O'Toole, S. Tavsamogulu, & D. Usher (Eds.), *Rebuilding the city: Property-led urban regeneration* (pp. 245–255). London: E & FN SPON.

Bondi, L., Davidson, J., & Smith, M. (2005). Introduction: Geography's 'emotional turn'. In L. Bondi, J. Davidson, & M. Smith (Eds.), *Emotional geographies* (pp. 1–16). Aldershot: Ashgate.

Castro Fernández, B., & Lois, R. C. (2006). Se loger dans le passé. La récupération emblématique de l'Hostal des Rois Catholiques de Saint-Jacques-de-Compostelle en hôtel de luxe. *Espaces et Sociétés, 126*, 159–179.

Castro Fernández, B. (2010). *El redescubrimiento del Camino de Santiago por Francisco Pons-Sorolla*. Santiago de Compostela: Xunta de Galicia.

CETUR (2007–2010). *Observatorio estatístico do Camiño de Santiago*. Santiago de Compostela: Universidade de Santiago de Compostela/Xunta de Galicia/Centro de Estudos e Investigacións Turísticas.

Coelho, P. (1987). *O diário de um mago*. São Paulo: Planeta do Brasil.

Coleman, S., & Eade, J. (Eds.). (2004). *Reframing pilgrimage: Cultures in motion*. London: Routledge.

Collins-Kreiner, N. (2010). The geography of pilgrimage and tourism: Transformations and implications for applied geography. *Applied Geography, 20*(1), 153–164.

Cosgrove, D., & Daniels, S. (1988). *The iconography of landscape*. Cambridge: Cambridge University Press.

Dalda, J. L. (2007). Planes y políticas urbanas. La experiencia urbanística de Santiago de Compostela desde 1988. *Urban, 12*, 102–125.

Delaney, D., & Leitner, H. (1997). The political construction of scale. *Political Geography, 16*(2), 93–97.

García Cantero, G. (2010). Ruta jacobea, jus commune y jus europeum. *Revista de Derecho UNED, 7*, 307–324.

Gaspar, S. (1996). *A Xeración Nós e o Camiño de Santiago*. Santiago de Compostela: Xunta de Galicia.

Hall, P. (1988). *Cities of tomorrow: An intellectual history of urban planning and design in the twentieth century*. Oxford: Blackwell.
Hobsbawm, E., & Ranger, T. (Eds.). (1983). *The invention of tradition*. Cambridge: Cambridge University Press.
Ivakhiv, A. (2003). Nature and self in new age pilgrimage. *Culture and Religion, 4*(1), 93–118.
Kerkeling, H. (2006). *Ich bin dann mal weg – Meine Reise auf dem Jakobsweg*. München: Piper Verlag.
Kong, L. (2004). Religious landscapes. In J. S. Duncan, N. C. Johnson, & R. H. Schein (Eds.), *A companion to cultural geography* (pp. 365–381). Malden/Oxford/Carlton: Blackwell.
Lefebvre, H. (1974). *La Production de l'espace*. Paris: Anthropos.
Lois, R. C. (2011). *El Camino de Santiago hoy: Debate sobre la multiculturalidad* [Keynote address delivered at the 9th International Conference of Jacobean Associations, in Alzira].
Lois, R. C. (2013). The Camino de Santiago and its contemporary renewal: Pilgrims, tourists and territorial identities. *Culture and Religion: An Interdisciplinary Journal, 14*(1), 8–22.
Lois, R. C., & Santos, X. M. (1999). El Camino de Santiago. In V. Bote (Ed.), *La actividad turística española en 1998* (pp. 597–603). Madrid: Asociación Española de Expertos Científicos en Turismo.
Lois, R. C., & Somoza, J. (2003). Cultural tourism and urban management in north-western Spain: The pilgrimage to Santiago de Compostela. *Tourism Geographies, 5*(4), 446–461.
López Alsina, F. (1993). El Camino de Santiago como eje de desarrollo urbano en la España medieval. *ICOMOS, 2*, 50–60.
López Carreira, A. (2005). *O reino medieval de Galicia*. Vigo: A Nosa Terra.
López Ferreiro, A. (1898–1909). *Historia de la Santa Apostólica Metropolitana Iglesia de Santiago*. Santiago de Compostela: Seminario Conciliar Central.
López Silvestre, F. (2004). *El discurso del paisaje. Historia cultural de una idea estética en Galicia* (pp. 1723–1931). Santiago de Compostela: Universidade de Santiago de Compostela.
Lopez, L., & Lois, R. C. (2011). *Peregrinos y testamentos: Fuentes históricas para la geografía* [Presentation delivered at the Regional Conference of the IGU, in Santiago de Chile].
Maddrell, A. (2009). A place for grief and belief: The witness Cairn, Isle of Whithorn, Galloway, Scotland. *Social & Cultural Geography, 10*(6), 675–693.
Marston, S. A. (2000). The social construction of scale. *Progress in Human Geography, 24*(2), 219–242.
Moralejo, S. (Ed.). (1993). *Santiago, Camino de Europa. Culto y cultura en la peregrinación a Compostela*. Santiago de Compostela: Xunta de Galicia.
Nolan, D. (2010). *A food lover's pilgrimage to Santiago de Compostela. Food, wine and walking along the Camino through Southern France and the North of Spain*. Camberwell: Lantern.
Ojeda, J. F. (1999). Naturaleza y desarrollo. Cambios en la consideración política de lo ambiental durante la segunda mitad del siglo XX. *Papeles de Geografía, 30*, 103–117.
Rey, O. (2006). *Los mitos del Apóstol Santiago*. Santiago/Vigo: Consorcio de Santiago/Nigra Trea.
Roger, A. (1997). *Court traité du paysage*. Paris: Gallimard.
Rossi, A. (1966). *L'architettura della città*. Padova: Marsilio Editori.
Sánchez Albornoz, C. (1979). *El Reino de Asturias: Orígenes de la Nación Española. Estudios críticos sobre la historia del Reino de Asturias*. Oviedo: Real Instituto de Estudios Asturianos.
Sànchez Pérez, J. E. (1992). *Geografía política*. Madrid: Síntesis.
Santos, X. M., & Lois, R. C. (2011). El Camino de Santiago en el contexto de los nuevos turismos. *Estudios turísticos, 189*, 95–116.
Santos, X. M. (1999). Mitos y realidades del Xacobeo. *Boletín de la Asociación de Geógrafos Españoles, 27*, 103–117.
Santos, X. M. (2002). Pilgrimage and tourism at Santiago de Compostela. *Tourism Recreation Research, 27*(2), 41–50.
Santos, X. M. (Coord.). (2005). *Galicia en cartel. A imaxe de Galicia na cartelaría turística*. Santiago de Compostela: Universidade de Santiago de Compostela.

Santos, X. M. (2006). El Camino de Santiago: Turistas y peregrinos hacia Compostela. *Cuadernos de Turismo, 18*, 135–150.
Soria y Puig, A. (1991). *El Camino de Santiago*. Madrid: Ministerio de Obras Públicas y Transportes.
Stump, R. W. (2008). *The geography of religion. Faith, place, and space*. Lanham: Rowman & Littlefield Publishers.
Turner, V. (1969). *The ritual process: Structure and anti-structure*. Ithaca: Cornell University Press.
Turner, V. (1974). *Dramas, fields and metaphors: Symbolic action in human society*. Ithaca: Cornell University Press.
Urry, J. (2007). *Mobilities*. Cambridge: Polity.
Vila Vázquez, J. I. (2011). Une analyse critique des «flagship projects» urbains: Le cas de la Bibliothèque Nationale de France. In M. J. Piñeira & N. Moore (Eds.), *New trends in the renewal of the city* (pp. 105–122). Santiago de Compostela: Instituto de Estudos e Desenvolvemento de Galicia.
Villares, R. (2003). A cidade dos "dous apóstolos" (1875–1936). In E. Portela (Ed.), *Historia da cidade de Santiago de Compostela* (pp. 465–542). Santiago de Compostela: Concello de Santiago de Compostela/Consorcio de Santiago/Universidade de Santiago de Compostela.
Yorgason, E., & della Dora, V. (2009). Geography, religion, and emerging paradigms: Problematizing the dialogue. *Social & Cultural Geography, 10*(6), 629–637.

Chapter 36
Religious Contents of Popular Guidebooks: The Case of Catholic Cathedrals in South Central Europe

Anton Gosar and Miha Koderman

36.1 Introduction

Popular guidebooks usually provide potential tourists with an inventory of places and objects of interest and include practical information on transportation, accommodation and costs. The evolution of this genre was, according to Jafari (2003: 268), closely shaped by geography and history books, and has over the course of century resulted in the numerous publishing houses, which specialize in the field of tourist guidebooks. One of the main characteristics of mainstream guidebooks is their impersonal, objective and systematic approach to providing reliable information and guidance to the general public.

When presenting buildings of worship and other religious sites, the guidebooks usually give a short description of places and then focus on works of art (paintings, sculptures, frescoes, mosaics, etc.). The latter are often subjects of relatively detailed description; together with the information on artist and the period in which the work was completed. The guidebooks, however, as a rule ignore the fact that potential reader may not be familiar with the religious contents of the work, and can furthermore come from a completely different cultural, historical, religious and ethnic background. As tourism is a widespread global phenomenon, these intercultural differences should be taken into consideration when describing the elements of religious heritage. Visiting religious buildings and objects is not, as Reinschede (1992: 57) argues, an exceptional feature of Christianity, but rather a worldwide

A. Gosar (✉)
Faculty of Tourism Studies, University of Primorska, Obala 11a, 6320 Portorož, Slovenia
e-mail: anton.gosar@fhs.upr.si

M. Koderman
Science and Research Centre of Koper, University of Primorska,
Garibaldijeva 1, 6000 Koper-Capodistria, Slovenia
e-mail: miha.koderman@zrs.upr.si

S.D. Brunn (ed.), *The Changing World Religion Map*,
DOI 10.1007/978-94-017-9376-6_36

phenomenon of religious history and can be found in every major religion, as well as in a number of smaller religious communities and in every cultural part of the world.

According to Timothy and Boyd (2003: 31), visitors of religious sites can be classified in two fundamental groups: those whose primary purpose is to gain a religious experience (pilgrim) and the potentially far larger group of those whose major motivation is visiting an element of the world's religious heritage (secular tourist). This classification suggests that there are many variations in people's motivation for visiting religious heritage paces – many tourist desire to visit religious structures and sites not only of their own faiths but out of interest in, or curiosity about, historic sites. Others may, as Olsen and Timothy (2006: 6) discuss, visit religious sites because they have an educational interest in learning more about the history of a site or understanding a particular religious faith and its culture and beliefs, rather than being motivated purely by pleasure-seeking or spiritual motives.

In regards to sacred buildings Vukonić (1996: 62–63) distinguishes two basic functions. First is their original function, where believers can satisfy their religious needs, and the second as building of greater or lesser cultural, historical, or artistic value in which both religious and non-religious tourists will seek these values just as they do in a museum or a gallery. Vukonić states that the second, profane function of sacred buildings in tourism almost overshadows their religious function, since the religious function is restricted to only one (religious) segment of tourist demand. He continues by stating that sacred buildings of special interest have often, wholly or in part, lost their religious function.

Among examples of Roman Catholic sacred sites in Europe, Vukonić (2006: 248–249) discusses the Notre Dame Cathedral in Paris, France, which is visited by 12 million people every year and has lost its dominant function as being a place of prayer. Another example of how a sacred space is increasingly becoming a profane tourist space, is Gaudi's Sacrada familia in Barcelona, Spain, although the same could be said of Westminster Abbey in London, England, Chartes Cathedral in France, Cologne Cathedral in Germany and almost all the churches in Rome. This change does not mean, however, that there are less visitors who come to these places for purely religious reasons or that they have stopped coming altogether, but rather points to a significant increase in secular reasons for the consecration of both sacred Catholic sites and individual sacred buildings.

Today, sacred sites present an additional and important value apart from the historical, cultural and architectural aspect to the destination and also the attractiveness of places to visit. One cannot overlook the fact discussed above, that visitors have turned some sacred buildings (variable from one object to another) into a major tourist destination of national or (in some cases) global importance. This brings us back to our initial thesis that guidebooks should provide potential tourist with impersonal and objective information on relevant tourist attractions. As Christianity is the most widespread religion in the world with an estimated two billion adherents (Vukonić 2006: 237), the Christian and Roman Catholic beliefs are embedded in to the so called "Western culture," can an author of the international guidebook bestseller ignore this basic fact, respect the aforementioned principle and provide (publish) an unbiased and non-Christian-centered guide? This question is the focus of our research.

36.2 Methodological Explanations

The main objective of this contribution is to study and evaluate the descriptions of primary religious objects in the selected countries of south-central Europe which were published in some of the most world-renowned and prominent tourist guidebooks. We selected Austria, Croatia, Hungary and Slovenia due to their geographical vicinity, common history since the times of the Austro-Hungarian Empire and the predominant religion they share (Roman Catholic Christianity; also shown in Fig. 36.1). The selection of the guidebooks was undertaken according to the three criteria: (1) they had to be issued (in English version) by the same publisher for all four counties; (2) each of the guidebooks had to be dedicated only to one of the four countries (not focused on the entire region) and (3) all of the guidebooks had to be as up to date as possible (not older than 5 years).

Based on the above mentioned criteria guidebooks from publishers *DK Eyewitness*, *Lonely Planet* and *Rough Guides* were selected (three guidebooks for each of the discussed countries). As already mentioned, the decision was made to study only the textual materials which relate to the primary religious subjects in each of the chosen countries. For example, the descriptions of the main national cathedrals were subjects of our evaluation. The latter included *St. Stephen's*

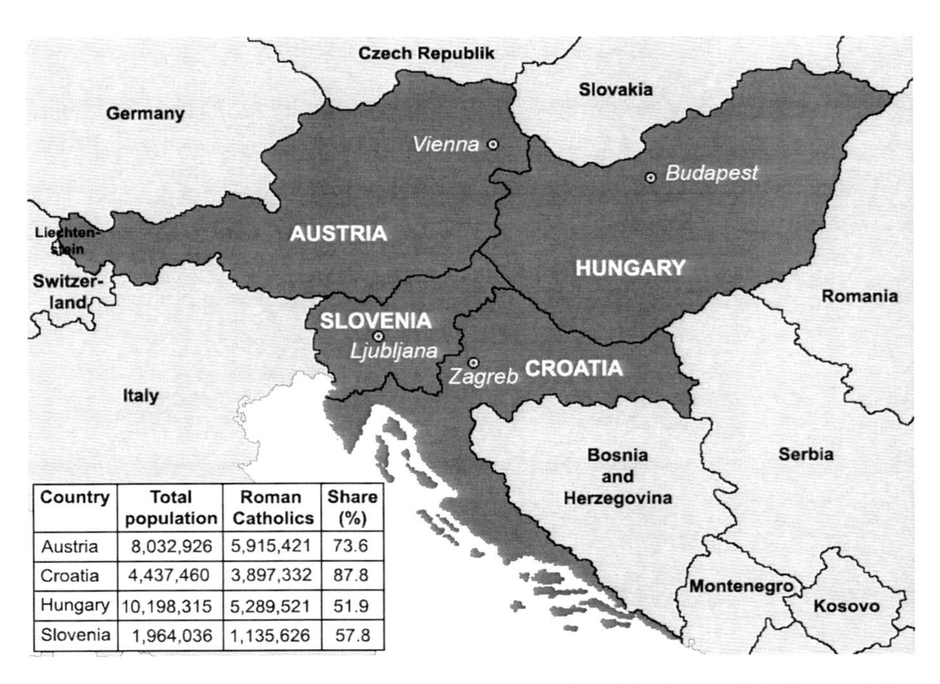

Country	Total population	Roman Catholics	Share (%)
Austria	8,032,926	5,915,421	73.6
Croatia	4,437,460	3,897,332	87.8
Hungary	10,198,315	5,289,521	51.9
Slovenia	1,964,036	1,135,626	57.8

Fig. 36.1 Location of selected countries showing distribution of the Roman Catholic population (Map by Dick Gilbreath, University of Kentucky Gyula Pauer Center for Cartography and GIS; commissioned by the editor. Data from Statistics Austria 2012; Croatian Bureau of Statistics 2012; Hungarian Central Statistical Office 2012a; Statistical Office of the Republic of Slovenia 2012a)

Cathedral (or *Stephansdom*) in Vienna, Austria, *Cathedral of the assumption of the Blessed Virgin Mary* (or *Kaptol*) in Zagreb, Croatia, *St. Stephen's Basilica* in Budapest, Hungary, and *St Nicolas's Cathedral* in Ljubljana, Slovenia. Despite being the principal places of worship for Catholics, they also *per se* represent some of the main tourist attractions in each country.

In order to place the guidebook descriptions of the selected cathedrals into a broader perspective, we also present basic statistical data on international tourist arrivals for each of the studied capitals and countries. The data include the details on the regions of tourists coming to the selected country. In this way, we can make provisional inferences regarding how many tourists come from countries with traditional non-Christian population (Africa and Asia), although they can in any case be found in all world regions.

36.3 Main Findings of the Evaluation

36.3.1 St. Stephen's Cathedral (Stephansdom), Vienna, Austria

Among the studied countries, Republic of Austria is the most developed in terms of economic and tourist indicators. As Table 36.1 shows, the country attracted over 22 million international visitors in the year 2010, what makes it one of the most visited places in the world. Austria's capital, Vienna, is the cultural, economic, and political center of the country and home to almost a quarter of its population. Once the capital of the Austrian and later Austro-Hungarian Empire, the city has a long tradition of history and culture, which is still present in numerous palaces, museums, galleries and other heritage sites. They are visited by approximately 3.6 millions (or 16.7 % of all) of international tourists every year.

Table 36.1 International tourist arrivals in Austria (and Vienna) from 2008 to 2010, by region

Austria						
International arrivals	2008 (Number)	Share (%)	2009 (Number)	Share (%)	2010 (Number)	Share (%)
Europe	19,900,490	90.7	19,512,749	91.4	19,926,153	90.6
Asia	808,390	3.7	772,282	3.6	907,914	4.1
Africa	39,957	0.2	38,794	0.2	44,965	0.2
Americas	672,610	3.1	616,730	2.9	708,515	3.2
Australia and Oceania	139,384	0.6	118,112	0.6	129,818	0.6
Other foreign countries (not specified)	374,578	1.7	296,772	1.4	286,901	1.3
Total	21,935,409	100.0	21,355,439	100.0	22,004,266	100.0
Vienna	3,532,871	16.1	3,349,738	15.7	3,682,503	16.7

Data source: Statistik Austria, Tourismus in Österreich 2008, 2009, 2010

Fig. 36.2 St. Stephen's Cathedral or "Stephansdom" is one of the most recognizable structures in Vienna and attracted over 380,000 visitors in 2010 (Photo by M. Koderman)

One of the city's most visible landmarks is the St. Stephen's cathedral (or Stephansdom in German). The cathedral is the mother church of the Archdiocese of Vienna and therefore the most important religious building in Austria's capital. St. Stephen's cathedral is also one of the city's most visited tourist destinations as it attracted over 380.000 visitors in the year 2010 (Wien Kultur 2012).

St. Stephen's cathedral (Fig. 36.2) is consequently exposed and recommended to the potential visitors in all three guidebooks which were analyzed in this research. Sentences, such as *Lonely Planet's* "The most bellowed and recognizable structure in Vienna" (Haywood et al. 2011: 58) or *DK Eyewitness'* "St Stephen, is the very soul of the city" (Czerniewicz-Umer et al. 2010: 58) clearly speak of its importance and popularity among locals and tourists. All three guidebooks initially present brief technical information on the cathedral (opening hours, address, contact numbers, etc.) and continue with more or less detailed description on the exterior an interior of the cathedral. In this element, the *Rough Guide* guidebook (Bousfield et al. 2008) prevails with 874 words dedicated to the cathedral, while *Lonely Planet* (Haywood et al. 2011) and *DK Eyewitness* guide (Czerniewicz-Umer et al. 2010) follow with 438 and 277 words respectively.

Although the name of the patron saint and other works of art symbolizing important Catholic persons and events are mentioned in all three guidebooks, their descriptions fail to give any explanations of the historical background. Two examples of such inadequate descriptions from *Lonely Planet* and *Rough Guide* guidebooks – both guidebooks are clearly designed for the usage of the broader public, regardless of their cultural or religious belief – are shown in the further text. *Rough Guide* (Bousfield et al.) presents, among others, the following description on the interior of the cathedral:

> At the far end of the north choir aisle, however, you should just be able to make out the winged Wiener Neustädter Altar, a richly gilded masterpiece of late-Gothic art, whose central panels depict Madonna and Child, the Birth of Christ, the Adoration of the Magi, the Coronation of Mary and latter's death, all sculpted in high relief out of wood. (Bousfield et al. 2008: 77)

This description is, of course, correct and probably welcomed by everyone who is visiting the cathedral without a guided tour, but the question arises whether the self-exploring visitor is familiar with the profound meaning of the *Madonna and Child* sculpture or the depiction of the *Adoration of the Magi*. The Magi, also known to the Catholics as the three wise men or three kings, were, according to Christian belief, a small group of royal leaders who visited Jesus of Nazareth after his birth. They are closely connected with the celebrations of Christmas and are an important part of the Christian tradition. A person with no previous knowledge of Christian religious heritage would, when faced with the current work of art and the above presented description, without doubt appreciate an explanatory sentence or two regarding this theme. In this way, one could gain a deeper understanding the artwork and, if further interested, obtain more information from Christian literature and other relevant sources.

The presented case of Christian-centric descriptions is by far an isolated example. Haywood et al. in the latest issue of *Lonely Planet*'s guide for Austria presents a group of sculptures located on the exterior of the Stephansdom's facade which refer to one of the main events in Christian history – the crucifixion Jesus Christ:

> Look closely at the decorations and statues on the exterior of the cathedral: at the rear the agony of the Crucifixion is well captured, while the glorious tilled roof shows dazzling chevrons on one end and the Austrian eagle on the other. (Haywood et al. 2011: 59)

The term *crucifixion* – a method of painful execution – in the above mentioned context refers to the event in which Jesus Christ was supposedly executed. Although the expression might sound clearer to the common (non-Catholic) public due to the fact that the events connected with the birth and the death of Jesus Christ later became globally referential as the start of the internationally accepted civil calendar – the Gregorian calendar (also called the Western or the Christian calendar), it could leave members of other so called high or major religions (Hinduism, Buddhism, Islam) dissatisfied.

36.3.2 Cathedral of the Assumption of the Blessed Virgin Mary, Zagreb, Croatia

The Republic of Croatia ranks third among the studied countries in number of tourist arrivals with over 9.1 million international visitors in 2010. In the same year, the country's capital and the largest city in Croatia, Zagreb, attracted 447,000 foreign tourists (or 4.9 % of total), as shown in Table 36.2. This fact presents a clear deviation from other three discussed capitals and countries, in which international tourists

Table 36.2 International tourist arrivals in Croatia (and Zagreb) from 2008 to 2010, by region

Croatia						
International arrivals	2008 (Number)	Share (%)	2009 (Number)	Share (%)	2010 (Number)	Share (%)
Europe	8,866,000	94.2	8,831,000	94.6	8,560,000	94.0
Asia	217,000	2.3	237,000	2.5	251,000	2.8
Africa	12,000	0.1	11,000	0.1	12,000	0.1
Americas	223,000	2.4	188,000	2.0	208,000	2.3
Australia and Oceania	97,000	1.0	68,000	0.7	80,000	0.9
Total	9,415,000	100.0	9,335,000	100.0	9,111,000	100.0
Zagreb	451,000	4.8	408,000	4.4	447,000	4.9

Data source: Croatian Bureau of Statistics, Statistical Yearbook of the Republic of Croatia, 2011

visiting the capital accounted for at least 16 % of all the visitors in the country. In this instance Zagreb unsuccessfully rivals with old historic cities situated along the Adriatic coast (such as Dubrovnik, Split, and Zadar), that traditionally represent primary destinations for international visitors coming to Croatia.

Despite the dominance of the so called sun, sea, sand destinations in Croatia, Zagreb can be considered as an important tourist center, mainly due to its rich cultural heritage and favorable geographic position in the contact area of the Dinaric, Adriatic and Pannonic regions. Visitors are especially drawn to the historical part of the city, located north of Ban Jelačić Square and composed of the Gornji Grad and Kaptol districts. The Kaptol, a medieval urban complex of churches, palaces and museums, is especially popular among tourists visiting Zagreb and usually presents the principal place of their interest. This is mainly due to the Cathedral of the Assumption of the Blessed Virgin Mary (or Katedrala Marijina Uznesenja in Croatian), which dominates the city's urban landscape and is one of city's most prominent landmarks (similar to the St. Stephen's Cathedral in Vienna). The cathedral and its surrounding buildings are also the seat of the Archdiocese of Zagreb, a metropolitan see of the Catholic Church in Croatia.

The Cathedral of the Assumption of the Blessed Virgin Mary is extensively recommended in all three studied guidebooks. The *DK Eyewitness* guidebook (Zoppé and Venturini 2011: 152), for example, clearly states that "... this is the most famous monument in the city," while the *Rough Guides* guidebook (Bousfield 2010: 77–78) uses carefully selected adjectives (such as *the filigree spires* or *ivy-cloaked turrets*) to describe its beauty and grandeur. In general, all three guidebooks present similarly long descriptions dedicated to the cathedral, which include basic explanations on history, interior and technical information. In this aspect, the *DK Eyewitness* guidebook (Zoppé and Venturini 2011: 152) provides the most thorough presentation (405 words) and prevails over the *Rough Guide* guidebook (Bousfield 2010: 77–78) with 355 words and *Lonely Planet* (Marić and Mutić 2009: 73) with 243 words.

While the *Lonely Planet* (Marić and Mutić 2009: 73) and *Rough Guide* (Bousfield 2010: 77–78) guidebooks avoid giving detailed descriptions on religious works of art in the cathedral, our examination of the studied guidebooks registered insufficient explanations on this subject in the case of the *DK Eyewitness* guidebook (Zoppé and Venturini 2011: 152). This guidebook offers (among other) the following report on the Gothic and Renaissance works in the interior of the cathedral:

> These works include a statue of St Paul (13th century), wooden statues of the saints Peter and Paul from the 15th century, a triptych entitled Golgotha (1495) by Albrecht Dürer and a 14th century Crucifixion by Giovanni da Udine. (Zoppé and Venturini 2011: 152)

The information on the authors of the works and the period of their origins, presented in the *DK Eyewitness* guidebook, is useful and contributes to the visitor's common appreciation of the works of art. However, a general description on the early Christian leaders and apostles *saints Peter* (also referred to as Simon Peter) and *Paul* (or Paul the Apostle) might be advantageous for visitors unfamiliar with the Catholic background. The term *Golgotha* (also called *Calvary*) – the site outside Jerusalem's walls where the crucifixion of Jesus Christ occurred – would similarly deserve a short explanation, same as the term *Crucifixion* (as already mentioned in the case of St. Stephen's Cathedral in Vienna).

36.3.3 St. Stephen's Basilica, Budapest, Hungary

Hungary, known for a number of geothermal springs and the largest lake in Central Europe (Lake Balaton), is yearly visited by around 9.5 million international tourists. The country's capital and largest city, Budapest, is considered one of the most beautiful cities in Europe and has several sites included on the UNESCO World Heritage list. It consequently received more than 22 % of all foreign tourists visiting Hungary in 2010, as shown in Table 36.3.

Table 36.3 International tourist arrivals in Hungary (and Budapest) from 2008 to 2010, by region

Hungary						
International arrivals[a]	2008 (Number)	Share (%)	2009 (Number)	Share (%)	2010 (Number)	Share (%)
Europe	7,867,028	89.3	8,099,298	89.4	8,571,251	90.1
Asia	322,219	3.7	326,401	3.6	315,527	3.3
Africa	20,356	0.2	21,538	0.2	22,668	0.2
Americas	541,174	6.1	547,653	6.0	535,903	5.6
Australia and Oceania	63,468	0.7	63,420	0.7	65,083	0.7
Total	8,814,245	100.0	9,058,310	100.0	9,510,432	100.0
Budapest	2,173,769	24.7	1,971,958	21.8	2,167,131	22.8

Data source: Hungarian Central Statistical Office (2012b)

[a]Based on number of international tourist arrivals at frontiers (excluding same-day visitors), this data differs from other studied countries, where the number of tourists is based on registered international tourist arrivals at collective tourism establishments

Among the city's main attractions is also the St. Stephen's Basilica (or Szent István-bazilika in Hungarian), named in honor of Stephen, the first King of Hungary. With 96 m (315 ft) in height, the basilica is one of the highest edifices in the capital (together with the parliament building) and, therefore, an obvious landmark of the city. In contrast to the previously presented St. Stephen's cathedral in Vienna and the Cathedral of the assumption of the Blessed Virgin Mary in Zagreb, the basilica is not as described in the three studied guidebooks as the latter two. Although recommended as one of the principal places of tourist interest in Budapest by *Lonely Planet* guidebook (Bedford et al. 2009: 74), the authors provide a rather short description of only 167 words. Both, *DK Eyewitness* guide (Olszańska et al. 2010: 86–87) and *Rough Guide* (Longley et al. 2010: 80) present almost twice the number of words on the same object (325 and 316, respectively).

All three guidebooks also present a brief history of the basilica and inform potential reader about the basic technical information (address, telephone numbers, opening hours and entry fees). Due to the relatively short description of the basilica, the explanations on works of art in and on the building are limited (the *DK Eyewitness* guide (Olszańska et al. 2010: 86–87) is dominant in this element). Similarly, only two short descriptions of the religious contents were noticed in all three guidebooks. Both were found in the above mentioned guidebook and referred to *the four Evangelists* and *the 12 Apostles* (Olszańska et al. 2010: 86–87). Another important and insufficient description of partially religious contents was noted in the *Lonely Planet* (Bedford et al. 2009: 74) and *Rough Guide* (Longley et al. 2010: 80) guidebooks, where no explanation on the basilica's Patron, St. Stephen (the first King of Hungary) was given.

36.3.4 St Nicolas's Cathedral, Ljubljana, Slovenia

The smallest of the studied countries, the Republic of Slovenia has a favorable strategic position among four European geographic units: the Alps, the Dinarides, the Adriatic Sea, and the Pannonian Plain. Although Slovenia serves to tourist mostly as a country-of-transit from Central Europe to the beaches of the Adriatic, its high level of landscape diversity and a number of historical and cultural sites attracted over 1.8 million foreign tourists in 2010 (Table 36.4). The centrally located capital, Ljubljana, is one of the most popular destinations for international tourists (visited by around 20 % of all visitors coming to the country).

Ljubljana's historic centre, marked by Baroque, Vienna Secession and Neoclassical architecture, offers some unique points of interest to potential visitors. Although overshadowed by the Ljubljana castle and the Triple Bridge at the Prešeren square, St Nicolas's Cathedral (also called Stolnica Sv Nikolaja in Slovene) can be considered one of them. Located on Cyril and Methodius Square near the Ljubljana Central Market, it is an easily recognizable structure with its green dome and twin towers. It is the most important sacred building in the country and serves as the seat of Archdiocese of Ljubljana.

Table 36.4 International tourist arrivals in Slovenia (and Ljubljana) from 2008 to 2010, by region

Slovenia						
International arrivals	2008 (Number)	Share (%)	2009 (Number)	Share (%)	2010 (Number)	Share (%)
Europe	1,771,053	90.5	1,646,406	90.3	1,680,403	89.9
Asia	93,279	4.8	99,861	5.5	101,098	5.4
Africa	3,760	0.2	2,911	0.2	3,396	0.2
Americas	65,701	3.4	55,336	3.0	61,941	3.3
Australia and Oceania	23,898	1.2	19,417	1.1	22,268	1.2
Total	1,957,691	100.0	1,823,931	100.0	1,869,106	100.0
Ljubljana	365,693	18.7	344,222	18.9	372,359	19.9

Data source: Statistical Office of the Republic of Slovenia 2012b

Fig. 36.3 Although overshadowed by Ljubljana Castle and the Triple Bridge at Prešeren Square, St Nicolas's Cathedral (in the background) is among the 10 most popular tourist attractions in Ljubljana (Photo by M. Koderman)

Compared with the studied sacred edifices in Vienna, Zagreb and Budapest, Ljubljana's St Nicolas Cathedral (Fig. 36.3) is the least exposed religious building. The descriptions in the inspected guidebooks were 282 words in *Rough Guide* – Longley 2010: 59–60; 226 words in *DK Eyewitness* guide – Bousfield and Stewart 2012: 50–51 and 137 words in *Lonely Planet* – Fallon 2010: 77 about the cathedral, which included a brief technical, historical and artistic presentation. Surprisingly and despite short descriptions, the *Rough Guide* (Longley 2010: 59–60) and the *DK Eyewitness* (Bousfield and Stewart 2012: 50–51) guidebook offer basic explanation of the cathedral's patron saint, St Nicolas, as presented below:

> Dedicated to St Nicolas, the patron saint of fishermen and sailors – many of whom lived in the suburb of Krakovo – the present building, designed by Andrea Pozzo from Rome and completed in 1706, stands on the site of a thirteenth-century basilica. (Longley 2010: 59)

> Built in the site of an earlier church by leading Jesuit architect Andrea Pozzo in 1707, this cathedral is dedicated to St Nicholas, patron saint of fishermen and sailors. (Bousfield and Stewart 2012: 50)

While some other religious terms appeared in both above mentioned guidebooks without any explanations (such as *Corpus Christi, the Crucifixion* and *the Apostles*), the *Lonely Planet* guidebook presented no terms of religious nature.

36.4 Conclusion

The analyses of the description of four Catholic Cathedrals in South Central Europe in 12 popular guidebooks (*Lonely Planet, DK Eyewitness Guides, Rough Guides*) clearly demonstrates that religious contents in the available travel literature, written in English, are minimal. In each of the three guidebooks studied per city, the nation-state's major cathedrals of Austria (Vienna), Hungary (Budapest), Croatia (Zagreb) and Slovenia (Ljubljana) are pointed out as the prime attractions of the country described. The church construction and the artistic uniqueness are exposed, in particular if regionally/worldly outstanding architects, painters and/or sculptors were involved. Their works and its artistic significance are placed in the foreground of descriptions, whereas the basic information of the religious content of the work presented is completely missing. Frescoes and other painted works of arts, stained glass artistic expressions, and wood/stone sculptures are placed in the artistic and societal time frame (gothic, renaissance, baroque, etc.) and even in the artist's production phase, but guidebooks in general fail to provide stories/descriptions of the presented religious stories (Bible source, saint's life, etc.). Interested lay and/or non-Christian tourists (readers of guidebooks) would have to make their way through extensive religious literature to find out about the content (and its importance for the religious community) in the works of arts of these cathedrals, for example, the Adoration of the Magi in the Viennese cathedral or in the case of St. Nicholas, the patron of the cathedral of Ljubljana.

Increasing Asian visitors to the sites of the Western Culture have in particular placed the discussed phenomenon in the foreground of discussion. If the tourism industry would focus only on visitors from the East Asia (yearly about 150,000 to Vienna, 70,000 to Budapest, 12,000 to Zagreb and at least 20,000 to Ljubljana – as presented in the tables), they would have to be enlightened into details of the Christian faith, or at least about stories the Christian monuments of the nation-state capitals that are presented as outstanding works of religious art. Contemporary popular guidebooks are written in general with the presumption that readers (tourists) are well informed about the scriptures of the Bible and the life of Christian/Catholic saints. The outstanding Christian, or, even better, Roman-Catholic – centered guidebooks studied here are of little help to the increasingly tourism-minded visitors to Europe of Islamic, Hindu or Buddhist faiths.

References

Bedford, N., Dunford, L., & Fallon, S. (2009). *Hungary*. Oakland/London/Footscray: Lonely Planet.
Bousfield, J. (2010). *The rough guide to Croatia*. London: Rough Guides.
Bousfield, J., & Stewart, J. (2012). *Slovenia*. London: Dorling Kindersley Ltd.
Bousfield, J., Humphreys, R., Walker, N., & Williams, C. (2008). *The rough guide to Austria*. New York/London/Delhi: Rough Guides.
Croatian Bureau of Statistics. (2011). *Statistical yearbook of the Republic of Croatia*, Godina 43. Zagreb.
Croatian Bureau of Statistics. (2012). 14. Stanovništvo prema vjeri, po gradovima/općinama, popis 2001. Zagreb.
Czerniewicz-Umer, T., Egert-Romanowska, J., & Kumaniecka, J. (2010). *Austria*. London: Dorling Kindersley Limited.
Fallon, S. (2010). *Slovenia*. Oakland/London/Footscray: Lonely Planet Publications.
Haywood, A., Sieg, C., & Christiani, K. (2011). *Austria*. London/Footscray: Lonely Planet.
Hungarian Central Statistical Office. (2012a). *Religious affiliation in Hungary (2001)*. Budapest.
Hungarian Central Statistical Office. (2012b). *4.5.3. Number and spending of foreign visitors in Hungary by countries (2006)*. Budapest.
Jafari, J. (Ed.). (2003). *Encyclopedia of tourism*. London/New York: Routledge.
Longley, N. (2010). *The rough guide to Slovenia*. New York/London/Delhi: Rough Guides.
Longley, N., Burford, T., Hebbert, C., & Richardson, D. (2010). *The rough guide to Hungary*. London: Rough Guides.
Marić, V., & Mutić, A. (2009). *Croatia*. Oakland/London/Footscray: Lonely Planet Publications.
Olsen, D. H., & Timothy, D. J. (2006). Tourism and religious journeys. In D. J. Timothy & D. H. Olsen (Eds.), *Tourism, religion and spiritual journeys* (pp. 1–21). London/New York: Routledge.
Olszańska, B., Olszański, T., & Turp, C. (2010). *Hungary*. London: Dorling Kindersley Limited.
Reinschede, G. (1992). Forms of religious tourism. *Annals of Tourism Research, 19*, 51–67.
Statistical Office of the Republic of Slovenia. (2012a). *8. Prebivalstvo po veroizpovedi in tipu naselja, Slovenija, popisa 1991 in 2002*. www.stat.si/popis2002/si/rezultati/rezultati_red.asp?ter=SLO&st=8. Accessed 6 June 2012.
Statistical Office of the Republic of Slovenia. (2012b). *Prihodi in prenočitve turistov po: vrste nastanitvenih objektov, vrste občin, države, meritve, leto*. (http://pxweb.stat.si/pxweb/Dialog/varval.asp?ma=2164505S&ti=&path=../Database/Ekonomsko/21_gostinstvo_turizem/02_21645_nastanitev_letno/&lang=2). Accessed 13 July 2012.
Statistics Austria. (2012). *Bevölkerung nach dem Religionsbekenntnis und Bundesländern 1951 bis 2001*. Wien. www.statistik.at/web_de/. Accessed 12 July 2012.
Statistik Austria. (2009). *Tourismus in Österreich 2008*. Wien.
Statistik Austria. (2010). *Tourismus in Österreich 2009*. Wien.
Statistik Austria. (2011). *Tourismus in Österreich 2010*. Wien.
Timothy, D. J., & Boyd, S. W. (2003). *Heritage tourism*. Harlow: Prentice Hall.
Vukonić, B. (1996). *Tourism and religion*. Oxford/New York/Tokyo: Pergamon.
Vukonić, B. (2006). Sacred places in the Roman Catholic tradition. In D. J. Timothy & D. R. Olsen (Eds.), *Tourism, religion and spiritual journeys* (pp. 237–253). London/New York: Routledge.
Wien Kultur. (2012). Website. www.wienkultur.info/page.php?id=98. Accessible 13 May 2012.
Zoppé, L., & Venturini, G. E. (2011). *Croatia*. London: Dorling Kindersley Limited.

Chapter 37
Sacred Crossroads: Landscape and Aesthetics in Contemporary Christian Pilgrimage

Veronica della Dora, Avril Maddrell, and Alessandro Scafi

37.1 Introduction

Pilgrimage is one of the most inherently geographical forms of religious expression. Its typologies and practices vary across the great world religions: from the *hajj* to Mecca, a highly-regulated institutional requirement focused on a specific sacred centre, to Hindu semi-structured and non-compulsory journeys to sites associated to legendary events and gurus, and Shinto followers completing the circuit of 88 sacred sites at Shikoku. While pilgrimage can take different forms, it can be essentially described as a simultaneously physical and inner journey in search of spiritual meaning. The pilgrim is usually focused on geographical localities (typically shrines or other sites important to his or her faith), yet is after an absolute. Through the transformative process of the journey, the pilgrim is continuously set in a liminal state of instability: between travel and place, earth and heaven, present and past, but also, and perhaps more significantly, between a process and its outcome.

Dwelling on the tension between the macro-scale of the earth as a place for global redemption and the topographical scale of the sites connected to the life of Jesus and to the lives of saints, Christianity has been characterized by a long and

V. della Dora (✉)
Department of Geography, Royal Holloway University of London,
Egham (Surrey) TW20 0EX, UK
e-mail: veronica.delladora@rhul.ac.uk

A. Maddrell
Department of Geography and Environmental Sciences, University of West England,
Bristol BS16 1QY, UK
e-mail: avril.maddrell@uwe.ac.uk

A. Scafi
Medieval and Renaissance Cultural History, The Warburg Institute, University of London,
London WC1H OAB, UK
e-mail: alessandro.scafi@sas.ac.uk

S.D. Brunn (ed.), *The Changing World Religion Map*,
DOI 10.1007/978-94-017-9376-6_37

rich pilgrimage tradition (Delano-Smith and Scafi 2005; della Dora forthcoming). The origins of Christian pilgrimage can be traced back to the early days of Christianity and the veneration of martyrs' tombs. However, it was not until the fourth century that pilgrimage practices assumed significant proportions, that is, after Constantine physically marked key sites associated to the life of Christ through his ambitious imperial building campaign (Maraval 2002). While Christians have enthusiastically embraced sacred journeys to these (and other) sites, such practices have not gone uncontested. Since the outset, their legitimacy was debated by the Church Fathers and by those who recalled Paul's words: 'God who made the world and everything in it, being the Lord of heaven and Earth, does not dwell in shrines made by man' (Acts 17: 24). Gregory of Nyssa (c. 335 – after 394), for example, characteristically argued that 'when the Lord invites the blessed to their inheritance in the kingdom of heaven, He does not include a pilgrimage to Jerusalem amongst their deeds' (Greg Nys, *On Pilgrimages*, trans. Schaffer 1885), whereas Athanasius of Alexandria decried 'the rush for martyrs' bones', 'the invention of holy places,' and the 'ecstatic devotions' he saw at the Egyptian shrines (quoted in Frankfurter 2005: 435).

Diverging views on the validity of Christian pilgrimage have multiplied over the centuries, with the various denominations attributing different value and theological meanings to it. As a result, a wide range of "spiritual geographies" and pilgrimage practices co-exist within the Christian faith. In the medieval Roman Catholic West, post-biblical holy places were mostly linked to the cult of martyrs and saints: the architectural structures built around these holy places had the purpose of marking sites of burial, martyrdom, and conservation of relics. In modern and contemporary Roman Catholic Christianity, however, an eschatological understanding of the biblical narrative unconnected with any specific locations and a topographical approach that developed the idea of holy places on earth did and do coexist as part of a fundamental dichotomy. For Roman Catholics, as well as for any Christian, God is everywhere, yet He can be sought in particular places (Smith 1987; Markus 1994; Delano-Smith and Scafi 2005). Catholic pilgrimage thus tends to focus on journeys to places associated with the shrines of saints, holy relics, healing and revelation.

By contrast, the Reformed churches, grounded in a theology of direct access, that is, anyone can approach God from any place, have traditionally rejected pilgrimage practices as corrupt and unnecessary. Nonetheless, over the last 10–20 years Protestant Christians have increasingly re-engaged with pilgrimage; this has tended to be with more emphasis on journeying and individual experience of God (even if participating in a communal activity) and with less emphasis on the sacred quality of specific shrines or liturgical rituals. At the other end of the spectrum, Orthodox pilgrimage focuses primarily on the shrine and the physical act of veneration of relics and icons, thus largely downplaying the role of the journey to reach it (Maddrell and della Dora 2013). The contrast is reflected in the Latin and Greek words for pilgrimage: *peregrinatio* places emphasis on the act of journeying, whereas *proskynema* (literally "bowing down") stresses the act of veneration at a shrine (Gothóni 1994). However, it should be noted that individuals and groups can blur these boundaries, as in the case of those who seek a new spiritual experience or perspective by "experimenting" with the practices of another denomination or faith group, for example, Protestants and non-Christians walking the Camino de Santiago de Compostela in Spain, or Buddhists undertaking Shinto pilgrimages in Japan.

Regardless of whether their focus is on the act of journeying through space or on the significance of the shrine(s), in each instance Christian pilgrims move through and interact with landscape. They also attribute different values and functions to it. While most literature on pilgrimage has usually focused on place (the shrine) or performance (the journey to reach it), the concept of landscape has largely remained out of these debates. Yet, we argue, in all the three above-mentioned Christian denominations landscape and aesthetics play, in different ways and to different extents, an important role in pilgrims' experience.

This short photographic essay explores the significance of landscape in contemporary Christian pilgrimage. It engages with three case studies, each associated with a different Christian denomination, but all characterized by stunning landscape sceneries: the Cave of Saint Benedict at Subiaco, Italy; a Protestant-led ecumenical pilgrimage in the Isle of Man; the Greek Orthodox rock-monasteries of Meteora in Thessaly, Greece. Because of their scenic qualities, these sites also attract other categories of visitors, including international tourists and excursionists, who in turn superimpose a further layer of meaning on the landscape, leading, in some cases, to its commodification. The essay is structured as a topographic journey through these different sacred landscapes. Each case study approaches landscape from a different denominational and geographical perspective, destabilizing traditional binaries such as nature and culture, meaning and performance, tourism and pilgrimage.

37.2 Landscape and Pilgrimage

Before we embark on our journey through sacred landscapes, it is important to address three questions:

1. What do we mean by landscape?
2. How do contemporary cultural geographers approach landscape?
3. Why is landscape a significant framework for studying pilgrimage?

We should start by saying that among all geographical concepts landscape is probably the most ambiguous. The term has traditionally been employed by human geographers to define "the appearance of an area, the assemblage of objects used to produce that appearance, and the area itself" (Johnston et al. 2000: 429). Landscape thus refers simultaneously to the land and to the way we perceive and represent it. It designates a framed view of specific sites and the scenic character of whole regions. It is simultaneously *a thing* and *"a way of seeing"* (Cosgrove 1985, 2003). Landscape therefore comes into being only through the encounter between territory and a viewer—in other words, there is no landscape without a subject looking at (or imagining) it. As a concept, landscape entangles a sense of "holism," of integration of its constitutive elements; a sense which has never ceased to charm geographers—from Alexander von Humboldt and Carl Sauer to our days.

Since the late 1980s, new cultural geographers have moved away from the traditional notion of landscape as an ensemble of material forms expressive of a culture in a given geographical area (Sauer 1925) towards an iconographic approach emphasizing the textual nature of landscape as a cultural image (Cosgrove and Daniels 1988; Duncan 1990), as well as the power relationships embedded in the act of producing and gazing at the land (Mitchell 1996; Mitchell 2002; Rose 1993). Other scholars, however, have called for approaches "less exclusively socially constructed and less visualist" (Carter 1987, 1992), emphasizing instead the corporeal and performative dimension of the encounter between landscape and subject (Foster 1998; Merriman 2006), as well as the emotional, subjective, even co-constitutive (Jones 2005; Wylie 2002) nature of such encounters, and thus their restorative potential.

Conradson (2007) has usefully reframed therapeutic *landscapes* as therapeutic *environments*. Within studies of pilgrimage, discourses of 'healing' have similarly become quite broad. They have also been concerned with the whole person rather than limited to the miraculous eradication of specific medical conditions (Williams 2007; Gesler 2003; Harris forthcoming). This broader sense of healing potentially operates through a number of potential processes at play through pilgrimage practices which may improve a pilgrim's sense of well-being and renewal in what is described as 'self-transformation' (Winkelman and Dubisch 2005: xx). The therapeutic benefits of walking and travel have led to a focus on mobilities in studies of western pilgrimage (Coleman and Eade 2004; Frey 1998; Maddrell 2011; Slavin 2003; Solnit 2002). The social dimensions of co-presence and community (aspects of what Turner and Turner 1978 referred to as *communitas*) and the embodied-affective experience of journeying through landscape, time-space for reflection and worship are part of the co-constitutive experience of pilgrimage. As the case studies below illustrate, both the immersion in beautiful natural landscapes and fellowship can in some cases be sources of spiritual insight and healing as much as specific acts of worship or ritual (also see Coleman and Eade 2004; Frey 1998; Gemzöe 2005).

As a concept blending vision and movement, cultural meaning and performance, landscape offers a unique lens to explore pilgrimage practices and their embodied narratives. Unlike place, landscape transcends the local and "fixed," self-contained dimension of the shrine. Conversely, unlike space, it has an unquestionably visual and material presence that exceeds pure performance (Cosgrove 2006). As with any human landscape, the landscapes looked at and traversed by pilgrims can thus be conceived of as palimpsests, "as surfaces with multiple inscriptions that build up over time and mark the presence of different cultural groups" (Anderson 2010: 20). Small chapels, crosses, roadside memorials, pilgrims' names inscribed on trees and in the rock, footprints transiently imprinted on the soil mark the pathways to the sacred.

In Subiaco, some visitors have scratched a name on a wall, even scribbled a date on a medieval fresco (Fig. 37.1). Others have left a mark of their passage in the visitors' book of the monastery. Each inscription is the visual expression of a specific set of beliefs. From a new cultural geography perspective, the landscape traversed by the pilgrim is also a text open to different interpretations, as well as an embodied and affective terrain. The pilgrim's act of traversing and gazing at the landscape can

Fig. 37.1 Marks on medieval fresco in Subiaco (Photo by Alessandro Scafi)

be differentiated from the tourist's by the role of his or her beliefs in interpreting that which is looked upon and experienced. Likewise, pilgrims from different denominations can read different stories in the same landscape (see Bowman 1991). Yet, pilgrims' landscape encounters go beyond mere interpretation. The pilgrim's gaze is not simply projected on a blank screen. The land speaks back to him or her. And, as we shall see in the next pages, it often does so in unexpected ways. Aesthetic and spiritual experiences are similar in that they entail a sense of wonder, a sense of the sublime and transcendent, but for the pilgrim these experiences emanate from or are channeled into worship of the divine.

Let us turn now to a site whose spectacular natural scenery has been enjoyed by both tourists and pilgrims. This site has been seen as blessed, because a saint led his early contemplative life there. But how has landscape played a role in the pilgrim's quest of this sanctified ground?

37.3 Landscape, Pilgrimage, and Aesthetics: Sacro Speco

Our journey starts from Italy and the monastery of the *Sacro Speco*, in the Subiaco valley, south east of Rome. The monastery in Subiaco enshrines the "holy cave" (the *Sacro Speco*) of St Benedict (Giumelli 1982). At the end of the fifth century, a

Fig. 37.2 The monastery of the Sacro Speco in the Subiaco valley southeast of Rome (Photo by Alessandro Scafi)

youthful Benedict, coming from a well-established family, was studying in Rome, but did not like what he saw around him and retreated to a cave near Subiaco. There he lived as a hermit for 3 years, before he organized his first monastic community (Gregory the Great 2001). Benedict's small and obscure cave in the Subiaco valley became the cradle of the Benedictine order. Subiaco is a place where people go on pilgrimage in order to experience the same encounter with God that St Benedict had experienced on that very spot (Figs. 37.2 and 37.3).

The first pilgrims to seek out Benedict's cave were the shepherds of the valley. According to tradition, Benedict preached to them and to the inhabitants of the neighbouring regions in a cave called "Grotto of the Shepherds," just below his own cave, the *Sacro Speco*. Following in the footsteps of these earlier visitors, popes, cardinals and royals have subsequently visited Subiaco. In the visitors' book of the monastery (the earliest signature dates to 1829 and visitors have been registered ever since) we find signatures of priests, friars, nuns, diplomats, politicians, prime ministers, as well as common people. Subiaco has been a site also for artists and scholars including art historians, ethnographers, archaeologists, and musicians. Whoever the visitor, landscape has always played a crucial role in facilitating the spiritual experience in Subiaco.

The reason is the stunning setting of the Monastery of St Benedict, carved in the rock over the saint's former cave. In the fourteenth century, the spectacular position of the sanctuary against the mountain was described by Petrarch as "the edge of paradise" (Petrarca 1975). In 1461 Pope Pius II compared the shrine stuck

Fig. 37.3 The "Sacro Speco" of St Benedict, with the seventeenth-century white marble statue of the saint by Antonio Raggi (Photo by Alessandro Scafi)

against the rock to a "swallow's nest" (Piccolomini 1984). Many local guides intended for pilgrims to place strong emphasis on the importance of landscape to facilitate spiritual insights. In his work on the Subiaco Pilgrimage, for example, the Benedictine monk Alberico Panella, describes with enthusiasm the beauty of the Holy Mountain:

> Not far from the town of Subiaco, where the river Aniene flows, still small and playful, Mount Talèo rises majestic over all the other heights that ring it, under such a clear sky and among so much variety of natural marvels that it makes it a delightful place to stay for those who wish to restore their tired soul, lifting it to meditations on paradise. ... Divine Providence has gathered here whatever beautiful and sublime is found in Nature: clarity of sky, rich woods, gigantic rocks, aromatic herbs, fragrant flowers with a thousand tints. Looking at this alone gives great pleasure. The charm is increased by the warble of little birds and the roaring of the river, which washes the slopes of the Holy Mountain with its clear and fresh waters with a thousand turbulences, now clear and calm, now rapid and foaming. (Panella 1875)

Similarly, Oderisio Bonamore, another Benedictine who wrote a description of the holy sites in Subiaco, starts his pilgrim guide with enthusiasm for the natural site:

> Here we are at the Holy Cave! Situated in a mountainous desert, camped, as it were, in mid air, like an eagle nest, it appears at first sight as an imposing building, attached to the rock, or better to a wall of rocks, projecting and overhanging sheer from the valley. (Bonamore 1884)

In the history of Subiaco, landscape has been experienced as a challenge, as the holy site was difficult to reach and only motivated pilgrims ventured to climb the mountain on foot. Nowadays the visit is easier, since the monastery can be reached by car or bus. Tourists, not only pilgrims, can enjoy the spectacular scenery of the Holy Cave. A survey conducted with visitors to the monastery of the Sacro Speco in 2010 has shown that 41 % of the respondants described themselves as "pilgrims;" the same percentage thought of themselves as "tourists," and the remaining 18 % refused to be labeled either as a "pilgrim" or as a "tourist" proposing other definitions. Interestingly, 71 % stated that they had a religious affiliation; however, 13 % of them declared that they were not practicing that faith. 21 % wrote to be non religious, not belonging to any confession.

The Holy site of the *Sacro Speco* has always been there for pilgrims. In his account of Benedict's life, Pope Gregory the Great wrote that "in that cave of Subiaco, where [Benedict] initially lived, it is still the case even today that, if the faith of those who petition so demands it, [Benedict] will shine forth miracles" (Gregory the Great 2001). In other words, the place becomes holy through the faith of the devoted visitor. Still today pilgrims come regularly to the monastery, some of them to live and pray with the monks for a short period of time.

As mentioned above, however, almost half of the respondents described themselves as "tourists." And yet quite a few visitors would rather be seen as pilgrims than "tourists," or at least both "tourists and pilgrims." The term "tourist" seems not to be too popular. A Greek woman wrote that she preferred the term "visitor," "because I consider tourism a consumerist activity, which Subiaco was not for me;" but interestingly her religious affiliation was defined as "not applicable." An Italian woman said that she would like to be a pilgrim, that she has got the soul of a pilgrim, but also that her experience in Subiaco was one of a tourist, as often happened in her life. An English woman, who admitted to having come to Subiaco as a tourist, preferred to consider herself as a "researcher." Locals, in their turn, insisted they were neither pilgrims nor tourists. Groups of people from the neighbourhood declared themselves as "atheists" or "agnostics" enjoying the landscape of the Holy Mountain almost everyday. "We come here to exercise, to enjoy the place, which we feel as our own. It is much better than going to the gym."

The importance of landscape in the Subiaco experience was emphasised by all the respondents, both pilgrims and tourists (Fig. 37.4). As one noted, "Subiaco was a lovely day out from Rome, so the aim was always "a day in the country," with the landscape playing a vital role in that experience." For some landscape contributed to a strong sense of transcendence: "The landscape is like a frame of a painting already rich in spiritual potency. Landscape crowns and stimulates a place of unique evocative power;" "The landscape gives me great tranquillity, serenity and inner peace." "It is my first experience in Subiaco, an overwhelming landscape that leads you to meditation and silent reflection;" "Landscape, I would say, almost paradisiacal, that excites inner tranquillity and peace of mind." For others landscape provoked worship of God: "Landscape is always a way to celebrate the marvels of God's love. Subiaco is a special place where to love and being loved. We all are loved and we all can love;" "I love it. It's beautiful!" However, as one English woman wrote, "It is the other way round. You see the landscape before you see Subiaco. The impact of landscape precedes the impact of Subiaco."

Fig. 37.4 View of the valley from the monastery (Photo by Alessandro Scafi)

Tourists and pilgrims share a common sensitivity to landscape. Our western world is not as secular as generally assumed and, surely, landscape plays a significant role as a natural bridge between the past to present, in view of the future. An Italian teenager, who wrote in the box allocated for religion "Lazio football club," nevertheless declared that in Subiaco "the landscape is extremely evocative, making the monastery a "timeless place.""

37.4 Landscape, Pilgrimage, and Performance: Praying the Keeills on the Isle of Man

The second part of our journey takes us to the Isle of Man. In 2006 an annual week of events known as "Praying the Keeills" (PTK) was initiated by churches in the Isle. These Ecumenical pilgrimage walks were motivated by a desire to prompt the (largely Protestant) local Christian community to engage with their shared faith heritage and to explore the keeills as a spiritual resource for the contemporary Christian community. At the same time they aimed at creating a different social context and dynamic by "taking the church outdoors" and into the landscape (Figs. 37.5 and 37.6). While the walks were carefully planned to cater for a variety of degrees of individual mobility, physical movement was given a central place in the programme, in the spirit of John Bunyan's Pilgrim's Progress (1678–1684), whereby the outer embodied physical journey was a means to inner reflection and spiritual mobility.

The keeills themselves are the remains of tiny medieval chapels, and historians and archaeologists estimate there were 200–250 scattered across the Manx

Fig. 37.5 Walking through the landscape (Photo by Avril Maddrell)

Fig. 37.6 Worship at PTK 2011 (Photo by Avril Maddrell)

landscape. Many are thought to have been built over by later parish churches or other buildings, leaving the visible remains of about 35 located principally in rural areas. These locations include cliff tops, glens, river valleys, farmland and even what is now a golf course (see www.prayingthekeeills.org/).

While many of the keeills were located to serve local communities, a number can be found at places of pre-Christian symbolic and sacred significance, for example, on or adjacent to Bronze or Iron Age forts or burial chambers or Druidic sites; in turn, some are overwritten by Viking ritual practice. This sense of a palimpsest landscape and inter-leaving of cultural practices and beliefs can be seen at Chapel Hill, Balladoole, named for Keeill Vael, site of a Bronze Age burial and Iron Age hill fort, as well as a later Viking ship burial.

This sense of historical continuity provides a sense of converging paths to sacred places and it is easy to see why many of these aesthetically powerful sacred sites were found to be inspirational. For example, Spooyt Vane waterfall, the location of St Patrick's keeill, was thought to have been an earlier Druidic site (Fig. 37.7). As noted in the 2012 programme: 'The ancient keeill sites are often found in areas of particular natural beauty and in such surroundings it is a common experience to sense something of the presence of God .... They are what our Celtic forebears would have described as being "Thin Places" where we can draw

Fig. 37.7 Spooyt Vane waterfall and keeill site, thought to have been a Druidic site (Photo by Avril Maddrell)

closer to God' (Praying the Keeills programme leaflet, May 2012). This sense of using the keeills as a spiritual crossroads, as a link between past and present faith practice and a prompt to setting aside time-space for reflection is further expressed in the program: "Prayer and meditation were very important to those who worshiped in or around the keeills, as they should be to us. Praying the Keeills, organized by local churches, gives an opportunity to step aside from the business of life, and rediscover what so many have lost" (Praying the Keeills program leaflet, May 2012).

"Sacred Space" was used as a thematic framework for the 2012 week of walks. Within such framework, even the solid keeills landmarks were conceptualized as part of a dynamic and fluid sacred space, as the daily notes for reflection on the program explained:

> What is "sacred space?" We know that single definitions do not bring fullness to concepts, so we begin by acknowledging that sacred space is discovered in many differing shapes, forms and geographical locations .... Although we may think of a sacred space as fixed, in reality it is mobile and always on-the-move because it is dependent upon the lively presence of the divine. Thus we might define sacred space as any and every place that breathes "alive" with the Holy. (Praying the Keeills programme leaflet, May 2012)

For one woman that sense of the holy was experienced through walking, talking and worship on the beach:

> I have enjoyed exercising my body while being among people who are friendly enough to converse. The worship on the beach was especially poignant as it interacted with the landscape… (PTK 4 2011, Female, 46-55 years of age, Methodist)

"Crossroads" was introduced as a theme for reflection on the last day's walk of the 2012 PTK pilgrimage. This walk around Maughold included an assemblage of sacred sites at Ballafayle: a known keeill site (with no visible remains), a Bronze Age burial cairn and a seventeenth century Quaker burial ground, the latter symbolizing of hard-won religious freedom from the rule of the Established church. As one participant noted of a previous PTK visit to this site:

> The landscape was central to the experience. My favourite keeill … is the one at Ballafayle – Neolithic burial chamber, a Celtic keeill and the Quaker grave of William Callow. (PTK, 2010, 15, Male, 66–75 years of age, Christian)

Participants were encouraged to reflect on the significance of "crossroads" at the outset and during meditation and prayers en route, crossroads being linked metaphorically to The Way of the Cross, as well as being meeting places, places of sharing and signposts, places of choice and decision. To date the keeills have not been associated with miraculous revelation or healing, as in the case of some other Christian pilgrimage sites such as Lourdes. Nonetheless, the wider, arguably more Protestant, sense of looking for and following divine "signs" along life's journey was echoed in the image and accompanying text of a photograph posted on the PTK 2012 blog, which read "We followed a sign" (Fig. 37.8).

Reflections on the nature and purpose of crossroads were underscored by the practical support of the walk leader's spouse who met walkers at two crossroads along the way of the day's unseasonably hot walk, offering fresh water and the

Fig. 37.8 "We followed a sign." A pilgrim took this photo as a reminder of the role of signs in many Biblical and pilgrimage narratives (Photo from http://www.prayingthekeeills.org/image_collection_216045.html, used with permission)

option to off-load bags or to ride in the car if necessary (Maddrell, Field Diary 2012). This in turn echoed the previous night's torchlight procession at Port St Mary, where "Thin Places" were linked to both a sense of God speaking through the landscape and to "Thin People," that is, those people who, like stained glass windows, let spiritual "light" shine through them, refracting colour and warmth like the saints depicted in stained glass windows (Maddrell, Field Diary 2012). This was expressed in the closing blessing:

> May the blessing of light be on you
> Light without and light within
> May it warm your heart
> Till it glows like a great peat fire
> So that strangers may come and warm themselves at it
> As well as friends.
> Any may the light shine out of your eyes
> Like a candle set in the windows of a house.
> (PTK, Friday worship sheet, Port St Mary, 2012; taken from an Irish blessing)

Prehistoric and Viking markers in the landscape signaled how the Isle of Man, located in the center of the Irish Sea, was an important signpost and "crossroads" for navigators at a time when sea travel dominated migration, trade and raids. Similarly Bronze and Iron Age cairns and standing stones acted as local signposts in the landscape (see Maddrell 2011). Evangelism was part of those exchanges over the sea, the Christian faith brought from Ireland, Wales and possibly Scotland, the keeills with their surrounding burial grounds being material evidence of that faith. The keeills themselves are a sign of the historical pre-Reformation foundation of the Christian church on the Isle of Man.

Today, a sense of spiritual journey is created in the small locale of the Isle of Man, through re-engaging the practice of pilgrimage, by visiting keeills in little known or private places, and beautiful places which provide a sense of being "transported" into the sublime. A combination of the embodied and social practice of walking in a group, with historical landscape analysis and spiritual reflection, worship and prayer mesh to offer a physical, educational, social and affective experience, and for many participants a sense of spiritual encounter and enrichment.

> Breathtaking beauty gave the feel of the 'thin' places on earth closer to God. The first [walk had a] … lasting impact, the peace, physical exhaustion, understanding we are such small specs on this earth … (PTK 14 2011, Female, 46-55 years of age, Anglican)
>
> The whole experience was a very moving and spiritual one. We met wonderful people, went to spectacular places and learnt much in the process … (PTK 3 2011, Male, 36-45 years of age, Methodist)

One pilgrim's account conveys the sense of transcendence experienced through the intermeshing of landscape, walking, history and prayer:

> The keeills week took me to beautiful previously unexplored parts of IoM [Isle of Man]. Sitting in a place of prayer surrounded by beautiful woodland and carpets of spring flowers was sheer delight. An awe-filled mystic experience each day being a new landscape, vista, and different type of weather. To have the mists surrounding St Runius surrounded by ancient graves, the dappled sunshine as we prayed at Glen Mooar up above Spooyt Vane, then down to the wild beach below. From hilltop and Viking burial grounds and grassy fields to the woodlands and waterfalls to the beaches, all enveloped in the glorious May splendour of new green and wildflowers. Magic (PTK 2010 18, Female, 66-75 years of age, Roman Catholic)

37.5 Landscape, Pilgrimage and Tourism: Meteora

The last part of our journey takes us to the plain of Thessaly in central Greece. As we pass the village of Trikkala and head North-West, our gaze bumps into a giant rocky wall, a brutal interruption on our visual horizon. Mircea Eliade (1959) would have named this an axis mundi, a point of encounter between earth and heaven. As we drive closer, the dark silhouette turns into a bizarre ensemble of rocky pillars. A nearer view reveals all sorts of shapes: pyramids, columns, obelisks, monoliths, masses. Some of the rocks rise for a couple of dozens metres, others reach up to nearly 400. We have reached the so-called Meteora (Fig. 37.9).

Meteora (literally, "the suspended") is one of the main monastic complexes and pilgrimage destinations in the Orthodox world. Its six still-functioning Byzantine monasteries are built on natural sandstone rock pillars, at the northwestern edge of the Plain of Thessaly. Physical geography alone makes Meteora one of the most evocative landscapes in the world. Yet, its distinctiveness lies in the combination of rock and Byzantine monastic tradition.

Fig. 37.9 Holy Monastery of Aghia Triada (Photo by Veronica della Dora)

Meteora's spiritual landscape has been carved out of its unique geology through ascetic discourse and practice. Hermits first moved to Meteora around the eleventh century. Seeking a retreat from the expanding Turkish occupation, they found the inaccessible rocky pillars to be an ideal refuge. Since the fourteenth century over 20 monasteries were built. Most of them were pillaged and destroyed over the last 200 years of Ottoman rule and during the Second World War (Nicol 1963). Today only six of them survive. These foundations host miraculous icons and relics, as well as active monastic communities who every morning perform their worship services as the world below is still asleep.

As a UNESCO world heritage site Meteora attracts up to a 1,000 tourists per day. As Greece's second largest monastic complex, it also attracts a vast number of pilgrims. Orthodox pilgrimage to Meteora focuses on the monasteries, and more specifically on the clergy and on the *sacra* therein contained. Unlike the western Christian pilgrimage traditions discussed above, Orthodox *proskynema* downplays physical distance from home and the transformative act of journeying. It rather centers on the physical veneration of sacred objects (relics and icons) at the shrine, on the participation to church services, and, on a more ordinary basis, on the possibility to meet with a spiritual guide from the monastery (Rahkala 2010). Hence pilgrims to Meteora include both Orthodox faithful traveling from other parts of Greece and of the world and, as with Subiaco, local villagers. The former venerate relics and icons, light candles, write down names to be commemorated by the monks. The latter usually drive up to the monasteries to attend to Sunday's liturgies (which take place early in the morning) and to seek counsel from a spiritual advisor. In both cases, pilgrimage is not simply an inspirational matter. What matters is physical interaction at the shrine, rather than the process of reaching it. Unlike Praying at the Keeills, here landscape thus serves as a background to, rather than as the focus of, pilgrimage (Fig. 37.10).

Fig. 37.10 Pilgrims descending from the monastery of Roussanou (Photo by Veronica della Dora)

Fig. 37.11 Fridge magnets on sale in the village of Kalambaka featuring Meteora and other commodified landscapes (Photo by Veronica della Dora)

A Greek Orthodox monk and an Egyptian Copt priest heading to one of the monasteries both referred to landscape in dismissive terms: "We are not here for the landscape; tourists are after landscape! We are here for the saints." As opposed to venerating the relics or taking part to church services, the two pilgrims perceived looking at landscape as a distraction. Likewise, a Finnish Orthodox priest set himself (and pilgrims in general) apart from tourists, because, he argued, "the tourist's approach always implies distancing, whereas the pilgrim's implies participation"- and as a concept landscape intrinsically implies visual distancing, separation.

Landscape in Meteora is indeed constantly framed by tourists' cameras, gazed through the windows of their coaches, or simply from the monasteries, which are themselves structured as viewing platforms. Landscape is also consumed through all sorts of souvenirs: postcards, fridge magnets, posters, etc. (della Dora 2012) (Fig. 37.11).

Yet, landscape also wraps the viewer. In Meteora, its presence is so insistent and overwhelming that it both exceeds the tourists' act of "Enframing" and the Orthodox pilgrims' focus on the shrine. While the focus of the latter remains the shrine, their journey to reach the monastery does not occur within an empty space. In Meteora *proskynema* cannot be totally divorced from moving through and gazing at the surrounding landscape. International Orthodox pilgrims still stop at panoramic spots and take pictures of the scenery (Fig. 37.12), though charging it with different values from tourists (for example, linking it to the ascetic lives of the founding fathers, or to higher spiritual truths):

> The landscape of Meteora is hard at first to even grasp. The buildings themselves and the now saints that dedicated their lives to this are amazing. (M2010, pilgrim, male, 46–55 years of age, Greek-Orthodox).
>
> Spiritually uplifting. The over 700-year experience of continuous monastic life was more important than the landscape, but I understand that this is what brought the monks here. (M2010, pilgrim, male, 36–45 years of age, Russian-Orthodox).

Fig. 37.12 Tourists at Megalo Meteoron Monastery capturing landscape through their cameras (Photo by Veronica della Dora)

For local pilgrims, landscape is part of their own identity: it is "lived" rather than "looked at" (Lowenthal 2007). Yet at the same time it is perceived as embedding a therapeutic quality. Setting the pilgrims aside from the everyday, it acts as a preparatory stage for their *proskynema* at the shrine and the meeting with their spiritual advisor, thus actively participating to spiritual healing processes.

> Just by looking at the Meteora makes me at peace. (M2010, local pilgrim, female, 46–55 years of age, Greek Orthodox)
>
> You can feel God near you in a quite and peaceful environment. (M2010, local pilgrim, male, 36–45 years of age, Greek Orthodox)

Conversely, tourists are often struck by the mystic atmosphere of the Byzantine church and the magic of the outdoor environment in general, something that exceeds words and the lens of their camera.

> The nature actually "demands" us to think about our lives. (M2010, tourist, male, 26–35 years of age, Roman Catholic)

The distinction between pilgrim and tourist is often blurred. As in Subiaco, further categories of visitors come to Meteora and are equally struck by the landscape. For these visitors, Meteora is but an elevated spot from which to put the world in perspective; it allows them to raise above earthly cares; to look at their lives from a distance, even if for a brief moment. For them Meteora is a pause.

> I have come here for the beauty of the rocks and for the calmness of the monasteries. Here you feel on the top of the world. (M2010, visitor, female, 36–45 years of age, Greek Orthodox)

Nature here has a therapeutic quality: I get away from the noise and my hectic life in Thessalonica. Here I feel regenerated. I feel better. (M2010, student, female, 26–35 years of age, Greek Orthodox)

Everyone in the coffee shop is talking about the elections nowadays. It's crazy! It causes me headache. I come up here to breath some oxygen. (M2010, local villager, male, 46–55 years of age, Greek Orthodox)

37.6 Conclusions: Intersecting Pathways

Pilgrimage sites and practices attract a range of participants with different motivations, interests and aspirations. Each pilgrimage is shaped by and attracts participants largely from its denominational tradition. However, increased openness to learning from different denominational practices and other faiths, as well as increased numbers of cultural tourists, has resulted in increased heterogeneity of visitors to pilgrimage sites and events. Drawing on the three case studies discussed above, accounts by Christian believers demonstrate both common and varying degrees of interaction between worship and landscape, while responses by non-believers illustrate the sense of transcendence many experience in and through these places and practices regardless of faith status.

Acknowledgements This chapter is based on data from an AHRC-ESRC Religion and Society Scheme funded research project (AH/HOO9868/1) on Christian pilgrimage and landscape. We would like to express our thanks to the many interviewees and respondents in the Isle of Man, Meteora and Subiaco who generously shared their experiences with us. We would also thank Sarah Evans for her precious suggestions and help with copyediting.

References

Anderson, J. (2010). *Understanding cultural geography: Places and traces*. London/New York: Routledge.

Bonamore, O. (1884). *Il Sacro Speco e Santa Scolastica*. Venezia: L'Immacolata.

Bowman, G. (1991). Christian ideology and the image of a Holy Land: The place of Jerusalem pilgrimage in the various Christianities. In J. Eade & M. Sallnow (Eds.), *Contesting the sacred: The anthropology of pilgrimage* (pp. 99–121). New York/London: Routledge.

Bunyan, J. (1678–1684). *The pilgrim's progress from this world to that which is to come*. London: George Virtue.

Carter, P. (1987). *The road to Botany Bay: An exploration of landscape and history*. London: Faber and Faber.

Carter, P. (1992). *Living in a new country: History, travelling, and language*. London: Faber and Faber.

Coleman, S., & Eade, J. (2004). *Reframing pilgrimage*. London: Routledge.

Conradson, D. (2007). The experiential economy of stillness: Places of retreat in contemporary Britain. In A. M. Williams (Ed.), *Therapeutic landscapes* (pp. 33–47). Aldershot: Ashgate.

Cosgrove, D. (1985). Prospect, perspective and the evolution of the landscape idea. *Transactions of the Institute of British Geographers, 10*, 45–62.

Cosgrove, D. (2003). Landscape and the European sense of sight-eyeing nature. In K. Anderson, M. Domosh, N. Thrift, & S. Pile (Eds.), *Handbook of cultural geography* (pp. 249–268). London: Sage.
Cosgrove, D. (2006). Modernity, community and the landscape idea. *Journal of Material Culture, 11*, 49–66.
Cosgrove, D., & Daniels, S. (Eds.). (1988). *The iconography of landscape*. Cambridge: Cambridge University Press.
Delano-Smith, C., & Scafi, A. (2005). Sacred geography. In Z. Török (Ed.), *Sacred places on maps* (pp. 121–140). Pannonhalma: Pannonhalmi Fıapátság.
della Dora, V. (2012). Setting and blurring boundaries: Pilgrims, tourists, and landscape in Mount Athos and Meteora. *Annals of Tourism Research, 39*, 951–974.
della Dora, V. (forthcoming). Paths to the holy: Early Christian pilgrimage and sacred landscapes. In A. M. Yesin (Ed.), *Early Christian and Byzantine architecture* (Cambridge world history of religious architecture series). Cambridge: Cambridge University Press.
Duncan, J. (1990). *The city as a text: The politics of landscape interpretation in the Kandyan Kingdom*. Cambridge: Cambridge University Press.
Eliade, M. (1959). *The sacred and profane*. London: Harcourt Brace Jovanovich.
Foster, J. (1998). John Buchan's 'Hesperides': Landscape rhetoric and the aesthetics of bodily experience on the South African Highveld, 1901–1903. *Ecumene, 5*, 323–347.
Frankfurter, D. (2005). Urban shrine and rural saint in Alexandria. In J. Elsner & I. Rutherford (Eds.), *Pilgrimage in Graeco-Roman and early Christian antiquity* (pp. 435–450). Oxford: Oxford University Press.
Frey, N. (1998). *Pilgrim stories: On and off the road to Santiago. Journeys along an ancient way in modern Spain*. Berkeley: University of California Press.
Gemzöe, L. (2005). The feminization of healing in pilgrimage to Fatima. In J. Dubisch & M. Winkelman (Eds.), *Pilgrimage and healing* (pp. 25–48). Tucson: University of Arizona Press.
Gesler, W. M. (2003). *Healing places*. Lanham: Rowman & Littlefield.
Giumelli, C. (Ed.). (1982). *I monasteri benedettini di Subiaco*. Mila: Silvana editoriale.
Gothóni, R. (1994). *Tales and truth: Pilgrimage on Mount Athos past and present*. Helsinki: University Press.
Gregory of Nyssa. (1885). On pilgrimages. In P. Schaff (Ed.), *Nicene and post-Nicene Fathers* (Series 2, Vol. 5). Text available at: www.tertullian.org/fathers2/NPNF2-05/Npnf2-05-33.htm. Accessed 13 July 2012.
Gregory the Great. (2001). *Vita di San Benedetto* (G. Rigotti, Ed.). Alessandria: Edizioni dell'Orso.
Harris, A. (2014). Lourdes and holistic spirituality: Contemporary Catholicism, the therapeutic and religious thermalism. *Culture and Religion, 14*, 23–43.
Johnston, R., Gregory, D., Pratt, G., & Watts, M. (Eds.). (2000). *Dictionary of human geography*. Oxford: Blackwell.
Jones, O. (2005). An ecology of emotion, memory, self and landscape. In J. Davidson, L. Bondi, & M. Smith (Eds.), *Emotional geographies* (pp. 205–218). Aldershot: Ashgate.
Lowenthal, D. (2007). Living with and looking at landscape. *Landscape Research, 32*, 637–659.
Maddrell, A. (2011). 'Praying the Keeills'. Rhythm, meaning and experience on pilgrimage journeys in the Isle of Man. *Landabréfið, 25*, 15–29.
Maddrell, A., & della Dora, V. (2013). Crossing surfaces in search of the holy: Landscape and liminality in contemporary Christian pilgrimage. *Environment and Planning A, 45*, 1105–1126.
Maraval, P. (2002). The earliest phase of Christian pilgrimage in the Near East (before the 7th century). *Dumbarton Oaks Papers, 56*, 63–74.
Markus, R. (1994). How on earth could places become holy? Origins of the Christian idea of holy places. *Journal of Early Christian Studies, 2/3*, 257–271.
Merriman, P. (2006). A new look at the English landscape: Landscape architecture, movement and the aesthetics of motorways in early postwar Britain. *Cultural Geographies, 13*, 78–105.

Mitchell, D. (1996). *The lie of the land: Migrant workers and the California landscape*. Minneapolis: University of Minnesota Press.
Mitchell, W. T. J. (Ed.). (2002[1994]). *Landscape and power*. Chicago: University of Chicago Press.
Nicol, D. (1963). *Meteora: The rock monasteries of Thessaly*. London: Chapman & Hall.
Panella, A. M. (1875). *Il pellegrinaggio sublacense*. Parma: Fiaccadori.
Petrarca, F. (Ed.). (1975). *De vita solitaria* (A. Bufano, Ed.). Turin: UTET.
Piccolomini, A. S. (Ed.). (1984). *Commentarii* (A. van Heck, Ed.). Vatican City: Biblioteca Apostolica Vaticana.
Rahkala, M. J. (2010). *Pilgrimage as a lifestyle: A contemporary Greek nunnery as a pilgrimage site*. Helsinki: University of Helsinki Press.
Rose, G. (1993). *Feminism and geography: The limits of geographical knowledge*. Oxford: Polity Press.
Sauer, C. (1925). *The morphology of landscape*. Berkeley/Los Angeles: University of California Press.
Slavin, S. (2003). Walking as spiritual practice: The pilgrimage to Santiago de Compostela. *Body and Society, 9*(3), 1–18.
Smith, J. Z. (1987). *To take place: Towards a theory in ritual*. Chicago: Chicago University Press.
Solnit, R. (2002). *Wanderlust: A history of walking*. London: Verso.
Turner, V., & Turner, E. (1978). *Image and pilgrimage in Christian culture*. Oxford: Oxford University Press.
Williams, A. (2007). *Therapeutic landscapes*. Aldershot/Burlington: Ashgate.
Winkelman, M., & Dubisch, J. (2005). Introduction: The anthropology of pilgrimage. In J. Dubisch & M. Winkelman (Eds.), *Pilgrimage and healing* (pp. ix–xxxvi). Tucson: University of Arizona Press.
Wylie, J. (2002). An essay on ascending Glastonbury Tor. *Geoforum, 33*, 441–454.

Chapter 38
Just Like Magic: Activating Landscape of Witchcraft and Sorcery in Rural Tourism, Iceland

Katrín Anna Lund

38.1 Introduction

"You have to excuse me if I burst into a laughter tomorrow, I don't really believe in things like this," said Anita, an elderly Italian lady I was having a meal with at Café Riis, Hólmavík, a town of just under 400 inhabitants in the Strandir region, northwest Iceland. Anita and I had earlier in the day met on the bus and both with the purpose of joining a trip around the region, titled "Landscape and Folklore," that had been scheduled the day after. Anita's statement surprised me. The tour had clearly been introduced as a journey into the landscape of magic introducing hidden people, trolls and ghosts, and I knew that partially it had been designed because Anita had been interested in it. Originally, it had been created as a hiking trip but Anita, although fit for her age, was not capable of walking much distances and thus the director of the Museum of Sorcery and Witchcraft in Hólmavík, the designer and guide for our tour, called Siggi, had redesigned and extended the original trip into a bus journey. I was going to join the tour as a researcher interested in landscape and folklore especially in looking at how routes, paths and roads, are used to narrate landscape in different ways. This summer of 2012 was my second year of doing fieldwork in the area for a bigger project focusing on the creation of a place as a tourism destination. Anita, on the other hand, was there as a tourist who had been travelling around Iceland for about a month, exploring its peculiarities, as she explained it. The day after we went on the trip and despite her remarks the trip did not bring out a reason for her to burst into laughter. Rather, the trip introduced her to different worlds that emerged in the landscape as we passed through it.

K.A. Lund (✉)
Department of Geography and Tourism, Faculty of Life and Environmental Sciences, University of Iceland, Reykjavik, Iceland
e-mail: Iceland.kl@hi.is

S.D. Brunn (ed.), *The Changing World Religion Map*,
DOI 10.1007/978-94-017-9376-6_38

She realized that it was not a question of believing or not, but rather to be able to delve into the magical landscape that she journeyed through allowing her to follow its narratives and gain an entrance into different temporal dimensions, as I will illustrate below.

The Strandir region has in recent years been growing as a tourist destination. The region is a sparsely populated coastal area in the North-west of Iceland (Fig. 38.1). Lowland is limited and the extreme contrast where mountains and the shore, stretching by the Arctic Ocean, meet, sometimes directly, is a recurring reminder of how nature is constantly in the making; "where land and water, solidity and liquidity transform and intermingle" (Mclean 2011: 609) (Fig. 38.2).

Its peripheral characteristics as well as location provide it an aura of remoteness, spatially and temporally, a sense of having been left behind, but simultaneously protected from the hustle and bustle of any modern invasions. Nevertheless, historically Strandir has always been caught up between diverse mobilities, entangled in constant and dynamic motions of people ideas and things in an erratic natural environment (for example, Urry 2000; Sheller and Urry 2004; Cresswell 2011). Economically, it has relied on fisheries and agriculture, but like in most other rural regions in Iceland these traditional economic backbones have been declining and thus the region has experienced almost constant outmigration after the second half of the twentieth century. During the last decade a slow but steady growth in tourism has been initiated which has become an influential factor in re-stabilizing the image of the area, hence as a tourism destination, emphasizing the cultural history of the region. This is a history which throws light on people who endured in a harsh environment, constantly dealing with the unpredictable doings of nature. A special focus was created into the seventeenth century, a period during which the area became notorious for the practice of witchcraft. This era is presented in the Museum of Sorcery and Witchcraft which was established in 2000, the starting point for the ethnographic account in this chapter.

My intention is to show how the establishment of the Museum has activated the regional landscape and mobilized it into different dimensions, spatially and temporally as a magical landscape. In order to do so I shall start by introducing how magic became associated with the region historically before showing how it is interwoven into the creation of the landscape introducing it as a tourism destination in contemporary context. This will reveal how the region does not exist as an entity marked by boundaries, but rather is constantly moving as a fluid image, a place of "in-betweenness" that extends and goes beyond boundaries as conventionally thought of, much as the fluid nature that characterizes the region. Furthermore, following Ingold and Hallam (2007), places become because of the "movements of people towards, around and away from them" (2007:8). Thus it can be argued that a place is a continuous happening (for example, Casey 1996; Massey 2006) created through the relational and simultaneous motions of people and its materiality. In this context I want to argue that places become through how they are improvised as people move through them "with each step a discrete entity, part of a fluid dance rather than a metronomic repetition of sameness" (Edensor 2010: 72). Thus people create their destinations as they travel through them. To reveal how, I intend to

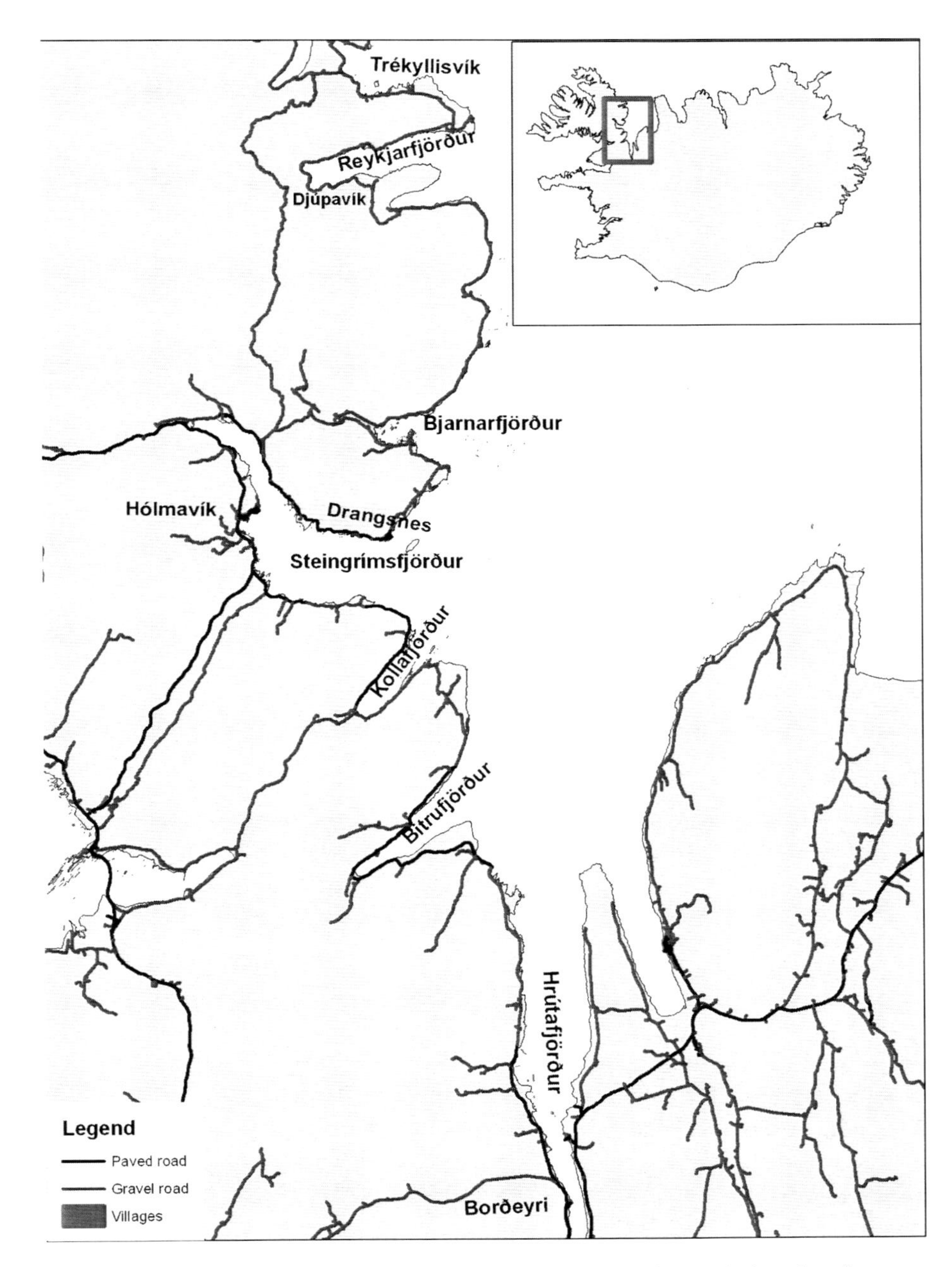

Fig. 38.1 Map of Strandir (Map by Friðþór Sófus Sigurmundsson for Katrín Anna Lund)

Fig. 38.2 Strandir (Photo by Claus Sterneck, www.claus-in-iceland.com, used with permission)

follow the narratives of the "ethnographic story" (Winthereik and Verran 2012) which the tour "Landscape and Folklore" provided me with, as well as my fellow travelers. The intention is to demonstrate how the region emerges as it is improvised through the steps of the guide and the visitors as they journey through it together, creating a magical landscape (Lund 2010). But that is not how the journey ends as my second example reveals because through embodying a magical destination while travelling around it tourists might not get entirely away from it because the place travels with them across geographical boundaries leaving them in a continuing "in-betweenness."

38.2 Magical Landscape

As a destination associated with magic Strandir can be described as a "place in play" according to Sheller and Urry (2004), a place that has been, at least partially, invented as such and provided with an image. I say partially because in the mind of some Icelanders, especially older generations, Strandir as a region has always been associated with occult practices, a correlation that stems from the seventeenth Century historical accounts of witchcraft accusations and burnings that brought Strandir into focus as place of murky performances (Jónsson 2008; Rafnsson 2003). During the later part of the twentieth Century this historical image had slowly been vanishing into distant and forgotten past, a process of fading memory that suddenly was changed in the year of 2000 when the Museum of Sorcery and Witchcraft was formally established in the area. The idea for establishing the museum originated from a report written by an ethnologist, Jón Jónsson, who is born and bred in the region. The report stressed the potential role of small scale tourism in the region by detailing many ideas of possible projects such as exhibitions on the history of the region both near and in more distant past. Jón underlined that these were not his ideas. His role as an author was to collect ideas from the people in the region, organize them and frame them within a discourse of sustainable development of cultural tourism. At the end of a long list of ideas the possibility to present the history of sorcery and witchcraft in the area is only briefly mentioned. This short paragraph, turned out to be the inspiration for the establishment of the Museum which Jón with a support from few local men initiated, financed by the County of Strandir. The exhibition is based on Jón's ideas, but a professional stage designer was hired to install it and Sigurður Atlason, or Siggi, was hired as its director.

Although some inhabitants had doubts at the beginning that this dark era of history was brought into focus again, the Museum has proved to be a positive innovation for tourism in the region. The museum throws a light on how in the name of God the authorities oppressed poor peasants, a reality which is in living memory at a distance but nevertheless real, not the least for the region's inhabitants. It provides a picture of how when religion is used as a mechanism of power and activated as such it brings forth cruelty and malice, a situation that for most people living in the Western World is a part of history but still a part of living reality for so many.

Located in Hólmavík, a village of just under 400 inhabitants and the central municipality for the region the Museum provides the initial steps for the visitor into the surrounding landscape. The Museum is located in the same building as the Tourist Information Centre in which the air of magic is also looming about. The souvenirs sold at the Information Centre are especially made for the area imprinted with runes and other magical signs, postcards featuring zombies and ravens have been especially made for the Centre and different types of books about magic, witchcraft and sorcery are on sale (Strandagaldur: Museum of Icelandic Sorcery and Witchcraft 2012). In this manner the Information Centre extends the display which takes place in the Museum itself; a display that can, in short, be described as a play with the imagination, the starting point for further improvisation as the visitor travels into the region.

The Museum's display allows the visitor to plunge into the world of witchcraft and sorcery through a presentation that is akin to what Alpers (1983) and Hetherington (1999) have referred to as "cabinets of curiosities" allowing each object on display to be speak for itself rather than making connections between objects for a linear story. Each cabinet features an item or items central for a particular magical act accompanied by relevant magical stave and instruction about how to practice it. Details such as where, why and by whom a particular magical act was practiced, or an exercise to connect the act to a particular past reality, is left out allowing the visitor to put the act into context in the way he or she wants to imagine it. To give an example it is worth mentioning the display of the Necropants (*Nábrók*). Necropants are pants of human skin that are the center of the display and on the floor below it a bed of coins. Beside the cabinet a card is located showing the magical stave, called *nábrókarstafur*, used for the charm and the one which signifies it. Below it is a short text which illustrates the way in which to perform the magic.

> If you want to make your own necropants (literally; nábrók) you have to get permission from a living man to use his skin after his dead. After he has been buried you must dig up his body and flay the skin of the corpse in one piece from the waist down. As soon as you step into the pants they will stick to your own skin. A coin must be stolen from a poor widow and placed in the scrotum along with the magical sign, nábrókarstafur, written on a piece of paper. Consequently the coin will draw money into the scrotum so it will never be empty, as long as the original coin is not removed. To ensure salvation the owner has to convince someone else to overtake the pants and step into each leg as soon as he gets out of it. The necropants will thus keep the money-gathering nature for generations. (Strandagaldur: Museum of Icelandic Sorcery and Witchcraft, Stave for Necropants 2012)

The visitor is left to further create a narrative context. Hence, the Museum's display is created through a play with magic, as it is put forward by the Museum and improvised by visitors. The improvisation continues as the visitor leaves the museum and playfully continues the creation as he or she travels through the destination. As a destination full of magic, Strandir is a place in play and a place of play. It is a place in play in how it is performed through continuous improvisation by people moving to it, from it and through it. It is a place of play in how the museum deliberately invites the visitor to take a part in the play encouraging him or her to continue the improvisation outside the museum's walls.

As a tourism destination it can thus be argued that Strandir has been created through play which indicates continuity with its murky past of witchcraft and sorcery, although the context is entirely different. Today the play with magic is a play that continues in order to provide the region with a distinctive identity. With a reference to Leví-Strauss, Michael Jackson points out that all magic is akin to play in how its tendency is to "miniaturize, simplify, and rearrange" (1998: 31) sometimes a daunting reality. Magic, like play is about reordering reality through acts that transform it and make it more manageable and tangible and in this context the magic performed in the seventeenth century can be compared to the transformation of Strandir into magical landscape in how it involves a creative play with circumstances. The origin of performing magic in Iceland can be traced as far back as to ancient Germanic mythology. In the Edda poems runic spells are frequently mentioned and magical acts were entwined in the cosmos (Rafnsson 2003: 9). Thus it may be assumed that knowledge of runes were common and the practice of magic a part of everyday life. The introduction of Christianity, in the form of Catholicism from the year 1000, did not affect people's daily practices in any great deal but during the middle of the sixteenth century the influence of the Lutheran Church was soon to win over in Iceland bringing in orthodox emphasis about suffering and salvation. Instead of trying to affect one's circumstances one should pray the lord as a humble sinner and be grateful for suffering (Rafnsson 2003). The oppositions between sacred and evil were clearly defined and play with magic was defined as evil. The power of the Lutheran Church implemented a new order through which the practice of magic became defined as witchcraft and sorcery, a criminal act.

To be accused of witchcraft and sorcery was to be doomed; one became an outcast of society and life unbearable. And as Jackson (1989) has pointed out, admitting to crime, however impossible it is to verify it, is to take the control in one's hand; or to leave circumstances unbearable to live with. In other words, to bring the game to a personal end and leave it for other people to continue with. Approximately three centuries later the creators of the Museum of Sorcery and Witchcraft pick up the sequence and continue the play, but in entirely new circumstances, now in order to bring forth economical changes for the region. It is interesting to note in this context that in a newspaper interview 2003 the bishop for the Lutheran State Church in Iceland disapproved of the Museum by accusing those behind it for making witch burning and murky forces in to an entertainment for tourists (Hin miklu öfl eru raunveruleg 2003: 22–23). Hence, although time and circumstances have changed in a more secular political environment, the Church's tolerance towards magic remains intact.

In an isolated region like Strandir, it can easily be imagined how peasants suffering poverty in often harsh, unforgiving and unpredictable nature desperately tried to control their circumstances by implying changes through magic as the only available instruments. This is not only imaginable in the contemporary world but it also seems to have been a widespread assumption that people in Strandir practiced their relations with supernatural powers more than people in other areas in Iceland did. Still, cases of witchcraft accusation during the seventeenth century were not isolated with Strandir. Nevertheless, it seems that in people's mind there had always been some association between occult practices and the people of Strandir, probably due to its peripheral location (for example, Rafnsson 2003). This old association is brought into light with the establishment of the Museum of Sorcery and Witchcraft and in that context magic is used in attempt to innovate and promote the region as a tourism destination. The play continues through how the Museum orders and displays its items in order to maintain the creation while encouraging tourist to take part in the play and continue to shape the region through further improvization.

Ingold and Hallam (2007) point to the difference between innovation and improvisation, as acts of creativity, in how the former is usually associated with end product while the latter "characterizes creativity by way of its processes" (2007: 2). In an agreement Stuart Mclean argues that what usually understood as creativity "often seems to amount to little more than a purely practical facility for problem solving" (2009: 213). Thus, when speaking of process of creativity we tend to "read it backwards, in terms of its result, instead of forwards, in terms of the movements that gave rise to them" (Ingold and Hallam 2007: 2–3). In this context we may look at the Museum as an innovation by reading it backwards, by looking at the process from when the idea was originally born to what it produced in the form of a museum; the Museum as an end product of innovative act. This way of looking would however leave out the process of how it activates the surrounding landscape as magical and, as such, as a tourism destination as well as erasing any indications of the region's mobility. Hence, the point of departure for the journey into the landscape of magic will be the Museum of Sorcery and Witchcraft.

38.3 The Journey

We met at the Museum of Sorcery and Witchcraft at 9 AM (Fig. 38.3). This was a welcoming, sunny and warm Saturday morning and when I arrived, Siggi, the guide and the director of the Museum told me that I had already performed magic by bringing this weather with me from my hometown, Reykjavík. This was a pleasant start of the day. Soon Anita arrived and then an American couple who we had met the day before and had decided to join us. The journey started inside the Museum. Siggi, dressed up as a sorcerer, performed two of the magical acts featured in the museum and then set up a ceremony, in which we all participated in, to quell a zombie. The ritual was acted out where a zombie is featured breaking up from underneath the museum floor and involved items and attributes such as skin of a seal, carved piece of wood, cross, cup of menstrual blood from Siggi's great grandmother, shouting, bell ringing, spitting and humming (Fig. 38.4). Needless to say, the act was successful.

Immediately a timeless and mystical past was made tangible and explicit through an imaginative act, bringing together bodies, human and non-human, earthly and divine, in different temporalities bringing forth an atmosphere of "in-betweenness." Following Deleuze and Guattari (1987), Anderson (2009) describes how atmospheres are "generated by bodies – of multiple types – affecting one another as some form of "envelopment" is produced" (2009: 80). As such atmospheres are fluid, in a constant motion filling "the space with a certain tone of feeling like a haze"

Fig. 38.3 Exhibit in the Museum of Sorcery and Witchcraft, Hólmavík, Iceland (Photo by Katrín Anna Lund)

Fig. 38.4 A zombie crawling through the museum floor (Photo by Katrín Anna Lund)

(Böhme 1993: 114). Thus, an atmosphere has generated which will set the tone for the journey we are about to undertake, like a haze in a constant motion producing countless combinations of colors, tones and shapes. The atmosphere of "in-betweenness" is thus an ambiguous one in which one is situated

> ...between presence and absence, between subject and object/subject and between the definite and the indefinite – that enable us to reflect on affective experience as occurring beyond, around and alongside the formation of subjectivity. (Anderson 2009: 77)

The atmosphere shapes a place. How we, on the other hand, reflect on and experience the ambiguity of the atmosphere is depending on how we carry on the improvisation or move with the atmosphere and, simultaneously, allow it to move us. In the case of our journey, the starting point is in the Museum of Sorcery and Witchcraft.

After the ceremony at the Museum the journey continued and now by mini-bus. On the way Siggi narrated the surroundings, sometimes the immediate ones such as spelling out some facts about Hólmavík, but also the non-immediate, for example, by pointing out cliffs and rocks as homesteads of elves and hidden people. The first stop of the journey was on the other side of the geographical boundaries of the region but, nevertheless, a significant place for this part of the country, the northwest peninsula of Iceland, since the merchant town for the hidden people is located there. We could see the rock appearing in the mist of the morning haze (Fig. 38.5). We stepped out of the bus and walked towards the coast to gain a closer perspective. A graceful cliff rose up from the ground at the other side of the fjord and one could easily see how this was an ideal place for a harbor.

Fig. 38.5 The city of the hidden people (Photo by Katrín Anna Lund)

Siggi spoke to us about the world of the hidden people and told how it is parallel to our own, still absent and hidden from human everyday reality, although not entirely separated. He explained how hidden people are generally perceived as human-like but still more beautiful and not the least more powerful beings that is manifested in how any negative nuances with them can lead human beings into trouble. These nuances are usually made evident through showing disrespect to the world of the hidden people, for example, by cutting grass on a hill in which they might live in or blowing up rocks close to their home. Any disturbance to their personal life is not tolerated by them and often they might in return put some spells on the person involved often resulting for example in accidents. This is often thought to be the case when frequent accidents happen during road construction and there are examples of roads being moved for such reasons. At this point I looked to Anita to see if she was about to burst into laughter as she had told me she would do, but she was not. She was serious and gazed into the distance towards the rock on the other side of the fjord. She asked Siggi if the hidden people had modernized in the same was as our human society has done. Siggi answered and told her that temporalities of the two worlds are not comparable despite the co-existence. The human and the hidden people share the earth and the environment, but in different spatial and temporal dimensions.

The journey continued back over the regional boundaries we had crossed earlier. On the way we followed an old dirt track which had served as the first road to the region, and to Hólmavík, when it was built in 1938. The road itself, which is in disuse today, made it feel like we were going back in time. Following the road also allowed us to enter different temporalities, natural and cultural, human and non-human.

Fig. 38.6 The trolls in Kollafjörður (Photo by Katrín Anna Lund)

We visited an old church, heard story about ghosts and experienced a truly magical moment when an eagle slowly hovered beside us before taking a turn up to the mountain. Siggi, our guide, continued the narratives and when we entered the region of Strandir the first thing we did was to visit a stoned trolls, a man and a woman, who had attempted to separate the Westfjord peninsula from the mainland by digging a tunnel. It had taken them too long and as soon as the day light hit them they turned into stone and materialize today as two giant rocks sitting on the coast by Kollafjörður (Fig. 38.6, see Fig. 38.1).

The journey took 10 h and at the end of it we had not only visited elves, hidden people, ghosts, trolls and other inhabitants of the region. We had passed by where fossils of a pre-historic arctic fox may be found to get a glimpse into an earthly history dating back further than any written history accounts for. We ate mussels, cooked by Siggi the sorcerer, on the beach on a fire made from driftwood which comes from Siberia and thus extended the journey to other parts of the globe. We visited the Sorcerer's Cottage which is an extending exhibit piece from the Museum of Sorcery and Witchcraft located in Bjarnarfjörður (see Fig. 38.1) before bathing in a natural warm spring containing water blessed by a thirteenth century Icelandic bishop. This is just to name few of the sites visited. On the way back to Hólmavík, Siggi pointed out to us a rather insignificant stone that sits in a visible distance from the road. "This is the Stone of Selkolla" he said and told us the moral tale of Selkolla, a baby which had disappeared as it was transported by two young people travelling on foot over the moors to be baptized. While on their way the young ones realized that they had feelings for each other and went behind the stone to make love whilst leaving the baby on the other side of it. When they came back the baby had disappeared, but few years later a giant female figure with a head akin to a seal's head started to make appearances, pestering the surrounding farmers. It was believed that this was the baby having been brought up by local trolls. In the end it was bishop who blessed the water, mentioned above, who exorcised it away.

The journey had taken us into assemblage of entangled heterogeneous temporalities whose materiality filters out through the journey's narratives (Edensor 2011) providing the region with motions that extends it beyond all boundaries shaping an atmosphere of in-betweenness. However, it can also be said that the journey itself is one that treads boundaries; boundaries of what is human and non-human, life and death, nature and culture, sacred and profane playing with the tension of how "presence and absence emerge and entwine" (Rose and Wylie 2006: 475). How the landscape appears as magical happens through the play that is improvised in the course of the journey. The Museum of Sorcery and Witchcraft where the journey begins sets the rules for the play but not merely because, as an innovation, its display is focused on the play with magic, sorcery and witchcraft, but rather because of how it performs its magic through a creative play in which the audience takes a direct part. The play shapes the atmosphere of an in-betweenness and the journey persists to tread the boundaries as it moves out of the museum. Furthermore the play continues, when the actual journey has been brought to an end, extending the atmosphere over geographical boundaries as the visitors leave the place, having embedded the magical atmosphere. In a mobile world the travel continues "virtually, communicatively and imaginatively" (Lean 2012: 153) although from a physical distance as I shall further reflect on below.

38.4 An End to the Journey?

In April 2012, when I was visiting Strandir for fieldwork, I asked Siggi if it was possible for me to go into the Museum to take some photographs for the research project which he allowed. Although I did not tell him, I also felt I needed to refresh my memory about the display. As I was entering the Museum area I noticed a handwritten piece of paper that had been framed and was hanging the wall (Fig. 38.7). I could not remember having seen this before and out of curiosity I walked up to it for a closer look. It was a letter to Siggi from a Swedish visitor to the Museum. Something that looked like a necklace was in the frame with it; a leather band with a stone and a Helm of Ave, the sign of lucky charm, carved on to it.

The letter says:

> Hello,
>
> In June I was at your fantastic museum together with a young German journalist (Alva). Perhaps you remember. I had a tooth ache … I bought a talisman to give me more luck. But guess what happened …!? Everything went wrong. I had the necklace for about fourteen days and all was unhappiness. I took it off; and felt free. Then I thought perhaps it was because I had tooth ache when I took it on…? So I tried it on once again a day when the luck was on my side. But, the same thing happened – everything turned into black and sorrow.

Hello,
in June I was at your fantastic museum
together with a young german
journalist (Alva) Perhaps you remember.

Fig. 38.7 The letter and the Talisman (Photo by Katrín Anna Lund)

> So now I send this talisman back (as you said I could do). I don't want to give it to anyone either, I don't want to risk anything.
>
> I was really fond of the necklace and I hope it is possible to get another one – which is fond of me …
>
> Best wishes and regards

How the letter was responded to is not revealed in the display of it, but below it a little note is attached saying:

> 2007 became a happy year for me probably because of the lucky charm you sent me (instead of the talisman I chose myself – if you remember…)
>
> Now I wish you a happy 2008!
>
> All the best from Annita Sander up in the north of Sweden.

The improvisation continues and the magical landscape extends itself over geographical boundaries. The region Strandir, as a tourism destination, continues to be a place in play and plays of play through how it is embodied as a magical place. The Swedish tourist bought a talisman in the Museum which was supposed to bring her good luck. She had been feeling lousy when she bought it and thus not connecting appropriately to the aura of the lucky charm and all goes wrong. Nevertheless, the narratives of the journey had not come to an end and the play continues because the magic is still lurking around somewhere in between in the shape of a talisman and makes its presence through it. Hetherington and Lee (2000) discuss the blank figure, also known in anthropological accounts as the trickster, a figure that under normal circumstances is absent but may suddenly make its presence in any shape and form stirring up the order of everyday life bringing about a state of in-betweenness. In this case the talisman acts as the trickster that brings the landscape of magic into presence and makes it tangible and real although from a physical distance.

38.5 Conclusions

This takes me back to the beginning of this chapter as it refers directly to Strandir and its creation as a tourism destination; as a magical place. As I pointed out above, Strandir, as an isolated and remote region, was in the past associated with occult activities, an image that was enforced during the era of witch hunting in the seventeenth century. Although this image had been fading away during the later part of the twentieth century it was still present in its absence; a blank figure lurking about. During the process of innovating the Museum of Sorcery and Witchcraft in order to create Strandir as a tourism destination, its presence is made explicit in order to mobilize and affect the region's appearance. And the magic continues to mobilize the place, through an ongoing play, as it is improvised in ever changing contexts, even in different corners of the world.

We started the journey by meeting Anita from Italy, telling me that she might burst into laughter due to her lack of believe in the supernatural or other worldly beings. But she did not laugh although the journey introduced our proximate relations with a vastly vivid world of non-human beings in many ways akin to the one Tylor defined as religion in primitive culture; or animism in 1871 (Lambek 2002). Ingold (2006) describes animacy as the

> …dynamic, transformative potential of the entire field of relations within which beings of all kinds, more or less person-like or thing-like, continually and reciprocally bring one another into existence. (2006: 10)

Thus animism is not "a way of believing *about* the world" (Lambek 2002) but rather a "condition of being in it" (2006: 10). It is about how we situate us amongst humans as well as non-humans as we improvise our journeys, no matter how mundane they are, and listen to the narratives that emerge and embrace us, bringing forth present absences in which we enmesh. The world through which we journeyed, described in this chapter, entwined multiplicity of paths that brought us into different spatial and temporal directions momentarily allowing us to delve into it. It is this entanglement that simultaneously moves us and the world. However, often it is the case that motions do not integrate and even go against each other which often results in how we try to manage our surroundings by slowing down, and even put the motion at illusionary standstill. This brings us to the seventeenth century practice of magic in Strandir where poor peasants tried to affect their nearest surroundings, the harsh reality of everyday life, for example, by bringing in extra money or even just to change the weather condition practicing magic. This they did only to meet the power of the authorities who imposed their power in order to organize the world to their own benefits: a world in which there was no space for magic which became defined as playing with supernatural forces which for the authorities indicated a threat. However, standstill is always an illusion. The magical landscape of Strandir remained, although absent in its presence. This should remind us that when mapping a religion on to the world leaves out the practices and activities of how the world is experienced, through how we live it and continuously create it, as we persist to improvise it as we journey, embraced by it.

Acknowledgements The research this chapter is based on is funded by the project Chair in Arctic Tourism. Destination Development in the Arctic (2010–2012), hosted by Finnmark University College, Alta, Norway, and financed by the Norwegian Ministry of Foreign Affairs and the University of Iceland Research Fund. I want to thank my colleagues, Gunnar Þór Jóhannesson and Guðrún Þóra Gunnarsdóttir, my co-workers on the project and I am grateful to Gunnar for reading this chapter at earlier stages and providing comments. I am indebted to all the people in Strandir we have provided us with invaluable informations, especially Sigurður Atlason at the Museum of Sorcery and Witchcraft. His support has really made magic!

References

Alpers, S. (1983). *The art of describing: Dutch art in the seventeenth century*. Harmondsworth: Penguin.

Anderson, B. (2009). Affective atmospheres. *Emotion, Space and Society, 2*, 77–81.

Böhme, G. (1993). Atmosphere as the fundamental concept of a new aesthetic. *Thesis Eleven, 36*, 113–126.

Casey, E. S. (1996). How to get from space to place in a fairly short stretch of time: Phenomenological prolegomena. In S. Feld & K. Basso (Eds.), *Sense of place* (pp. 13–52). Santa Fe: School of American Research Press.

Cresswell, T. (2011). Mobilites I: Catching up. *Progress in Human Geography, 35*(4), 550–558.

Deleuze, G., & Guattari, F. (1987 [1980]) *A thousand plateaus* (B. Massumi, Trans.). London: Continuum.

Edensor, T. (2010). Walking in rhythms: Place, regulation, style and the flow of experience. *Visual Studies, 25*(1), 69–79.

Edensor, T. (2011). Entangled agencies, material networks and repair in a building assemblage: The mutable stone of St Ann's Church, Manchester. *Transactions of the Institute of British Geographers, 36*(2), 238–252.

Hetherington, K. (1999). From blindness to blindness: Museums, heterogeneity and the subject. In J. Law & J. Hassard (Eds.), *Actor network theory and after* (Sociological review monographs, pp. 51–73). Oxford: Blackwell.

Hetherington, K., & Lee, N. (2000). Social order and the blank figure. *Environment and Planning D: Society and Space, 18*, 169–184.

Hin miklu öfl eru raunveruleg. (2003, October 19). *Fréttablaðið*, pp. 22–23.

Ingold, T. (2006). Re-thinking the animate, re-animating thought. *Ethnos, 71*(1), 9–20.

Ingold, T., & Hallam, E. (2007). Creativity and cultural improvisation. An introduction. In E. Hallam & T. Ingold (Eds.), *Creativity and cultural improvisation* (pp. 1–24). Oxford: Berg.

Jackson, M. (1989). *Paths toward a clearing: Radical empiricism and ethnographic inquiry*. Bloomington: Indiana University Press.

Jackson, M. (1998). *Minima ethnographical: Intersubjectivity and the anthropological project*. Chicago: University of Chicago Press.

Jónsson, M. (2008). *Galdrar og siðferði í Strandasýslu á síðari hluta 17. Aldar*. Hólmavík: Strandagaldur.

Lambek, M. (Ed.). (2002). *A reader in the anthropology of religion*. Oxford: Blackwell.

Lean, G. L. (2012). Transformative travel: A mobilities perspective. *Tourist Studies, 12*(2), 151–172.

Lund, K. (2010). Slipping into landscape. In K. Benediktsson & K. A. Lund (Eds.), *Conversations with landscape* (pp. 97–108). Farnham: Ashgate.

Massey, D. (2006). Landscape as a provocation – Reflections on moving mountains. *Journal of Material Culture, 11*(2), 33–48.

McLean, S. (2009). Stories and cosmogonies: Imagining creativity beyond "nature" and "culture". *Cultural Anthropology, 24*(2), 213–245.

McLean, S. (2011). Black goo: Forceful encounters with matter in Europe's muddy margins. *Cultural Anthropology, 26*(4), 589–619.

Rafnsson, M. (2003). *Angurapi: Um galdramál á Íslandi*. Hólmavík: Strandagaldur.

Rose, M., & Wylie, J. (2006). Guest editorial: Animating landscape. *Environment and Planning D: Society and Space, 24*, 475–479.

Sheller, M., & Urry, J. (Eds.). (2004). *Tourism mobilities: Places to play, places in play*. London: Routledge.

Strandagaldur: Museum of Icelandic Sorcery and Wirchcraft. (2012). Available at: www.galdrasyning.is/index.php?lang=en. Accessed 15 Aug 2012.

Strandagaldur: Museum of Icelandic Sorcery and Witchcraft, Stave for Necropants. (2012). Available at: www.galdrasyning.is/index.php?option=com_content&view=article&id=212%3Anabrokarstafur&catid=18&Itemid=60&lang=en. Accessed 15 Aug 2012.

Urry, J. (2000). *Sociology beyond societies: Mobilities for the twenty-first century*. London: Routledge.

Winthereik, B. R., & Verran, H. (2012). Ethnographic stories of generalisations that intervene. *Science Studies, 1*, 37–51.

Chapter 39
Hindu Pilgrimages: The Contemporary Scene

Rana P.B. Singh and Martin J. Haigh

39.1 Introduction

Experiencing the power of place through acts of pilgrimage is a central feature of Hinduism (Jacobsen 2012). For Hindus, pilgrimage is a sacramental process that both symbolizes the participation of the pilgrim in the spiritual realm and actively establishes a two-way relationship between the pilgrim and the divine. Many pilgrimage places draw devotees through their reputation for granting some specific spiritual, social or material blessing, usually expressed in terms of purification and the healing of soul, mind and body (Stoddard 1997). However, Hindu pilgrimage is also a social duty, a rite of passage, and a way of gaining favor, which "equally involves searching for spiritual experience in special places and learning that these material places lie outside the spiritual, mystical, true reality" (Sopher 1987: 15). The liminal *"faithscape"'* that is so created encompasses sacred places, sacred time, sacred meanings, and sacred rituals. The focal points for Hindu pilgrimage travel are called *tirthas*. The word "tirtha" means a 'ford' or river-crossing and, by extension, these are places that allow passage between the mundane and spiritual realms (Bhardwaj and Lochtefeld 2004). Each Hindu pilgrimage is a '*tirthayatra*' (*tirtha* journey) and the geographical manifestation of each '*tirthayatra*' evokes a new kind of landscape that, for the devotee, overlays sacred and symbolic meaning upon a physical and material base. Hindu pilgrims often conceive their sacred journeys as an earthly adventure that combines spiritual seeking and physical tests (Sax 1991). If touring is an outer journey in geographical space, then pilgrimage is the geographical expression of an inner journey. If touring is something largely oriented

R.P.B. Singh (✉)
Department of Geography, Banaras Hindu University, Varanasi, UP 221005, India
e-mail: ranapbs@gmail.com

M.J. Haigh
Department of Social Sciences, Oxford Brookes University, Oxford OX3 0BP, UK
e-mail: mhaigh@brookes.ac.uk

S.D. Brunn (ed.), *The Changing World Religion Map*,
DOI 10.1007/978-94-017-9376-6_39

Fig. 39.1 Allahabad, Kumbha Mela 2001: Pilgrims' camp, where lived over a million pilgrims for a month (Photograph by Rana P.B. Singh, February 2001)

to pleasure seeking (and/or the satisfaction of curiosity), then pilgrimage is something that combines spiritual and worldly aspirations in places where the immanent and the transcendent mesh. Today, most Hindu sacred places are dominated by hybrid spaces that blend the religious and the mundane in complex, often contradictory, forms. Each "*sacredscape*" of sacred spaces, religious ritual performances, and religious functionaries (Vidyarthi et al. 1979) is embedded within the socioeconomic-environmental attributes of the mundane world and so creates the wholeness of a geographical *"faithscape"* (cf. Singh 2013: 69).

In India pilgrimage tourism is big business, part of a gigantic $18 billion, 300 million participant, "*religious tourism and hospitality market*" (Wright 2007). In 2008, it generated around US $100 billion, which is expected to increase to US $275.5 billion by 2018 (Mishra et al. 2011). Overall, India claims more than 562 million domestic and 5 million annual "foreign" tourist visitors, many of them from the diasporas (Kanjilal 2005). Religious tourism provides half share of the total domestic tourists in India. The numbers of people involved are vast; a recent Kumbha Mela festival held in 2001 (Fig. 39.1) brought 68 million visitors to Allahabad (Prayag) and the recent one held in 2013 estimated to around 75 million visitors. Tourism is India's largest service industry worth around 6 % of GDP (almost 9 % of total employment) and it is a major growth engine for the Indian economy (Mishra et al. 2011).

However, pilgrimages knit together the diverse Hindu population at many different integrative levels, socially and culturally (Bhardwaj 1973: 228). Collectively, they develop the complex web of pilgrimage routes and places that defines the sacred geography of India (Eck 2012). Although they are outward expressions of Hindu religious beliefs, driven by each pilgrims deeper quest for union of the human and divine, collectively, they reflect Hinduism's vitality, resilience, and syncretism.

Today, there is a rising tide of pilgrimage tourism in India, which may be related to an increased desire among Hindus to assert their identity. Partly, this is a reaction to the new militancy of Islam, perhaps partly to increasing prosperity, but also, partly, it is consequence of the sectarian politics of "Hindutva," conservative Hindu nationalism and the rivalry between secular parties, such as Congress and "identity" parties, such as, in north India, the (high caste dominated) Bharatiya Janata Party (BJP) and (lower caste dominated) Bahujan Samaj Party (BSP), which promotes a concept of Hindu cultural nationalism based on Hindu scriptures (Narayan 2009) and which led to the destruction of the Babri Mosque, Ayodhya, on December 6, 1992. Of course, expanding religions often place their shrines atop those of the religion they would supplant. Many British churches cover pagan Holy Wells, a mosque stands over Lord Krishna's birthplace in Mathura, and the Babri Mosque covered the birth place of Lord Rama. Such historical contestations are very easily exploited for political gain (Singh 2011b). Here, the result was another round of inter-communal disturbances throughout the country. However, a side effect has been that large numbers of Hindus have become more conscious of their religious heritage. The result has been increased participation in traditional rituals, celebrations, the construction of new temples and, of course, pilgrimages. Meanwhile, Hinduism, which is itself very diverse, remains broadly tolerant of diversity and there are examples of regional level Hindu pilgrimages, such as Sabarimalai in Kerala (South India) in which Christians and Muslims freely participate (Sekar 1992).

39.2 The Pilgrim's Progression

The motivations for pilgrimage are complex. Schmidt (2009) classifies them into several types: devotional, healing, obligatory or socially required, ritual cycle, whether related to the calendar of stages of human life or "wandering," that is, free-form. Bhardwaj suggests that pilgrimages to the highest level shrines are made more for spiritual gains while pilgrimages to lower level shrines tended to seek more material goals (Bhardwaj 1973). Respecting this view, we propose a typology of five classes arrayed along a spectrum. At the one extreme are: (1) *Tourists* – those who are there to see the sights, take a picture, buy a souvenir, eat some food, but who have no major spiritual or emotional engagement with the sacred messages of the site. (2) *Pilgrims of Duty* – people who travel to the sacred not necessarily through belief, but out of respect to their Social Dharma. It is something they must do and be seen to be doing by their community. Their pilgrimage is not especially spiritual, but it is expected of them; it is a display of social conformity. (3) *Pilgrims of Need* – Spiritual Supplicants – people who travel on a pilgrimage in order to gain some result in the material world. They are believers, but their mind is troubled by rough weather in the ocean of material life. In the Indian Himalaya, Uttarakhand's

Fig. 39.2 Golu Devata Temple, Chitai, Kumaun Himalaya – pilgrim 'thank you' gifts of temple bells (Photograph by Martin Haigh)

Chital temple, which is devoted to Shri Golu Dev, the Kumauni God of Justice, is covered in *manautis* – requests for success in legal disputes, examinations, interviews, etc., – all backed up with promises of gifts, usually a new temple bell (Fig. 39.2), if the wish is granted (Agrawal 1992). (4) *Pilgrims of Hope* – Spiritual Tourists are those who seek spiritual uplift from association with the Supreme, they have spiritual goals and seek things that are mainly outside the mundane world, but they are part-timers. They access the liminal mainly to leaven otherwise worldly lives. (5) *Pilgrims of Union* – true Spiritual Seekers for whom all experience is a spiritual journey, who follow *moksha dharma*, a path that seeks escape from the material world and the Hindu cycle of rebirth.

Most, Hindu pilgrimages are performed on auspicious occasions, sacred times that are often defined in terms of astronomical-astrological correspondences, which underpin their associated qualities of sacredness (*pavitrika*) and merit-giving capacity (*punya-phala*). These special occasions very often coincide with the timing of sacred festivals and share the belief that, at such times, the spiritual benefits of a particular *tirtha* are most powerful. This practice, of course, can lead to the development of mass pilgrimages like those of the Kumbha Mela and Panchakroshi Yatra. Of course, the many and varied regional traditions of Hinduism, together with the rival claims of each *tirtha*, contain many such occasions and festivals. However, at a pan-India level there appear about 31 key dates (Table 39.1).

Table 39.1 Hindu festive dates for pilgrimages and the Roman dates (CE), 2012–2018

Festival	Hindu Date/*Tithi*	2012	2013	2014	2015	2016	2017	2018
Makara Samkranti: always on the 14th January; the Sun leaves the house of Capricorn and enters Aquarius								
Pausha Purnima	Pausha, L-15, F	9 Jan	27 Jan	16 Jan	5 Jan	24 Jan	12 Jan	2 Jan
Magha Amavasya	Magha, D-15, N	23 Jan	11 Feb	30 Jan	20 Jan	8 Feb	27 Jan	17 Jan
Vasant Panchami	Magha, L – 5	28 Jan	15 Feb	4 Feb	24 Jan	13 Feb	1 Feb	22 Jan
Magha Purnima	Magha, L-15, F	7 Feb	25 Feb	14 Feb	3 Feb	22 Feb	10 Feb	31 Jan
Maha Shivaratri	Phalguna, D – 14	20 Feb	10 Mar	1 Mar	17 Feb	8 Mar	25 Feb	14 Feb
New Samvata starts, Vasant Navaratri -1	Chaitra, L-1	23 Mar	11 Apr	31 Mar	21 Mar	8 Apr	29 Mar	18 Mar
	--- *Samvata* ---	2069	2070	2071	2072	2073	2074	2075
Rama Naumi	Chaitra, L – 9	1 Apr	20 Apr	8 Apr	28 Mar	15 Apr	5 Apr	25 Mar
Mahavira Jayanti	Chaitra, L-15, F	6 Apr	25 Apr	15 Apr	4 Apr	22 Apr	11 Apr	31 Mar
Akshaya Tritiya	Vaishakha, L-3	24 Apr	13 May	2 May	21 Apr	9 May	29 Apr	18 Apr
Buddha Purnima	Vaishakha, L-15, F	6 May	25 May	14 May	4 May	21 May	10 May	29 Apr
Ganga Dashahara	Jyeshtha, L – 10	31 May	18 Jun	8 Jun	28 May	14 Jun	4 Jun	22 Jun
Ratha Yatra	Ashadha, L – 3	22 Jun	10 Jul	29 Jun	18 Jul	6 Jul	25 Jun	14 Jul
Guru Purnima	Ashadha, L-15, F	3 Jul	22 Jul	12 Jul	31 Jul	19 Jul	9 Jul	27 Jul
Naga Panchami	Shravana, L – 5	24 Jul	11 Aug	1 Aug	19 Aug	7 Aug	27 Jul	2 Aug
Ganesh Caturthi	Bhadrapada, D – 4	5 Aug	24 Aug	13 Aug	1 Sep	21 Aug	11 Aug	30 Aug
Krishna Janmasthmi	Bhadrapada, D – 8	10 Aug	28 Aug	17 Aug	5 Sep	25 Aug	15 Aug	3 Sep
Lolarka Shashthi	Bhadrapada, L – 6	21 Sep	11 Sep	31Aug	19 Sep	7 Sep	27 Aug	15 Sep
Ananta Chaturdashi	Bhadrapada, L – 14	29 Sep	18 Sep	8 Sep	27 Sep	15 Sep	5 Sep	23 Sep
Pitri Visarjana – 14	Ashvina, D – 14	15 Oct	4 Oct	23 Sep	11 Oct	30 Sep	19 Sep	8 Oct
Navaratri, NR 1	Ashvina, L – 1	16 Oct	5 Oct	25 Sep	13 Oct	1 Oct	21 Sep	10 Oct
Lakshmi Puja, NR 8	Ashvina, L – 8	22 Oct	12 Oct	2 Oct	21 Oct	9 Oct	28 Sep	17 Oct

(continued)

Table 39.1 (continued)

Festival	Hindu Date/*Tithi*	2012	2013	2014	2015	2016	2017	2018
Dashahara, NR 10	Ashvina, L – 10	24 Oct	14 Oct	4 Oct	23 Oct	11 Oct	30 Sep	19 Oct
Kojagiri/Purnima	Ashvina, L-15, F	29 Oct	18 Oct	8 Oct	27 Oct	16 Oct	5 Oct	24 Oct
Dipavali/Divali	Karttika, D-15, N	13 Nov	3 Nov	23 Oct	11 Nov	30 Oct	19 Oct	7 Nov
Surya Shashthi	Karttika, L – 6	19 Nov	8 Nov	29 Oct	17 Nov	6 Nov	26 Oct	13 Nov
Prabodhini Ekadashi	Karttika, L – 11	24 Nov	13 Nov	3 Nov	22 Nov	11 Nov	31 Oct	19 Nov
Karttika Purnima	Karttika, L-15, F	28 Nov	17 Nov	6 Nov	25 Nov	14 Nov	4 Nov	23 Nov
Margasirsa Purnima	Margasirsa, L-15,F	28 Dec	17 Dec	6 Dec	25 Dec	13 Dec	3 Dec	22 Dec
Lunar Eclipse	Full Moon (F)	–	–	–	4 Apr	10 Feb	9 Jul	31 Jan, 27 Jul
Solar Eclipse	New Moon (N)	–	–	–	–	–	–	–

Source: Singh (2009a: 390–391)

Lunar Month: *D* Dark Fortnight (Waning), *L* Light Fortnight (Waxing), *F* Full Moon, *N* New Moon

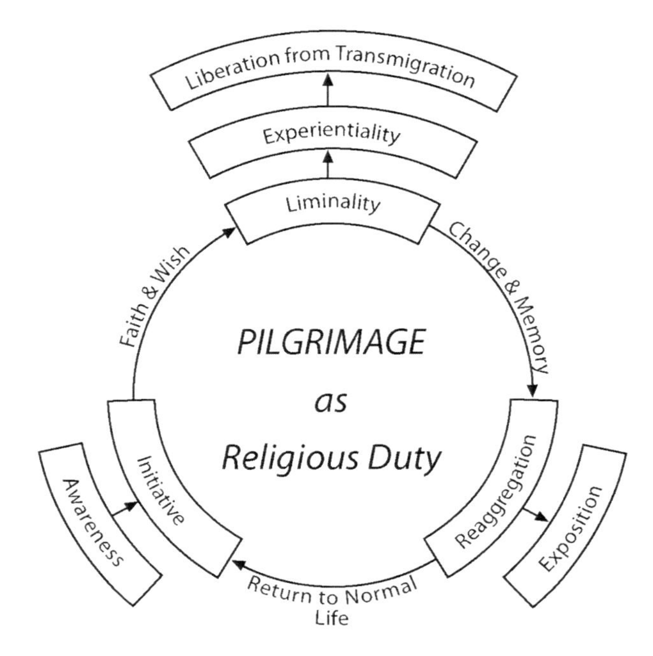

Fig. 39.3 Pilgrimage as religious duty: 'Pilgrimage Mandala' (After Singh, 2011a, p. 17)

Most Hindu Pilgrims in India aspire toward transcendence of the mundane and often express their experiences in purely spiritual terms, perhaps as a dialogue with their Deity or greater Self. Such deep pilgrimages, whether motivated from a sense of duty, hope or devotion, are rites of passage involving a cycle *(mandala)* of three stages: *initiation* (from awareness to start), *liminality* (the journey and its experiences), and *reaggregation* (the homecoming) as well as the upward path of liberation *(moksha)* (Fig. 39.3). This cycle is a sacramental process that involves the pilgrim in the liminal dimensions of each *faithscape.* It seeks to develop a two-way relationship between the pilgrim and the divine, however conceived, and it offers two means of departure, one back to the mundane world and one to the spiritual realms (Singh 2011a). Singh's model of Hindu perspectives on pilgrimage charts (a minimum of) four layers, interconnecting through sacred space and sacred time that link the individual believer to the Ultimate (Fig. 39.4). These two process systems bind sacred space, time, territory and religious functionaries together into a sacrosanct spatial organization (Caplan 1997).

39.3 Making Sense of Pilgrimage Places

The most ancient parts of the Vedas, the *Rig Veda*, (ca. Thirteenth Century BCE), attach four chief connotations to the notion of *tirtha.* It is: (1) a place where one can receive power (*Rig Veda*, 1.169. 6; 1.173.11); (2) a place of purification where

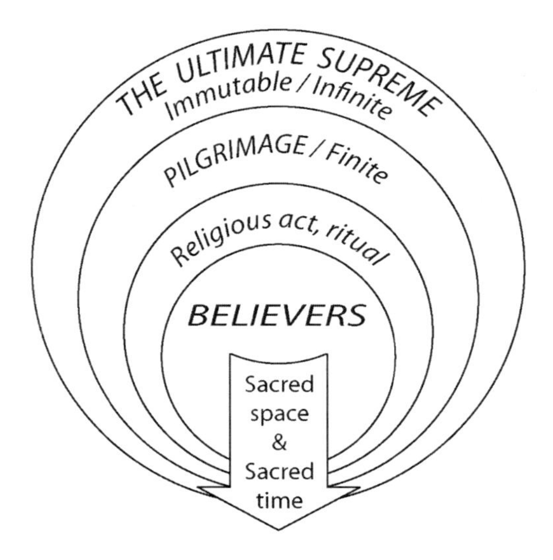

Fig. 39.4 Hindu outlook of Pilgrimage (After Singh, 2011a, p. 18)

people can dip in sacred waters (*Rig Veda* 8.47.11; 1.46.8); (3) a sacred site where God is immanent (*Rg Veda*, 10.31.3); and (4) the location of some divine pass-time or activity (*Satapatha Brahmana*, 18.9). A huge, if not quite as ancient, literature describes the many and varied blessings awarded by taking a holy dip at different pilgrimage *tirtha*. These include the *Puranic Tirtha Mahatmya* texts dating from the first millennium or so CE, which dominate the *Skanda* and *Padma Purana*s. In addition to spiritual gains, Hindu pilgrimages have always been concerned with gaining social status and the relief of worldly cares (Haigh 2011). In these texts, many of the blessings described concern the relief of sins or the fulfillment of wishes for health, wealth, success etc. (Jacobsen 2012). Hence, the *Brahma Purana* (70.16-19) classifies *tirthas* into another four categories: (1) sites related to gaining blessings from specific deities; (2) sites associated with the propitiation of mythological demons who performed malevolent works and sacrifices there; (3) sites associated with the lives of important spiritual leaders; and (4) human-perceived sites, which are not believed to be "chosen," but merely discovered and revered by humans.

A huge literature has developed to catalogue these sacred sites and several geographers have sought to make order of this complexity by classification, seeking norms, generalizations, and rankings (Singh 2011a). Today, the most popular pan-Hindu pilgrimage destinations are Kashi (Varanasi), Allahabad (Prayag), and Gaya (Vidyarthi 1978), which together are eulogized as the three ladders to heaven, while for overseas visitors especially, the "Braj Mandala," an area around the temple towns of Mathura and Vrindavan (Uttar Pradesh), is also important. Among pan-Indian pilgrimage places, there are the seven sacred cities; they include Mathura, Dvarka, Ayodhya, Haridvar, Varanasi, Ujjain and Kanchipuram . Also

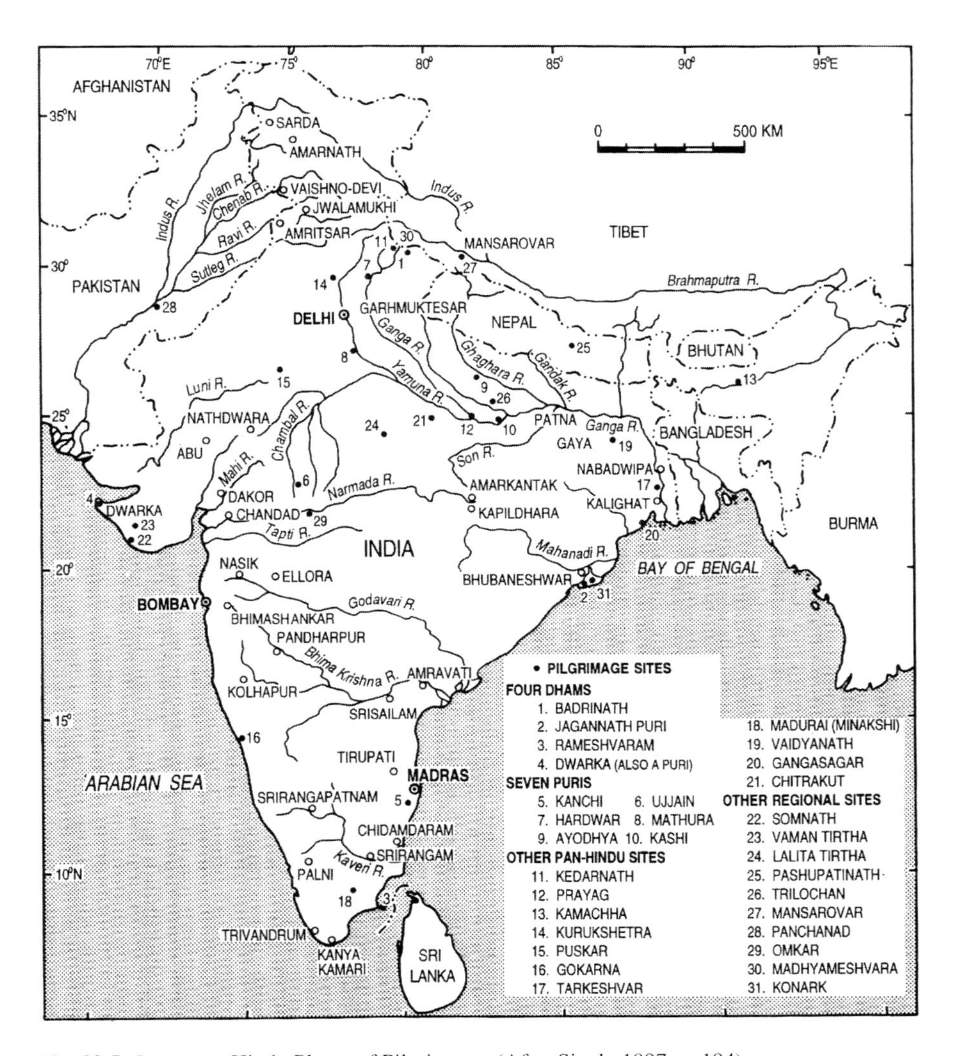

Fig. 39.5 Important Hindu Places of Pilgrimages (After Singh, 1997, p. 194)

scattered across India, are 12 important Jyotir Linga *tirthas* of Lord Shiva, 4 abodes of Lord Vishnu (the Preserver God), 1 major temple devoted to Lord Brahma (The Creator God) at Pushkar, and 51 special sites sacred to the Goddess (Motivation and Power) (Fig. 39.5). According to the *Kalyana Tirthank* list, 35 % of all sacred places are associated to the Lord Shiva, followed by Lord Vishnu (16 %), and the Goddess (12 %). In addition, there are layer upon layer of more local and regional shrines, for example, in Kumaun, the western part of India's Uttarakhand State, pilgrims from many states flock to the Shaivite temple complex at Jageshwar (Agrawal 2010), while in the local area around Kumaun's capital, Almora, the temple named for the Kumauni mountain Goddesses, Nanda Devi and Su Devi, takes local pride of place during festivals. Feldhaus (2003) has explored the intricacies of such interconnected regional identities and "sacredness" in western India. Of course, in recent years, this pattern has become still more complicated with the rise of the new pilgrimage shrines of World Hinduism, such as the BAPS Svaminarayan Akshardham in Delhi, ISKCON's Mayapurdham in rural Bengal, and Aurobindo's Auroville complex in Puduchery, which attract people from all over the world (Brooks 1989). In truth, the number of Hindu *tirthas* in India is so vast and the practice of pilgrimage so ubiquitous that some argue that India itself might be better conceived as sacred space organized into pilgrimage centers and hinterlands, many of which are better defined as cultural and mental space than physical spaces (Bhardwaj 1973).

Elsewhere, in the "new Indias" of more recently established Hindu communities, immature but ontologically similar patterns can be seen developing in places as diverse as Bali, Mauritius, the West Indies, Europe and the Americas (Prorok 2003). For example, the most sacred Hindu place in Mauritius is Ganga Talao (Grand Bassin), a crater-lake temple complex high in the island's mountainous heart, which, since 1897, has been seen as having a spiritual connection to the Ganga River. Today, during the festival of Maha Shivaratri, as many as 40,000 pilgrims leave their homes to walk many miles into the mountains to collect its Ganga water as an offering to Lord Shiva (Eisenlohr 2006; Seewoochum 1995).

Singh explored the cultural geography of Varanasi's *Panchakroshiyatra* of *Ashvina Malamasa* (18 September-16 October 2001). Surveying a sample of 432 from its 52,310 pilgrims, Singh (2009a) found that most travelled as small, typically family, groups (3–6 persons) (*sim*, Gujarat: Sopher 1968, and the UK: Chauhan et al. 2012). In Varanasi, most participants (66 %) were female and most came from the local area. However, there were also cohorts from Bengal and from the international diaspora (see Singh 2009a, b). Over half of the pilgrim-tourists were middle-aged, 40–60 years and 20 % came from the lower classes, including peasantry and menial servants. Their educational status was also low with 57 % of the local pilgrims claiming to have an education between primary school and graduation (Grades 5–10), compared to 70 % of the pilgrim-tourists from further afield. People of Brahmin caste predominated because undertaking such rituals helps reinforce their professional image and religious status. Together, the Brahmin and Merchant castes shared a little over half of the total (see Singh 2009a, b*)*. Similar results are found in a survey of 500 pilgrims to the Shaivite Jageshwar shrines of Kumaun (Agrawal 2010), which was also dominated by Brahmin participation and by married people, both who have more responsibilities. Agrawal (2010) also found that rural respondents and those from lower income groups were more inclined toward God than more wealthy urban people.

Bhardwaj (1973), again following Stoddard (1966), found a weak correlation between a catchment-area-based hierarchy of pilgrimage sites and the caste of the pilgrim visitors. This finding suggests that visitors to pan-Hindu or supra-regional shrines are more often from higher castes than those visiting regional or local shrines (Bhardwaj 1973). Undoubtedly, part of the reason for this high caste prevalence is socioeconomic but part is the greater familiarity of higher castes with the Sanskrit texts. Bhardwaj (1973) also mentions a counter trend caused by the "*Sanskritization*" of the lower castes. Since India's independence in 1947, an aspect of upward mobility among the lower castes has been their adoption of symbols and religious activities formerly associated with higher castes. The Sanskrit law books and *Purana*s commend pilgrimage as meritorious for poor people, members of the low castes, and women. However, even today, very low caste Hindus rarely make pilgrimages (Morinis 1984).

The emotional qualities, experiential learning and religious understandings of pilgrims in Rajasthan have been explored by Gold (1988: xiii) who inquired: "what people did and what they said about what they did, and what they said about what they did, to each other and to me." Although it tackles a very atypical "international" pilgrimage, Bhakti Caitanya Swami (2007) has recorded an evocative 8–9 h video documentary that shares the route, practices, ethos and experience of a month-long *Braj Mandala Parikrama* made by devotees from the International Society for Krishna Consciousness (ISKCON). However, much more research could be done to explore the different levels of emotional experience gained by pilgrims of different types, who so clearly see their surroundings with very different eyes. For example, on a mundane level, Radhakund, "the most Sacred Place in the Universe" (Bhakti Caitanya 2010), is a small ordinary tank filled with water of questionable biological quality, albeit one surrounded by a bustle of small temples and religious shrines. However, for a devotee, it is sublime, a place filled with pure and holy water that is the liquid essence of the Goddess Radha and whose waters preserve the seeds of eternal devotion for Lord Krishna. In transcendental vision, its surroundings of unpainted bricks and stone fade to lush forest and in its place appear four bridges across which heavenly maidens (*Gopis*) walk toward a be-jeweled central pavilion, where Radha relaxes among her women friends. Such visions contrast sharply with more mundane experience, such as that charted by the travelogue of Dev Prasad, an IT engineer from Bangalore, who describes a thoroughly modern pilgrim's progress through the lands of Krishna (Prasad 2010).

Together, the landscape with its sacred symbolic geography creates a "*faithscape*." To cross such a sacred landscape is to seek to be transformed and realize a new spiritual identity. The act of pilgrimage is itself a ritual and a crossing of a threshold to a new way of being (Singh 2011a). After a positive pilgrimage experience, the transformed pilgrim returns to his/her everyday life and shares their experiences with others, who themselves become inspired to be pilgrims themselves. This forms a cycle known as the "*pilgrimage mandala*" (see Fig. 39.3).

39.4 Making Sense of Pilgrimage Routes: The Pilgrimage Mandala

Topographically, holy *tirthas* may be classified into three groups: (i) *Water-sites*, associated mostly with a sacred bath on an auspicious occasion, (ii) *Shrines*, related to a particular deity or sect and mostly visited by pilgrims that belong to, or are attached to, that particular deity or sect, and (iii) *Kshetra* (sacred grounds), areas usually shaped by the form of cosmic *mandala,* the travelling of which brings special merit. Typical cases include Varanasi (Singh 2009a, b), Ayodhya (Bakker 1986), Mathura's Braj Mandala (Entwistle 1987), but arguably not the Kathmandu Valley (Stoddard 2009).

For the pilgrim, Braj Mandala, the holy land of Lord Krishna, is a place where pilgrims hope to experience the interpenetration of the spiritual world and material worlds (Entwhistle 1987). No single mundane place is the goal. Instead, pilgrimage trails manifest as a nested series of circular "*Parikrama*" (meaning 'path surrounding') or "*pradakshina*" (meaning "path to the right") *mandalas*. These Parikrama trails lead through a shifting array of places that have spiritual associations, in this case, with the romantic play, *lilas*, of the divine couple, Radha-Krishna, and their associates. The hope of the Parikrama participant is to transcend the material world to the spiritual place where the sacred transcendental pastimes of the divine couple continue eternally. Their ambition is to achieve an insight that no longer is trapped by the mundane, often squalid, realities of Mathura, Vrindavan and its surrounding villages, but one that shares the higher plane in an idyllic forest paradise, "Goloka Vrindavan," where these activities continue, and their goal is to become a personal participant.

Singh tracks 56 pilgrimage routes through that most sacred Hindu city, Varanasi (Kashi), and describes their associated numerical/cosmic associations, varieties of divinities, and rituals (Singh 2009a, b). Here, the *Panchakroshi Yatra* pilgrimage trail describes a cosmic circuit centered on the temple of Madhyameshvara and with the shrine of Dehli Vinayaka (Ganesha) at its radius (Fig. 39.6). There are 108 *tirthas*, shrines and sacred places, along its 88.5 km (54.9 mi) route including 56 devoted to Shiva (*linga*). These archetypically indicate the division of time into the 12 stations of the zodiac and space into the 9 planets of Hindu mythology, the eight cardinal directions and the centre. Singh (2011a) tracks the origins of this pilgrimage to the mid-sixteenth century and descriptions in Puranic mythology. Significantly, this journey tales place in a month outside of normal time, the intercalary month of leap year, known as *Malamasa*. Similar cosmic interpretations of such holy journeys are reported from West Bengal (Morinis 1984) and many other places.

The Kumbha Mela, the world's largest pilgrimage gathering, owes its origin/location and timing to two traditions. The first establishes its timeframe through astrological calculations while the second emerges from the *Puranas*. This tradition tells of a battle between gods and demons during which four drops of sacred nectar (*amrita*) fell at each of the *mela* sites (Feldhaus 2003). The Kumbha Mela is held four times every 12 years and its location rotates between Allahabad at the

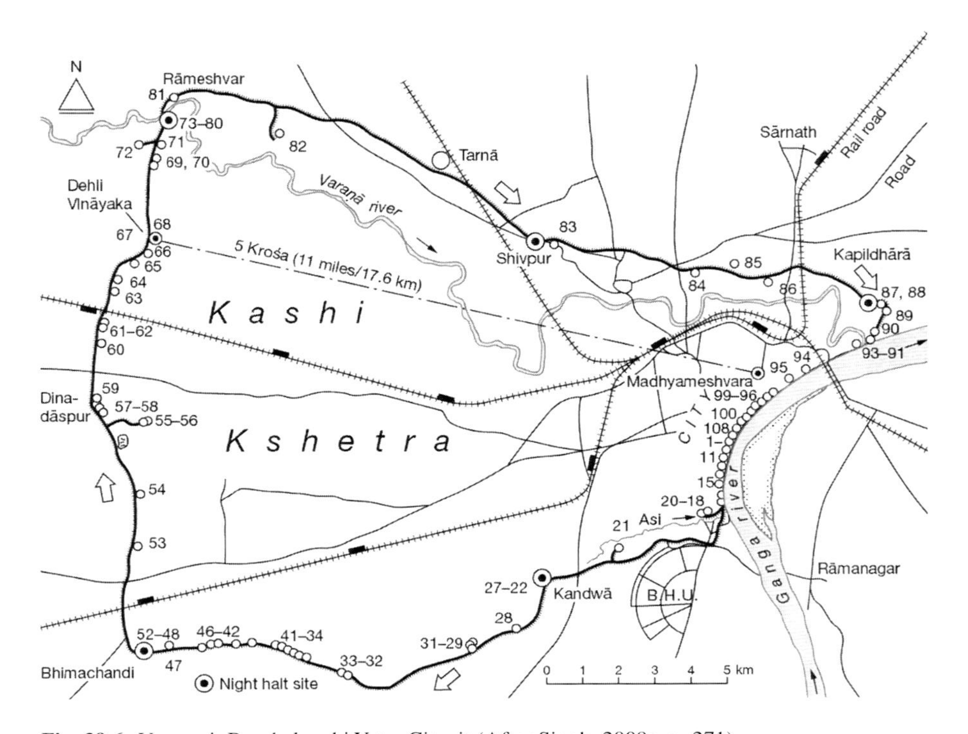

Fig. 39.6 Varanasi: Panchakroshi Yatra Circuit (After Singh, 2009a, p. 271)

confluence of the rivers Ganga, Yamuna and mythical Sarasvati (see Fig. 39.1), Nasik on the Godavari River, Ujjain on the Shipra River, and Haridvar on the Ganga River [Singh 1987: 316–318; Fig. 39.7). Taking a holy dip in these rivers at the Kumbha Mela is considered to bring great merit and to cleanse both body and spirit.

The Kumbha Mela has great antiquity, the Chinese Buddhist pilgrim, Hsuan-tsang, described a Magha Mela at Allahabad in 643 CE, but the modern festival was shaped by the ninth century philosopher Shankaracharya (Dubey 2001). Shankaracharya, leader of the non-dualist Advaitist School of Vedanta, called Hindu ascetics, monks and sages to meet at these places to exchange of philosophical views and build mutual understanding, while the laity followed, in increasing numbers, to benefit from association with these assembled sages.

Gaya is eulogized as the most sacred place for ancestral rituals that aim to help ancestral spirits (who, through *karma* or premature death, are trapped in limbo) become free and reach the realm of the ancestor spirits. This site, which attracts over a million pilgrims each year, is mentioned in the ancient *Rig Veda* (1.22.17), while the rites in the Vishnupad Temple are described in texts dating from the fifth century CE and claim continuity of tradition since the eighth century CE. Presently, 84, from an original 324 sites, related to ancestral rites are identifiable in nine sacred

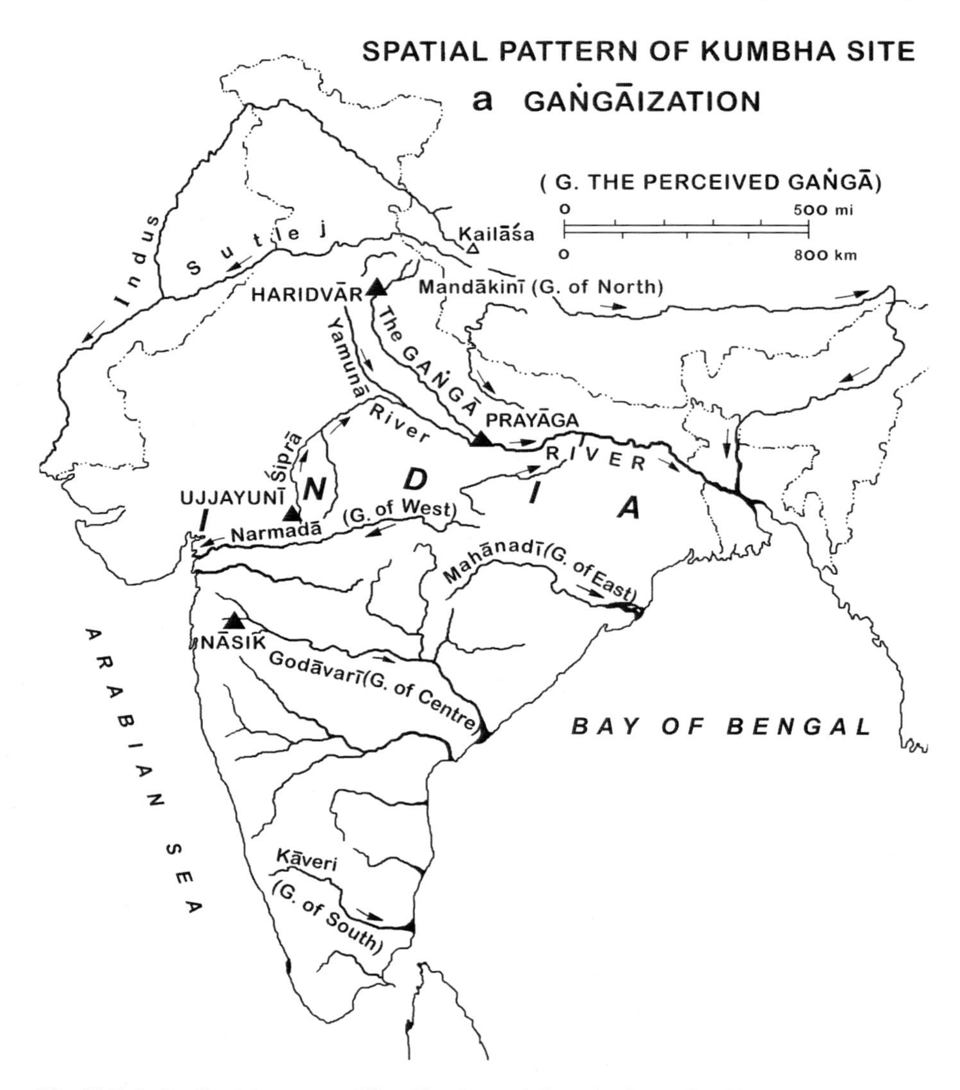

Fig. 39.7 India: Spatial patterns of Kumbha sites and Gangaization (After Singh, 1987, p. 317)

clusters, but modern pilgrims rarely visit more than 45, while more than three quarters perform their ancestral rites at the Phalgu River, the Vishnupad Temple and its associated sacred centers (Vidyarthi 1978). The cosmogony of Gaya's faithscape contains three territorial layers: Gaya *Mandala*, Gaya *Kshetra* (literally "field"), and Gaya *Puri* (township), through which shrines and rituals chart a complex of interweaving themes concerning birth, fertility, life and death (Singh 2009b). Lord

Vishnu's footprints, enshrined in the Vishnupad temple, serve as the *axis mundi* for the Gaya *mandala* whose cardinal and solstitial points are marked by hills, including Pretashila ("the hill of the ghosts").

39.5 Hindu Pilgrimage: Hermeneutics

Victor Turner's pioneering studies focused on the pilgrimage journey, and the emergence of the status-less social union of pilgrims he called "*communitas*" (Turner 1973), while Emile Durkheim's school saw pilgrimage as reaffirming social structure. In Hinduism, the truth is one, the other or both together. The Hindu concept for right action is *"Dharma," dharma* is the foundation of Hindu behavioral ethics. But there two major types of *dharma – social dharma*, the dharma of duties and the rites prescribed by community, and culture – and *Moksha dharma*, which concerns appropriate action for those seeking to detach themselves from the Hindu cycle of rebirth. In Hinduism, the path taken by those seeking liberation, *moksha*, is commonly socially antinomian and, by tradition, involves formal detachment from social responsibilities (Singh and Aktor 2015). In Vaishnavism, the avatar Lord Rama and his associates provide the model of *social dharma*. He was the ideal King, Sita was the ideal wife, and Hanuman was the model servant and devotee etc. By contrast, another *avatar*, the wild ascetic, Dattatreya, set aside all social conventions of dress, behavior, and even social interaction itself, along with all other distractions that stood in the path of the search for spiritual liberation. The coexistence of these two aspects is one reason why those who study pilgrimage in materialistic terms are seriously constrained.

Pilgrimage is a rite of passage that operates in a *liminal*, that is, a transitional, space between the material world and a transcendental reality. Turner and Turner assert that pilgrimage is "exteriorised mysticism while mysticism is an interior pilgrimage" (Turner and Turner 1978: 33–34). In Hinduism, the role of much religious discipline, including pilgrimage, is the removal of the veil of illusion that divides the material world from the spiritual reality that provides its context. Among Hinduism's many descriptors for the material world is the term *maya,* which also includes the root *ma*, the name for the Goddess as Mother, the *mater (*Latin: Mother*)* in materiality, and, hence, embedded within the material. This is why pilgrimage research that attempts to study the visible without knowing of what lies beneath it are doomed to failure (Llewellyn 2001). Pilgrimage is an aspect of the pilgrim's consciousness, which may be constructed at different levels. Bhardwaj (1973) classifies Hindu pilgrimage sites by suggesting that travelers to the highest level shrines tend to seek spiritual merit, while those to lower level shrines have more mundane purpose. In Rajasthan, Gold also found that pilgrimages to local shrines, *jatra* (a Rajasthani version of the Sanskrit *yatra*), were for material purpose while, *yatra,* pilgrimage to distant shrines, sought spiritual "merit" (Gold 1988; Morinis 1984; Llewellyn 2001).

39.5.1 Tourist Resource Management

Hindu pilgrimage places often have an ambivalent approach to the modern pilgrim tourist. Most welcome the revenue pilgrims bring, but many are reluctant to become reduced to a "sight-seen" rather than a religious experience and, ancient sites like the Jagannath Temple, Puri, Orissa, or Padmanabhaswamy Temple, Tiruvananthapuram, Kerala, which energetically exclude perceived outsiders (Prasad 2010). New pilgrimage places tend to be more inclusive, for example, Bhagwan Swaminarayan Sanstha's Akshardham, a vast and magnificent new temple complex in New Delhi which aims to provide a showcase for both Hindu and Indian tradition as well as to raise awareness of the Swaminarayan sect and provide a focus of pride for its devotees (Singh 2011c). Akshardham is part of a new breed of Hindu pilgrimage site, often emerging in places where such developments were long suppressed by foreign rulers and which represent a modern re-visioning of the great ancient temple complexes built through the royal patronage of ancient Hindu kings (Jaffrelot 2010).

However, elsewhere the spiritual environments of some ancient holy places are being degraded into tourist spectacles. For example, two of most popular pilgrimage destinations are Vaishno Devi (in Kashmir, in the north) and Tirumala-Tirupati (in the south). Here, the estimated annual income of the temple trust at each place exceeds Rs 5 billion (ca. US $100 million) per annum. Unfortunately, little of this vast income is reinvested in maintaining the serene and sacrosanct environments of these holy places.

Despite its complexity, variety and potential, relatively few geographers to date have explored the complex relationships between Hindu religion, culture, spirituality, and tourism, especially heritage and religious tourism. Nevertheless, the pilgrimage, pilgrimage sites and their related festivities in combination are driving motivations for domestic and international tourist travel and the source of much interest in both heritage and its conservation. Of course, the new attention to the conservation of sacred sites also bears witness to the resurgence, redefinition and modernization of a previously repressed Hindu culture. Equally, it is affected by the commercial responses to modern sensibilities and by contemporary cultural developments. For example, much of the recent increase of religiously motivated travel to sacred sites is blended with a modern, middle-class, urbane, and New Age spirituality which brings travelers who have non-traditional demands upon the infrastructures of heritage sites (Timothy and Olsen 2006). The process is magnified by the increasing impact of the Westernized Indian diaspora at Hindu sites along with their reverence for heritage and expectations for hotels, transport, etc. Currently, around three-fourths of the expenditures of the modern pilgrimage tourism traveler goes to supporting infrastructure, while local stakeholders receive only marginal benefits. Recent studies of Vrindavan and Tirumala-Tirupati suggest an immediate need, not only to develop comprehensive environmental management policies for such pilgrimage places, which should draw the religious institution 'enterprises' into some kind of regulatory framework for environmental improvement, but also include strategies to build stakeholder participation and meet the needs of this community (Shinde 2009).

39.6 Postscript

As globalization accelerates, the expansion of pilgrimage tourism has encouraged "heritage-making" ("*heritagization*" or "*patrimonialization*" in French) within an international framework. Four of the chief pilgrimage cities of India are now part of a Green Pilgrim Cities Initiative (GPCI), viz., Dwarka, Somnath, Ambaji (in Gujarat), and Amritsar (in Punjab); while Varanasi (Uttar Pradesh) is in the process of nomination (Anon 2011). The GPCI is affiliated with interfaith Alliance of Religions and Conservation (ARC), which has worked for the environmental conservation of sacred sites and pilgrimage routes for more than 20 years. Adoption of the GPCI framework has encouraged these Indian cities to awaken, activate and start educating their communities for the conservation of their sacred places and the need for eco-friendly pilgrimage. However, the national government of India has yet to institute its "heritage act" and, because India was founded as a secular-based system, it finds it hard to legislate on matters in the religious domain. A strategy for managing the development of the economic enterprises associated with pilgrimage-tourism may be another matter.

Meanwhile, in India, while pilgrimage-tourism remains centered on devotion-based informal activities in pilgrimage centers, the two key features within religious travel remain. Religious-tourism (*dharma*) and spiritual-tourism (*moksha*), while intertwined, have different infrastructural needs, require different services, and have different driving forces, organizers, managers and modes. In practical terms, understanding these differences is a necessary prerequisite for the effective development of strategies for sustainable development within the overall framework of India's national development and within this framework institutions and charitable trusts have a vital role to play (Shinde 2011).

References

Agrawal, C. M. (1992). *Golu Devata: The God of justice of Kumaun Himalaya*. Almora: Uttarakhand, Shree Almora Book Depot.

Agrawal, C. M. (2010). *Jageshwar: Abode of Lord Shiva*. Delhi: Indian Publishers and Distributors.

Anon. (2011). *Green pilgrim cities network: Leaving a positive footprint on the earth*. Bath: ARC (Alliance of Religions and Conservation).

Bakker, H. T. (1986). *Ayodhya* (Groningen oriental series 1). Groningen: Egbert Forsten.

Bhakti Caitanya, S. (2007). *Vraj Mandala Parikrama* (Parikrama Series (4 DVD Set)). Vrindaban: ISKCON.

Bhakti Caitanya, S. (2010). *Radhakund: The most sacred place in the universe* (Parikrama Series (DVD)). Vrindaban: ISKCON.

Bhardwaj, S. M. (1973). *Hindu places of pilgrimage in India: A study in cultural geography*. Berkeley: University of California Press.

Bhardwaj, S. M., & Lochtefeld, J. G. (2004). Tirtha. In S. Mittal & G. Thursby (Eds.), *The Hindu world* (pp. 478–501). New York: Routledge.

Brooks, C. R. (1989). *The Hare Krishnas in India*. Princeton: Princeton University Press.

Caplan, A. (1997). The role of pilgrimage priests in perpetuating spatial organization within Hinduism. In R. H. Stoddard & A. Morinis (Eds.), *Sacred places, sacred spaces* (Geoscience and man, Vol. 34, pp. 209–233). Baton Rouge: Louisiana State University, Department of Geography and Anthropology.

Chauhan, S. C., Sita, R. das, Haigh, M., & Rita, N. (2012). Awareness vs. intentionality: Exploring education for sustainable development in a British Hindu community. *Sustainable Development* 18 (Article published online: 7 Sept 2010).

Dubey, D. P. (2001). *Prayaga: The site of the Kumbha Mela*. New Delhi: Aryan International.

Eck, D. L. (2012). *India: A sacred geography*. New York: Harmony (Random House Group).

Eisenlohr, P. (2006). *Little India: Diaspora, time, and ethnolinguistic belonging in Hindu Mauritius*. Berkeley: University of California Press.

Entwistle, A. W. (1987). *Braj: Centre of Krishna pilgrimage* (Groningen oriental series 3). Groningen: Egbert Forsten Publisher.

Feldhaus, A. (2003). *Connected places. Region, pilgrimages, and geographical imagination in India*. New York: Palgrave Macmillan.

Gold, A. G. (1988). *Fruitful journeys: The ways of Rajasthani pilgrims*. Berkeley: University of California Press.

Haigh, M. J. (2011). Interpreting the Sarasvati Tirthayatra of Shri Balarāma. Itihas Darpan. *Research Journal of Akhil Bhartiya Itihas Sankalan Yojana (New Delhi), 16*(2), 179–193.

Jacobsen, K. A. (2012). *Pilgrimage in the Hindu tradition: Salvific space*. London: Routledge.

Jaffrelot, C. (2010). *Religion, caste and politics in India*. Delhi: Primus Books.

Kanjilal, G. (2005). *India tourism (through the inner eyes)*. Mumbai: Orchid. Available at: http://www.gourkanjilal.com/books-authored/index.htm. Accessed June 2011.

Llewellyn, J. E. (2001). *The Kumbha Mela homepage*. Springfield: SW Missouri State University.

Mishra, P. K., Rout, H. B., & Mohapatra, S. S. (2011). Causality between tourism and economic growth: Empirical evidence from India. *European Journal of Social Sciences, 18*(4), 518–527. Available from: http://www.eurojournals.com/EJSS_18_4_02.pdf

Morinis, E. A. (1984). *Pilgrimage in Hindu tradition. A case study of West Bengal*. Delhi: Oxford University Press.

Narayan, B. (2009). *Fascinating Hindutva*. New Delhi: Sage.

Prasad, D. (2010). *Krishna: A journey through the lands and legends of Krishna*. Ahmedabad: JAICO.

Prorok, C. V. (2003). Transplanting pilgrimage traditions in the Americas. *Geographical Review, 93*(3), 283–307.

Sax, W. S. (1991). *Mountain goddesses. Gender and politics in a Himalayan pilgrimage*. Oxford: Oxford University Press.

Schmidt, W. S. (2009). Transformative pilgrimage. *Journal of Spirituality and Mental Health, 11*(1), 66–77.

Seewoochum, C. S. (1995). *Hindu festivals in Mauritius*. Quatre Bornes: Editions Capucines.

Sekar, R. (1992). *The Sabarimalai pilgrimage and Ayyappan cultus*. New Delhi: Motilal Banarasidass.

Shinde, K. A. (2009). *Environmental governance for religious tourism in pilgrim towns: Case studies from India: Vrindavan and Tirumala-Tirupati*. Köln: Lambert Academic Publishing.

Shinde, K. A. (2011). Religious travel industry in India: Prospects and challenges. In World Tourism Organization (Ed.), *Study on religious tourism in Asia and the Pacific* (pp. 295–312). Madrid: UNWTO.

Singh, R. P. B. (1987). Towards myth, cosmos, space and mandala in India. *National Geographical Journal of India, 33*(3), 305–326.
Singh, R. P. B. (2009a). *Banaras, making of India's heritage city*. Newcastle upon Tyne: Cambridge Scholars Publishing.
Singh, R. P. B. (2009b). *Cosmic order and cultural astronomy. Sacred cities of India*. Newcastle upon Tyne: Cambridge Scholars Publishing.
Singh, R. P. B. (2011a). Holy places and pilgrimages in India: Emerging trends & bibliography. In R. P. B. Singh (Ed.), *Holy places and pilgrimages: Essays on India* (pp. 7–56). New Delhi: Shubhi Publications.
Singh, R. P. B. (2011b). Politics and pilgrimage in north India: Varanasi between communitas and contestation. *Tourism, an International Interdisciplinary Journal, 59*(3), 287–304.
Singh, R. P. B. (2011c). Pilgrimage and religious tourism in India: Countering contestation and seduction. In R. P. B. Singh (Ed.), *Holy places and pilgrimages: Essays on India* (pp. 307–334). New Delhi: Shubhi Publications.
Singh, R. P. B. (2013). *Hindu tradition of pilgrimage: Sacred space and system*. New Delhi: Dev Publishers.
Singh, R. P. B., & Aktor, M. (2015). Hinduism and globalization. In S. D. Brunn (Ed.), *The changing world religion map*. Dordrecht: Springer (in press).
Sopher, D. E. (1968). Pilgrim circulation in Gujarat. *Geographical Review, 58*(3), 392–425.
Sopher, D. E. (1987). The message of place in Hindu pilgrimage. In R. L. Singh & R. P. B. Singh (Eds.), *Trends in the geography of pilgrimages. Homage to David E. Sopher* (Research Pub. 35, pp. 1–17). Varanasi: National Geographical Society of India.
Stoddard, R. H. (1966). *Hindu holy sites in India*. Ph.D. dissertation, Department of Geography, University of Iowa, Iowa City (University Microfilms, 1973).
Stoddard, R. H. (1997). Defining and classifying pilgrimages. In R. H. Stoddard & A. Morinis (Eds.), *Sacred places, sacred spaces: The geography of pilgrimages* (Geoscience and man, Vol. 34, pp. 41–60). Baton Rouse: Louisiana State University, Department of Geography and Anthropology.
Stoddard, R. H. (2009). Pilgrimage places and sacred geometries. In J. M. Malville & B. N. Saraswati (Eds.), *Pilgrimage: Sacred landscapes and self-organised complexity* (pp. 163–177). Delhi: DK Printworld for IGNCA.
Timothy, D. J., & Olsen, D. H. (Eds.). (2006). *Tourism, religion and spiritual journeys*. London: Routledge.
Turner, V. (1973). The center out there; Pilgrim's goal. *History of Religions, 12*(3), 191–230.
Turner, V., & Turner, E. (1978). *Image and pilgrimage in Christian culture. Anthropological perspectives*. New York: Columbia University Press.
Vidyarthi, L. P. (1978). *The sacred complex in Hindu Gaya* (2nd ed.). New Delhi: Concept Publishing Co.
Vidyarthi, L. P., Jha, M., & Saraswati, B. N. (1979). *The sacred complex of Kashi. A microcosm of Indian civilisation*. New Delhi: Concept Publishing Co.
Wright, K. J. (2007). *Religious tourism: A new era, a dynamic industry*. Leisure Group Travel. E-zine 2007/2008 Winter Issue, 1. Available from: http://www.grouptravelblog.com/comments.php?id=1362_0_1_0_C. Accessed 26 Feb 2010.

Chapter 40
A World Religion from a Chosen Land: The Competing Identities of the Contemporary Mormon Church

Airen Hall

And inasmuch as ye shall keep my commandments, ye shall prosper, and shall be led to a land of promise; yea, even a land which I have prepared for you; yea, a land which is choice above all other lands. (I Nephi 2:20)

One cannot even be sure if the object of our consideration is a sect, a mystery cult, a new religion, a church, a people, a nation, or an American subculture; indeed, at different times and places it is all of these. (Ahlstrom 1972: 508)

40.1 Introduction

Is the Church of Jesus Christ of Latter-day Saints (also known as the LDS Church or the Mormon Church) a world religion? Or is it only an American religion? Can it be both? How does a religious movement with such strong theological and historical ties to the place of its founding balance those bonds with an ever-increasing worldwide membership? This chapter explores how the LDS Church manages the tensions between its roots and its ambitions.

Palmyra, a small town in central New York State, served as the backdrop for Joseph Smith's early prophetic ministry. The visions that moved him to found what is now the Church of Jesus Christ of Latter-Day Saints took place in the vicinity; Smith officially organized the Church in nearby Fayette. Today, Palmyra is one of the most significant pilgrimage sites for Latter-day Saints (Fig. 40.1). Tens of thousands of visitors come every year to tour Joseph Smith's childhood home, wander through the grove of trees where he claimed to see God and Jesus, and climb to the top of the hill where he found the metal plates he said were the source for the Book of Mormon. That hill is also the site of the Hill Cumorah Pageant, an annual

A. Hall (✉)
Department of Theology, Georgetown University,
3700 O St NW, Washington, DC 20057, USA
e-mail: aehall01@gmail.com

S.D. Brunn (ed.), *The Changing World Religion Map*,
DOI 10.1007/978-94-017-9376-6_40

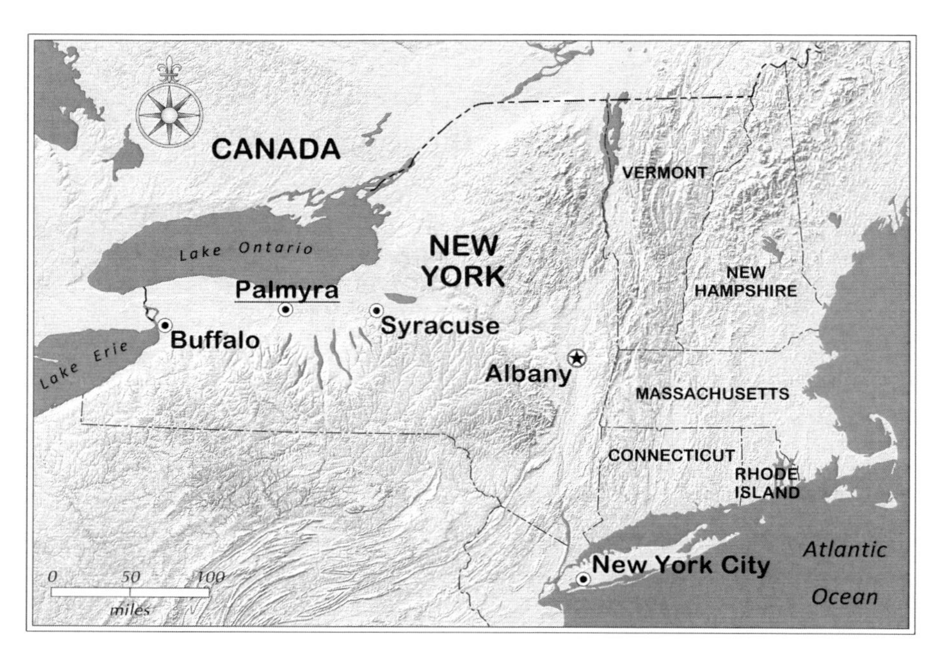

Fig. 40.1 Location of Palmyra, New York (Map by Dick Gilbreath, University of Kentucky Gyula Pauer Center for Cartography and GIS; commissioned by the editor)

spectacle with a cast of hundreds which the Church bills as the largest outdoor theatrical production in the United States. The pageant runs for a single week in July each year, attracting tens of thousands of audience members and considerable media attention (see www.hillcumorah.org/Pageant/).

As the birthplace of the Latter-day Saint movement, the Palmyra area holds a special place in LDS history and theology. Many Latter-day Saints describe the area as sacred, often pointing to Smith's vision of God and Jesus as a reason for this judgment. Though Latter-day Saint temples are usually only constructed in areas with a large or growing LDS population, former Church President Gordon B. Hinckley chose to build a temple in Palmyra, despite a lack of members in the area. He dedicated the completed building in 2000 as a structure with deep symbolism for members beyond that of most temples (Fig. 40.2).

Palmyra is meant to be a welcoming site to visitors from all over the United States and the world. Visitors to Palmyra during the pageant's run are greeted by a host of international flags as they approach the Hill Cumorah (Fig. 40.3). Inside the Visitors' Center, which sits adjacent to the Hill, visitors can see a collection of copies of the Book of Mormon printed in dozens of languages. The pageant's publicity team eagerly points to the diversity of pageant participants, listing the various countries from which cast members hail. This list changes every year with the all-volunteer cast, but previous years have included cast members from Dubai, Italy, and the United Kingdom.

Fig. 40.2 LDS temple in Palmyra, New York (Photo by Airen Hall)

Fig. 40.3 Flag atop Hill Camorah in Palmyra, New York, site of LDS pageant about Mormon history (Photo by Airen Hall)

Yet the reality is that the vast majority of pageant participants, like visitors to Palmyra in general, are Americans. Even many of the visitors or cast members who travel from other nations are actually Americans who simply live abroad. Moreover, the pageant itself is as much a declaration of American patriotism as it is a presentation of Mormon beliefs. On several occasions the narration refers to America in prophetic terms. During the scene when Jesus visits early inhabitants of the Americas after his death in Jerusalem, he tells them he will someday establish a "great nation," a "free nation" on the continent. This statement is not a direct quote from the Book of Mormon account of Jesus's coming to America, but an extrapolation of the pageant's author, Orson Scott Card. Card, an active Latter-day Saint, has become known in recent years as much for his conservative political commentary as his wildly successful science fiction novels and stories. Card has said that he wrote the script in consultation with Church leaders, who encouraged him to "shape a clear and coherent story that would present the book's most important themes for an audience of nonmembers" (Card 1993). Evidently, American patriotism was one of those themes.

Identifying the United States of America as a blessed or chosen nation is nothing new for Latter-day Saints (or for Christian settlers of the Americas for centuries). The American roots of the LDS Church run deep. America, the "land which is choice above all other lands," remains a geographical and ideological anchor for the LDS tradition. However, the Church now claims a strong international presence, with more than half of its 14 million members living outside the United States of America, the country of its birth. Church leaders make much of this worldwide expansion and emphasize the global nature of the Mormon religion. As the Church gains continued mainstream acceptance and attention in the public sphere in the United States, but also gains worldwide membership, it becomes a compelling case study for exploring the tensions facing a globalizing body and issues of how space and place are intertwined with identity formation and maintenance.

Palmyra stands as a geographic emblem of these tensions, a pilgrimage site which both lays claim to a global church while maintaining the American identity of the movement. In what follows, I examine the idea of the LDS Church as a world religion and the ties the Church has to the Americas, then return to how these two strands of LDS identity come together in Palmyra to help build what Alberto Melucci calls "collective identity."

40.2 An American Religion or a Global Church?

In 1984, Rodney Stark proposed a theory that has become a kind of cliché in Mormon studies: that Mormonism was poised to become the first new "world religion" since Islam (Stark 1984). When he revisited this thesis in 1996, he affirmed his projections for future LDS membership and his theory (Stark 2005). Certainly such a future is in line with the hopes the Church of Jesus Christ of Latter-day Saints has for itself, as expansion is a major concern of the group. According to the church's website (www.mormonnewsroom.org/facts-and-statistics/), over 50,000 missionaries for the Church are currently serving full-time missions with the primary goal of making new converts. Most of these missionaries are young men and women between the ages of 19 and 23, visible symbols of the Church's emphasis on wholesome living and family values in their white shirts, ties, and black name cards.

At the time of this writing, the Church professes a worldwide membership of approximately 14 million ("LDS Facts and Statistics," 2013); a little over six million of those live in the United States (for a map illustrating the distribution of Church membership, see Shumway's chapter, Chap. 64, this volume). This means that while the United States certainly claims more Mormons than any other single country, more members of the Church live outside the borders of the tradition's birth nation. Nearly 3.5 million members live in South America, with another million in Asia. These numbers are not uncontested, as I will examine below, but they demonstrate the Church's sense of itself as a global religious body.

One good indicator of the Church's growth around the world is the location of proposed temples. For Latter-day Saints, temples are sacred buildings where members engage in essential rituals for salvation. Temples are quite different from meeting-houses, where local congregations worship on a weekly basis. There are currently 136 temples operating around the world, and generally the locations of these temples are based on membership needs in a particular area (for a map illustrating the distribution of temples, see Shumway's chapter, Chap. 64, this volume). When an area has a large or rapidly growing Latter-day Saint population, Church leaders identify that area as needing a temple. Accordingly, there are 13 temples in Utah (the state nearly two million Mormons call home) and only 3 temples in the whole continent of Africa (where only about 330,000 Mormons live). However, there are 30 temples which leaders have announced will be built; many are already under construction. Of these new temples, two will be located in Africa (in South Africa and the Democratic Republic of Congo) and two in Asia (in the Philippines and Japan). Seven will be in South America. The Church leadership clearly views these areas as sites of growth.

In recent decades, leaders of the Church have frequently referred to the "global church," emphasizing the international expansion and nature of the group. Twice a year, Church leaders speak to all members of the Church at a General Conference which is broadcast around the world. At one General Conference, former Church President Howard W. Hunter emphasized that Mormonism is "a global faith with an all-embracing message" and insisted that "Mormonism, so-called, is a world religion, not simply because its members are now found throughout the world, but chiefly because it has a comprehensive and inclusive message based upon the acceptance of all truth, restored to meet the needs of all mankind" (Hunter 1991).[1] Taking another tack several years later, Dieter Uchtdorf, the Second Counselor of the First Presidency of the Church, has pointed to sheer statistics, saying, "This conference is being broadcast to 68 countries and translated into 55 languages. This is truly a global Church, with members spread across the nations of the earth" (Uchtdorf 2002). Uchtdorf, who is German, goes further than most high-ranking Church leaders (who are primarily American citizens) in emphasizing the global nature of the Church. His position is that the LDS Church is explicitly *not* an American church, but "a

[1] All General Conference talks can be accessed as Ensign articles at the Church's website.

universal church for all people everywhere" (Uchtdorf 2008). Most other Church leaders do not take quite this same approach, though it is common for current leaders to emphasize the newly global character of the modern church.

Former President Gordon B. Hinckley frequently spoke about the worldwide expansion of the Church. In 2003, he compared the Church to the British Empire, listing the many countries where the Church had members and then concluding that "that the sun never sets on this work of the Lord as it is touching the lives of people across the earth" (Hinckley 2003). Four years later, he proclaimed that "a great miracle is taking place right before our eyes," the miracle being that the growth of the church was God's fulfillment of promises made to the prophet Daniel (Hinckley 2007). In the most recent General Conference of the Church, current Church President Thomas S. Monson opened the proceedings with the declaration that, "We come together as one, speaking many languages, living in many lands, but all of one faith and one doctrine and one purpose. From a small beginning 182 years ago, our presence is now felt throughout the world" (Monson 2012).

Beyond the pronouncements of high-ranking leaders, the Church is involved in a public relations campaign designed to emphasize the diversity of the membership, the "I'm a Mormon" campaign. This campaign combines more traditional advertisements with a revamped Church website which showcases profiles created by members. The advertisements have been spreading across the United States over the past 2 years; this year they will begin running in Australia, as well. The website boasts over 50,000 member profiles. Such a campaign demonstrates the Church's investment in projecting a global image not only to current members, but also to outsiders who may stereotype the church as consisting only of white, middle-class Americans.

Since Rodney Stark predicted that the Church of Jesus Christ of Latter-day Saints would develop into a new world religion, nearly 30 years have passed and the Church has indeed grown worldwide and made renewed efforts to brand itself as having an international presence. Yet many scholars have questioned calling Mormonism a "world religion" and have preferred to label it a "global religion." That is, the LDS Church may have a presence in countries around the world, but it has retained a certain distance between itself and the cultures of those countries. Douglas Davies, a British scholar of Mormon studies, proposes that a "world religion" must adapt more fully to the different places it travels, taking on new incarnations in new settings (Stack 2005). Accordingly, the LDS Church does not quite qualify.

Sociologist Rick Phillips makes an even stronger case against viewing the LDS movement as a new world religion. Arguing that the membership claims of the LDS Church are "greatly inflated;" he insists that it is a "predominantly North American church with tendrils in other continents" (Phillips 2006: 23 and 60). To prove his point, he examines discrepancies in reported membership between the Church rolls and national census date in various countries.

For instance, in 2002, the Church claimed over 525,000 members in Chile, but a national census from the same year showed only a little over 103,000 self-reported members. Similarly, in Mexico, the Church claimed over 882,953 members in 2000, but the Mexican census of that year counted only 205,229 (Phillips 2006: 55).

These divergences show a serious gap between Church rolls and self-identification as Mormon, even accounting for limitations of the study. Phillips insists that he does not mean to call into question the veracity of the Church's claims; he says that each person on the church rolls was baptized and confirmed a member, so in one sense their numbers are accurate. Yet he wants to caution that the discrepancies raise questions about how "*social scientists* should define who is or is not a Mormon, and how we should think about the significance of Mormonism's international expansion" (Phillips 2006: 62). Some estimates suggest that by 2020, over half of all Latter-day Saints on church rolls will live in Latin America, but he points out that "conclusions about the future of Mormonism using these projections must take into account the disparity between LDS membership claims and the number of self-identified Latter-day Saints in Latin American nations" (Phillips 2006: 61).[2]

Studies like Phillips' and distinctions between "world religions" and "global religions" like Davies' point to the complicated nature of the LDS effort to both portray itself and actually become a worldwide religious movement. As the Church grows, global expansion is a point of pride and even proof, as missionary success is seen as evidence of God's favor. Yet the Church remains firmly anchored in a North American identity.

Indeed, the American roots of the LDS movement are as strong as ever. In 2001, even while pointing to the fact that the Church is a "global organization" with "members in more than 150 nations," former Church president Gordon B. Hinckley spoke strongly of the deep significance of America (Hinckley 2001).

> Great are the promises concerning this land of America. We are told unequivocally that it "is a choice land, and whatsoever nation shall possess it shall be free from bondage, and from captivity, and from all other nations under heaven, if they will but serve the God of the land, who is Jesus Christ" (Ether 2:12). This is the crux of the entire matter—obedience to the commandments of God.

Hinckley was speaking in the immediate aftermath of 9/11 attacks, thus his focus on America was precipitated by current events. Yet other disasters in other nations have not encouraged leaders to speak of those nations as uniquely blessed or chosen, even when advocating for Church members to provide aid and support (in the case of the earthquake in Haiti, for instance, or the tsunami in Japan). The American continent – and the United States in particular – hold a deep theological meaning for the Latter-day Saints.

The sacrality of America emerges in LDS scriptures, as seen in the quote which opened this chapter and in other passages within both the Book of Mormon and the Doctrine and Covenants. A typical reading of I Nephi, the first book in the *Book of Mormon*, involves interpreting visions which Nephi has as depicting the "discovery" of America by Christopher Columbus, the arrival of the Pilgrims, and the American Revolution (Flammer 1988; Romney 1979). Much is made of the idea that God directly intervened to ensure that the United States of America would be a nation in which a chosen prophet (Joseph Smith) could establish God's true Christian church.

[2] Ryan Cragun is another sociologist who finds that LDS growth becomes more complicated when specific locations are analyzed separately from aggregate numbers. See R. T. Cragun. (2010).

More than that, Joseph Smith taught that God had told him that the United States would be the place where a new Zion will be built and for several decades leaders encouraged converts to the Church to immigrate to America to join with other Latter-day Saints.[3]

Church leaders have long pointed to the particular role the U.S. is meant to play in God's plan for humanity. Former Apostle Mark E. Petersen passionately proclaimed that the "United States then is God's country. It is His government set up specifically to provide freedom of speech and religion in this country so that our Church could be established here and so that we, as the missionaries of the Church, may travel the world over on American passports and have the protection of this great government as we preach the gospel to the world." He went on to add that he has "always felt that the Stars and Stripes, our glorious flag, is God's own banner because it represents His country" (Petersen 1970).

Former Church president Ezra Taft Benson took this religiously-infused patriotism even further, saying "I reverence the Constitution of the United States as a sacred document. To me its words are akin to the revelations of God, for God has placed His stamp of approval upon it" (Benson 1978). Such language is not only a product of particular leader's political leanings, but of Cold War America, in which religiosity became a significant marker of virtue in opposition to secular communism and fascism. Latter-day Saint leaders have always been products of their time. Moreover, Benson's position on this subject has a scriptural basis from an LDS perspective; in the *Doctrine & Covenants*, one of Joseph Smith's revelations quotes God as taking direct responsibility for the United States Constitution: "And for this purpose have I established the Constitution of this land, by the hands of wise men whom I raised up unto this very purpose, and redeemed the land by the shedding of blood" (D&C 101:80).

In recent years, Church leaders are less likely to specifically emphasize the chosen or blessed nature of America over other nations, but there are more subtle ways in which they grant America a privileged position. At General Conferences directed at the entire church, leaders speak about American patriotism or American Mormonism in a manner which assumes that the primary audience is America. Senior Apostle Boyd K. Packer re-told the story of the early Latter-day Saints as a story of patriotic triumph, emphasizing the great "loyalty" the first Mormons had to the United States government (Packer 2008). Stories about Mormon loyalty to other governments are rare. In the April 2012 conference, Elder Dallin H. Oaks of the Quorum of the Twelve congratulated members for their "Christian love, service, and sacrifice" and for their "caring for the poor through personal efforts and through supporting Church welfare and humanitarian contributions," citing a "nationwide study" which concluded that active members of The Church of Jesus Christ of Latter-day Saints "volunteer and donate significantly more than the average American and are even more generous in time and money than the upper [20 %] of religious people in America" (Oaks 2012).

[3]Article of Faith 10. Now, Church leaders counsel converts to stay in their home countries and build up the Church locally.

Outside the Church, various scholars have pointed to the idea of a Mormon culture which is strongly tied to the American West (Meinig 1965; Yorgason 2010). The idea that Mormonism is a culture inextricably bound up with the region of the inter-mountain west lends credence to the notion that the LDS Church is an American institution at its core, no matter how many members there are in other nations. Alan Wolfe sums up these concerns with a single question: "Can the culture that originally sustained the religion survive a diaspora that includes Saints living outside the original Great Basin as well as accept the multicultural character of the modern world?" (Wolfe 2005: 149). Some LDS leaders and scholars have tried to downplay this concern. Eric Eliason, for example, proposes that the fact of a "Mormon homeland" is part of what makes the Latter-day Saint tradition a "world faith," comparing it to the Punjab for Sikhism (Eliason 2001: 12). Still, from a purely social scientific standpoint, the reality of an LDS culture that is dependent on a particular history in a particular setting is not easily pushed aside (Wood 2008; Uchtdorf 2008). Lee Trepanier and Lynita K. Newswander suggest that, in fact, "Mormonism is the most American of all of the American religions" (Trepanier and Newswander 2012: 111).

Beyond the claims of Church leaders or the arguments of scholars, though, is the simple fact that the Church has its historical origins in the United States of America. Joseph's Smith prophetic mission took place in an American context; his key revelations and most of the major events of early church history took place in the United States. This becomes particularly significant when one understands the role of history in the Latter-day Saint tradition; it is generally acknowledged that Church history is understood in theological terms, that history is theology for Latter-day Saints (Davies 2000; Givens 2004; Mitchell 2004). Thus, a history so deeply embedded in America results in a theology that also takes on an American flavor.

40.3 Pilgrimage to Palmyra, the "Birthplace of the Restoration"

Palmyra, the "birthplace of the Restoration," is a site of Church history which is resonant with theological meaning for Latter-day Saints as a place where God spoke with a living prophet and began the process of establishing the true Christian gospel on the earth. In that sense, no site could more strongly emphasize the special nature of America as "a land which is choice above all other lands." Palmyra is a tangible evidence of God's blessing on America, a physical location that provides the groundwork for the whole of the LDS movement. As one local church leader said at a talk given during a church meeting, "Welcome to God's first choice" ("Journal," 2010). The patriotic themes of the Hill Cumorah pageant are not only reasonable under this view, but perhaps even essential. If God chose America to be the nation where the true Church would be founded and flourish, then America is indeed a promised land.

Moreover, Palmyra is a site for the celebration of the Book of Mormon as a scriptural record. Smith claimed to find the metal plates buried in the side of the Hill Cumorah and the first edition of the Book of Mormon was published in Palmyra.

Fig. 40.4 LDS Pageant in Palmyra, NY; Jesus Visits America (Photo by Airen Hall)

Fig. 40.5 Pageant cast members distribute the Book of Mormon to theater goers (Photo by Airen Hall)

The Hill Cumorah pageant re-enacts a series of Book of Mormon stories and the founding of the LDS Church; the key scene in both the Book of Mormon and the pageant is the visitation of Jesus Christ to the Americas immediately following his death (Fig. 40.4). Thus, Palmyra not only affirms God's choosing of the Americas by highlighting Joseph Smith's visions, but also by emphasizing a previous divine appearance. Both visitations are essential and unique points of LDS theology.

Yet Palmyra is not a site that invites visitors to look only to the past. The Church pours money into the area, funding historical renovations, the grand production of the pageant, and more primarily in order to create a space for evangelization and public relations. The audience for the Hill Cumorah pageant may still primarily be LDS, but the Church hopes to use the production to attract non-Mormon visitors. Cast members go out amidst waiting audience members before the shows carrying a stack of comment cards (Fig. 40.5). They try to speak with every single visitor, asking all non-member attendees to write in their contact information in order to receive free gifts from the Church (such as a copy of the Book of Mormon or a DVD of a Church-made film) and/or invite the missionaries into their homes. Such cards are also available at all the Church-owned historical sites, like the building where the Book of Mormon was published or Joseph Smith's childhood home. Full-time missionaries staff all those sites and serve as tour guides and have done so since the 1970s (Pykles 2010: 156). The Church specifically calls a public relations director for the Hill Cumorah pageant and seeks out media attention for the production.

Palmyra is a place, then, where the Church attempts to draw in new members and offer an image of the Church's identity, thus defining the future of the movement. The international flags fluttering around the Hill Cumorah and the carefully arranged displays of the multi-lingual Book of Mormons are signs of what the Church hopes to become as much as what it already is. Palmyra is a site where the Church can tell its own story on its own terms. It is a place meant to act as a bridge between past and future, a moment in the present that brings memory and vision together.

40.4 Collective Identity and the Mormon Experience

To consider the significance of a place like Palmyra for contemporary Mormonism, I turn to Italian sociologist Alberto Melucci. Melucci who has proposed a concept of "collective identity" which he uses in discussing social movements in a globalizing world (Melucci 1996a, b). Considering how a pilgrimage site might relate to the building of religious community is nothing new. Pilgrimage sites have long been viewed as places where religious communities come together. In their classic works on the subject, Victor and Edith Turner offered a vision of pilgrimage in which participants come together in *communitas*, a "nonrational fellowship," in which "unity and homogeneity" supersede "the disunity and heterogeneity … in the mundane sphere" (Turner and Turner 1978: 39). In this vision, stratification is superseded by shared religious experience; individuals come together in a mystical union.

Such a view has several pitfalls, however. While many pilgrimage experiences may incorporate an element of this type of unity, many others do not. Pilgrimages can be divisive, creating or sustaining social differences. Even amongst a group of people doing similar activities, individuals have personal motivations, experiences, and understandings of pilgrimage which make an overriding of heterogeneity difficult to fully achieve. The Turners may have described an ideal for certain pilgrimages, but their model cannot be a totalizing framework (Badone and Roseman 2004; Reader 2006). Melucci's notion of collective identity serves as a constructive contrast to *communitas*, because it denotes a meaningful coming together while never requiring a dissolution of self or implying an idealistic (and improbable) unity. Collective identity emphasizes the reciprocal dynamic between autonomous action and collective action (Melucci 1996a).

For Melucci, no identity can be formed without a strong sense of history and time. Melucci argues that "the present becomes a crucial dimension, not as a point-like, instantaneous dimension but rather as the possibility of forging in the here and now the connection between the past and the future, between memories and projects"

(Melucci 1996a: 85–86). Latter-day Saint pilgrimage sites are primarily sites of Church history: Joseph Smith's birthplace in Vermont, the places in central New York where Smith had his early visions and first organized the Church, and sites where early followers gathered in Ohio, Missouri, and Illinois. While the future of the LDS Church may be as a global tradition, the history of the Church is embedded in the geography of the United States. Thus, these sites of pilgrimage are able to serve as physical anchors for the Church, functioning as those places where an understanding of the present is forged in the remembering of the past and the gaze toward the future. At such places, the American identity of the Church is solidified and celebrated, even as the Church makes a concerted effort to build a second identity as a worldwide religious body.

40.5 Conclusion

The Church of Jesus Christ of Latter-day Saints is positioned at a crossroads. On one hand, the Church is experiencing real growth around the world and is actively pursuing further growth. Current Church leaders clearly want to claim an identity as a global church. The Church is making concerted efforts to emphasize the diversity of its membership and to grow its missionary program worldwide by making use of new technologies and media. Still, there is no denying the significance of the tradition's American roots. The pilgrimage sites of the Church, of which Palmyra is only one, are places which where the Church proclaims a particular commitment to an *American* identity, even while celebrating worldwide expansion. In spatially anchoring a globalizing religious tradition in the physical geography of America, a site like Palmyra both exemplifies and resolves the tension between these two identities – both the American roots and the global vision are evident, and bringing them together offers an opportunity for creating a collective identity that transcends either. As the Church continues to grow, such sites will only increase in significance.

As Sydney Ahlstrom noted 40 years ago in the second quote which opens this chapter, the Latter-day Saint tradition is difficult to pin down. Scholars have been debating whether or not Mormonism qualifies as a world religion for decades and may continue to debate the question for decades to come. This chapter has referenced a sampling of the relevant considerations to this complicated issue: cultural assimilation, sheer membership numbers, identifications by Church leaders, and the evidence offered by LDS pilgrimage sites.

The evidence seems to be sufficient to conclude that Mormonism is an American religion, a tradition infused from its earliest beginnings with American values and rooted even now in American geography. Some may view this conclusion as a definitive statement against considering Mormonism a world religion, but that is not a necessary deduction. Religious traditions always come out of geographically and culturally specific locations. Hinduism, for instance, is clearly a product of India, deeply and inextricably tied to Indian culture and geography. Similarly, Judaism emerges from a particular experience of a particular group of people in the Middle

East. Any good scholar of religion or history recognizes that the origins of all traditions are bound up in geography, politics, economics, and a host of other factors.

Deciding the issue on pure numbers is tempting. Yet merely noting that a certain percentage of Latter-day Saints live inside or outside the United States is not especially informative, as highlighted by Rick Phillip's research. Membership numbers are subjective and the strength of LDS communities is not consistent from state to state, let alone nation to nation. Moreover, who can draw the line on the magic number at which a tradition qualifies as a "world religion"? Is it ten million? Twenty? Thirty?

Perhaps the question comes down to issues of history and longevity. At barely over 180 years old, the LDS tradition simply is not old enough to have gained the credibility that the weight of hundreds of years provides. Decade after decade provides the layers of texture the designation of "world religion" seems to require. Again, though, there is no perfect determiner of exactly when a tradition is "old enough." At what point does a tradition gain enough historical traction?

In the end, the debate over whether the Church of Jesus Christ of Latter-day Saints is "only" an American religion or if it has achieved the status of "world religion" is more significant than the answer to that question. It is the tension between roots and goals, between past and present, which both drives collective identity and makes Mormonism a fascinating subject of study and invites further research on the topic. Questions to consider include: how will growing LDS populations in other nations affect international policy and politics? How might perceptions of foreigners that the LDS Church is primarily American affect the Church's proselytization goals? As more non-American members reach higher leadership positions, how will Church doctrines and practices shift to accommodate cultural variety? Will the experiences of members outside the United States result in new pilgrimage sites which draw American Mormons outside their own nation to better understand or more actively live out their faith? Such explorations have the potential to be provocatively fruitful.

References

Ahlstrom, S. E. (1972). *A religious history of the American people*. New Haven: Yale University Press.

Badone, E., & Roseman, S. R. (2004). Approaches to the anthropology of pilgrimage and tourism. In E. Badone & S. R. Roseman (Eds.), *Intersecting journeys: The anthropology of pilgrimage and tourism* (pp. 1–23). Chicago: University of Illinois Press.

Benson, E. T. (1978, November). Our divine constitution. *Ensign*. Retrieved May 25, 2012, from www.lds.org/ensign/1987/11/our-divine-constitution

Card, O. (1993). *A storyteller in Zion: Essays and speeches*. Salt Lake City: Bookcraft.

Cragun, R. T. (2010). The secular transition: The worldwide growth of Mormons, Jehovah's Witnesses, and Seventh-Day Adventists. *Sociology of Religion, 71*(3), 349–373.

Davies, D. J. (2000). *The Mormon culture of salvation*. Aldershot: Ashgate.

Eliason, E. (2001). Introduction. In E. Eliason (Ed.), *Mormons & Mormonism: An introduction to an American world religion* (pp. 1–21). Chicago: University of Chicago Press.

Flammer, P. M. (1988). A land of promise choice above all other lands. In M. S. Nyman & C. D. Tate Jr. (Eds.), *First Nephi, the doctrinal foundation* (pp. 217–229). Provo: Religious Studies Center, Brigham Young University.

Givens, T. (2004). *The Latter-day Saint experience in America*. Westport: Greenwood Press.
Hinckley, G. B. (2001, November). The times in which we live. *Ensign*. Retrieved May 25, 2012, from www.lds.org/ensign/2001/11/the-times-in-which-we-live
Hinckley, G. B. (2003). The State of the Church. *Ensign*. Retrieved May 25, 2012, from www.lds.org/ensign/2003/11/the-state-of-the-church
Hinckley, G. B. (2007, November). The stone cut out of the mountain. *Ensign*. Retrieved May 25, 2012, from www.lds.org/ensign/2007/11/the-stone-cut-out-of-the-mountain
Hunter, H. W. (1991, November). The gospel – A global faith. *Ensign*. Retrieved May 25, 2012, from www.lds.org/ensign/1991/11/the-gospel-a-global-faith
"Journal". (2010). Author's fieldwork journal, Sunday, July 11, 2010.
"LDS Facts and Statistics". (2013). www.mormonnewsroom.org/facts-and-statistics/. Accessed 20 Mar 2013.
Meinig, D. W. (1965). The Mormon culture region: Strategies and patterns in the geography of the American West, 1847–1964. *Annals of the Association of American Geographers, 55*(2), 191–220.
Melucci, A. (1996a). *Challenging codes: Collective action in the information age*. Cambridge: Cambridge University Press.
Melucci, A. (1996b). *The playing self: Person and meaning in the planetary society*. Cambridge: Cambridge University Press.
Mitchell, H. (2004). Being there: British Mormons and the history trail. In S. Coleman & J. Eade (Eds.), *Reframing pilgrimage: Cultures in motion* (pp. 27–46). London: Routledge.
Monson, T. S. (2012, May). As we gather once again. *Ensign*. Retrieved May 25, 2012, from www.lds.org/ensign/2012/05/as-we-gather-once-again
Oaks, D. H. (2012, May). Sacrifice. *Ensign*. Retrieved May 25, 2012, from www.lds.org/ensign/2012/05/sacrifice
Packer, B. K. (2008, November). The test. *Ensign*. Retrieved May 25, 2012, from www.lds.org/ensign/2008/11/the-test
Petersen, M. E. (1970). The Church and America. April. Retrieved May 25, 2012, from www.lds.org/pa/display/0,17884,4892-1,00.html
Phillips, R. (2006). Rethinking the international expansion of Mormonism. *Nova Religio: The Journal of Alternative and Emergent Religions, 10*(1), 52–68.
Pykles, B. C. (2010). *Excavating Nauvoo: The Mormons and the rise of historical archaeology in America*. Lincoln: University of Nebraska Press.
Reader, I. (2006). *Making pilgrimages: Meaning and practice in Shikoku*. Honolulu: University of Hawaii Press.
Romney, M. G. (1979, September). America's promise. *Ensign*. Retrieved May 25, 2012, from www.lds.org/ensign/1979/09/americas-promise
Stack, P. F. (2005, July). Mormonism is global, but is it a 'world religion'? *Salt Lake City Tribune*. Retrieved May 25, 2012, from www.sltrib.com/faith/ci_2900770
Stark, R. (1984). The rise of a new world faith. *Review of Religious Research, 26*, 18–27.
Stark, R. (2005). The rise of a new world faith. In R. L. Nielson (Ed.), *The rise of Mormonism* (pp. 139–146). New York: Columbia University Press.
Trepanier, L., & Newswander, L. K. (2012). *LDS in the USA: Mormonism and the making of American culture*. Waco: Baylor University Press.
Turner, V., & Turner, E. (1978). *Image and pilgrimage in Christian culture*. New York: Columbia University Press.
Uchtdorf, D. (2002, November). The global church blessed with the voices of prophets. *Ensign*. Retrieved May 25, 2012, from www.lds.org/ensign/2002/11/the-global-church-blessed-by-the-voice-of-the-prophets
Uchtdorf, D. (2008). The church in a cross-cultural world. In R. L. Nielson (Ed.), *Global Mormonism* (pp. 294–306). Provo: Religious Studies Center, Brigham Young University.
Wolfe, A. (2005). *The transformation of American religion: How we actually live our lives of faith*. Chicago: University of Chicago Press.

Wood, R. S. (2008). A babe upon its mother's lap: Church development in a developing world. In R. L. Nielson (Ed.), *Global Mormonism* (pp. 65–83). Provo: Religious Studies Center, Brigham Young University.
Yorgason, E. R. (2010). *Transformation of the Mormon culture region*. Urban: University of Illinois Press.

Further Reading

Bushman, C. L. (2006). *Contemporary Mormonism: Latter-day saints in modern America*. Westport: Praeger.
Davies, D. J. (Ed.). (1996). *Mormon identities in transition*. London: Continuum.
Jordan, M. (2007, November 19). The new face of global Mormonism. *The Washington Post*. p. A01.
Mauss, A. (1994). *The angel and the beehive: The Mormon struggle with assimilation*. Chicago: University of Illinois Press.
Neilson, R. L. (Ed.). (2008). *Global Mormonism*. Provo: Religious Studies Center, Brigham Young University.
Nelson, G. (Ed.). (2010). *On Sunday: Mormon portraits of a global church*. New York: Mormon Artists Group.
The Book of Mormon. (1981). Salt Lake City: The Church of Jesus Christ of Latter-day Saints.
The Doctrine and Covenants. (1981). Salt Lake City: The Church of Jesus Christ of Latter Day Saints.

Chapter 41
Religious Nationalism and Christian Zionist Pilgrimages to Holy Landscapes

Tristan Sturm

41.1 Introduction

This paper contributes to recent literature on theories produced largely in sociology and anthropology on the topic of discourses and performatives on the relationship between religion and nationalism, not all of which can be called "religious nationalism." Most nationalisms are inflected with religious discourse to give (Secor 2001; Woltering 2002; Mullin 2010):

- a moral core;
- further cement common belief systems among a community of common values;
- act as a legitimizing power for state leaders especially for rallying support in religiously conservative regions;
- act as a powerful political lobbying body;
- use the power of religion to elicit a common heritage and, therefore, create common socio-cultural ground; and
- as we see in the case of Islamic movements in much of the Middle East over the last two decades and especially following the "Arab Spring," religion can provide the counter geopolitical discourse to perceived Western-based models of secular nationalism.

Explicit and well known examples of these functional, instrumental, and facilitative uses of religion are currently employed in Turkey, Ukraine, Russia, and the United States among many others. Such a tally is clearly not exhaustive of the ways

T. Sturm (✉)
School of Geography, Archaeology and Palaeoecology, Queen's University Belfast, Elmwood Ave, Belfast BT9 6AY, UK
e-mail: tristan.sturm@gmail.com

S.D. Brunn (ed.), *The Changing World Religion Map*,
DOI 10.1007/978-94-017-9376-6_41

religion and nationalism are fused, but it is this taxonomy that illustrates the American Christian Zionist case study for this chapter.

Christian Zionists, in brief, are made up of socially conservative evangelicals and Pentecostals who believe the wars in the Middle East are portending an imminent End Times scenario which will be centered in the modern state of Israel where a clash between Satan's Russian and Arab led armies meet those Westerners who are non-Christian but, nevertheless, serve Christ's army. In a nutshell, this scenario largely defines the eschatological beliefs of the over 20 million American Christian Zionists, without which, their national allegiance to the territory of Israel and the perceived religiously homogenous nation of Jews (for which they cannot belong but does not presuppose the performance of this religious nationalism), hence Christian *Zionism*, would not be possible. Jews are seen as the Chosen People of Earth and are, therefore, to be unwaveringly supported as God's army soldiering toward the apocalypse, while Christians are understood as the Chosen People of Heaven and a post-millennial Earth. This outsider diaspora nationalism is, I argue, a form of religious nationalism where religious beliefs provide the possibility and core discourse for their nationalism for the territory of Israel and Jews with Israeli citizenship and in the diaspora. This, of course, differs from nationalism employing religious discourse for inflection as it is often understood as merely a functional device.

This story of Christian Zionist nationalism is interesting and begs explanation precisely because it disrupts expectations of ethno-national-religious correspondence and because it shows an unexpectedly convergent expression of religious territoriality, thus contributing to theoretical observations that identities can be multiple and competing, a "multiplicity" of subject and in this case, dual nationalistic, positions. This paper is divided into several sections below: the first section, "religious nationalism," explores the various explorations by nationalism scholars of religion while navigating the theoretical position following the work of Talal Asad (1991); second, "Territory and apocalypse," provides a brief history of the idea of the apocalypse in America as it relates to the New World and also sketches-out a brief history of post-Civil War American national interests in Palestine, illustrating a historical vacillation between millennial thought and a search for healing origin myths in relation to America; and in the third section entitled, "Post 1967 making of Judeo-Evangelical nationalism," I outline in depth how Christian Zionists gained political power in the U.S., fomented an outsider diasporic religious nationalism for Israel, and how much of this identity today is contingent on the performance of Islam as an evil Other. Through this history I provide discursive evidence from Christian Zionist leaders that indeed their national allegiance is shifting from the U.S. to Israel. These contrasted historical moments distinguish between the "religious nationalism" of modern Christian Zionists, a nationalism founded on circulating and performed religious doctrine, and "nationalisms inflected with religion" or given further sentimental communal feelings through religious idioms.

41.2 Religious Nationalism

There is almost a religious devotion among all nationalisms to a territory or homeland and more often than not there is religious discourse embedded within performances of nationalism. As John Agnew (2006: 185) notes, "much nationalism and imperialism have found purpose and justification in religious difference and in proselytizing." Countless sentiments of devotion comprise nationalism, but territory and a geopolitical imagination of it are key, especially in contrast to common enemies and neighboring territories. Because of the nationalist identification of "internal" and "external," "our nation" and "their nation," Agnew argues (2008), these binaries make nationalism the most territorial of all major ideologies between socialism, liberalism, and nationalism. Herb and Kaplan argue that "territory becomes a vital constituent of the definition and identification of the group living within it" (Herb and Kaplan 1998: 2; cf. Sturm and Bauch 2010).

The "nation" is often defined as a group of people who feel they share a common set of myths concerning a territory, sharing common experiences of danger, destiny, historical struggles, and cultural affinity in relation to common places. However, I argue consistent with theories that the "nation" is a "social text," that the nation does not exist outside the performance of such binding myths. Instead these common performances are what can be termed as the verb, "nationalism," the doing of the "nation" (Krishna 1996; Calhoun 2001; Agnew 2008).

If we take this limited taxonomy provided in the introduction of what binds nationalism, then there does emerge a type of nationalism (that is, nationalism as understood as the performance of these traits of cohesion, heritage, and a destiny of the "nation") among Christian Zionists cemented in selective interpretations of the Old Testament that imagine Israel as a redemptive national territory. While not all nationalist expressions are religious, almost any can be adopted into a religious discourse and be performed as such, just as any religion can adopt national allegiances as part of the ritualized performance of their religion (McAlister 2008). Christian Zionists practice a particular form of diasporic nationalism that challenges notions of nationalist exclusivity. Christian Zionists can be best described as performing a type of nationalism: an *ethno-religious* nationalism that contributes to the making of a more accurately termed "Judeo-Evangelical" religious identity rather than the inflected use of "Judeo-Christian" in recent American politics as an attempt to exclude Islam from the foundations of the West (Silk 1984; Schultz 2011; Sturm 2010). This unique brand of nationalism emerges from American social, economic, ethnic, and racial anxieties, and forces us to reconsider how nationalism, religion, and space can be conceived together.

41.2.1 Religious Nationalism?

Much of the work on nationalism has tended to ignore religion or explain it as a function of nationalism. In this view, nationalism is a modernist project that replaces religion by emphasizing socioeconomic factors or cultural or political modernity

(Gellner 1994; cf Asad 1999). That said, there are many examples of synergy between religion and nationalism, some mobilizing strong nationalisms such as Ireland and Poland, and other relatively weak movements such as in Russia and Ukraine (Bruce 2003). However, care must be taken in fusing these two terms together as "religious nationalism" because there are often many reasons for, expressions of, and geographically specific types of nationalism, the use of religious signifiers being but one of them, often of secondary influence, epiphenomena, or used as a guise for political means (Agnew 2008).

In this respect, Roger Friedland (2001: 138) writes that religion can be used to constitute the "state and territorially bounded population in whose name it speaks." Friedland (2001: 138) has suggested that there is an alternative type of nationalism, in form and content, which he calls "religious nationalism," where nationalism is defined as "discursive practice by which the territorial identity of a state and the cultural identity of the people whose collective representation it claims are constituted as a singular institutional fact." Religion in this conception fuses state, territory, and culture in an analytical way. Friedland's "religious nationalism," however, hinges on the state, as if religious expression can only be funneled through the state's political voice or that religious expression matters only if it is sanctioned by the state. Rather than "religious nationalism" Friedland's observation might better be termed "religious statism" (Brubaker 2006).

Friedland's argument is then underinclusive, ignoring other non-state-centric forms of nationalism. But for the purposes of this scholarship, Friedland's conception has relevance to nationalism. Christian Zionists perceive themselves as the highest representatives of the nation, its moral mores, and transcendently sanctioned critics, albeit a subgroup and minority of the American "nation." There are still shared perceptions of origin and myth that are utilized. A more nuanced and somewhat less state-centric case can be made as well, where the lack of nationalist congruence with the state may bear and nurture a type of religious nationalism.

Although nationalism certainly has a "spiritual principle," as Ernest Renan (1996 [1882]: 52) classically observed, the attempt to formulate a theory for religious nationalism is one fraught with problems.[1] There have been many attempts to fuse religion and nationalism as it relates to overlapping language, analogous histories, that religion is part of nationalism, or that there is a particular form of religious nationalism. Some have suggested the term to be oxymoronic, as Brubaker (2006: 23) provocatively concluded on the possibility of a religious nationalism, "nationalist politics is carried out in the name of the nation, religious politics in the name of God." Despite this, it is argued here that this fusion takes place most convincingly not in how national discourse is inflected by religious language or how religious discourse is inflected with nationalist language, but rather, following Talal Asad's (1999) demand, that we again understand the category of religion as one of *practice*

[1] Although Juergensmeyer (1993: 1–2) argues that religious nationalism has replaced Cold War identities, like many scholars at this time looking for a new world order, based on political ideologies: "the new world order that is replacing the bipolar powers of the old Cold War is characterized… by the resurgence of parochial identities based on ethnic and religious allegiances."

rather than solely as *belief*. Therefore, what we assume to be nationalist language is also religious practice through prayers, sermons, and pilgrimages to mention only a few performances. As McAlister (2008: 875) explains, it is not that "everything is religion, it is just that religion can be virtually anything." Nationalism too can be religious practice.

While I take my theoretical impetus from Asad to explain religious nationalism, other scholars of religion and nationalism, like Hobsbawm and Ranger (1983), have taken a functionalist perspective arguing that rational and deliberate decisions, made most often by social elite, guide the precepts of nationalism. In Hobsbawm and Ranger's (1983: 1) words, national traditions are politically motivated entrepreneurial inventions, most times consciously constructed through "a set of practices, normally governed by overtly or tacitly accepted rules and of a ritual or symbolic nature, that seek to inculcate certain values and norms of behaviour by repetition, which automatically implies continuity with the past." Therefore, not only are traditions invented, they can be quite nascent, relying on repeated narratives and representations that enter modern practice and memory concerning a place. I take issue, however, with Hobsbawm and Ranger's use of the term "invented" because it suggests traditions are rational, manipulative, and orchestrated. I argue instead that traditions, however mythical, operate as emergent and performative forms of discourse that can be used and abused for political ends by virtue of the loose networks of normative performances. That Christian Zionists would make dogma an "Judeo-Evangelical" tradition has less to do with it being consciously and deliberately "invented" and more to do with what Derrida called the "play" between cultural performances.

Christian Zionism is then not a natural or inevitable perennial outcome either. Anthony Smith (2003) makes the argument that proto-nationalisms based on religious and ethnic groupings, what he calls "ethnies," pre-existed and pre-disposed the Western world to modern nationalism. Smith outlines four kinds of religion based proto-nationalist cultural resources: (1) myth of ethnic election; (2) attachment of terrains as sacred; (3) yearning to recover the spirit of a golden age; (4) regenerative powers of sacrifice to ensure glorious destiny. All of the above relate to Christian Zionism it that their discourse and performance of religious nationalism stress that both Jews and Christians have been Chosen to act in a grand play to reclaim sacred territory given to them in a Covenant with God and will result in an imminent return of Christ who will redeem them in Holy glory. That said, most nationalisms imagine themselves in these terms which suggests that religion is often a legitimating factor in nationalist identifications. Therefore, while often employed to make the argument for a "religious nationalism," Smith's analytical division is not particularly convincing. I therefore deny that there is a natural continuity as presented by Smith, nationalisms are emergent for a variety of reasons from many different historical experiences.

Smith's notion of proto-nationalisms based on sacred texts and religious practices is not argued here. Rather, I maintain that religious myths are often symbols of heritage and destiny and are pulled into the performances that are co-constitutive of space and national identity. So while modern nationalism has a vast array of mythical heritage events, religion is often a central legitimating claim. It is only within this vein of practice that is informed by such discourse that Smith's (2003: 254) charge is taken up to or serve the period of nationalism with "sacred foundations" and "deep cultural resources" in the nation's commitment to authenticity, however contrived and selective. Nationalism is then constructed from pre-existing culture but through its performative making it also changes that culture (Derrida 1977; Butler 1997).

41.2.2 Diaspora and Extra-territorial Nationalism

A significant amount of recent literature has focused on diasporic or extra-territorial forms of nationalism (Anderson 1983; Brubaker 2005). Diasporas are often the most extreme and unwavering nationalisms (Herb 1999: 20). Christian Zionists are not a classical diaspora-based "nation," nevertheless, they share in common a basic definition as a group of people living outside of a territory for which they claim *irridenta* rights, restoration commitments, and heritage claims, based on myths and memories of a perceived "homeland" which they imagine set out in the Bible. Conventionally, diasporas refer to having a sustained connection to a homeland and keeping ethnic and cultural community in place. It is commonly argued that there are three elements of any diaspora: (1) dispersion through space; (2) orientation to a homeland; (3) boundary-maintenance within a larger polity (Safan 1991). These definitional limits taken, Brubaker (2005) rightly argues that rather than categorize "a diaspora" as a cultural fact, a diaspora should be approached through stances, projects, claims, idioms, and practices of those who identify themselves as "dispersed in space." Therefore, to add to the tri-partite definition above, it is important to understand diasporas through "categories of practice," that is for groups to identify themselves in such a way, as opposed to imposing upon them preconceived "categories of analysis." In the final section that follows, I provide a case study to illustrate this by fleshing out through Christian Zionist discourses how they imagine themselves to be a religious nationalism diaspora despite living in America and also performing a "civic" secular constitution-based nationalism. It is this imagination and identification that allows Christian Zionists to have a diasporic national self-image.

Foon (1986) argues that people can have a loyalty to two different ethnic groups and have two nationalisms because they serve two different functions. Much modern cultural theory would suggest that even harboring two conflicting ideas of "nation" is not necessarily caustic or competing (Gupta and Ferguson 1992). Following Gupta and Ferguson (1992: 16), there is no isomorphic parallel overlap between identity, religion, ethnicity, nation, and place. They write, "if we question a pre-given world of separate and discrete 'peoples and cultures,' and see instead a difference-producing set of relations, we turn from a project of juxtaposing preexisting differences to one of exploring the construction of difference to one of exploring the construction of differences in historical process." Gupta and Ferguson (1992: 12) seek to make this case clear through the example of America, a diverse state and set of nationalisms, which calls into question these very categories. In other words, while there is an assumed overlapping of ethnicity, nationalism, territory, and religion, identities are more complicated, less austerely categorized, and territorially blurred of the bounded assumptions we wish to impose on them (Agnew 1994).

Gabriel Sheffer (2003: 232–233) distinguishes between "total," "dual," and "divided" diaspora loyalties. For Christian Zionists, their religious nationalism is committed to God first, and America second, and, therefore, rejects Sheffer's analytical gradient because of the narrow definition of loyalty as that which is attributed to the state. But in no way is the echelon static or stable, it is the result of performative contingencies that have defined their modern identities. In a recent poll, 42 % of Christians saw themselves as Christian first and American second (Pew Global Attitudes Project 2006: 3). As such, Christian Zionist love for Israel is a religious commitment based on future history that sees a moral, religious, economic, and political decline in the United States and an ascendency of these attributes in Israel until the culmination of the Rapture. For example, John Hagee, a major Christian Zionist figure who heads a megachurch in San Antonio Texas and Chairs the charitable organization and Israeli lobby *Christians United for Israel*, has been clear concerning the moral, political, and religious position of the United States: "the laws of God transcends the laws of the United States government and the U.S. State Department" (quoted in Mearsheimer and Walt 2007: 150). In other words, in so far as the United States plays a role in this script, it is to support and protect Israel. In his 2006 book Jerusalem Countdown: A Warning to the World, Hagee suggests former President George W. Bush's support for Israel "fulfills a biblical injunction to protect the Jewish state" which is leading to "a pivotal role in the second coming" (Hagee 2006).

41.2.3 Ethnic vs. Civic Nationalism

Understanding the distinction between "civic" and "ethnic" nationalism is essential to grasping the meaning of the ethno-religious nationalism seen in the Christian Zionist movement. Both Brubaker and Smith make distinctions between civic and ethnic nationalism (Smith 2003; Brubaker 1992). The concept of American civic nationalism—otherwise referred to as the "American creed" which is based on democratic and enlightenment beliefs upon which the American Constitution is founded—is alive and well and continues to bind Christian Zionists to the American state and people. Nevertheless it is thought to be under attack by a broad spectrum of culturally conservative Americans giving way to the so called "culture wars." For Christian Zionists this eroding of this civic nationalism is thought to be a sign of the End Times.

Like French civic nationalism, it is assimilationist not pluralist, and, therefore, while based demographically on immigration it has a nativist expectation for conversion to the American creed and this is an especially poignant observation when that creed is imagined to be founded in a mythologized Jewish and Christian synergy (Kazin and McCartin 2006). This mix of Protestantism and Enlightenment thought is what has been called "civil religion" in the United States (Bellah 1976). However, what Lieven (2004) calls the American nationalist "Antithesis" or ethnic nationalism, a sometimes competing form of American civic nationalism, he argues co-exists with civic nationalism, but has its roots in ethno-religious beliefs and Jacksonian ideals. While this latter form of nationalism is often subordinate to civic nationalism, it can rise to prominence in times of crisis. Imaginations of an imminent apocalypse, being an embattled minority, and the belief that Israel and Jews being persecuted by the surrounding world provide such crisis mentality particularly when juxtaposed against a common enemy, which as we will explore below, has often been Islam (Sturm 2010).

It is this form of ethno-religious nationalism that Benedict Anderson predicted to be "the wave of the future" for nationalism, one consisting of "ethnic and racial stereotypes, xenophobia, sectarian 'multiculturalism' and the more brutal forms of identity politics" (Anderson 1996: 12–13). For Lieven (2004: 6), in his condemnation for America's support for Israel's radical right, it is this ethno-religious nationalism that cements "America's attachment to Israel, ethno-religious factors have become dominant, with extremely dangerous consequences for the war on terror." What distinguishes ethno-religious nationalism from other cultural identities, is not only its territorial and political dimensions and ideals, but also its commitment to the pursuit of authenticity (however mythical) of the heritage and destiny of the imagined national community.

41.3 Territory and Apocalypse

Visions of an imminent apocalypse have provided just such binding territorial exceptionalism many times throughout history, from Munster in 1534–35 to the English Puritans who migrated to New England. The latter marks the beginnings of what historian, Stephen Stein (quoted in Boyer 1992: 68), calls "the Americanization of the apocalyptic tradition."[2] The Puritans serve here as a brief example of a millennial attachment to place. As Avihu Zakai explains of the Puritan reverence with America's shores, their

> …attitudes towards [New England] space according to eschatology and apocalyptic visions, or according to prophetic imagination… [led to a] desacralization of England as a scared place in providential history, [this] reveals their geoeschatologic and geoapocalyptic awareness that England was not elect but rather represented apostasy within the course of the history of salvation. And geoeschatolic and geoapocalyptic consciousness gives evidence as well of the sacralization of an alternative place within the eschatology and apocalyptic drama of salvation and redemption. (Zakai 1992: 72)

[2]Although Thomas Jefferson denigrated the Book of Revelation as "merely the ravings of a maniac" and made his own Bible, cleansed of what he thought was pseudoepigraphia, leaving only the gospels. Quoted in Boyer, note 8, p. 68; see also Bercovitch 1978 and Cherry 1998.

Within this quote we can discern that England was, prior to migration, seen as sacred space but that sacredness was eventually moved to New England (Zakai 1992: 74). The New England Puritan interpretation of America's shores, as Sacvan Bercovitch (1978: 41–42) succinctly argued, turned "geography into eschatology." This story of cleansing the Puritans and Yankees in the waters of the Atlantic is a powerful one in American national identification, and one that continues to set America apart not merely by distance from Europe but by moral absolutes as to the uniqueness of being "born-again" as American and Christian (Hughes 2003).

41.3.1 Rediscovering Sacred Territory in Mid-nineteenth Century Palestine

Christian Zionist territorial identification with Israel differs from the post-Civil War American peripatetic interest and fascination with Palestine, however. One mid-nineteenth century Methodist Episcopal bishop and pilgrim to Palestine, Henry White Warren, for example, reminisced that "This [Palestine] is the first country where I have felt at home" (Quoted in Davis 1996: 16). Palestine became at this time not only an origin myth but also a moral guide, a "Fifth Gospel," as it was colloquially referred to (Vogel 1993: 11).

It is important to first flesh out this story of the nineteenth century "Holy Land Mania," to contrast the modern Christian Zionist reverence for Israel with the post-American Civil War American obsession with Palestine (Obinzinger 1999). Nineteenth century American pilgrimage to Palestine's landscapes was one of self-imagination and renewal after the deep and divisive scars of the American Civil War (Quoted in Vogel 1993: 67).[3] The thought was that if Palestine could be restored as sacred space, so too could America. As Obinzinger argues, "travel to Palestine allowed Americans to read sacred geography.... While the persistent preoccupations with the Bible and biblical geography stood at the ideological core of American colonial expansion, actual travel to Palestine allowed Americans to contemplate biblical narratives at their source in order to reimagine—and even to reenact—ethno-religious national myths, allowing them, ultimately, to displace the biblical Holy land with the American New Jerusalem" (Obinzinger 1999: 5). Palestinian landscapes were, therefore, the medium for *American* national self-definition.[4]

---

[3] Prophecy was important to the American imagination of Palestine, but was not crucial to pilgrimage identities. For example, one travel writer remarked that Palestine is "overwhelming evidences of the truth of scriptural prophecy."

[4] Of course Perry Miller's popular description of American settlement being viewed as a new Jerusalem, it is important to point out that exuberant celebration of renewal this was short lived and that second generation immigrants were more critical of their "sins, crimes, misdemeanors, and nasty habits," tapping into the "attuned to the narrative of decline." In other words, America never neatly fit the perfect and expectant model of Jerusalem and Israel the way these modern places have come to be revered by Christian Zionists (Jendrysik 2008: 4).

In the nineteenth century both Palestine and America were thought of as the twin cradles of civilization, where Palestine marked the alpha—the beginning—American marked the omega—the millennial destiny (Stephanson 1995). On the other hand, what makes the contemporary American Christian Zionist experience so interesting is that rather than reaffirming an *American* national identity, pilgrimage to Israel becomes a critique of that identity. Christian Zionists today increasingly are coming to see America as the moral antipode of a Jewish Israeli nation that fuses future redemption with a present (but adopted) nation.

Christian Zionists are largely disillusioned with the assumption of God's divine providence attached to America as a new terrestrial space to live out a God given holy life. Playing on Hal Lindsey's (1971) best-selling book of doom, *The Late Great Planet Earth*, Mark Hitchcock (2010) for example, has entitled his recent book with a more specific national focus, *The Late Great United States*. In it Hitchcock details the inevitable decline of the United States, the ascendant prominence of Israel, and the imminent prophecies to take place there. Victor Mordecai, a Messianic Jew who is well known in evangelical circles for his adamant anti-Islamic thought, argued that "Christianity is going out of style in America in favor of ethnic diversity represented by many immigrant groups and other faiths." There is a "fading Judeo-Christian ethic in American," as the title of his article laments, America is in decline then not just economically or politically but at its ethical core. Mordecai's core values are biblically based, this is what he means by a "Judeo-Christian" foundation, one that has seen, as he points out, Muslims outnumber Jews 4 to 1 in America. Those who are to blame for this undercutting of America's foundation via Muslim immigration are those who support progressive politics. He writes, "Obama represents a synthesis of Muslim and ultra-left radicalism of the 1960s, both of which are inimical to the values of Judeo-Christian America and Israel" (Mordecai. 2010: 12; cf. Dittmer 2010). As Grace Halsell also observed of the disenchantment of the American West for apocalyptic forms of Christianity, "since the 'frontier' of America is gone, they seek to recreate it elsewhere" (Halsell 1986: 113–114).

Israel is now the Alpha and the Omega. Furthermore, Palestine in the nineteenth century was interpreted not in alterity, but in continuity that linked an origin story to Palestine rather than to Europe and, therefore, the material landscapes of the Bible. The difference between viewing Palestine as the performance of a "Fifth Gospel" is the literal emphasis of a new homeland that often rejects America in favor of Israel. The modern phenomenon is an attempt to make Americans into Israelis, not Palestine into America. This is made possible by accepting an apocalyptic vision of the future and a religious adoption of Jews as legitimate biblical actors. Prior to the early 1970s, evangelical imaginations of Israel and Jews were often fraught with anti-Semitism that saw Jews serving a functional role in the End Times (Barkun 2010; Weber 2004; Boyer 1992; Cohen 1979). It was not until American evangelical culture saw the territorial maximalism of Israel taking place with the occupation of the West Bank and Gaza Strip, Jews as the Chosen People of the earth with a prophetic role to immigrate to Israel, a populist envisioning of America as founded on a "Judeo-Christian" tradition, and Islam as an emergent enemy of the West was an American Christian Zionist diaspora nationalism for Israel possible. As one recent Christian Zionist defined the historical distinction, unlike the nineteenth century Christian pilgrims to Palestine who were "friends of the Jews," today we are "full partners in fulfilling His eternal promises to re-gather His beloved people" (Bühler 2008: 53).

41.4 Post-1967 Making of Judeo-Evangelical Nationalism[5]

With the 6-Day War of 1967 and growing Arab hostility against Israel throughout the mid-twentieth century, Christian Zionists increasingly focused on the role of Israel within the prophetic End Times scenario.[6] The 1967 war that saw the unlikely defeat of Arab forces by a much smaller Israeli army and the occupation of the West Bank and the Gaza Strip, had a convincing affect on American evangelicals that major prophetic events were taking place in the Middle East. These events marked the beginning of their national shift and territorial identity.

Cohn-Sherbok points to America's bicentennial year, 1976, as a time when several issues coalesced to reinvigorate Christian Zionism, which included the outgrowing of mainstream Protestantism, the election of born-again and southern Jimmy Carter, and the election of Menachem Begin as Prime Minister of Israel (Cohn-Sherbok 2000: 165). American Christian Zionism became more involved in federal politics as well for many reasons that culminate coincidentally at this time (Bruce 1988). It is in part the elevating of the decline of perceived American Christian values that brought voters to the polls on issues of abortion, prayer in schools, intelligent design, homosexuality, the spread of "liberal humanism," and the more recent example of stem cell research.

The values voter was originally mobilized by Televangelists, Pat Robertson, progenitor of the Conservative Coalition and voice of the 700 Club, Jerry Falwell, the founder of the Moral Majority, and Tim LaHaye, founder of the American Coalition for Traditional Values among them. These three men in the revival of southern and conservative Christian voter bloc coalesced around the "return" of so-called "Judeo-Christian" values and also renewed American Christian interest in pilgrimage to Israel and Palestine.[7]

---

[5] Much of this brief history is inspired by Shalom Goldman's authoritative text (Goldman 2010; see also, Sandeen 1970; Boyer 1992; Marsden 1980).

[6] Koeing (2006), for example, correlates (selective) American catastrophes to moments in history when American support for Israel waned or when America pressured Israel toward a resolution over the Palestinian occupied territories, arguing that God was punishing America for not supporting Israel (see Clark 2007: 252–55).

[7] Institutionally, the movement found a voice in the American political landscape in the early 1970s, but its language and ideology emerged much earlier (cf. Lichtman 2008; Dochuk 2011). That said, the interest in Jews and Judaism was a relatively recent post-World War II emphasis employed by the American conservative political right. Revisionist histories by cultural conservatives and especially culturally and theologically conservative Evangelicals has become a mainstay

Central to this story is the beginnings of the "Judeo-Evangelical" relationship that brought together political and religious leaders in America and Israel, their congregations, and tourism money within the emerging state of Israel (Chafets 2008).[8] The mid-twentieth century roots of this Judeo-Evangelical relationship can be traced to Oral Roberts and Billy Graham's both of whom were particularly supportive of the nascent state of Israel. In particular, Roberts laid the foundation for future evangelical leaders to approach and be approached by every Israeli Prime Minister beginning with David Ben Gurion. Ben Gurion, while known as a proud secular Zionist who claimed to enter a synagogue only once in 40 years, met and allegedly "prayed" with Roberts in 1959 (Segev 1998; Goldman 2011). And yet despite his secular ethos, recognizing the political and economic support of Christian Zionists, Ben Gurion helped facilitate the Sixth World Conference of Pentecostal Churches to take place in Jerusalem in 1961, to which he addressed by courting their beliefs "[in Israel] today we are privileged to see the fulfillment of the prophecy and promise of the Bible" (Goldman 2011: n.p.). Jerry Falwell's relationship with Menachem Begin, Prime Minister of Israel from 1977 to 1983, has often been credited with being the relationship that led post-World War II American evangelicals to acknowledge the occupation of the West Bank and Gaza Strip as prophetic events. Begin was the first Israeli PM to openly support American evangelical tourism to the Holy Land and Falwell's ministries brought them by the thousands (Harding 2000). In 1980 Falwell became the first non-Jew to be awarded the Jabotinsky medal for Zionist excellence (Cohn-Sherbok 2000: 162).

The relationship between the rise of Israeli right wing politicians and the Christian Right in America has coterminous parallels related to the settlement movement and the 1967 and 1973 Israeli wars (Bruce 1998; Lienesch 1993). Ehud Spirnzak (1991) points out that Israel's radical right began to gain political clout in 1967 by defining the occupied territories as part of biblical *Eretz Yisrael.* The messianism of the Israeli radical right, largely under leadership of Rabbi Rav Kook and his *Gush Emunim* settlement movement, was made possible by defining the secular state of Israel as a messianic means toward territorial maximalism (Deuternonmy 7: 1–23) (Nyroos 2001).[9]

---

of reinventing the American tradition. See for example, Catherine Millard's book, The Rewriting of America's History (1991), which as its title suggests, "is dedicated to the glorious truth that this nation was established upon biblical principles: its founders were men of Christian nobility" (Millard 1991: iv).

[8] On the hyphenation, "Judeo-Evangelical" see Chafets (2008). As for ethnic inclusivity (see Noll 2008; Ariel 2000; Ehrenhaus and Owen 2004; Diamond 1996).

[9] Almost 900 years later Rabbi Abraham Isaac Kook, the early twentieth-century thinker respected by most Orthodox Jews and revered by Religious Zionists, had similar sentiments. Like Maimonides, Kook saw Christianity as akin to idolatry, writing that "with Christianity and its concepts one should share nothing, not even what seems good or beneficial… It is only by distancing oneself from Christian concepts, and by implementing the absolute refusal to gain any benefit from that world of ideas, that our own intellects and sense of self will become purer and stronger" (Goldman 2011).

The Israeli and Palestinian landscape for *Gush Emunim* took on a millennial significance that differed from the German romantic landscape vision of Labour Zionist settlers (Friedland and Hecht 2000). The former believed that possession, both visual and physical, was essential to the performance of a prophetic national identity. As Weizman eloquently explains "for most settlers, the landscape was not initially much more than a pastoral view, but for the ideologists of Gush Emunim, its topographical features were cast as national metaphors. A constructed way of seeing sought to re-establish the relation between terrain and sacred text" (Weizman 2007: 135). For Christian Zionists, the proxy possession of landscape through their perceived national peers is essential to the production of their own Judeo-Evangelical nationalism. Christian Zionists have supported the "national religious" settler movement, whether they like it or not, in the quest for territorial maximalism.[10]

This definition gave agency to many orthodox Jews who had previously thought that Israel could only be founded by the hand of God. Israel's capture of the Golan Heights, Gaza, and the West Bank in 1967 suggested to many that God's hand was working through secular agents. For this burgeoning movement, settling these newly won territories was a self-fulfilling prophecy that would result in the return of the messiah (Sprinzak 1991). The Israeli settler movement or "national religious" have facilitated the Judeo-Evangelical relationship as it relates to the occupation of territory and the exclusion of Palestinians from it. These expanded borders that now include Palestine have come to redefine the Israeli nation and, therefore, challenging those borders by suggesting a two-state solution to the Israel-Palestine conflict, for example, is understood to be synonymous with not only challenging the definition of the nation itself but more importantly God's prophetic work.

This national "cartographic anxiety" is a motivating factor for most Christian Zionist pilgrims and is a rallying call for their political and economic efforts in Israel and America (Krishna 1996). As a poignant example, the televangelist pastor John Hagee openly contributes millions of dollars in Christian Zionist donations to settler movements. Pat Robertson, voice of the 700 Club, dispensationalist, and one-time presidential hopeful, in a 1991 interview on the 700 Club with Gershon Solomon, leader of the Temple Mount Faithful, a group focused on the rebuilding of the Third Temple, declared that "we will never have peace until the mount of the House of the Lord is restored" (quoted in Ariel: 153–154). The *International Christian Embassy in Jerusalem*'s founding principle, for example, was to affirm Israeli sovereignty over the West Bank later spinning off the group in response to the Oslo Accords' returning some territorial control to Palestinians (Weber 2004).[11]

---

[10] John Hagee estimates that in 2009 alone, he contributed $58 million. Hagee's support for settlements has resulted in a sports arena being named after him in the Jewish West Bank settlement of Ariel (Rutenberg et al. 2010).

[11] A common meme among ex-pat Christian Zionists living in Israel, for example, is that Ariel Sharon's coma was God's punished for removing settlers from Gush Katif.

41.4.1 Judeo-Evangelical Tradition and the Exclusion of Islam

There have been many examples where powerful government representatives from Presidents to members of Congress have not only tried to redefine American history in favor of a "Judeo-Christian" foundation narrative, but have also tried to influence American foreign policy in favor of a "Judeo-Evangelical" relationship with Israel specifically with regard to territory. The West Bank, for example to James Inhofe, Oklahoma Senator, former Chair of the Senate Committee on Environment and Public Works, and Christian Zionist, in a speech to the Senate in 2002 entitled, "Seven Reasons Why Israel is Entitled to the Land" makes it clear that Israel has *irredenta* rights to all land West of the Euphrates river up to the Mediterranean Sea that may include western Iraq, parts of north western Saudi Arabia, and all of Jordan, Syria, Lebanon, and the Sinai Peninsula (Rossing 2004: 52–53). More recently Michele Bachmann, Sarah Palin, Rick Perry, and Mike Huckabee stated that Israel should not give up Palestinian occupied territories because it was biblically promised to Jews (Lizza 2011; Kalman 2009). On a recent pilgrimage to Israel, Huckabee told the *New Yorker* reporter, Ariel Levy, that a two-state solution would not happen "on the same piece of real estate" (Levy 2010: n.p.). Rather the Palestinians, whom he denies legitimate nationality, are strictly an Arab problem. He was also quoted expressing his "Judeo-Evangelical" nationality saying, "I worship a Jew!... I have a lot of Jewish friends, and they're kind of, like, 'You evangelicals love Israel more than we do.' I'm, like, 'Do you not get it? If there weren't a Jewish faith, there wouldn't be a Christian faith!'" This territorial identification with Israel and the national ethno-religious identification with Jews is one set against a racialized exclusion of Arabs and Islam.

In matters of international relations the so-called "Judeo-Christian tradition" has been used to refer to "Western civilization" by, for example, Samuel Huntington (2003) and Bernard Lewis (1987). In both of these cases, it is used specifically to mark a difference between Islam and the "West." Slavoj Zizek picks up this later division and calls it into question by posing a "Jewish-Muslim civilization." He argues that because "we usually speak of the Jewish-Christian civilization—perhaps, the time has come, especially with regard to the Middle East conflict, to talk about the Jewish-Muslim civilization as an axis opposed to Christianity" (Zizek 2006: np). Bernard Lewis, however, anticipated this hyphenation 20 years before Zizek: "the term 'Judaeo-Islamic' is at least as meaningful and as valid as 'Judaeo-Christian' to connote a parallel and in many ways comparable cultural tradition" (Lewis 1987: x). For Maimonides, Goldman (2011: np) reminds us, the similarity between Islam and Judaism was closer than that between Christianity and Judaism because of perceived idolatry of the Trinity concerning the latter, "Islam, though a 'mistaken' religion in his view, is monotheistic, and thus cannot be construed as idolatrous."

European Christians often saw Jews as those who were the "outsiders within" and saw Muslims as the "outsiders outside," the latter's presence defined the territorial limits of a Christian Europe (Buchanan and Moore 2003: 7). The most recent and obvious catalyzing event to demonize Arabs Muslims was 11 September 2001. This event seemed to provide evidence for Samuel Huntington's then failing "clash of civilizations" thesis as well as the dualistic foundations of apocalyptic Christianity.[12] This is symptomatic of a wider practice of how religious categories are grafted onto racial ones generally (Sturm 2010).[13] Muslims through immigration have become the threat to the Judeo-Christian tradition inside and are the threat outside. In this way, post-Cold War prophecy adopted, informed, and inflamed these binary qualifications between American capitalism and Soviet communism during the Cold War and American Christians and Arab Muslims in the post-Cold War period, rendering both sides into divisive geopolitical halves.

The national division of insider and outsider can take dramatic shifts in times of geopolitical change (Agnew 2003). But most interesting is a new religious identification that sees Jews as not just a question of a shared text and values or tradition, something that in and of itself is difficult for most American Protestants to relate to, but also the genealogy that often took on a tone of family, blood, and race that excludes some in favor of others. Jews became rewritten into God's contemporary interventions in history and the proof was the reestablishment of Israel and his clear territorial sovereignty and protection of it (Goldman 2011). Jews became once again the people of the Book and took on a transcendent role for the redemption of all Christians. In other words, Jews became insiders.

Central to this definition of inside/outside is where Christian Zionist national allegiances are placed. Common among Christian Zionists is the assumption that dissention or criticism of Israel is synonymous with criticizing God: Israel is God's work. Indeed, today, for Christian Zionists, America and Americans still holds a moral voice and is respected as homeland largely represented by civic or constitutional nationalism. This civic nationalism, as illustrated above however, takes second place to the imagined moral light of Israel and Jews. The foreign policy of Israel is infallible and inerrant while American foreign policy is sinful if it challenges Israeli policy. Israel always trumps America. William Koenig, a former 3rd-party presidential candidate on the Christian Right, has recently published two books in which he argued that both Hurricane Katrina and 11 September 2001 were God's responses to wavering American support for Israel. Koenig's book, *Eye to Eye* (2006), argues that all major catastrophes on American soil in recent history were God's vengeance for the US asking Israel to hammer out a peace plan with Palestine.

---

[12] I say failing because of the vast amount of literature celebrating the 200th anniversary of Kant's perpetual peace (Habermas 1997).

[13] Race is thought to be descended from particular biblical individuals from which their characteristics are generalized across whole "races" and "nations." For example, Arabs are thought to be the heirs of Abraham's cast-out son Ishmael, while Isaac is taken as the patriarch of the Jews. Abraham's cast-out son Ishmael is thought to have begat Arabs, and his son Isaac, Jews for which the former is not only subservient to the latter as the preferred son, but was also given the Covenant to Israel. Furthermore, the descendents of Noah's son Ham are thought to be Africans and his son Japheth are thought to be Europeans and, therefore, ancestral Americans. These generalizations are then used to make stereotypical character judgments, justify prejudice assumptions concerning social standing, and associate whole continents of people as either "evil" or "good."

Koenig's book succinctly captures what he believes as an undeniable correlation rather than coincidence and selective history:

> What do these major record-setting events have in common? The ten costliest insurance events in U.S. history; The twelve costliest hurricanes in U.S. history; Three of the four largest tornado outbreaks in U.S. history; The two largest terrorism events in U.S. history All of these major catastrophes and many others occurred or began on the very same day or within 24-hours of U.S. presidents Bush, Clinton and Bush applying pressure on Israel to trade her land for promises of "peace and security," sponsoring major "land for peace" meetings, making major public statements pertaining to Israel's covenant land and /or calling for a Palestinian state. Are each one of these major record-setting events just a coincidence or awe-inspiring signs that God is actively involved in the affairs of Israel? In this book, Bill Koenig provides undeniable facts and conclusive evidence showing that indeed the leaders of the United States and the world are on a collision course with God over Israel's covenant land. 'And it shall come to pass in that day, that I will seek to destroy all the nations that come against Jerusalem.' (Zechariah 12:9; Koenig 2006: back cover)

The feeling among Christian Zionists is that America is in moral, economic, and political decline. At times, America itself becomes the enemy of Judeo-Evangelical nationalism. Unlike the nineteenth century American pilgrims seeking to cleanse the morality of their American nation in Palestine, Christian Zionists today understand Israel as being the new millennial promise while America is irrevocably lost. Israel is both Alpha and Omega.

41.5 Conclusion

This Christian Zionist national identification to Israel and Jews, I argue, is grounded in an eschatological narrative of the future. Susan Harding (2000) argues that Christian Zionists read history backwards, from the future to the present. They know what is *going to happen*, just as the mythical histories presented as the heritage of the Judeo-Christian Tradition present *what happened*. Christian Zionism is an ethno-religious national identity that is framed by anticipation for a "future history" as one scholar of eschatology put it (Schussler 1985: 40).

Christian Zionists are therefore "spreading [reremembered] time out in space" (Boyarin 1994: 19). Daniel Boyarin is arguing in this quote that time and space cannot be separated out and that memory, whether past or future recollection, is always spatialized. Any memorialized territory requires for its performance to be effective that historical space and historical time be brought together. As argued above, inserted into this calendar of "future history" and "future memory" is the establishment of Israel in 1948 and its expansion in the 1967 6-Day War. Read as a miracle created by the hand of God by Christian Zionists, these events are considered the most important signs portending Christ's return. All of this will manifest in a territory and place: Israel. Their *future is a foreign country*.[14] Time and space here

[14]This is an inversion of David Lowenthal's, The past is a foreign country (1985), to suggest the dislocation Christian Zionists feel not only between the present and past but also the present and the future. Furthermore, it is to suggest that their national allegiance is shifting from the America

merge as prophetic time manifests itself in prophetic space. And the hyphenation of Judeo and Christian is key to developing a future, dualist Manichean memory dividing what is inside—"ours"—and outside—"theirs"—in often simplistic geopolitical divisions drawn up between Jews and Christians on the one hand, and everybody else—especially Muslims—on the other.

Despite this alterity toward Islam, Christian Zionists challenge the Eurocentric model that clear boundaries can be placed around ethnic and religious groupings. "Us" and "them" binaries are challenged as Israel is linked to an American Venn diagram where Christian Zionists move toward the Israeli half of the circle concerning ethnic and religious elements of nationalism and, therefore, away from American civic nationalism. In other words this hybridity resists naming[15] and decenters representation by complicating our usual isomorphic overlapping of religion, nationalism, and territory.[16] Therefore a hybrid Jewish/Christian and Israeli/American identity is never fully consumed by any one categorization. Christian Zionism is then not wholly Western nor Other. Furthermore, this is not to say that there is a complete internal coherence to Christian Zionism (Spivak 1988). Rather, as Clifford challenged cultural studies in the early 1990s, culture is always "travelling." Therefore, modern American Christian Zionism is "not your father's Christian Right" as McAlister observed, it is less isolationist and staunchly American and has become more dynamic as religious interests have merged with national ones in Israel (McAlister 2003: 775).

This "Judeo-Evangelical nationalism" is important precisely because it seeks to politically redefine the relationship between Judaism and Christianity, and Israel and America. As Boyer explains in his magisterial and commanding book on American cultural expectations of the apocalypse, such expectations are not merely reflections of other realities rather "apocalyptic cosmologies have functioned dynamically, helping to mould political and social ideology and thus influencing the course of events" (Boyer 1992: 78). Indeed as Mearsheimer and Walt (2007: 137–8) have recently made clear of the political, cultural, and economic implications of Christian Zionists:

> By providing financial support to the settler movement and by publically inveighing against territorial concessions, the Christian Zionists have reinforced hard-line attitudes in Israel and the United States and have made it more difficult for American leaders to put pressure on Israel. Absent their support, settlers would be less numerous in Israel, and the U.S. and Israeli governments would be less constrained by their pressure in the occupied Territories as well as their political activities. Plus, Christian tourism (a substantial portion occurring

---

to a foreign country, Israel. Lowenthal, however, takes his title from the prologue of L. P. Hartley's The Go-between (2002 [1953]: 17) which reads, "The past is a foreign country: they do things differently there."

[15] "Judeo-Evangelical" nationalism remains a hopelessly imperfect distinction, not least because excludes Pentecostals.

[16] Gilroy argues similarly, "the specificity of the modern political and cultural formation I want to call the Black Atlantic can be defined, on one level, through [a] desire to transcend both the structures of the nation state and the constraints of ethnicity and national particularity. These desires are relevant to understanding political organizing and cultural criticism" (Gilroy 1993: 19).

under the evangelical auspices) has become a lucrative source of income for Israel, reportedly generating revenues in the neighborhood of $1 billion each year.

Christian Zionists are motivated by an apocalyptic vision that co-constitutionally performs Israel and their Judeo-Evangelical nationalism, with politically consequential implications for the future of America and Israel/Palestine. This hyphenation between Jews and Christians is a spatial relationship, one that poses that if the Israel can be possessed through law, colonialism, and performative definition, then so too can the credentials of truth and faith be possessed, validated, and confirmed. In other words the possession of territory equates to a possession of truth. Such broad Christian Zionist eschatological expectations are often performed with little variation where deviation from eschatologically discursive norms as to an alternative future is anathema to maintaining deep attachments to particular places and spiritual convictions. In the case of Christian Zionists, convictions are based on assuming biblical infallibility, but are also cross-referenced with historical, material, and popular evidence (Sturm 2010). Such conviction helps produce Israel, and is fed by a positive feedback loop that reinforces their particular religious national identity.

References

Agnew, J. (1994). The territorial trap: The geographical assumptions of international relations theory. *Review of International Political Economy, 1*(1), 53–80.

Agnew, J. (2003). *Geopolitics: Re-visioning world politics* (Vol. 2). New York: Routledge.

Agnew, J. (2006). Religion and geopolitics. *Geopolitics, 11*(2), 183–191.

Agnew, J. (2008). Nationalism. In J. Duncan, N. C. Johnson, & R. H. Schein (Eds.), *A companion to cultural geography* (pp. 223–238). Oxford: Blackwell.

Anderson, B. (1983). *Imagined communities: Reflections on the origin and spread of nationalism*. New York: Verso.

Anderson, B. (1996). Introduction. In G. Balakrishnan (Ed.), *Mapping the nation* (pp. 12–13). New York: Verso.

Ariel, Y. (2000). *Evangelizing the chosen people: Missions to the Jews in America, 1880–2000*. Chapel Hill: University of North Carolina Press.

Asad, T. (1999). Religion, nation-state, secularism. In P. van der Veer & H. Lehmann (Eds.), *Nation and religion: Perspectives on Europe and Asia* (pp. 178–196). Princeton: Princeton University Press.

Barkun, M. (2010). The new world order and American exceptionalism. In J. Dittmer & T. Sturm (Eds.), *Mapping the end times: American evangelical geopolitics and apocalyptic visions* (pp. 119–131). Farnham: Ashgate.

Bellah, R. N. (1976). Civil Religion in America. *Daedalus, 96*(1), 1–21.

Bercovitch, S. (1978). *The American jeremiad*. Madison: University of Wisconsin Press.

Boyarin, D. (1994). Introduction. In D. Boyarin (Ed.), *Remapping memory: The politics of timespace* (pp. vii–xiv). Minneapolis: University of Minnesota Press.

Boyer, P. (1992). *When tomorrow shall be no more*. Cambridge: Belknap.

Brubaker, R. (1992). *Citizenship and nationhood in France and Germany*. Cambridge: Harvard University Press.

Brubaker, R. (2005). The 'diaspora' diaspora. *Ethnic and Racial Studies, 28*(1), 1–19.

Brubaker, R. (2006, July 6–8). *Religion and nationalism: Four approaches*. Presented to the conference, Nation/religion, Konstanz, pp. 1–23.

Bruce, S. (1988). *The rise and fall of the Christian right: Conservative protestant politics in America 1978–1988*. Oxford: Clarendon.
Bruce, S. (2003). *Politics and religion*. Oxford: Polity Press.
Buchanan, A., & Moore, M. (2003). Introduction: The making and unmaking of boundaries. In A. Buchanan & M. Moore (Eds.), *States, nations, borders: The ethics of making boundaries* (pp. 1–18). Cambridge: Cambridge University Press.
Bühler, J. (2008). *The history of Christian Zionists: An ancient and noble tradition now flourishes. The Lion of Judah 2008 Feast of Tabernacles conference program*. Jerusalem: International Christian Embassy Jerusalem.
Butler, J. (1997). *Excitable speech: A politics of the performative*. New York: Routledge.
Calhoun, C. (2001). *Nationalism*. Minneapolis: University of Minnesota Press.
Chafets, Z. (2008). *A match made in heaven: American Jews, Christian Zionists, and one man's exploration of the weird and wonderful Judeo-Evangelical alliance*. New York: Harper Perennial.
Cherry, C. (1998). *God's new Israel: Religious interpretations of American destiny*. Chapel Hill: University of North Carolina Press.
Clark, V. (2007). *Allies for Armageddon: The rise of Christian Zionism*. New Haven: Yale University Press.
Cohen, N. W. (1979). Antisemitism in the gilded age: The Jewish view. *Jewish Social Studies, 41*(3), 187–210.
Cohn-Sherbok, D. (2000). *Messianic Judaism*. London: Continuum.
Davis, J. (1996). *The landscape of belief: Encountering the holy land in nineteenth-century American art and culture*. Princeton: Princeton University Press.
Derrida, J. (1977). Ltd., Inc. abc.... *Glyph, 2*(1), 162–254.
Diamond, S. (1996). Right-wing politics and the anti-immigration cause. *Social Justice, 23*(1), 154–168.
Dittmer, J. (2010). Obama, son of perdition? Narrative rationality and the role of the 44th president of the United States in the end-of-days. In J. Dittmer & T. Sturm (Eds.), *Mapping the end times: American evangelical geopolitics and apocalyptic visions* (pp. 73–98). Farnham: Ashgate.
Dochuk, D. (2011). *From Bible Belt to Sunbelt: Plain-folk religion, grassroots politics, and the rise of evangelical conservatism*. New York: Norton.
Ehrenhaus, P., & Owen, S. (2004). Race lynching and Christian evangelicalism: Performances of faith. *Text and Performance Quarterly, 24*(3), 276–301.
Foon, C. S. (1986). On the incompatibility of ethnic and national loyalties: Reframing the issue. *Canadian Review of Studies in Nationalism, 13*(1), 1–11.
Friedland, R. (2001). Religious nationalism and the problem of collective representation. *Annual Review of Sociology, 27*(1), 125–152.
Friedland, R., & Hecht, R. (2000). *To rule Jerusalem*. Berkeley: University of California Press.
Gellner, E. (1994). *Encounters with nationalism*. Oxford: Blackwell.
Gilroy, P. (1993). *The black Atlantic: Modernity and double consciousness*. New York: Verso.
Goldman, S. (2010). *Zeal for Zion: Christians, Jews, and the idea of the promised land*. Chapel Hill: University of North Carolina Press.
Goldman, S. (2011, January 21). What do we mean by 'Judeo-Christian'? *Religion Dispatches*. Retrieved December 15, 2011, from www.religiondispatches.org/archive/politics/3984/what_do_we_mean_by_%E2%80%98judeo-christian%E2%80%99_/
Gupta, A., & Ferguson, J. (1992). Beyond 'culture:' Space, identity, and the politics of difference. *Cultural Anthropology, 7*(1), 6–23.
Habermas, J. (1997). Kant's idea of perpetual peace, with the benefit of two-hundred years. In J. Bohman & M. Lutz-Bacmann (Eds.), *Perpetual peace essays on Kant's cosmopolitan ideal* (pp. 113–154). Cambridge: MIT Press.
Hagee, J. (2006). *Jerusalem countdown: A warning to the world*. Lake Mary: Frontline.
Halsell, G. (1986). *Prophecy and politics: The secret alliance between Israel and the U.S. Christian Right*. Chicago: Lawrence Hill Books.
Harding, S. (2000). *The book of Jerry Falwell: Fundamentalist politics and language*. Princeton: Princeton University Press.

Hartley, L. P. (2002 [1953]). *The go-between*. New York: The New York Review of Books Press.
Herb, G. H. (1999). National identity and territory. In G. H. Herb & D. H. Kaplan (Eds.), *Nested identities: Nationalism, territory, and scale* (pp. 9–30). Lanham: Rowman & Littlefield.
Herb, G. H., & Kaplan, D. H. (Eds.). (1998). *Nested identities: Nationalism, territory, and scale*. Lanham: Rowman & Littlefield.
Hitchcock, M. (2010). *The late great United States: What Bible prophecy reveals about America's last days*. Sisters: Multnomah Books.
Hobsbawm, E., & Ranger, T. (Eds.). (1983). *The invention of tradition*. Cambridge: Canto Press.
Hughes, R. T. (2003). *Myths America lives by*. Urbana: University of Illinois Press.
Huntington, S. (2003 [1996]). *The clash of civilizations and the remaking of world order*. New York: Simon and Schuster.
Jendrysik, M. S. (2008). *Modern jeremiahs: Contemporary visions of American decline*. Lanham: Lexington Books.
Juergensmeyer, M. (1993). *The new cold war? Religious nationalism confronts the secular state*. Berkeley: University of California.
Kalman, M. (2009, August 19). Huckabee's first 2012 campaign stop: Israel. *Time.com*. Retrieved December 15, 2011, from www.time.com/time/politics/article/0,8599,1917389,00.html
Kazin, M., & McCartin, J. A. (Eds.). (2006). *Americanism: New perspectives on the history of an ideal*. Chapel Hill: University of North Carolina Press.
Koenig, W. R. (2006). *Eye to eye: Facing the consequences of dividing Israel*. Phoenix: About Him.
Krishna, S. (1996). Cartographic anxiety: Mapping the body politic. In H. Alker Jr. & M. Shapiro (Eds.), *Challenging boundaries: Global flows, territorial identities* (pp. 193–215). Minneapolis: University of Minnesota Press.
Levy, A. (2010, June 28). Prodigal son: Is the wayward Republican Mike Huckabee now his party's best hope? *The New Yorker.* Last assessed 23 Sept 2014. http://www.newyorker.com/magazine/2010/06/28/prodigal-son
Lewis, B. (1987). *The Jews of Islam*. Princeton: Princeton University Press.
Lichtman, A. J. (2008). *White protestant nation: The rise of the American conservative movement*. New York: Grove.
Lienesch, M. (1993). *Redeeming America: Piety and politics in the new Christian Right*. Chapel Hill: University of North Carolina Press.
Lieven, A. (2004). *America right or wrong: An anatomy of American nationalism*. Oxford: Oxford University Press.
Lindsey, H. (1971). *The late great planet earth*. New York: Bantam.
Lizza, R. (2011, August 15). Leap of faith. *The New Yorker*. Last assessed 23 Sept 2014. http://www.newyorker.com/magazine/2011/08/15/leap-of-faith-4
Lowenthal, D. (1985). *The past is a foreign country*. Cambridge: Cambridge University Press.
Marsden, G. M. (1980). *Fundamentalism and American culture*. Oxford: Oxford University Press.
McAlister, M. (2003). Prophecy, politics, and the popular: The left behind series and Christian fundamentalism's new world order. *The South Atlantic Quarterly, 102*(4), 773–798.
McAlister, M. (2008). What is your heart for?: Affect and internationalism in the evangelical public sphere. *American Historical Literature, 20*(4), 870–895.
Mearsheimer, J. J., & Walt, S. M. (2007). *The Israel lobby and U.S. foreign policy*. New York: Farrar, Straus and Giroux.
Millard, C. (1991). *The rewriting of America's history*. Camp Hill: Christian Publications.
Mordecai, V. (2010). Fading Judeo-Christian ethic in American. *Israel Today*, p. 137.
Mullin, C. (2010). Islamist challenges to the 'liberal peace' discourse: The case of Hamas and the Israel-Palestine 'peace process.'. *Millennium: Journal of International Studies, 39*(2), 525–546.
Noll, M. (2008). *God and race in American politics. A short history*. Princeton: Princeton University Press.
Nyroos, L. (2001). Religeopolitics: Dissident geopolitics and the 'fundamentalism' of Hamas and Kach. *Geopolitics, 6*(3), 135–157.

Obinzinger, H. (1999). *American Palestine: Melville, Twain, and the holy land mania*. Princeton: Princeton University Press.
Pew Global Attitudes Project. (2006, July 6). *Muslims in Europe: Economic worries top concerns about religious and cultural identity*. Washington: Pew Research Center for the People and the Press.
Renan, E. (1996 [1882]). What is a nation? In G. Eley & R. G. Suny (Eds.), *Becoming national: A reader* (pp. 41–55). Oxford: Oxford University Press.
Rossing, B. (2004). *The rapture exposed: The message of hope in the book of revelation*. New York: Basic Books.
Rutenberg, J., Mcintire. M., & Bronner, E. (2010, July 6). Tax-exempt funds aid settlements in west bank. *New York Times*. Retrieved December 15, from www.nytimes.com/2010/07/06/world/middleeast/06settle.html
Safan, W. (1991). Diasporas in modern societies: Myth of homeland and return. *Diaspora, 1*(1), 83–99.
Sandeen, E. R. (1970). *The roots of fundamentalism: British and American millenarianism, 1800–1930*. Chicago: University of Chicago Press.
Schultz, K. M. (2011). *Tri-Faith America: How Catholics and Jews held postwar America to its protestant promise*. New York: Oxford University Press.
Schussler, F. (1985). *Book of revelation: Justice and judgment*. Philadelphia: Fortress Press.
Secor, A. (2001). Islamist politics: Anti-systemic or post-modern movements? *Geopolitics, 6*(3), 117–134.
Segev, T. (1998 [1986]). 1949: *The first Israelis*. New York: Owl Books.
Sheffer, G. (2003). *Diaspora politics: At home abroad*. Cambridge: Cambridge University Press.
Silk, M. (1984). Notes on the Judeo-Christian tradition. *American Quarterly, 36*(1), 65–85.
Smith, A. D. (2003). *Chosen peoples*. Oxford: Oxford University Press.
Spivak, G. C. (1988). Can the subaltern speak? In C. Nelson & L. Grossberg (Eds.), *Marxism and the interpretation of culture* (pp. 271–313). Chicago: University of Illinois Press.
Sprinzak, E. (1991). *The ascendance of Israel's radical right*. Oxford: Oxford University Press.
Stephanson, A. (1995). *Manifest destiny: American expansion and the empire of right*. New York: Hill and Wang.
Sturm, T. (2010). Imagining apocalyptic geopolitics: American evangelical citationality of evil others. In J. Dittmer & T. Sturm (Eds.), *Mapping the end times: American evangelical geopolitics and apocalyptic visions* (pp. 133–154). Farnham: Ashgate.
Sturm, T., & Bauch, N. (2010). Nationalism and geography: An interview with Rogers Brubaker. *Geopolitics, 15*(1), 185–196.
Vogel, L. I. (1993). *To see a promised land: Americans and the holy land in nineteenth century*. University Park: The Pennsylvania State University Press.
Weber, T. P. (2004). *On the road to Armageddon: How evangelicals became Israel's best friend*. Grand Rapids: Baker Academic.
Weizman, E. (2007). *Hollow land: Israel's architecture of occupation*. London: Verso.
Woltering, R. (2002). The roots of Islamist popularity. *Third World Quarterly, 23*(6), 1133–1143.
Zakai, A. (1992). *Exile and kingdom: History and apocalypse in the Puritan migration to America*. Cambridge: Cambridge University Press.
Zizek, S. (2006). *A glance at the archives of Islam*. Lacanian Ink. Retrieved: December 15, 2011, from www.lacan.com/zizarchives.htm

Chapter 42
Spaces of Rites and Locations of Risk: The Great Pilgrimage to Mecca

Sven Müller

42.1 Introduction

The great Islamic pilgrimage to Mecca, Saudi Arabia, named the Hajj, is one of the largest pilgrimages in the world (Wilson 1996). Besides the Islamic creed, daily prayers, almsgiving, and fasting during Ramadan, the Hajj is one of the five pillars of Islam. The pilgrimage occurs from the 8th to 12th day of Dhu al-Hijjah, the 12th and last month of the Islamic calendar. The Hajj is a religious duty: every able-bodied Muslim is obliged to make the pilgrimage to Mecca at least once in their lifetime. However, substitution may take place. That is, relatives or friends may act as substitutes for disabled persons (Haneef 2005).

Although the Hajj is an Islamic pilgrimage, the roots of the Hajj trace back to the time before the seventh century of the Christian calculation of times. Back then, each year tribes from all around the Arabian Peninsula converged on Mecca, as part of the pilgrimage. The exact faith of the tribes was not important at that time. Since the foundation of Islam by the prophet Muhammad (CE 570–632 A.D.) in the seventh century, the rites of the Hajj are generally as known today. The Hajj is considered as a devotion to Allah and it is directly associated with the life of Muhammad and the life of Abraham. Most of the rites performed by the pilgrims during the Hajj are related to important incidents of the life of Abraham. They also symbolize the solidarity of Muslims worldwide (Bianchi 2004). These rites of the Hajj are performed at certain places in the holy city of Mecca and the region east of Mecca (Fig. 42.1).

Generally, the Hajj is performed as a combination of Umrah (the small pilgrimage to Mecca that consists of rites located only at the al-Haram Mosque in Mecca) and the Hajj itself – therefore, called sometimes "the greater Hajj" (Shariati 2005).

S. Müller (✉)
Institute for Transport Economics, University of Hamburg,
Von-Melle-Park 5, D-20146 Hamburg, Germany
e-mail: sven.mueller@wiso.uni-hamburg.de

S.D. Brunn (ed.), *The Changing World Religion Map*,
DOI 10.1007/978-94-017-9376-6_42

Fig. 42.1 The Mecca region of holy sites (21°25′00.00″ N 39°49′00.00″ E) (Map by Sven Müller with aerial photo from Google Earth, September 2012)

Here, we only consider this so called "greater Hajj." During the Hajj, male pilgrims are required to dress only in the *ihram*, a garment consisting of two sheets of white cloth plus a pair of sandals. Women are simply required to maintain their *hijab* – a normal modest dress, which does not cover the hands or face. The *ihram* is meant to show equality of all pilgrims, in front of God: there is no difference between a prince and a pauper. The *ihram* is also symbolic for holy virtue and pardon from all past sins. While wearing the *ihram*, a pilgrim may not shave, clip their nails, wear perfume, swear or quarrel, have sexual relations, uproot or damage plants, kill or harm wild animals, cover the head (for men) or the face and hands (for women), marry, wear shoes over the ankles, or carry weapons (Peters 1994).

Pilgrims generally travel to the Hajj in groups, as an expression of unity. In 1950 the number of pilgrims during the Hajj was less than 100,000. That number doubled 5 years later, and since 1970 there are constantly more than 500,000 pilgrims each year. In 1983, the number of pilgrims coming from abroad exceeded one million for the first time (Salahuddin 1986). Due to the rapidly increasing numbers, in 1988 the Organization of Islamic Conference passed a resolution to specify a pilgrims' quota for each country according to its population (Metcalf 1990). Since 1995, Saudi Arabia has hosted over one million pilgrims from abroad as well as more than 500,000 from within the Kingdom. The vast majority arrive by air (through the gateway city of Jeddah), with a small percentage arriving by land or sea. The pilgrim numbers from 1995 to 2009 are given in Table 42.1. Note, the numbers given are only for the registered pilgrims. There are round about one million non-registered

Table 42.1 Registered pilgrims, 1995–2009

Christian year	Islamic year	Registered pilgrims
1995	1416	1,865,234
1996	1417	1,942,851
1997	1418	1,832,114
1998	1419	1,831,998
1999	1420	1,839,154
2000	1421	1,913,263
2001	1422	1,944,760
2002	1423	2,041,129
2003	1424	2,012,074
2004	1425	2,164,479
2005	1426	2,258,050
2006	1427	2,378,636
2007	1428	2,454,325
2008	1429	2,408,849
2009	1430	2,313,278

Data source: Ministry of Hajj, Kingdom of Saudi Arabia

Table 42.2 Casualties due to crowd disasters

Islamic Year	Christian Year	Location	Casualties
1414	1994	Jamarat Bridge	266
1417	1997	Eastern Entrance of Jamarat Plaza	22
1418	1998	Eastern Entrance of Jamarat Plaza	118
1421	2001	Jamarat Bridge	35
1424	2004	Jamarat Bridge	249
1426	2006	Eastern Entrance of Jamarat Plaza	363

Data source: http://en.wikipedia.org/wiki/incidents_during_the_Hajj

pilgrims performing the Hajj yielding a total number of more than three million pilgrims in 2009 for example (Baxter 2010; Central Department of Statistics & Information, Kingdom of Saudi Arabia 2010).

Besides health issues (infectious diseases for example) that accompany such a huge gathering of humans (Memish et al. 2012), there have taken place several severe stampedes with hundreds of casualties (BBC News 2004). As Table 42.2 shows the most dangerous ritual is the stoning of the devil at the Jamarat Bridge located in the Mena valley. Since 2006 there is an ongoing improvement concerning the infrastructure of the locations of rites (particularly the Jamarat Bridge) and the operational level in order to avoid crowd disasters by modern crowd-control measures. This contribution outlines some of these improvements.

42.2 Locations of Rites

Upon arrival in Mecca the pilgrim, now known as a Hajji, performs a series of ritual acts symbolic of the lives of Abraham and his wife Hagar (Trojanow and Morrison 2007). The acts are located at different sites in the city of Mecca and the region east of Mecca. In the subsequent sections a detailed description of the locations and the corresponding rites is given.

42.2.1 The Kaaba

In contrast to most other Arabic settlements, Mecca did not emerge from an oasis. Oases were traditional places of farming, craft and trade. Rather the importance of Mecca immediately stems from the location of the Kaaba (Watt 2008). The Kaaba, also known as the Primordial House, is a cuboid-shaped building of 15 m (49.2 ft) in height. The building relates to the arrival of Adam on earth after the fall of humankind. God sent the Kaaba to Adam as a sign of mercy. After Adam has circuited the Kaaba seven times he went to the plain of Arafat, where he met Eva again after a quest of 200 years. The Kaaba has been gone with the Deluge (Hawting 2003; Wensinck 2008).

Later, Abraham rebuilt the Kaaba together with his son Ismael as the temple for God. It is said, that the black stone located in the eastern corner of the Kaaba is an original stone from the Kaaba of God. A few meters/feet east of the Kaaba the station of Abraham is located. This is a glass and metal enclosure with what is said to be an imprint of Abraham's foot. Abraham is said to have stood on this stone during the construction of the upper parts of the Kaaba, raising Ismael on his shoulders for the uppermost parts (Armstrong 2001a).

At the time of Muhammad, his tribe was in charge of the Kaaba, which was at that time a shrine containing hundreds of idols representing Arabian tribal gods and other religious figures. Muhammad earned the enmity of his tribe by claiming the Kaaba to be dedicated to the worship of Allah alone; all other idols evicted. By then, the Kaaba was re-dedicated as an Islamic house of worship, and henceforth, the annual pilgrimage was to be a Muslim rite, the Hajj, with visits to the Kaaba and other sacred sites around Mecca (Guillaume 1955).

42.2.2 The Hills of Safa and Marwah

When wandering in the desert Hagar, Abraham's second wife and Ismael's mother, began frantic search for water for her son Ismael. It is said that this search has taken place between the two small hills Safa (round about 200 m from the Kaaba) and Marwah (round about 120 m from the Kaaba). As she searched, the Zamzam Well

Fig. 42.2 The Mena valley from east to west. The Jamarat Bridge is in the background on the *left* (Photo by Sven Müller)

was revealed to her by an angel, who hit the ground with his heel (or brushed the ground with the tip of his wing), upon which the water of the Zamzam started gushing from the ground (Levenson 2004).

Both – the Kaaba and the two hills Safa and Marwah – are nowadays located within the complex of the al-Haram Mosque in Mecca. In remembrance of Abraham and his wife Hagar, the pilgrims perform the ritual Tawaf upon arrival at the city of Mecca. Therefore, they circuit the Kaaba seven times and try to touch the black stone each time. After this the pilgrims drink water from the Zamzam Well, which is made available in coolers throughout the Mosque. Then the pilgrims walk back and forth seven times between the hills Safa and Marwah. At the end of the first day (the 8th day of Dhu al-Hijjah) the pilgrims move to the Mena valley, where they spend the night in tents (Fig. 42.2). After sunrise the pilgrims move from the Mena valley to the plain of Arafat (Shariati 2005).

42.2.3 The Plain of Arafat and Muzdalifah

According to Islamic tradition, it was on Mount Arafat (454 m height above sea level) that Adam and Eve were forgiven by God for their transgression after offering their repentance. It is also known as the Mount of Mercy. The level area surrounding the hill is called the Plain of Arafat. The hill is the place where Muhammad stood and delivered the Farewell Sermon to the Muslims who had accompanied him for the Hajj towards the end of his life (Armstrong 2001b).

Fig. 42.3 Pilgrims on Mount Arafat (Photo by Sven Müller)

A main reason of the ritual of pilgrimage is the renewal of that prayer of repentance every year standing on the hill of mercy, the climax of Hajj. The pilgrims will spend the whole day (the 9th day of Dhu al-Hijjah) on Arafat supplicating to God to forgive their sins and praying for personal strength in the future (Fig. 42.3). After sunset the pilgrims move to the area of Muzdalifah (see Fig. 42.1). In the area of Muzdalifah the pilgrims stay overnight in the open. They pray and collect pebbles. The pebbles are needed for the next ritual: the stoning of the devil (Ramy al-Jamarat) at the Jamarat Bridge located in the Mena valley (Shariati 2005).

42.2.4 The Mena Valley

God demanded Abraham to sacrifice his own son Ismael. As Abraham was up to leave the Mena valley through a defile in the west in order to do so, the Devil appeared to him at a stone heap. The angel Gabriel prompts him to pelt the devil and so Abraham threw seven stones at the devil so that he disappeared from him. The devil appeared again at the middle stone heap and Abraham pelted the devil again with seven stones. Then the devil appeared a last time at a little stone heap and Abraham pelted him again using seven stones. All three stone heaps (*jamarat*) represent the devil: the first and largest represents his temptation of Abraham against sacrificing Ismael; the second represents the temptation of Abraham's wife Hagar to induce her to stop him; the third represents his temptation of Ismael to avoid being sacrificed. He was rebuked each time, and the throwing of the stones symbolizes those rebukes. Now, as God saw that Abraham was willing to sacrifice his own son, God intervenes. God had mercy on Abraham and replacing his son Ismael with a ram, which Abraham then sacrificed (Peters 1993). Because of God's mercy Abraham and Ismael (re)build he Kaaba in honor of God.

On days 3–6 (10th day to the 13th day of Dhu al-Hijjah) of the Hajj the pilgrims perform Ramy al-Jamarat – the symbolic stoning of the devil. Therefore, the pilgrims throw seven stones at each pillar. The pilgrims stay in tents in the Mena valley for the time of this ritual. Due to capacity reasons there exists a bridge called the Jamarat Bridge to enable pilgrims to throw stones at the three jamarat pillars from either the ground level or from the bridge. The pillars extend up through three openings in the bridge (the pillars are replaced by ellipsoid walls nowadays). At the end of the first day of this ritual (10th of Dhu al-Hijjah), pilgrims slaughter a sacrificial animal (at the slaughter house). They also might buy a sacrifice ticket instead. After this ritual the pilgrims shave their heads (women are allowed to cut off a strand of hair instead). This day is the most important holiday in the Islamic calendar (Shariati 2005). After the 5th or the 6th day of the Hajj the pilgrims leave the Mena valley to Mecca in order to perform a final *Tawaf*.

42.3 Crowd Disasters and Improvements to Security

Due to the tremendous amount of people performing the rituals and the fact that some rituals have to be performed at sites which are basically not designed for so many people given time constraints (preferred stoning times) there were severe crowd disasters (particularly stampedes) in the past (see Table 42.1). Although such incidences have taken place more or less all over the place (at the al-Haram Mosque and the Moaisim Tunnel in the north of the Mena valley, for example), the most dangerous location are the Jamarat Bridge and its surroundings (Johansson et al. 2008). Here, we focus on the Jamarat Bridge. Example: the behavior of the pilgrims, waiting until noon in a vast crowd in order to perform stoning at a most preferred time, puts an enormous load on the Eastern Entrance of the Jamarat Plaza in front of the Jamarat Bridge. The width is 45 m (147 ft), the crowd can exceed several hundred thousand which exceeds the safe capacity of the entrance (Selim and Al-Rabeh 2001; Al-Haboubi 2003). In such an occurrence the critical density for humans (typically four people per sq met or four people per 10 sq ft) is exceeded (Still 2001). In order to avoid crowd disasters, numerous improvements have taken place. In the following we discuss some selected improvements.

42.3.1 Infrastructure Improvements

Obviously, the capacity of the Jamarat Bridge was not sufficient. The old bridge was a 30-year-old structure that 135,000 people were attempting to cross per hour – using a single entrance ramp that was just 45 m (147 ft) wide. The bridge had a

Fig. 42.4 The new Jamarat Bridge. On the *right* is the eastern entrance (from Mena valley) and the two access ramps to the first level. Pilgrims approaching the bridge by bus or metro enter via the long, curved ramps (second and third level) from the *upper-left corner* (western entrance). Hence, there is a one-directional flow of pilgrims on the bridge (Source: Sven Müller)

flow-rate capacity of 100,000 an hour. It was a classic bottleneck (Selim and Al-Rabeh 2001; Al-Haboubi 2003; Helbing et al. 2007). Since the crowd disaster in 2006 (see Table 42.1) the site has undergone significant changes. The Jamarat Bridge has been transformed into a five-level, 950-m-long permanent structure (Fig. 42.4) that can handle 300,000 pilgrims per hour (950 m or about 3,116 ft). There are 11 entry points and 12 exits, and further access points, including helipads, where the authorities can intervene in case of emergency. For the first time, the bridge has been designed so that pilgrims on each floor are moving in the same direction. This construction fosters a strictly one-way route system. And the exit ramps are wider than the entrance ramps, which means crowds can disperse quickly. The transformation, which cost nearly 900 million Euro or 1.1 billion USD, was completed in November, in time for the 2010 Hajj (Dar Al-Handasah 2012).

42.3.2 Video Tracking System and Pedestrian Crowd Modeling

Based on the insights from the analysis of the crowd disaster described before, new tools and measures to detect and avoid critical crowd conditions have been proposed, and some of them have been implemented in order to reduce the likelihood of similar disasters in the future. With cameras covering all the inputs and outputs to and from a given area, authorities are able to add all the flows of pilgrims and determine the difference between inflows and outflows. From this information the increase or decrease of the crowd density inside of a given area can be calculated. Further on, one could compare the pilgrim flows under each camera to the available capacity of the corresponding section. By this, the security forces are able to assess the potential risk of over-utilization of different parts of the system. This information supports their decisions to stop or redirect flows of pilgrims depending on their spatial distribution (Johansson et al. 2008; Johansson 2009).

42.3.3 Route System and Pilgrim Scheduling

In order to align the pilgrim flows a new route system (Fig. 42.5) has been established (Haase 2006): in the proximate area of the Jamarat Bridge these routes are strictly one-way for the purpose to avoid crossing flows. Moreover, barriers are established at the plaza in front of the bridge in order to avoid large-area, stationary crowds of pilgrims (Fig. 42.6). Figure 42.7 displays exemplarily the assignment of camps to a path (forward and return) on a certain route. In general, there are several feasible assignments of a camp to different paths. Note, a camp might contain several thousand pilgrims. In order to enhance the control of the pilgrim flow an innovative mathematical model has been developed (Haase 2007). The objective of this mixed integer program is constant utilization of the routes and the bridge (-levels). At the same time we regard for the preferred stoning times of the pilgrims. To achieve this task we partition all pilgrims of a given camp into groups of 250 pilgrims each. For each group the program determines a respective route and a level of the Jamarat Bridge (for example, a path) as well as a stoning time period. We consider a constant average speed of the pilgrims for the sake of resolvability of the program. We expect the highest capacity utilization at locations proximate to the Jamarat Bridge.

We employ a mesoscopic simulation model for the validation of the results of the mathematical program. The expected walking times from the camps to the bridge are of particular interest. In order to simulate the expected speed of the pilgrims, we use empirical data and the fundamental diagram of traffic flow (Johansson 2009). The densities respective to the capacities are determined for 1,253 segments of the route system. Each segment is 90 ms (or 300 ft) in length. Due to the planning we could achieve mostly constant capacity utilization (Fig. 42.8). The one-way route system and the barriers in front of the Jamarat Bridge ensure a continuous flow of pilgrims. As a result of the simulation we are able to identify potential locations of risks: The Jamarat Bridge and its surroundings as well as tunnels in the Mena valley.

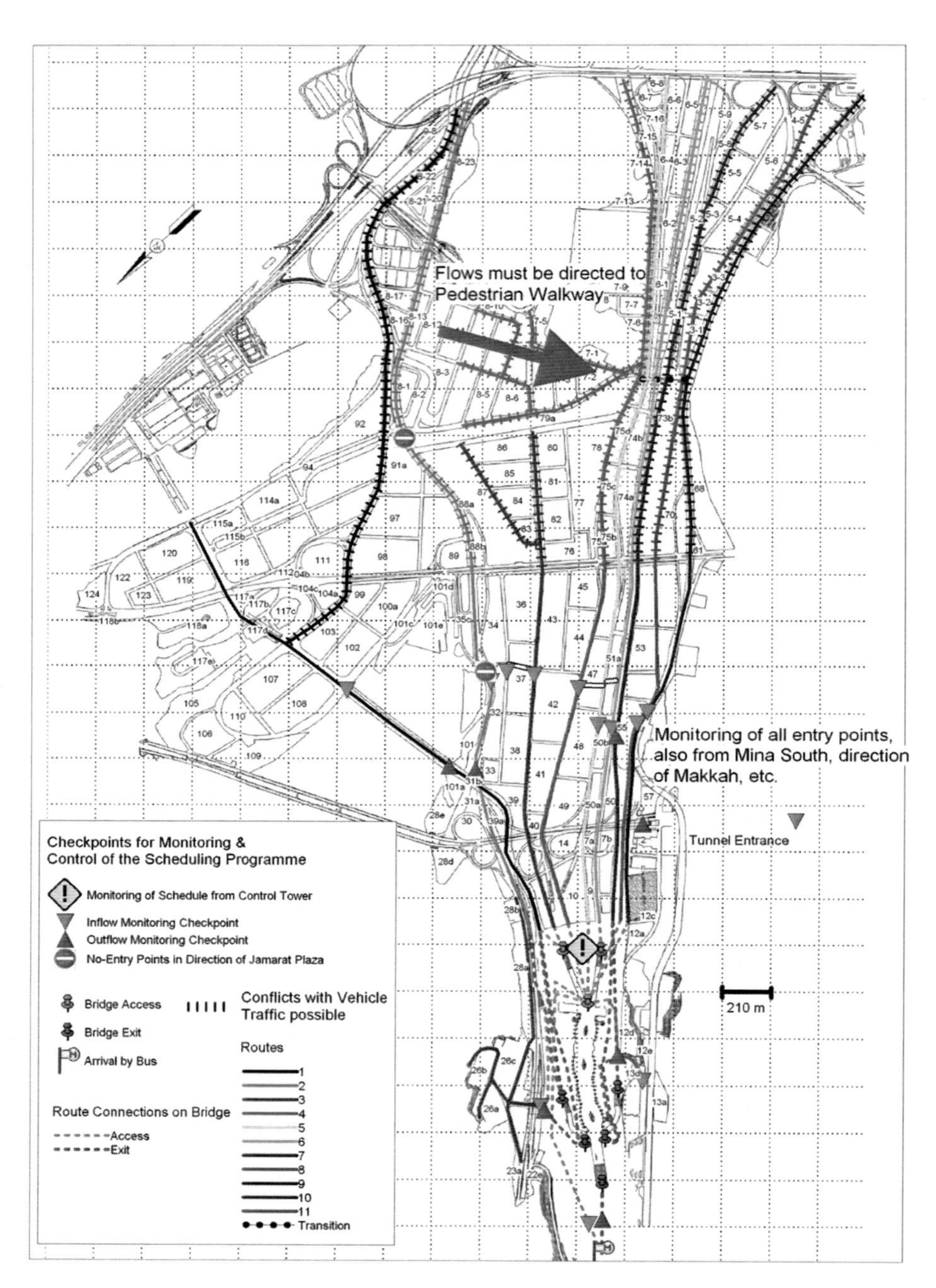

Fig. 42.5 One-way route system (Source: Sven Müller)

Fig. 42.6 Jamarat Plaza – eastern entrance. The Mena valley and tents are in the background; five access routes are in the foreground (the *left*-most ramp is only partly pictured). An egress route can be seen in the *upper-right* (Photo by Sven Müller)

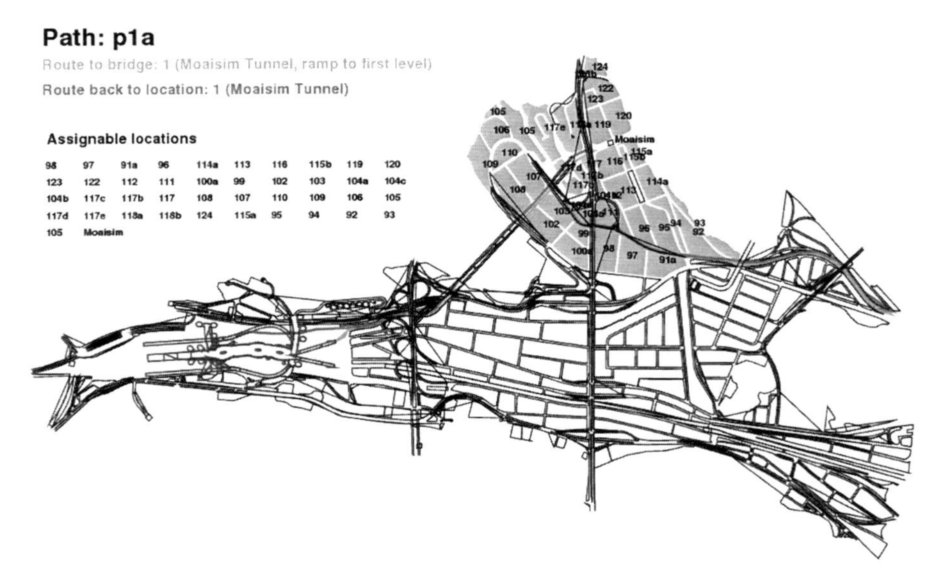

Fig. 42.7 Feasible assignments of locations (containing camps) to a path from the camp to the bridge using a specific route (access and egress) (Source: Sven Müller)

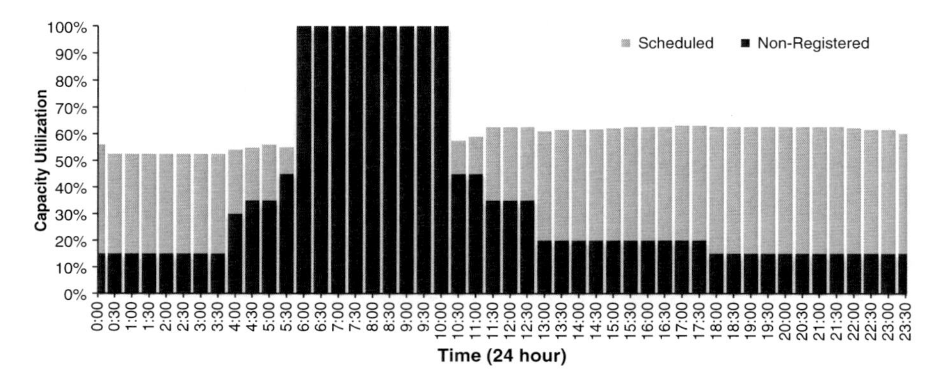

Fig. 42.8 Example of capacity utilization: Expected utilization of the first level of the Jamarat Bridge on the 10th day of Dhu al-Hijjah. During the most preferred stoning times (in the morning) no registered pilgrim is scheduled for stoning because of the uncertainty about the behavior of the unregistered pilgrims. It is assumed that most of the unregistered pilgrims favor stoning in this period (Source: Sven Müller)

42.4 Summary

Performing the Hajj is an exhaustive religious duty. Nowadays pilgrims might sometimes suffer from heat and thirst. However, this is not a life-threatening issue as it was back in the days. Today, due to the economic growth in Islamic countries and the reduced transport cost (particularly in air transport) the new life-threatening issue is the tremendous number of pilgrims. Moreover, it is expected that the number of pilgrims will further increase in the succeeding years. The coincidence of large crowds and constrained sites makes such sites dangerous locations. However, the actions so far yield a reduction of the risks: from 2006 to 2011 no major accidents or crowd disasters have taken place.

References

Al-Haboubi, M. (2003). A new layout design for the Jamarat area (Stoning the Devil). *The Arabian Journal for Science and Engineering, 28*(2), 131–142.

Armstrong, K. (2001a). *Islam: A short history*. Quezon City: Phoenix House.

Armstrong, K. (2001b). *Muhammad: A biography of the prophet*. Quezon City: Phoenix House.

Baxter, E. (2010). 15 % increase in Saudi pilgrims in 2010. *Arabian Business*. Retrieved July 19, 2010, from www.arabianbusiness.com/582928-15-increase-in-saudi-pilgrims-in-2010

BBC News. (2004). *Hundreds killed in Hajj stampede*. Retrieved September 17, 2012, from http://news.bbc.co.uk/2/hi/middle_east/3448779.stm

Bianchi, R. (2004). *Guests of God: Pilgrimage and politics in the Islamic world*. Oxford: Oxford University Press.

Central Department of Statistics & Information, Kingdom of Saudi Arabia. (2010). *The number of pilgrims for the Years From 1416H. (1995G.) to 1430H. (2009G.)*. Retrieved June 25, 2010, from www.cdsi.gov.sa/showcatalog.aspx?lid=26&cgid=1002

Dar Al-Handasah. (2012). *Jamarat Pedestrian bridge*. Retrieved September 17, 2012, from www.dargroup.com

Guillaume, A. (1955). *The life of Muhammad*. Oxford: Oxford University Press.

Haase, K. (2006). Scheduling of Hajjis Groups in Hajj 1427H. *Specialized Architectural, Engineering & Technical Reviewed Magazine issued by Ministry of Municipal & Rural Affairs. 10*, 49–61.

Haase, K. (2007). Scheduling and re-scheduling the departure and stoning times of the hajjis in 1428H. *Specialized Architectural, Engineering & Technical Reviewed Magazine issued by Ministry of Municipal & Rural Affairs, 11*, 86–92.

Haneef, S. (2005). *What everyone should know about Islam and Muslims*. New Delhi: Islamic Book Service.

Hawting, G. (2003). Kàba. In J. D. McAuliffe (Ed.), *Encyclopaedia of the Qur'an* (pp. 75a–80a). Leiden: Brill.

Helbing, D., Johansson, A., & Al-Abideen, H. (2007). The dynamics of crowed disasters: An empirical study. *Physical Review E, 75*(4), 046109.

Johansson, A. (2009). *Data-driven modeling of pedestrian crowds*. Saarbrücken: VDM Verlag Dr. Mueller.

Johansson, A., Helbing, D., Al-Abideen, H., & Al-Bosta, S. (2008). From crowd dynamics to crowd safety: A video-based analysis. *Advances in Complex Systems, 11*(4), 497–527.

Levenson, J. (2004). The conversion of Abraham to Judaism, Christianity and Islam. In H. Najman & J. Newman (Eds.), *The idea of Biblical interpretation: Essays in Honor of James L. Kugel* (pp. 3–40). Leiden: Brill.

Memish, Z., Stephens, G., Steffen, R., & Ahmed, Q. (2012). Emergence of medicine for mass gatherings: Lessons from the Hajj. *The Lancet Infectious Disease 2, 12*(1), 56–65.

Metcalf, B. (1990). The pilgrimage remembered: South Asian accounts of the Hajj. In D. F. Eickelman & J. Piscatori (Eds.), *Muslim travellers: Pilgrimage, migration, and the religious imagination* (pp. 85–107). London: Routledge.

Peters, F. (1993). *A reader on classical Islam*. Princeton: Princeton University Press.

Peters, F. (1994). *The Hajj: The Muslim pilgrimage to Mecca and the holy places*. Princeton: Princeton University Press.

Salahuddin, M. (1986). The political role of the Hajj. In Z. Khan & Y. Zaki (Eds.), *Hajj in focus* (pp. 41–54). London: Open Press.

Selim, S., & Al-Rabeh, A. (2001). On the modeling of pedestrian flow on the Jamarat Bridge. *Transportation Science, 25*(4), 257–263.

Shariati, A. (2005). *Hajj: Reflection on its rituals*. North Haledon: Islamic Publications International.

Still, K. (2001). *Crowd dynamics*. Ph. D. thesis, Department of Mathematics, University of Warwick

Trojanow, I., & Morrison, R. (2007). *Mumbai to Mecca: A pilgrimage to the holy site of Islam*. London: Haus Publishing.

Watt, W. (2008). Makka – The pre-Islamic and early Islamic periods. In P. Bearman, T. Bianquis, C. E. Bosworth, E. van Donzel, & W. P. Heinrichs (Eds.), *Encyclopaedia of Islam* (pp. 177–179). Leiden: Brill.

Wensinck, A. (2008). Kàba. In P. Bearman, T. Bianquis, C. E. Bosworth, E. van Donzel, & W. P. Heinrichs (Eds.), *Encyclopaedia of Islam* (Vol. 4, pp. 584–585). Leiden: Brill.

Wilson, C. (1996). *Atlas of holy places and sacred sites*. London: Dorling Kindersley Publishers.

Chapter 43
Finding the Real America on the TransAmerica Bicycle Trail: Landscapes and Meanings of a Contemporary Secular Pilgrimage

Thomas W. Crawford

> *Having traveled across this country many times, I was not able to feel, taste, and experience it in a deep way until I got on my bike and rode across it. (TransAmerica bicyclist 1986)*

43.1 Introduction

Since 1976, thousands of people have pedaled across the United States on the TransAmerica Bicycle Trail. The route, which can be ridden in either direction, begins at the Yorktown Victory Monument at in Virginia and ends at the mouth of the Columbia River in Astoria, Oregon. By directing the cyclist across the rural back roads of America, the Trail acts as a pathway that provides for a unique experience of the American landscape and satisfies a quest for the authentic experience of place interpreted here as the "Real America." Drawing from the pilgrimage literature, I theorize the transcontinental bicycle journey on the Trail as a form of secular pilgrimage. In doing so, I use empirical data to analyze ways in which the bicycle journey embodies several components of pilgrimage drawn from major pilgrimage theories including sacred space, rites of passage, bodies, liminality, *communitas*, and heterogeneity.

This chapter itself has had a long journey to publication whose context merits mention. During the summer of 1989 as a recent college graduate, I completed a solo crossing of the Trail. Five years later, my master's thesis "Riding into America on the TransAmerica Bicycle Trail: An Interpretation of Meaning" used survey data, interviews, and archival materials to describe and interpret the Trail experience (Crawford 1994). While the theme of the "Real America" was central to the thesis, at that time I had not yet engaged with the pilgrimage literature so that only weak

T.W. Crawford (✉)
Department of Geography, East Carolina University, Greenville, NC 27858, USA
e-mail: crawfordt@ecu.edu

S.D. Brunn (ed.), *The Changing World Religion Map*,
DOI 10.1007/978-94-017-9376-6_43

and implicit traces of pilgrimage were evident. I moved on to other topics for my dissertation and subsequent research and unfortunately never published materials from this work. The current chapter revisits the TransAmerica Bicycle Trail and makes pilgrimage explicit in an analysis of the same data used in the original thesis from 1994.

43.2 Pilgrimage Theory

During the 1990s, pilgrimage studies arguably experienced a renewed interest among geographers that was influenced by the wider adoption of critical theory perspectives (Bhardwaj 1997; Kong 2001; Collins-Kreiner 2010b). Others have reviewed and traced key developments in the pilgrimage literature so that below I only briefly sketch key theoretical ideas that are most relevant to an interpretation of the TransAmerica Bicycle Trail (hereafter the Trail) as a secular pilgrimage. In later sections, I explicitly connect these ideas to empirical data that describe the Trail and participating cyclists and provide a qualitative interpretation of cyclist narratives.

Mircae Eliade's theory of sacred space distinguishes between *sacred space* and *profane space* (Eliade 1959). Sacred space is set apart from the profane and results from adherents interpreting a historical or contemporary hierophanic irruption of the sacred or supernatural "real" at certain sites acting to make them sacred. Sacred space thereby comes to be viewed as worthy of devotion or esteem attracting the development of shrines and a circulation of pilgrims. The profane is the mundane, everyday lifeworld spaces from which pilgrims originate. Related to this idea, Victor Turner argued that pilgrimage sites are typically remote and "out there" (Turner 1973). Unlike the ordinariness of the profane, they are extraordinary and excentric to centers of population. Paradoxically, pilgrims journey to a remote "center out there" to experience in some meaningful way the centrality of the sacred that they wish to become more imprinted in their lives.

Turner also theorized pilgrimage as a ritual process involving a three-phase rite of passage (Turner 1969; Turner and Turner 1978). During the pilgrim journey and culminating at the sacred site, the pilgrim is "out of place" and in a liminal phase located between the pre and post-liminal phases associated with profane home spaces. *Liminality* is characterized as antistructural as the pilgrim is freed from norms and structures of regular everyday society. During this phase pilgrims may be more open to new experiences and may try on new identities. For groups of pilgrims, this antistructure can promote an intense sense of solidarity and social equality that Turner and Turner (1978) termed *communitas*.

Eade and Sallnow's (1991) edited volume *Contesting the Sacred* challenged these prior ideas by arguing that pilgrimage sites are sites for competing discourses regarding histories and meanings and also for competition over political and economic control of sacred space. The sacredness of a site is not simply accepted as an "irruption" with a magnetic pull attracting pilgrims. Instead of a singular grand narrative or meaning there are competing discourses from which pilgrims recover meaning. No place is intrinsically sacred; rather, pilgrim sites and circulations result out of social processes involving contestation such that pilgrimages are made and not simply revealed (Chidester and Linenthal 1995). As important as the sacred pilgrim sites themselves are the sites and contexts from which pilgrims originate.

> It is at the sites whence the pilgrims set out on their searches for the centre that pilgrims learn what they desire to find. At the centres where they go in expectation of fulfilling that desire pilgrims experience little other that than that which they already expect to encounter. (Bowman 1991: 120–121)

Other research examines the distinction between tourism and pilgrimage resulting in a blurring of the two as witnessed by terms such as "religious tourism," "tourist pilgrim," or "pilgrim tourist" (Singh 2005; Collins-Kreiner 2010a). MacCannell (1973) suggested that the tourist's quest for authenticity in travel experiences shares strong similarities with the pilgrim's quest for an authentic "center" of his/her belief system. Others have problematized the notion of authenticity. Urry (1990) developed the idea of the "tourist gaze" with respect to heritage and cultural tourism arguing that the gaze is a set of expectations for authenticity that tourists place on destination sites. This gaze is often reflected back by host communities to satisfy expectations in a commodified way that may destroy the very authenticity being sought.

Pilgrimage has been described as a phenomenon universal to nearly all major world religions (Park 1994). Given the importance of religion in U.S. history, it is perhaps ironic that formal religious pilgrimage sites are less prevalent in the U.S. compared to many other regions. Campo (1998) identifies several reasons specific to the U.S. context for this conundrum: (1) a disdain for the perceived archaic and ritualistic nature of pilgrimage, (2) an individual orientation rather than a community orientation (*communitas*) that is often central to pilgrimage, and (3) an increase in secularization. Despite this relative dearth of pilgrimage sites, pilgrimage is present in the U.S. and Campos identifies a typology of American pilgrimage landscapes: (1) *organized religion* – sites connected with formal organized religion, (2) *civil religion* – sites connected with values, symbols, practice of American civil religion, (3) *cultural religion* – sites connected with popular culture. While elements of formal religion (that is, Christianity, Judaism) may (or may not) be present with Campo's civil and cultural religion types, they share a difference from organized religion in that both are more adequately envisioned as forms of secular pilgrimage. The idea of secular pilgrimage has received much debate with "purists" pitted against those who interpret structural similarities, related to theoretical ideas discussed previously, between non-religious and religious motivated journeys (Palmer et al. 2012). Selected examples of secular pilgrimage include journeys to the Vietnam War Memorial (Dubisch 2004), Elvis Presley's Graceland gravesite (Alderman 2002), the New York City 9/11 site (Selby 2006), and New Zealand youth overseas journey experiences (Bell 2002).

Using Campo's (1998) typology of American pilgrimage landscapes, the Trail is positioned in the dual categories civil religion and cultural religion. No explicit attempt is made here to resolve the debate regarding the legitimacy of secular pilgrimage. Morinis defines pilgrimage as "a journey undertaken by a person in quest of a place or a state that he or she believes to embody a valued ideal" (Morinis 1992: 4). Adopting this definition, as I do here, clearly opens possibilities for the existence of secular pilgrimage. However, if one is to interpret a secular pilgrimage, it is important to clearly demonstrate the ways in which a particular case study aligns with or extends existing theories of pilgrimage. The remainder of this chapter attempts to do just that. First, I describe the actual route of the TransAmerica Bicycle Trail. Second, I summarize survey data results to describe characteristics of Trail cyclists and their trip behaviors. Third, I use qualitative methods to discursively analyze a set of reflective narratives handwritten by Trail cyclists to provide an interpretation of the Trail experience as a secular pilgrimage to a center termed the "Real America."

43.3 The Route

43.3.1 Founding Story

The idea for a national cross-county bicycle trail originated from two couples (Adventure Cycling Association 2012), Greg and June Siple and Dan and Lys Burden, in 1973 while engaged on an 18,000-mile bicycle trek from Alaska to Tierra Del Fuego (Burden 1973). Their idea was to organize a mass cross county bicycle tour with thousands of participating cyclists to be held in 1976 as a way to celebrate the U.S. Bicentennial with thousands of participants. To achieve this vision the couples formed the non-profit organization Bikecentennial whose official publication stated in its first issue that "Bikecentennial was created to unite the celebration of the introduction of bicycling to America and the founding of the nation" (Bikecentennial 1974: 1). Route selection was designed with cyclist safety in mind, but as important as safety was the desire for a route that allowed for an intimate experience of rural America as is evident in other Bikecentennial literature.

> … the bicycle has evolved as a means to see and experience the rapidly disappearing rural landscape. Thus cyclists are turning to the rural roads that their predecessors helped to develop, and in doing so are rediscovering a whole culture that still lives, for a time at least, at the very heart of America. (Bikecentennial 1979: 5)

During the summer of 1976, 4,000 cyclists participated in Bikecentennial's cross county tour, with approximately 2,000 cyclists, staggered into smaller, more manageable groups completing the entire route (D'Ambrosio 1986). Since its founding, Bikecentennial has continued as a non-profit organization that has developed and mapped a nationwide network of routes in addition to acting as a bicycle advocacy platform. The Trail as designed for the 1976 bicentennial celebration remains intact to this day with the exception of minor changes. In 1994 the organization changed its name from Bikecentennial to the Adventure Cycling Association.

43.3.2 TransAmerica Bicycle Trail

The Trail is approximately 4,000 miles (6,400 km) from Yorktown, Virginia to Astoria, Oregon and follows federal, state, and local roads (Fig. 43.1). It is not a specially constructed bike path and instead is dependent on existing roads through mainly rural areas most of which are lightly traveled by motor vehicles. To highlight the rural, small town nature of the route, GIS analysis using Census 2000 data identified 449 census-defined populated places located within 10 mi (16 km) of the route and summarized the distribution of place populations: 50th percentile (median) = 921, 75th percentile = 2,858, 90th percentile = 9,429. Only four places had a population over 100,000: Richmond, Virginia (197,790), Eugene, Oregon (137,893), Salem, Oregon (136,924), and Pueblo, Colorado (102,121). Arrayed along the route are several places of historic and environmental interest some of which are listed in Table 43.1.

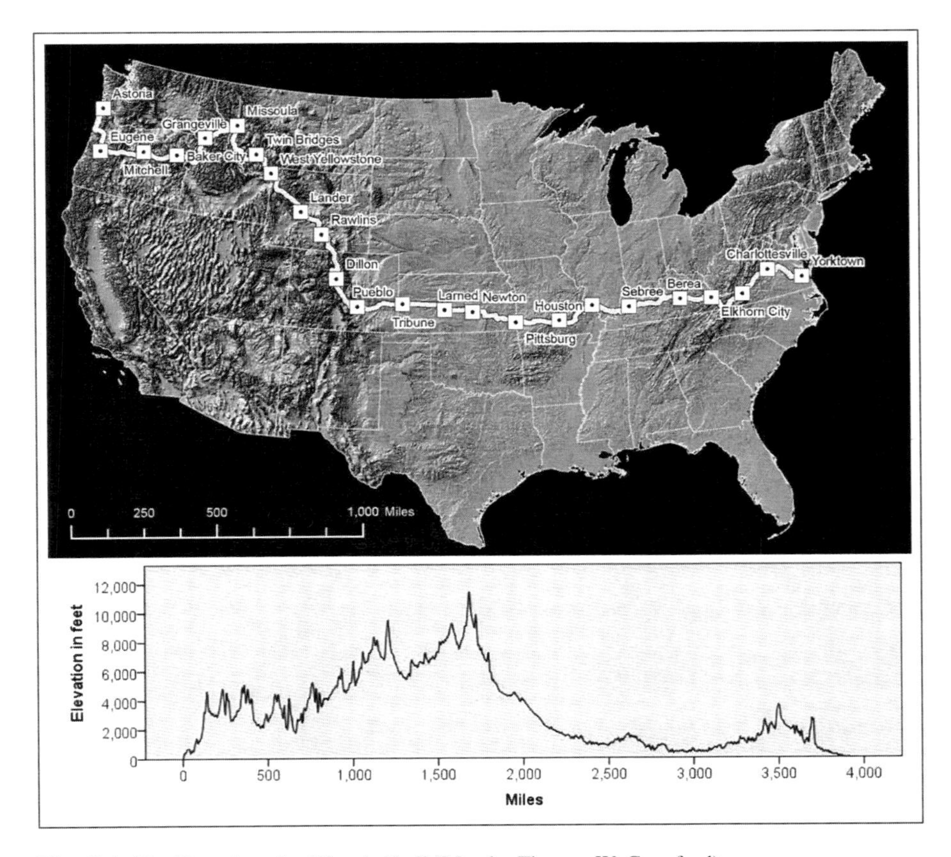

Fig. 43.1 The TransAmerica Bicycle Trail (Map by Thomas W. Crawford)

Table 43.1 Selected places on the TransAmerica bicycle trail

State	Place
Virginia	Yorktown
	Yorktown Victory Monument, Trail terminus
	Colonial Williamsburg
	Colonial capital of Virginia
	Monticello
	Home site of Thomas Jefferson
Kentucky	Abraham Lincoln Birthplace
	Replica of log cabin birthplace
Illinois	Mississippi River
	Bridge crossing at Chester, Illinois
Missouri	Ozark Mountains
	Ozark National Scenic Riverways national park
Kansas	Flint Hills
	Largest intact tallgrass prairie ecoregion
	Fort Larned National Historic Site
	Santa Fe Trail
Colorado	Royal Gorge
	"Grand Canyon" of the Arkansas River
	Hoosier Pass
	Trail's high point (11,542 ft.) at Continental Divide
Wyoming	Grand Teton National Park
	Yellowstone National Park
Montana	Big Hole National Battlefield
	Major battle of the Nez Perce War
Montana – Idaho	Lewis and Clark Trail
	Route of Lewis and Clark Expedition 1804–1806
Idaho – Oregon	Hells Canyon
	Snake River, North America's deepest river gorge
Oregon	McKenzie Pass
	First or last major pass of Trail, Cascade Range
	Astoria
	Mouth of Columbia River, Trail terminus

Source: Thomas W. Crawford

43.4 Survey Results

A survey was deployed during the summer of 1994 at three strategic locations along the route known to be frequented by Trail cyclists. The survey was formatted in a way to facilitate easy return through the mail. A total of 53 usable surveys were received. The sample of 53 was not randomly generated so that no claims of statistical representativeness can be made; rather, it was a sample of convenience. Based on an estimate that 250–500 cyclists completed the Trail during the time period of the survey, it is likely that the sample of 53 included between 10 and 20 % of cyclists who rode the Trail during 1994. International cyclists, largely European, travel the Trail each year. Survey results and subsequent qualitative analysis focus solely on domestic U.S. cyclists.

Demographically, Trail cyclists were predominately white, male and from urban or suburban settings (Table 43.2). GIS analysis revealed that approximately 75 % of cyclists lived within 30 miles of the largest 125 U.S. cities. Many were in their 20s, although almost one third of the sample was at least 40 years old. Cyclists had high education levels with 55 % holding a college or post-graduate degree. Regarding trip behaviors, most cyclists rode between 50 and 80 miles per day and were split evenly in terms of direction. A strong majority tent camped most nights and prepared their own meals. This form of "loaded touring" typically does not rely on supporting vehicles as cyclists load and transport belongings in panniers attached to racks on the bicycle frame (Fig. 43.2). Total trip expenses varied considerably. The greatest number of cyclists rode solo, although there were many who rode in pairs or larger groups.

Table 43.2 Demographic characteristics of 53 Trail cyclists

Residence	%	Marital status	%
Rural	7.5	Single	54.7
Suburban	45.3	Married	37.7
Urban	41.5	Divorced	7.6
Other	5.7		
Gender		Education	
Male	77.3	<High school	9.4
Female	22.7	High school	7.6
Race		Some college	26.4
Caucasian	96.2	College degree	39.6
Black	0.0	Graduate degree	15.1
Other	3.8	No response	1.9
Age			
15–19	1.9	Household income[a]	
20–29	41.5	<$31,000	24.5
30–39	24.5	$31,000–$61,999	33.9
40–49	17.0	$62,000–$92,999	17.0
50–59	9.4	>$93,000	18.9
60+	5.7	No response	5.7

Source: Thomas W. Crawford

[a]Shown in equivalent 2013 dollars

Fig. 43.2 Loaded bicycle tourist-pilgrim east of Lander, Wyoming (Photo by Thomas W. Crawford)

43.5 Qualitative Analysis of Cyclist Narratives

The analogy of a bicycle wheel (Fig. 43.3) is used here to provide a framework to help interpret significant themes evident in cyclist narratives. The wheel is comprised of a tire, rim, supporting spokes, and a central hub connecting the various elements. The tire affixed to the metal rim represents the cyclists contact with the external landscape, literally "where the rubber meets the road." The spokes, which connect the tire/rim to the hub act to strengthen and hold the wheel together. They are viewed here as a series of themes that connect cyclists' experiences of the Trail's landscapes to the central hub which represents the broader meaning and significance of the journey. Qualitative methodology involved careful analysis of over 100 cyclist narrative essays written from 1982 to 1993. These narratives were obtained and photocopied at the Missoula, Montana offices of the Adventure Cycling Association in 1994. Many cyclists visit the ACA offices as part of their journey, typically signing a logbook and chatting with ACA staff. Since the early 1980s, ACA staff members have distributed a form document in which cyclists were requested to write narratives about their journeys. Many completed the document on site. Others completed at journey's end and returned by mail. All narrative essays were coded and themes that were repeated by many cyclists emerged as important

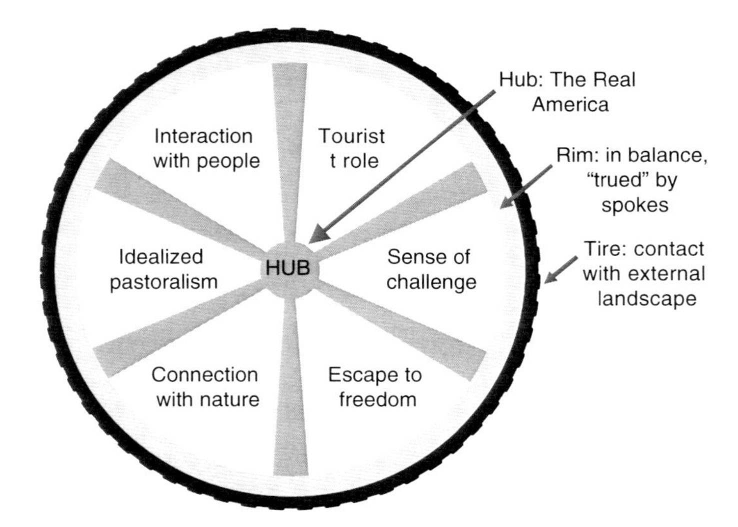

Fig. 43.3 Interpretative framework for the Trail as a secular pilgrimage to the "Real America" (Source: Thomas W. Crawford)

"spokes" interpreted as significant themes of the overall Trail experience. ***Selected quotes are cited in a format such that "86.7" indicates the year (1986) and the number of the individual cyclist's narrative from that year (7).***

43.5.1 Tourist Role

Pilgrimage scholar Victor Turner stated, "a tourist is half a pilgrim and if a pilgrim is half a tourist" (Turner and Turner 1978: 20). If the Trail journey is to be considered a secular pilgrimage, it is important to recognize that participants have dual identities as both touring cyclist and pilgrim. Lowenthal (1968) described how tourists, as *outsiders*, are more concerned with form; natives, as *insiders*, are more concerned with function. Trail cyclists are outsiders and undoubtedly are drawn to a curiosity and appreciation of aesthetic qualities of landscapes more so than permanent residents of these same landscapes who, as insiders, have stronger concerns for the landscape's practicality utility. Relph introduced the term *empathetic insideness* as a condition resulting from being physically in a place that demands "a willingness to be open to significances of a place, to feel it, to know and respects its symbols (Relph 1976: 54). Cyclist narratives revealed in many cases an attitude of empathetic insideness as they pedal through the Trail's largely rural places. Yet, it seems inappropriate to describe these tourist-pilgrims in terms of empathetic insideness since they are predominantly suburban or urban outsiders who are just passing through, albeit slowly. Extending Relph's concept to the term *empathetic outsideness* perhaps better reflects the tourist role of many Trail cyclists. Empathetic outsideness connotes an appreciation of the unique qualities and significances of places while simultaneously indicating that the cyclists are outsiders with no permanent role as inhabitants of places along the Trail.

Boorstin (1964) claimed that tourists encounter largely artificial or inauthentic settings that create images which satisfy the tourist's yearning for aesthetic satisfaction. Others have distinguished between landscapes of consumption versus landscape of production (Ateljevic 2002). Historical village tourism sites or Disneyland are examples of this form of inauthenticity. A notable feature of the Trail experience is that the vast majority of the route's landscapes and places are authentic in the sense that their forms and functions are not explicitly directed towards outsiders. There are of course exceptions such as historic Williamsburg or Colorado's Royal Gorge which is a natural landscape feature around which a theme park-style tourist venue has grown. The Trail passes through predominantly ordinary vernacular landscapes. It was designed in the mid-1970s for this very purpose so that the farmer in the field, the small town Main Street or courthouse square, and the impoverished homes of eastern Kentucky are authentic and not created for consumption by passing cyclists. They exist independent of the Trail.

Similar to Bowman's (1991) aforementioned observation regarding pilgrims' expectations, many Trail cyclists had clear and lofty expectations for their journeys. The trip requires substantial amounts of planning, money, and physical effort so that an expected return on this investment is, at a minimum, some sort of positive and memorable experience. For many, the trip had significance much deeper than simply going on vacation and fulfilled a long held dream to traverse the continent by bicycle. Essays revealed multiple cases expressing a desire for a "grand adventure" or an "experience of a lifetime" and an opportunity to encounter the "Real America" to be found on the back roads of rural American.

> We were about to embark after many weeks of preparation on the cross-county bike trip that would fulfill a ten-year dream of mine. (87.1)
>
> A retirement party for me. I had planned this for years … the trip across America is something I had wanted to do for a long time. (87.4)

43.5.2 Sense of Challenge and Personal Achievement

Many cyclists derived a strong sense of personal achievement from overcoming the challenges, both physical and mental, of the journey. Similar to Turner's (1969) ritualistic rite of passage, for many it was a personal test whose passage created a sense of self-actualization. One participant stated, "I wanted to set a difficult goal and achieve it … the process I knew would bring some good things from which I would learn and grow as an individual" (92.1). This and similar expressions suggest the tour as a pilgrimage involving a quest for *existential authenticity* (Steiner and Reisinger 2006), an encounter with one's authentic self that emerges through the trials of the self-propelled long distance journey. Many symbolized their achievement through a ritualistic "dipping of the tires" (Fig. 43.4). This "baptismal" rite involves

Fig. 43.4 Tire dipping at Yorktown, Virginia (Photo by Terry Smith, used with permission)

starting the journey with the bicycle's rear wheel partially submerged in an initiating water body. At journey's end, the front wheel is dipped in a terminating water body. Survey results from 1994 revealed that 43 % of the sample performed an initial dipping. Presumably they also performed a dipping at journey's end. Note that the eastern terminus at Yorktown, Virginia is situated on the York River in which many dipped their tires. Others desiring an ocean-to-ocean ritual extend their journey to Virginia Beach in order to access the Atlantic Ocean.

Contrary to those not familiar with long distance cycling, riding the Trail does not require extraordinary physical strength or ability, although it does involve some form of hardship and at times suffering (Table 43.3). In addition to the planning and financial resources, it requires determination and resolve to persevere through difficult situations such as stifling heat, strong headwinds, rain, equipment problems, pain and soreness, and at times mental anguish and boredom. Many cyclists commented on the hardships involved, without which the journey would be less meaningful.

> All adventures are marked by times of pain, misery, and discomfort. However, it is the overcoming of such conditions that makes bicycle touring so rewarding and makes the truly great days so memorable. (93.2)
>
> ... the discovery of untapped inner resources, the deep feelings of homesickness, the steepness of the Ozark hills, the six hour climb up McKenzie Pass in Oregon (it rained too) and the sense of satisfaction as we came home, our final day ... and we did it a pair of friends ... made a dream come true. (82.1)

Table 43.3 Trip characteristics of 53 Trail cyclists

Daily mileage	%	Lodging	
<50	0.0	Tent camping	88.7
50–69	68.0	Motel/hotel	7.5
70–89	30.1	Other	3.8
90–100	1.9	Meals	
Direction		Prepared own food	66.0
East-to-West	53.0	Restaurant, cafe, diner	34.0
West-to-East	47.0	Total expenses[a]	
Size of group		<$3,000	26.4
One (solo)	37.7	$3,000–$5,999	41.5
Two	28.3	$6,000–$9,999	30.2
Three	7.6	>$10,000	1.9
Four or more	26.4		

Source: Thomas W. Crawford
[a]Shown in equivalent 2013 dollars

Suffering as an aspect of the Trail, while inspired by different motives, shares commonalities with suffering endured by traditional religious pilgrims who may intentionally undertake physical suffering as an expression of piety, devotion, or penitence; for example, pilgrims to Marian shrines who may crawl on their knees for great distances approaching the sacred site. Suffering also engenders Turner's condition of *communitas* due to hardships that all Trail cyclists share in common often trading "war stories" among each other.

43.5.3 Escape to Freedom

For many Americans, the road trip is a particularly alluring expression of personal freedom. One author describes this allure in the following manner:

> Traveling across the country is an American dream. To climb aboard a bus or a train, or to hop onto a motorcycle or slip into a car and go, just go, is a basic romantic impulse. Movement is freedom, and to travel the breadth of the country is to proclaim that freedom in a bold and ebullient way. (O'neil 1980: 8–9)

The Trail journey is similar to other forms of vacation in that it provides a temporary respite from the burdens and rhythms of everyday life, an escape to freedom. However, riding the Trail is different from more common forms of vacation due to the unique sense of autonomy and sensory experience provided by the bicycle.

> The senses are exposed to so much more on a bicycle than in a car. The sensation of speed, visual and aural sensations as well as the feelings of one's own heart, lungs and legs working together as an incredibly efficient machine … the 'under you own power' independent philosophy …. (93.5)
>
> And traveling 5 to 20 mph affords time to savour, to think about and meditate on the smallest weed reaching out from the crack in the pavement, and the fenceless gate

creaking in front of an old abandoned church, the sound of birds fluttering by, and the fragrance of a freshly plowed field. (91.3)

Several cyclists expressed motives consistent with Turner's theme of liminality involving a temporary removal from the profane spaces and structures of everyday life.

I got the 'bug' to get away from it all in early '88 after a second failed marriage, after being fed-up with an all-consuming job and after reading a very good book about taking sabbaticals. (88.14)

Truly a 'go-work-go' lifestyle. Decided to change perspectives on life. Wanted to retreat to a more relaxed way of life so decided to leave my job and see America. (88.8)

The room where I worked was temperature controlled all year. There were many rules and regulations, spoken and unspoken, affecting behavior, work style and attitudes. Touring removes all those barriers, the climate is definitely not controlled and I am no longer in a hierarchical social structure. (93.6)

43.5.4 Connection with Nature

Connecting with nature was another important theme for many cyclists. This connection results partly from the openness and heightened sensory perception inherent to bicycle travel. The route also exposes cyclists to landscapes of great natural beauty including iconic national parks, national forests, and national wildlife and refuge areas.

I enjoyed seeing Yellowstone Park and the areas in Montana following Yellowstone the most. I don't believe I ever have seen anything that beats the natural beauty of that country. (82.3)

The naturally slow pace and generally quiet are that go with cycling make it a perfect medium in which to watch wildlife. How thrilling it is to round a bend and come face to face with a browsing deer, elk, or moose. (85.1)

For many cyclists, the connection with nature goes deeper than viewing wildlife and beautiful scenery. For many it induces a feeling of liminality and simplicity which can be described as a "back to nature" attitude. The daily rhythm of rising with the sun, cycling across terrain relief, and sleeping outdoors helps some cyclists to "appreciate the important things in life and to live by nature's clock, not man's." (93.3)

Somehow when I'm riding, and particularly alone, and for more than four days, the complexities of my life fade away. My needs simplify. Being somewhat removed from the modern technology which surrounds our lives, more at the mercy of the land, I find bike touring refreshing and exhilarating. It seems to turn life upside down, returning it to its more natural state. (90.4)

43.5.5 Idealized Pastoralism

Idealized pastoralism is among the most significant themes of the Trail journey. *Pastoral* can be defined as having to do with shepherding or farming, as being rustic, as being peaceful or simple. No single definition of the term is promoted here, yet I argue that at core it resonates with the theme of rurality embodied along much of the route and that the route founders intentionally designed for the original Bikecentennial celebration in 1976. According to Sears, tourism "demands a body of images and descriptions of those places – a mythology of unusual things to see – to excite people's imagination and induce them to travel" (Sears 1989: 3). To the suburban and urban oriented cyclists who make up a vast majority of participants, pastoral America is both unusual and mythical. The Trail's small towns, farms, and natural landscapes act as sacred spaces that are far removed from the profane spaces of most Trail cyclists. One cyclist noted, "Kansas had always been there, at least in my mind, a mythical place; an MGM creation called Oz, the geographic and psychological center of America" (Eddings 1986: 21). This and other sentiments resonate with Turner's notion of pilgrimage destinations being remote centers "out there." In order to find the "center" of America, many cyclists felt they had to pedal America through the country's rural back roads.

Strong images and memories of farming landscapes were evident in many cyclist essays. A cyclist from suburban Washington commented "I'll never forget the wheat harvest of Kansas" (93.1). Another cyclist stated:

> I was mainlining mid-America, with its farms, meadows, forests … my highs came from the mundane: a red tractor in a green field … a dog sleeping on a porch, the sound of grain rushing up an elevator. (Glickman 1992: 15)

Jakle (1982) identified the "myth of the small town" as a view held by many Americans. This view posits that many Americans envision a national identity rooted in a "Main Street" America of an idealized past. Essay materials suggest that a major part of the Trail's significance appears to be a validation of this myth.

> … the whole trip is about Americana, small town lifestyles, food, you name it. (94.17)
>
> Having been raised in Los Angeles all my life gave me such a narrow view of what the rest of the country was like. Having been there has given me a deep and lasting appreciation of life in rural America. I should never forget the small towns of America that I visited. (86.17)
>
> All across the country we consistently saw the warmth and goodness of small town America. (91.3)
>
> Small town America is still out there, don't dream it, do it! (90.3)

43.5.6 Interaction with People

Commenting on the original 1976 tour, a Bikecentennial staff member noted that "for many, Bikecentennial had turned into a grand pilgrimage based on an interchange between rider and resident (Burden 1976: 1). This and similar accounts from 1976 indicate a great outpouring of interest and assistance on the part of many local communities along the route that allowed groups to sleep in fields, schools, and churches. Over the years, communities are now accustomed to Trail cyclists so that levels of cyclist-resident interactions has probably diminished or at least are not so novel from the hosts' perspectives. Nonetheless, several cyclists reported local interactions as among the most meaningful aspects of their journeys.

> I wanted to meet the everyday ordinary people of this country and be able to put my trust in them. What I found was that time and time again you could open up yourself and trust the people across this country to help you when you most needed it. (89.2)
>
> Most importantly, it restored my faith not only in America, but in people and their ability to lend a hand when times were tough and just as often when times were good. People offered their yards for us to camp on, their water for use to clean and cook with, their bathrooms, food, shade, etc. (82.4)

In addition to the majority of references to positive impressions and interactions with local communities, there were a small number negative comments such as one from a cyclist from a large eastern metropolitan area who stated "most of the people are in the sticks, racists, ignorant, low-tech, backwards ... but they are friendly as hell" (94.22). However, the dominant impression from the cyclist narratives can be summarized as "... most people in this country are down-home, friendly and genuinely giving people." (94.20)

Communitas is evident in different ways with this particular theme of interaction with people. As expressed by cyclists above, the help and assistance provided by locals provided an increased "faith" and "trust" in the sense of goodness and community at host sites along the route. This can be interpreted as a form of *communitas* yet one that is different from the traditional interpretation of solidarity among fellow pilgrim travelers. Cyclists traveling in groups experience this more traditional form of intra-pilgrim *communitas* in various ways. Their common suffering as mentioned previously is one way. Mechanical assistance or words of encouragement and motivation are other ways. Drafting by riding in single file as a group with cyclists taking alternate turns at front to minimize wind drag is a quite literal expression of solidarity. Perhaps most poignant is the preparation and sharing of a common meal at the end of a long day's ride.

43.6 Pilgrimage to the Real America

Returning to the bicycle wheel as analogy and interpretive framework, the six themes described above act as spokes that strengthen and hold the wheel together centered on the central hub. Due to inherent diversity among participating cyclists, individual themes will necessarily vary in strength depending on the individual cyclist. Figuratively, those spokes with more tension represent themes of meaning that are more important to an individual. There are, therefore, many wheels, one for each participating cyclist and each with his/her unique set of spoke tensions. Thus, this framework explicitly recognizes the heterogeneity of cyclists' motivations, experiences, and meanings; a heterogeneity that is consistent with contemporary theories of pilgrimage (Eade and Sallnow 1991). For virtually all Trail cyclists, these themes will be present at varying degrees and contribute to the overall meaning and significance of the individual tour experience. The theme of contestation in the sense of inter-group conflict that figures prominently in pilgrimage scholarship was not evident in the narratives analyzed. This is likely due to the fact that no single institution or set of competing institutions exercise ownership and control over the multitude of places comprising the Trail.

To synthesize and distill results from the array of data involved into a broader interpretation of the TransAmerica Bicycle Trail is challenging given the role of heterogeneity. This broader interpretation is located at the hub of the wheel. Significantly, four of the six themes; escape to freedom, idealized pastoralism, connection with nature and interaction with people, all have strong connections to the theme of pastoralism. The strong majority of cyclists escape from urban or suburban environments to a state of liminality and the freedom of pastoral landscapes. By idealizing pastoralism, many cyclists are celebrating or at least encountering mythologized American pastoral landscapes, a valued ideal. Cyclists are also connecting with nature, another component of pastoralism. These pastoral and natural landscapes serve as sacralized space for the Trail's cyclist-pilgrims. Cyclists report the importance of interactions with local people and communities from pastoral places along the route. The remaining two themes, the role of the tourist and the sense of challenge, serve to emotionally charge the Trail experience and the potential for a rich and meaningful personal experience involving an encounter with self and an encounter with the "Real America."

Beyond an emotionally charged celebration of or encounter with pastoral landscapes of rural America, the central hub is argued to represent a secular pilgrimage to a place termed here the "Real America." Cyclist-pilgrims embarking on the TransAmerica Bicycle Trail are on a journey seeking to find this "Real America." Congruent with Morinis' definition of pilgrimage (Morinis 1992), they are on a quest for a place that they believe embodies a valued ideal. Given the continental scale of the journey, it is perhaps ironic to identify quest for place as a unifying central hub. Place is often associated with smaller spatial scales such that "place is a compelling focus of a field, it is a small world, the node at which activities converge" (Tuan 1974: 236). The Trail, paradoxically, is large in linear length, but Trail cyclists experience it through smaller increments of 50–100 mile daily rides which are punctuated by nodal stops arrayed along the breadth of the route. Some cyclists commented that the transcontinental journey in fact made America seem smaller to them and more "mine."

> The U.S. is much smaller now. I can conceive the breadth of the continent now, and the makes America more 'mine'! (83.1)

Significantly, narratives often explicitly focused on the idea of the "Real America" commenting on how the Trail enabled them to see this "real" America.

> The best part about touring is that it takes you away from the armchair and the books, away from the couch and the TV and it allows you to see America as it really is. (85.3)
>
> … the freedom and adventure to roam about from area to area and discover the America I had read about since grade school by wanted to discover first hand. (86.17)
>
> We wanted to see the Real America. (88.3)
>
> America is really out there: purple mountains, amber waves of grain and all … (91.2)

For the vast majority of Trail cyclist-pilgrims, the "Real America" they encounter is a very different place from their home settings and the actual America of the contemporary suburban and urban places from which they come. The "Real America" that they find on their journeys and that lives on in their memories is a collage of images consisting of a collection of idealized American icons such as small towns, Main Street, working farms, and nature's beauty, and encounters with local residents and fellow cyclists. Many narratives indicated an awareness that the reality of contemporary American places and societies differ greatly from the "Real America" found on the Trail. The passion and richness with which they describe their experiences shows them to cling fondly to memories and images that collectively form the "Real America" which is an idealized sense of place that many seek and find through various themes of meaning during their cross country bicycle pilgrimages.

43.7 Conclusion

Similar to traditional religious pilgrimage, bicycling the TransAmerica Bicycle Trail has the potential to be a transformative experience impacting the lives of pilgrims as they exit pilgrimage's liminal phase and enter the post-liminal phase upon returning to everyday profane spaces at journey's end. Several years after bicycling the Trail in 1986, one TransAmerica cyclist reflected on his journey in his online journal *Uphill and Against the Wind* (Tuzinowski 2009):

> All the memories and events leading me to ride the Trans America Trail came flooding back. I started to realize that the Trans America Trail has been part of my life for 26 years, either consciously or unconsciously... I made many decisions about life in those 90 days in 1986 ... I realized the Trans America Trail was about the journey, not the destination. The trip was not really about the bike at all, or even the goal to ride coast to coast, but about everyone and everything in-between. The trip was really about the experience, the people, the friends, and the memories along the way ... I know that reliving the journey all over again 20 years later pulled forward the fun times, and dulled the physical and mental agonies of the road. I did experience the highs and lows all over again in my mind, some days I was bored to tears and didn't want to write, some days I was so excited I couldn't wait to get to it. I experienced the whole gamut of emotions, including the anguish of the end of the road. The Trans America Trail was a life altering experience for me, and it can't ever be changed, erased or forgotten. If anyone has the opportunity to ride the Trans America Trail I highly recommend it, but remember, sometimes it's not all fun and games. (Dave Tuzinowski 2009)

Acknowledgements Thank you to Wil Gesler, Derek Alderman and students of the Spring 2011 ECU Honors College Seminar "The Power of Place." Greg Siple of the Adventure Cycling Association provided important assistance. A GIS layer of the Trail's route was graciously provided by Nathan Taylor and Carla Majernik of the Adventure Cycling Association. GIS support was provided by ECU's Center for Geographic Information Science. Tire dipping photos by Peter McLaren and a contributor wishing to remain anonymous were used with permission.

References

Adventure Cycling Association. (2012). *Our history in brief.* www.adventurecycling.org/whoweare/history.cfm. Last accessed 11 June 2012.

Alderman, D. H. (2002). Writing on the Graceland wall: On the importance of authorship in pilgrimage landscapes. *Tourism Recreation Research, 27*(2), 27–33.

Ateljevic, I. (2002). Circuits of tourism: Stepping beyond the 'production/consumption' dichotomy. *Tourism Geographies, 2*(4), 369–388.
Bell, C. (2002). The big 'OE:' Young New Zealand travellers as secular pilgrims. *Tourist Studies, 2*(2), 143–158.
Bhardwaj, S. M. (1997). Geography and pilgrimage: A review. In R. H. Stoddard & A. Morinis (Eds.), *Sacred places, sacred spaces. The geography of pilgrimages* (pp. 1–23). Baton Rouge: Geoscience Publications, Louisiana State University.
Bikecentennial. (1974, May). *BikeReport*. Missoula: Bikecentennial.
Bikecentennial. (1979, January–February). *BikeReport*. Missoula: Bikecentennial.
Boorstin, D. (1964). *The image: A guide to pseudo-events*. New York: Harper Colophon.
Bowman, G. (1991). Christian ideology and the image of a holy land: The place of Jerusalem in the various Christianities. In J. Eade & M. J. Sallnow (Eds.), *Contesting the sacred, the anthropology of Christian pilgrimage* (pp. 98–121). London: Routledge.
Burden, D. (1973). Bikepacking across Alaska and Canada. *National Geographic, 143*(5), 682–695.
Burden, D. (1976). We've pulled it off! *BikeReport*, November edition.
Campo, J. E. (1998). American pilgrimage landscapes. *Annals of the American Academy of Political and Social Science, 558*(1), 40–56.
Chidester, D., & Linenthal, E. T. (1995). Introduction. In D. Chidester & E. T. Linenthal (Eds.), *American sacred space* (pp. 1–42). Bloomington: Indiana University Press.
Collins-Kreiner, N. (2010a). The geography of pilgrimage and tourism: Transformations and implications or applied geography. *Applied Geography, 30*, 153–164.
Collins-Kreiner, N. (2010b). Geographers and pilgrimages: Changing concepts in pilgrimage tourism research. *Tijdschrift voor Economische en Sociale Geographie, 101*(4), 437–448.
Crawford, T. W. (1994). *Riding into America on the TransAmerica bicycle trail: An interpretation of meaning*. Unpublished master's thesis, Department of Geography, UNC-Chapel Hill.
D'Ambrosio, D. (1986). A summer to remember. *BikeReport* (a newspaper), June edition (pp. 11–15). Available online at http://www.adventurecycling.org/default/assets/resources/198606%5FASummerToRemember%5FD%27Ambrosio%2Epdf
Dubisch, J. (2004). Heartland of America: Memory, motion and the (re)construction of history on a motorcycle pilgrimage. In S. Coleman & J. Eade (Eds.), *Reframing pilgrimage. Cultures in motion* (pp. 105–132). London: Routledge.
Eade, J., & Sallnow, M. J. (Eds.). (1991). *Contesting the sacred: The anthropology of Christian pilgrimage*. London: Routledge.
Eddings, K. (1986). A last fling on the TransAmerica trail (part one). *BikeReport*, March edition, pp. 8–9.
Eliade, M. (1959). *The sacred and the profane*. New York: Harcourt, Brace and World.
Glickman, J. (1992). A string of roads. *BikeReport*, June edition, pp. 14–15.
Jakle, J. (1982). *The American small town. Twentieth-century place images*. Hamden: Archon Books.
Kong, L. (2001). Mapping 'new' geographies of religion, politics and poetics in modernity. *Progress in Human Geography, 25*(2), 211–233.
Lowenthal, D. (1968). The American scene. *Geographical Review, 58*, 61–88.
MacCannell, D. (1973). Staged authenticity: Arrangements of social space in tourist settings. *American Sociological Review, 79*, 589–603.
Morinis, A. (1992). Introduction. In A. Morinis (Ed.), *Sacred journeys. The anthropology of pilgrimage* (pp. 1–28). Westport: Greenwood Press.
O'neil, T. (1980). *Backroads America, a portrait of her people*. Washington, DC: National Geographic Society.
Palmer, C. T., Begley, R. O., & Coe, K. (2012). In defence of differentiating pilgrimage from tourism. *International Journal of Tourism Anthropology, 2*(1), 71–85.
Park, C. (1994). *Sacred worlds. An introduction to geography and religion*. London: Routledge.
Relph, E. (1976). *Place and placelessness*. London: Pion.
Sears, J. (1989). *Sacred places*. New York: Oxford University Press.

Selby, J. (2006). The politics of pilgrimage: The social construction of Ground Zero. In W. H. Swatos Jr. (Ed.), *On the road to being there. Studies in pilgrimage and tourism in late modernity* (pp. 159–196). Boston: Brill.

Singh, S. (2005). Secular pilgrimages and sacred tourism in the Indian Himalayas. *GeoJournal, 64*, 215–223.

Steiner, C. J., & Reisinger, Y. (2006). Understanding existential authenticity. *Annals of Tourism Research, 33*(2), 299–318.

Tuan, Y. F. (1974). Space and place: Humanistic perspective. *Progress in Geography, 6*, 211–257.

Turner, V. (1969). *The ritual process*. Chicago: Aldine.

Turner, V. (1973). The center out there: Pilgrim's goal. *History of Religion, 12*(3), 191–230.

Turner, V., & Turner, E. (1978). *Image and pilgrimage in Christian culture*. New York: Columbia University Press.

Tuzinowski, D. (2009). Uphill and against the wind. *Trans America E-W*. Online journal last accessed 11 June 2012 at www.tuz.net/cycling/NmHvZ/1571/thumbnail/index.html

Urry, J. (1990). *The tourist gaze*. London: Sage.

Part V
Education and Changing Worldviews

Chapter 44
Geographies of Faith in Education

Peter J. Hemming

44.1 Introduction

Over the last decade, religion has begun to take a more prominent role in the social sciences than was previously the case. Developments related to globalization, such as migration and international mobility have led to more religiously diverse populations in the West, while geopolitical events such as the 9/11 attacks and the subsequent "war on terror" have also highlighted the significance of religion for making sense of the world. Scholars have begun to realize that a purely secular lens is not always sufficient for understanding social processes in their entirety and that the importance of religion cannot be underestimated, even in so-called "secular" societies (for example, ESRC/AHRC Religion and Society Programme in the UK, www.religionandsociety.org.uk/). In the context of geography, Kong (2010) has usefully outlined a number of "global shifts," including urbanization and inequality, deteriorating environment, aging, and increased mobility, where an engagement with religion could help to better understand human responses to these key issues.

The study of education is no exception to this trend and interest in the significance of religion in educational contexts has also continued to grow. While social studies on education have, in the past, often been concerned with issues such as class, economy and culture, and later gender, ethnicity and globalization, it is reasonable to assert that less attention has been paid to religion. This has now started to change and bodies of literature can be found that explore key themes such as religion in the curriculum, faith schools, and religious identity and difference in education. In fact, these are the three areas that make up the main part of this chapter,

P.J. Hemming (✉)
School of Social Sciences, Cardiff University, Cardiff, Wales, UK
e-mail: hemmingpj@cardiff.ac.uk

S.D. Brunn (ed.), *The Changing World Religion Map*,
DOI 10.1007/978-94-017-9376-6_44

specifically the contribution that a geographical approach can bring to such issues and debates.

Education has not always been of major concern in geography, although, like religion, the topic is currently rising in prominence in geographical circles. However, there has been very little scholarly interest in the intersection between religion and education from geographers, despite concepts such as space and scale having much to offer debates in this area. In this chapter, I explore some of the relevant issues and literature relating to the three key areas mentioned in the last paragraph in order to highlight the significance of such debates to geography. I begin by introducing the recent body of work on geographies of education and some of the relevant tools and concepts that can help to enhance the study of religion in education.

44.2 Geographies of Education

A good place to start in developing a geographical approach to the study of religion in education is with a consideration of the 'new' geographies of education. This emerging sub-discipline is well represented by two reviews in *Progress in Human Geography*. The first, by Hanson Theim (2009), highlights the significance of geography for interrogating the political and economic aspects of educational processes, in contrast to Holloway et al. (2010), which emphasizes their social and cultural dimensions, particularly the role of young participants and their families. Despite these differences in perspective, both reviews, nevertheless, point to four key areas where geography has started to make inroads into grappling with the significance of education.

The first area relates to "geographies of educational provision and consumption," encompassing the role that education plays in communities of identity and social networks, the structuring of neighborhoods and housing decisions, the transnationalization of higher education markets, and experiences of education in the global majority world. The second area is concerned with both "formal and informal curricula and spaces of learning," including issues of citizenship and national identity, the development of social subjectivities, the nature of curriculum governance and the role of a range of actors in supporting or contesting these processes. The examples given later in this chapter fall mainly within these first two areas.

The third area identified by the two reviews is the issue of "knowledge spaces," particularly the role of higher education in economic and political processes, educational restructuring as a result of globalization, and the resultant experiences of students and their impact on the urban landscape. The final area relates to "institutional restructuring," including economically, politically and socially. This is concerned with the relationship between education and the neoliberal project, whether through its engagement with concepts such as rationalization and privatization, or through the shifting responsibilities for social reproduction and the challenges this might present for the project.

Although the two reviews successfully identify a number of thematic areas for which the "new" geographies of education have shown a certain amount of coverage, they only travel so far down the road of developing specific geographical approaches to education. Hanson Theim (2009) does point to the need for geographies of education to not only pay attention to the role of space in education ("inward-looking" studies), but also the ways that education can affect external macro processes and their spatial manifestations ("outward-looking" studies). Holloway et al. (2010), however, problematize this inward- and outward- categorization. They argue that many studies focusing on children, youth and families that could be classed as "inward" in this sense, actually reveal important links between educational and other wider spaces and discourses.

As a geographer writing from within education studies, Taylor (2009) has also made some interesting contributions to the development of a distinctly geographical approach to the study of educational processes. He utilizes the concept of operational scale to outline a range of research agendas that have the potential to focus on micro, meso and macro scales. These scales include learners, sites of learning, communities of learners, local authorities, central governments, and international contexts. Of course, there have been numerous debates in human geography about how scale should be theorized, and many of these have moved away from hierarchical notions, towards inter-linking and socially constructed understandings. More recent conceptions have questioned the usefulness of scale altogether, instead emphasizing networks and practices (for example, Moore 2008). But as MacKinnon (2010) has argued, this does not necessarily mean that scale as an analytical category should be discarded. Social actors may well employ hierarchical understandings of scale in order to make sense of the world, and this may result in real material effects.

The fluidity of scale is something that Taylor (2009) recognizes as a limitation of his model, while nevertheless, identifying a whole range of current research examples that fit within it, such as embodied geographies, school design, residential segregation, education markets, devolution and globalization. The contribution of Holloway et al. (2010) to this way of thinking about scale is to highlight how both formal and informal education micro-spaces may connect and relate to wider scales and spaces of reference. A model of scale that takes analytical scale seriously, but nevertheless recognizes the fluidity and socially constructed nature of the concept, as well as the connections and linkages between spaces and scales, is, therefore, central to the present chapter.

In the remainder of this chapter, I focus on key debates in the field of faith in education, drawing out some of the spatial and scalar implications, including the construction of public and private space, the relationship between local and national power, different understandings of community and community cohesion, and the way in which citizenship operates as a multi-scalar process. I explore the first of these with reference to the place of religion in the curriculum, as it operates as a contested space.

44.3 The Curriculum as a Contested Space

One of the key functions of schools and other educational institutions is the transfer of knowledge through various curricula. Although many countries have moved towards standardized national curricula in recent times, their content and form often remain fiercely contested. The place of religion in the curriculum is no exception, although it maintains a very different role depending on the particular national context in question. An examination of this issue reveals some interesting relationships, firstly between the role of religion in public and private space, and, secondly, between power at the national and the local level in educational processes.

The extent to which religion should be viewed as a public or a private affair is often an important factor in debates about its role in the curriculum. As a political theory, liberalism generally sees religion as something that should rightly be confined to private space, achieved through the separation of church and state (Locke 2003, cited in Martin 2010:61). In reality, though, many liberal democratic states do not always maintain a separation between church and state, and even if they do, religion and state politics are often intensely intertwined, even if they are not presented as such (Berger 1999). The place of religion is often, therefore, much more complex than its official status implies, as is illustrated in the examples below. Furthermore, as will be apparent, the debate over public and private space often plays out through the dynamic between local and national power.

44.3.1 Religious Education

While many state-funded education systems allow little or no official place for religion in the curriculum, others, such as in England and Wales, have a long tradition of including religious education (RE) in the standard provision offered to pupils. The place of religion in the English school curriculum very much reflects the historical role of the Churches in the national education system. The Church of England originally introduced schooling for the masses in the early nineteenth century, and after universal state education was established in 1870, a "dual system" was gradually developed through various Acts of Parliament from 1902–1944. This enabled the incorporation of existing church schools into the wider system, offering a pragmatic solution to a difficult logistical problem.

"Religious instruction" had always been a key component of the curriculum in church schools, and was also adopted by state schools as a means to uphold national standards of morality (Baumfield 2003). Yet the subject gradually lost its confessional basis in state schools, particularly following the 1960s and 1970s and the religious diversification that occurred as a result of new migration from the former colonies. Increasingly, the subject was instead taught from an educational perspective through a phenomenological focus on the various elements of the world religions. The subject was eventually renamed "religious education," remaining a compulsory part of the curriculum in the 1988 Education Reform Act.

It is at this point that one of the key geographical issues emerged, highlighting RE as a contested space. By the late 1980s, the subject had become an area of considerable controversy, particularly over the balance between different religions taught. This tended to reflect concerns regarding the emphasis given to Christianity, viewed by many as part of the national cultural and religious heritage, compared with other religions, and the recognition and respect their inclusion extended to religious minorities. At the time, central government succeeded in sidestepping this debate in the 1988 Education Reform Act by omitting RE from the national curriculum, but nevertheless ensuring that it remained a compulsory part of the whole curriculum.

This was achieved by devolving the power to set the RE programs of study to the local level – specifically to Standing Advisory Committees for Religious Education (SACREs). SACREs are groups of advisors, teachers and representatives from the major faiths in each particular local area. Although there is a requirement for all local syllabuses to show a certain amount of priority to Christianity, there is still substantial freedom for SACREs to develop RE content that reflects their locality. Crucially, it also allows central government to avoid becoming involved in controversial discussions over what should be taught in each local area. Localism, or devolving power to the local level, was, therefore, used strategically to sidestep difficult debates about the role of religion and RE in the public sphere.

In the last decade or so, various developments have placed the subject of RE under renewed scrutiny in England, both in terms of its place in the curriculum and its actual content. It is of interest that all of these developments occurred at the national level, potentially changing the balance of power between the state and the local areas. Firstly, the introduction of compulsory citizenship education in all secondary schools in England from 2002 onwards represented a potential competitor with RE for timetable space in an already crowded curriculum. This led to suspicions on the part of many RE teachers that the new subject could herald the demise of their own area of expertise, partly because of the substantial overlap between the two subjects on issues such as common values and respect for difference (for example, see Watson 2004).

Secondly, the publishing in 2004 of a non-statutory national framework for RE by the then QCA government curriculum organization (Qualifications and Curriculum Authority 2004), marked a shift away from the more locally-based SACRE system established by the 1988 Education Reform Act. Although the framework was presented as advisory only, inevitably it influenced many SACREs in devising their local syllabuses. The national framework was also significant in that it was the first time that secular philosophies and less well-known religions were suggested as important areas to cover in RE lessons, alongside Christianity and the other major world faiths. If teaching about religion in publically-funded schools was considered controversial by some, the inclusion of explicitly non-religious philosophies, such as atheism and humanism, in the RE curriculum further fuelled debates about the place of religion in the public sphere, and its relationship to secular functions and systems of thought.

44.3.2 Science and Sex Education

Religion and its relevance to the school curriculum is, however, much wider than the issue of RE. There have been many high profile cases in the U.S. of educational organizations and institutions attempting to introduce the teaching of creationism or "intelligent design" into science lessons, presented as a valid alternative to theories of evolution. School boards and state legislators have, in the past, come into conflict with the U.S. Supreme Court, which in 1987, outlawed religious content in science lessons on the grounds that this would be advancing a particular religion. Despite this ruling, states such as Louisiana and Tennessee have, in the last few years, been successful in passing state laws that permit the discussion of alternative theories alongside evolution in the school classroom (for example, see Flock 2012, www.washingtonpost.com/national/law-allows-creationism-to-be-taught-in-tenn-public-schools/2012/04/11/gIQAAjqxAT_story.html). These examples can be viewed in the wider context of the U.S. "culture wars," whereby education has become a key battleground between conservative and progressive forces.

Central to many of these debates, is the (conservative) argument that secular schools are not promoting totally neutral philosophies, but are teaching from an explicitly liberal position. Teachers could, therefore, be accused of subverting the religious beliefs of children and discriminating against particular religious positions (Collins 2006). In these situations, the rights of parents to raise their children in the religious values of their choice grate against the liberal ideal of religion being confined to the private sphere. Pike (2008) has also adopted a similar argument in his analysis of English citizenship education, arguing that it effectively teaches children to "believe" in liberal democracy, when these are values that particular religious communities may not share. Although this is a rather controversial argument, it is nevertheless aligned with the kind of debates that typically occur around such issues.

Alongside science, the subject of sex and relationship education is another area of the curriculum where religious considerations can be pertinent. Collins (2006) examines a Canadian case study, involving the Board of Surrey School District in British Columbia refusing to allow the use of three particular books in schools that portrayed same sex parents. This incident, and the 6 years of controversy that surrounded it, highlighted some of the ambiguity around the boundary between public and private space and the relation of religion to this divide. Whereas conservative parental groups maintained that religious values and interests should play a role in public decision-making, liberal lobbyists contended that values of dignity and respect of all persons, including sexual orientation, should be paramount, placing religion firmly in the private sphere. After various rulings at different levels of the Canadian legal system, the principle was established that religious beliefs and values *could* be considered in decisions regarding public education, but not at the expense of issues of equality and social justice. In other words, the place of religion was deemed to be more than merely confined to the private sphere, but not quite as much as some conservative lobbyists would have liked.

In both the examples of science and sex education above, power dynamics between local group interests, school boards, state legislators and various levels of law courts were central to the eventual resolution of the issues. Typically in these cases, local control and devolved decision-making are favored by conservative interests, in contrast to liberal and progressive concerns for a degree of state oversight to safeguard rights and combat discrimination (Collins 2006). In many ways, these examples mirror the case of English RE, and the power dynamics between the local and the national scale discussed earlier in the section. As Merrett (1999:599) argues, in these kinds of debates, "the politics of scale are used by the contending sides to determine the appropriate scale for social reproduction." Not only this, but the effective boundary between public and private space, and the place of religion in relation to this divide, is also constituted through these scalar politics between national and local interests.

44.4 The Faith Schools Debate

In addition to the curriculum, religion has often been intertwined with education through the provision of single faith institutions, including schools, colleges and universities. As mentioned in the previous section, universal education in England and Wales was predated by the establishment of a vibrant church school sector, from the start of the nineteenth century, before being incorporated into the state system. State funding for faith schools is a tradition that has continued in England and Wales, and has been replicated in many other countries, for example, the Netherlands. Yet despite this, the issue remains highly controversial, with arguments for and against such schools regularly rehearsed by politicians, lobby groups and the media (for example, see Jackson 2003).

Arguments in favor of state-funded faith schools include the following:

- religious needs are better catered for in faith schools;
- faith schools provide choice for parents and diversity of provision;
- faith schools represent justice and fairness for members of different religious groups;
- faith schools foster positive ethos and supportive religious values.

In contrast, arguments against state-funded faith schools include the following:

- faith schools limit pupil autonomy and promote indoctrination;
- faith schools use selection procedures that effectively disadvantage other schools by selecting the best pupils;
- faith schools work against educational aims and equal opportunities;
- the state should not fund religion as it should be a private matter.

Despite these wide-ranging arguments, the most vigorous debates occur over the issues of community, belonging and social division. It is to these issues that I now turn.

44.4.1 *Faith Schools and Institutional Belonging*

The idea of community is often very central to how faith schools view themselves, and this is illustrated well in the study by Valins (2003) on Jewish schools in England. Valins (2003) outlines how Jewish community leaders and school gatekeepers sought to form and institutionalize desirable religious identities among pupils through educational provision. This was achieved by the construction and maintenance of socio-spatial boundaries between the school community and the outside world. Particularly in Jewish Orthodox schools, faith-based education was viewed as a way to combat the perceived assimilation of Jews into mainstream British (secular) culture. The process of boundary construction and formation was not, however, straightforward but was a contested process, involving different school stakeholders.

Despite the emphasis on religious belonging from community and school leaders, parents of pupils who attended the school often had rather different priorities for what they hoped their children would obtain from their educational experiences. Rather than specific teaching about the Jewish religion, parents instead valued academic standards and a more general sense of Jewish ethos and identity. Even so, some parents continued to hold concerns about isolationism and the need for children to be prepared for dealing with the "outside world." Many parents spoke particularly highly of schools in Liverpool, Birmingham and Glasgow, which due to lower demand for places, also took children from non-Jewish families and were therefore seen as the "best of both worlds."

Close-knit educational communities of faith are often consistent with state discourses such as parental choice in countries like England, but this may not always be the case, and there may be some misalignment between the aims and objectives of the school on the one hand, and the state on the other. Kong (2005) examines how Islamic schools in Singapore negotiate their community identities within the multicultural context of a state with a secular and modernist vision. In Singapore, religious schools are considered a right for groups, but whereas state schools are constructed as modern and multicultural, Islamic schools are seen as the opposite. For the Muslim community, however, such schools are seen as important alternative "educators" for reproducing religion and a holistic education. In order to negotiate this tension, Islamic schools have also attempted to show that they are modern and forward looking by making links with other schools and investing in new facilities. As was the case in Valins's (2003) study, the construction of a school community of belonging is shown to be a contested process, one that often involves the involvement of social and political actors from various different scales of reference.

44.4.2 *Faith Schools and Community Cohesion*

The issue of faith schools and isolationism has also been central to debates about social and community cohesion. Of course, the idea of community cannot only refer to the state of belonging to a particular religious or social group, but can also be concerned with the wider geographically-based community or neighborhood (for example, see Clark 2009 for a review of the community concept). This type of community has received much attention over the last decade or so particularly in inner-city urban contexts, amongst concern about community fragmentation and division (e.g. see Putnam 2000). In England, the 2001 riots in northern towns and cities such as Bradford, Oldham and Burnley were of particular significance, when some of the reports into these events highlighted the segregated nature of the education system in these urban centers (for example, Ouseley 2001). In the case of faith schools, the question is whether such institutions work to promote or erode social cohesion between different faith, ethnic and cultural groups. While one argument suggests that children must experience interfaith and intercultural encounters in order to understand each other, another insists that values of tolerance and acceptance can instead be taught as part of the curriculum.

As Dwyer and Parutis's (2013) recent work highlights, often the argument hinges on the particular understanding of community and community cohesion that is employed. The paper explores political and policy discussions in England regarding the place of state-funded faith schools in the education system and their relationship to debates about social integration and division. The authors focus on two specific mechanisms introduced by the state in order to respond to criticism of the state-funded faith school sector. These included a universal requirement for all schools to "promote community cohesion" (and to be assessed on this by the school inspectorate), as well as the introduction of a new "admissions code" for faith-based schools. Here, the focus is on the first of these mechanisms, particularly how various constructions of community were central to understanding the dynamics of the debate.

In the case of the requirement for schools to "promote community cohesion," Dwyer and Parutis (2013) show how responses to this new regulation from faith schools were often based on a re-working of the community cohesion concept. This re-working emphasized the theologically based values of respect and tolerance taught to pupils as part of the faith ethos, rather than actual inter-faith contact or encounter as envisaged by the state. Although there were some examples of "linking events," whereby schools with culturally homogenous intakes would develop partnerships with other more demographically diverse schools, these were generally in the minority. The desire from the government in 2006 to introduce a requirement for all new faith schools to reserve 25 % of places for families different faith or non-faith backgrounds, also met with resistance from parts of the faith school sector, and the attempt was eventually abandoned (Dwyer and Parutis 2013). The understanding of "community" and "community cohesion" adopted by many (although by no means all) faith schools, therefore focused much more on the school religious community than the local neighborhood based community (see also Hemming 2011a), or alternatively drew on notions of a "global community" through the teaching of universally applicable values.

The idea of the community as a geographically based neighborhood has also started to take on much greater significance in the context of faith-based schooling in certain metropolitan areas. This is particularly the case regarding the admissions process, often accused of inadvertently allowing the selection of the most affluent or academically able pupils rather than operating purely by religious criteria (for example, see Allen and West 2011). The effect of this, so it is argued, is to further exacerbate social divisions of various kinds in the local neighborhood. Research by Butler and Hamnett (2012) found that, in the crowded educational market of East London, where good state schools are in rather short supply and private school fees are often unaffordable, parents increasingly see faith schools as an attractive option for schooling their children. In this situation, many parents are willing to go to great lengths to establish their religious credentials in order to meet challenging admissions rules or even relocate to take advantage of the "distance from school" or catchment criteria increasingly used alongside religious considerations to select pupils for oversubscribed faith schools.

Butler and Hamnett (2012) argue that while parents often emphasize school standards, quality, behavior, ethos and values at such schools, implicit to their enthusiasm for the faith school sector also comes from a desire to find the 'right' social and ethnic mix for their children's social and educational development. Often this involves pupils experiencing limited contact with the ethnic or religious 'other', but not *too* much contact (see also Byrne 2006). By subtly insinuating that faith schools are a better reflection of the neighborhood, compared with local state schools that might admit a larger proportion of ethnic minority or working class pupils, these parents are effectively constructing their own socially acceptable version of the "local community." They are also enacting a certain type of community cohesion that ensures contact and encounters with other families that share broadly similar values.

44.5 Religion, Citizenship and Identity

The final section of this chapter consists of a focus on the relationship between education, citizenship and religious identity. Although citizenship has traditionally been defined at the scale of the nation, geographical research has increasingly highlighted its multi-scalar nature, particularly the role of everyday practices in constituting citizenship (Dickenson et al. 2008). Gordon et al. (2000) have pointed to the central role that schools play in creating future citizens, particularly in reference to gender, including through formal, informal, physical and embodied space. But religion has, to date, received rather less attention in this context, compared with other social identities such as class, gender and ethnicity. Notable exceptions outside of geography include the work of Nesbitt (2004) and Smith (2005), among others, but there is much potential for geographers to add to these bodies of work.

In my own work, I have been interested in developing and exploring the concept of "religious citizenship." Building on Joppke's (2007) three dimensions of citizenship (status, rights and identity), I have defined religious citizenship as "the role of religion in devising criteria for access to state or community membership, the political rights and responsibilities attributed to particular religious groups within that membership, and the religious aspects of collective social/cultural identity that influence belonging" (Hemming 2011b:444). I have also emphasized the significance of everyday practices and discourses to the evolving nature of the dimensions outlined above.

Although the concept of religious citizenship is relevant and applicable at a variety of scales, in the case of my own research, I have made use of it in the context of education spaces. The idea that religion as an identity is important in its own right creates an alternative lens through which to view work on multiculturalism and education, whereby religion has often been subsumed under the umbrella of "culture" and "ethnicity." As my work has highlighted (Hemming 2011b), religion actually constitutes a further layer of stratification in schools, influencing which pupils are privileged and which are not. In the case of the primary (elementary) case-study school in my research (Rainbow Hill), these positions were often formed through everyday school practices, including the way in which various religions were recognized and accommodated, and the responses of various social actors to these arrangements. Religious citizenship, therefore, influenced the dynamics of belonging that pupils in school experienced and (re)negotiated.

One of the key ways in which schools can extend recognition to religious minorities is to make a commitment to marking or celebrating festivals, either as a formal part of the RE curriculum, or during more informal parts of the school day, such as whole school or class assemblies. Nesbitt (2004) highlights how multi-faith schools in England increasingly celebrate a range of non-Christian festivals, such as Diwali and Eid, alongside the more traditional Christian-based festivities at Christmas and Easter. In my research, Rainbow Hill also marked many of these non-Christian festivals, but often at a much smaller scale than the Christian ones. This created a situation where Christianity, if only in a cultural sense, maintained its primacy as the 'default' religion in school, effectively privileging certain pupils over others, either culturally or religiously.

In the case of religious accommodation and minority rights, the issues of food, dress and prayer needs were significant. As has been observed in previous studies, school lunchtimes can create a context for the reinforcement of difference and the negotiation of children's religious identities (Nesbitt 2004; Smith 2005). At Rainbow Hill, the school dinner service catered for children in school who only ate Halal meat or were vegetarians due to their religion. The school was also generally quite tolerant of dress needs, such as religious bangles and headscarves, and provided space in the school library for Muslims to pray during Ramadan. Yet this provision for religious minorities only went so far, with curriculum requirements and health and safety legislation taking priority in the case of arrangements for fasting and the wearing of religious jewelry during physical education. Some parents and pupils also complained about the lack of prayer space and problems they encountered with bringing prayer mats into school.

All of these examples highlight the limitations of recognizing and accommodating religious minorities through legal and cultural arrangements developed within a western Christian social context. But what was also interesting, was some of the responses to these arrangements from various social actors connected to Rainbow Hill, often reinforcing, contesting or renegotiating them. On the part of parents, while a few of the participants with minority faiths felt that more could be made of their religious festivals in school, many of the other parents were in favor of the status quo, pointing to England's status as a "Christian country." Similarly, whilst some parents felt that the school could go further in accommodating religious needs such as respecting Sabbath days, others expressed the view that the tolerance and respect extended to religious minorities was not always reciprocated in the case of Christian cultural and religious customs. Such examples highlight the role of particular constructions of the nation and national identity in constituting religious citizenship.

The most fascinating finding from this research study was the response of some of the pupils to the arrangements for recognizing and accommodating religious minorities at Rainbow Hill. These were often quite inventive, and included creating provision for prayer space where the institution had failed to do so. One way of achieving this was to construct mental prayer space by changing the words of the Christian prayers so that they were more appropriate for another religion, or praying to a different God in the child's head. Another was to use toilet cubicles as private spaces in order to satisfy daily prayer requirements. This last example clearly raises some important issues about the extent to which the school really was providing for pupils' religious needs, when children were forced to disregard the usual customs on cleanliness and purity in order that they could pray. All of these cases point to the way in which children were able to contest and renegotiate religious citizenship through everyday embodied practices.

In its entirety, the study further highlights the multi-scalar nature of citizenship processes and in this case religious citizenship. Although citizenship is a concept that is often discussed in relation to the state, it can be just as useful for making sense of the educational institution as a geographical space and the social processes that take place within and through it. Furthermore, the way in which religious citizenship was reinforced, contested and negotiated, brings to light the significance of a number of other scales, including the nation and national identity and embodied practices within informal spaces. This latter example in particular underlines the continued importance of hierarchical scale when considering issues of power, since it was only at this level that children were able to contest constructions of religious citizenship influenced by the state and the institution. However, this nevertheless implied a rather fluid and interconnected understanding of scale, as Holloway et al. (2010) maintain.

44.6 Conclusion

As the study of religion becomes increasingly important in geography and the social sciences in general, the role of faith in education is an area that holds much potential for further scholarly inquiry. The "new" geographies of education sub-discipline offers an initial framework for the development of a distinctly geographical approach to the topic, alongside contributions from the field of educational studies itself, through core concepts such as space and scale. In this chapter, I have focused on three areas of interest, including religion in the curriculum, the faith schools debate, and religious identity, difference and citizenship in education and explored some of the ways that geography can shed further light on these issues. Key themes that emerged from this analysis included the construction of public and private space, the relationship between local and national power, different understandings of community and community cohesion, and citizenship as a multi-scalar process.

The coverage of this chapter is, however, necessarily limited, both by the space available, but also by the scarcity of geographical research on religion and education. As a topic that spans a whole range of sub-disciplines in human geography, including social, cultural, urban and political geography, the potential for the development of this field is really quite substantial. Future research needs to engage with some of the major questions in the study of religion, such as secularization, post-secularization, atheism, radicalism and new spiritualities, in order to examine the ways in which these processes are impacting geographically on education. Similarly, many of the contemporary developments in education, such as neo-liberalism, decentralization and privatization may have consequences for both religious institutions and individual believers. Geography has a real contribution to make to these debates and it is important that it does so.

References

Allen, R., & West, A. (2011). Why do faith secondary schools have advantaged intakes? The relative importance of neighborhood characteristics, social background and religious identification amongst parents. *British Educational Research Journal, 37*(4), 691–712.

Baumfield, V. (2003). The dignity of difference: Faith and schooling in a liberal democracy (editorial). *British Journal of Religious Education, 25*(2), 86–88.

Berger, P. L. (1999). The desecularization of the world: A global overview. In P. L. Berger (Ed.), *The desecularization of the world: Resurgent religion and world politics* (pp. 1–18). Grand Rapids: William B. Eerdmans.

Butler, T., & Hamnett, C. (2012). Praying for success? Faith schools and school choice in East London. *Geoforum, 43*(6): 1242–1253.

Byrne, B. (2006). In search of a 'good mix:' 'Race', class, gender and practices of mothering. *Sociology, 40*(6), 1001–1017.

Clark, A. (2009). From neighbourhood to network: A review of the significance of neighbourhood in studies of social relations. *Geography Compass, 3*(4), 1559–1578.

Collins, D. (2006). Culture, religion and curriculum: Lessons from the 'three books' controversy in Surrey, BC. *The Canadian Geographer, 50*(3), 342–357.

Dickenson, J., Andrucki, M. J., Rawlins, E., Hale, D., & Cook, V. (2008). Introduction: Geographies of everyday citizenship. *ACME: An International E-journal for Critical Geographies, 7*(2), 100–112.

Dwyer, C., & Parutis, V. (2013). 'Faith in the system?' State funded faith schools in England and the contested parameters of community. *Transactions of the Institute of British Geographers, 38*(2): 267–284.

Flock, E. (2012). *Law allows creationism to be taught in Tennessee public schools*. Retrieved: May 25, 2012, from www.washingtonpost.com/national/law-allows-creationism-to-be-taught-in-tenn-public-schools/2012/04/11/gIQAAjqxAT_story.html

Gordon, T., Holland, J., & Lahelma, E. (2000). *Making spaces: Citizenship and difference in schools*. Basingstoke: Macmillan.

Hanson Theim, C. (2009). Thinking through education: The geographies of contemporary educational restructuring. *Progress in Human Geography, 33*(2), 154–173.

Hemming, P. J. (2011a). Building a sense of community: Children, bodies and social cohesion. In L. Holt (Ed.), *Geographies of children, youth and families: An international perspective* (pp. 55–66). London: Routledge.

Hemming, P. J. (2011b). Educating for religious citizenship: Multiculturalism and national identity in an English multi-faith primary school. *Transactions of the Institute of British Geographers, 36*(3), 441–454.

Holloway, S. L., Hubbard, P. J., Jöns, H., & Pimlott-Wilson, H. (2010). Geographies of education and the importance of children, youth and families. *Progress in Human Geography, 34*(5), 583–600.

Jackson, R. (2003). Should the state fund faith-based schools? A review of the arguments. *British Journal of Religious Education, 25*(2), 89–102.

Joppke, C. (2007). Transformation of citizenship: Status, rights, identity. *Citizenship Studies, 11*(1), 37–48.

Kong, L. (2005). Religious schools: For spirit, (f)or nation. *Environment and Planning D: Society and Space, 23*(4), 615–631.

Kong, L. (2010). Global shifts, theoretical shifts: Changing geographies of religion. *Progress in Human Geography, 34*(6), 755–776.

Mackinnon, D. (2010). Reconstructing scale: Towards a new scalar politics. *Progress in Human Geography, 35*(1), 21–36.

Martin, C. (2010). *Masking hegemony: A genealogy of liberalism, religion and the private sphere*. London: Equinox.

Merrett, C. D. (1999). Culture wars and national education standards: Scale and the struggle over social reproduction. *The Professional Geographer, 51*(4), 598–609.

Moore, A. (2008). Rethinking scale as a geographical category: From analysis to practice. *Progress in Human Geography, 32*(2), 203–225.

Nesbitt, E. (2004). *Intercultural education: Ethnographic and religious approaches*. Brighton: Sussex Academic Press.

Ouseley, H. (2001). *Community pride not prejudice*. Bradford: Bradford Vision.

Pike, M. A. (2008). Faith in citizenship? On teaching children to believe in liberal democracy. *British Journal of Religious Education, 30*(2), 113–122.

Putnam, R. D. (2000). *Bowling alone: The collapse and revival of American community*. London: Simon & Schuster.

Qualifications & Curriculum Authority. (2004). *Religious education: The non-statutory national framework*. London: QCA.

Religion & Society Programme. Website from www.religionandsociety.org.uk/. Accessed 25 May 2012.

Smith, G. (2005). *Children's perspectives on believing and belonging*. London: National Children's Bureau for the Joseph Rowntree Foundation.

Taylor, C. (2009). Towards a geography of education. *Oxford Review of Education, 35*(3), 651–669.

Valins, O. (2003). Defending identities or segregating communities? Faith-based schooling and the UK Jewish community. *Geoforum, 34*(2), 235–247.

Watson, J. (2004). Educating for citizenship: The emerging relationship between religious education and citizenship education. *British Journal of Religious Education, 26*(3), 259–271.

Chapter 45
Religion, Education and the State: Rescaling the Confessional Boundaries in Switzerland

Mallory Schneuwly Purdie and Andrea Rota

45.1 Introduction

Historically, Switzerland has been a bi-confessional country, Roman Catholic and Evangelical Reformed. For centuries, cantonal territorial boundaries corresponded with confessional boundaries. Since the foundation of the modern Swiss Confederation, in the mid nineteenth century, the autonomy of the cantons in their handling of the relationships between State and religious communities has been confirmed in all Federal Constitutions. This means that, to the extent that sovereignty is not limited by federal law, each canton can freely grant privileges to one Church or the other; and that each canton remained, until recently either predominantly Catholic or Protestant (Forclaz 2009).[1]

Since 1960, the religious landscape of Switzerland has experienced major evolutions. Firstly, researchers show that Switzerland is experiencing a vivid secularization process (Campiche et al. 1992, 2004; Stolz et al. 2011): institutional religiosity is drastically declining,[2] religious beliefs are less often transmitted and individuals grant less social significance to religion. Secondly, the Swiss religious landscape is

[1] In some cantons and territories this predominance was less strong, for instance in the territories of today's cantons of Geneva, Aargau, St. Gallen or Graubünden; however, these were the exception, not the rule.

[2] Among other indicators, we would like to mention the frequency of churchgoing and praying and the number of declared members.

M. Schneuwly Purdie (✉)
Institut de sciences sociales des religions contemporaines, Observatoire des religions en Suisse, Université de Lausanne – Anthropole, 1010 Lausanne, Switzerland
e-mail: mallory.schneuwlypurdie@unil.ch

A. Rota
Institute for the Study of Religion, University of Bern, Lerchenweg 36, 3012 Bern, Switzerland
e-mail: andrea.rota@relwi.unibe.ch

S.D. Brunn (ed.), *The Changing World Religion Map*,
DOI 10.1007/978-94-017-9376-6_45

no longer bi-confessional, but has pluralized: in addition to its historical religious congregations, Switzerland now also shelters numerous Muslim communities,[3] as well as Buddhist, Hindu, Jewish congregations and Christian Free Churches (Baumann and Stolz 2009; Stolz et al. 2012). Not only is the religious landscape evolving, but so are the 'religious products' available on the 'religious market' (Stolz 2006). Beliefs and practices are no longer monopolized by religious institutions, but along with the process of individualization, the social actors choose in what they believe, when and why they get involved in various spiritual practices such as yoga, reiki, qi gong or shamanic rituals (Stolz et al. 2012, 2014).

This contribution aims to examine the ways that territorial and symbolic boundaries are drawn on the basis of a specific case study: the religious education (RE) classes in Swiss public schools. In social sciences, RE has been mainly studied from a pedagogical and didactical angle (see for instance Alberts 2007). From our point of view, RE is not only a matter of education, but it also is a powerful instrument for analyzing and understanding the public presence of religions as well as their interactions with society and the contemporary secular State. As we will show, not only do individuals and religious institutions play an important role in the drawing of symbolic boundaries, but the State also appears to be a major actor in this process (Jödicke 2009; Jödicke and Rota 2010; Rota 2011, 2013). Along with the profound changes of the Swiss religious landscape, the structure and the purpose of RE have been profoundly reshaped and the State endorses a growing influence since it assumes the supervision of new RE classes addressed to all the pupils, indifferently as to their confession.

Our contribution is articulated around the following core question*: How does the emergence of the State as an actor in the field of RE reveal and shape the evolution of the confessional boundaries in the public sphere?* For the purpose of our study, we will focus on the French-speaking part of Switzerland. This region in the Western part of the country is composed of seven cantons (Fig. 45.1): Fribourg, Wallis, and Bern (French-German bilingual); Jura (Independent since 1979, previously part of the canton of Bern); Geneva, Neuchâtel, and Vaud. The region combines different confessional traditions, State-Church relations, and religious education models within a uniform linguistic and cultural framework.

In order to understand the transformations of RE and the drawing of confessional boundaries in Switzerland, it is necessary to put our analysis in geographical and historical perspective. Therefore, our paper will be structured in four points. Firstly, we shall present theoretical insights of the boundary work concept. Secondly, we shall briefly outline some fundamental features of the Swiss juridical system and discuss their implications on the organization of RE. Thirdly, we will cast some light upon three major thresholds in the history of RE allowing us to draw important consideration to the (de)construction of confessional boundaries. Finally, we shall draw on our conclusion.

---

[3] We are of the opinion that we cannot speak of 'the' Muslim community in Switzerland, which does not exist as a homogenous entity. On the contrary, researches have shown that there are many Muslim communities evolving side by side, each one mobilising the Islamic reference differently, along with other markers such as ethnicity, gender, age or social class (see Schneuwly Purdie 2011).

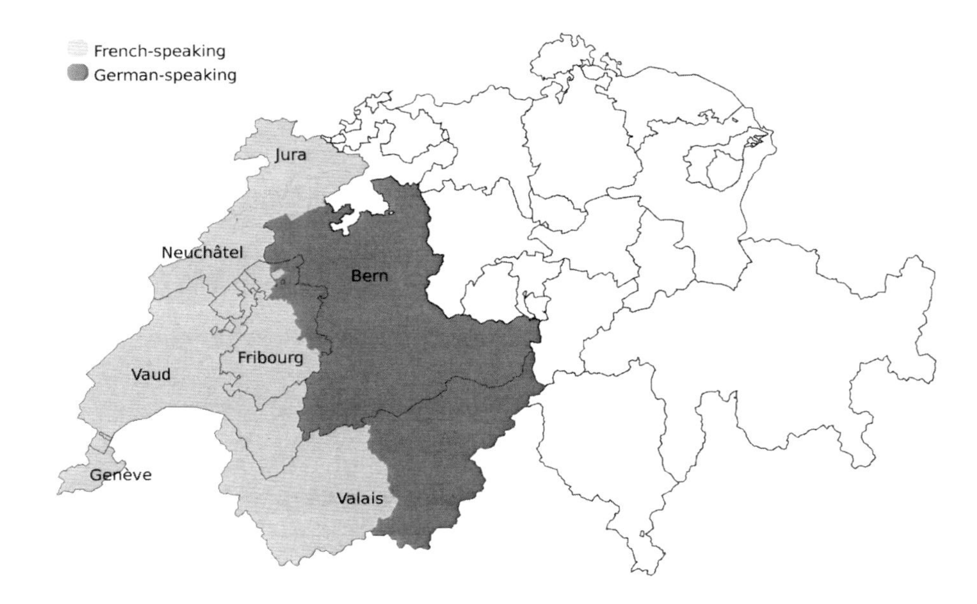

Fig. 45.1 French-speaking Switzerland (with bi-lingual cantons) (Map by A. Rota with data from the Swiss Federal Statistical Office, 2000)

45.2 Drawing Territorial and Symbolic Boundaries

Common notion in geography, the concept of boundary became popular in the social sciences in the 1990s. For social scientists, a boundary is a conceptual border that demarcates individuals into two (or more) distinctive groups. As defined by Alba, a boundary is "a distinction that individuals make in their everyday lives that shapes their actions and mental orientations towards others" (Alba 2005: 22). Among other forms, boundaries can be territorial or symbolic. Territorial boundaries outline the edges of a realm, a state, a region, a district or, in the Swiss federal context, a canton. They also designate the limits of specific forms of political management of religious matters valid within their geographical borders.

On the other side, symbolic boundaries are drawn upon the ways that individuals mobilize religious, cultural, national or ethnic (re)sources in their social interaction. As Lamont and Molnar define them, symbolic boundaries are "conceptual distinctions made by social actors to categorize objects, people, practices, and even time and space. They are tools by which individuals and groups struggle and come to agree upon a definition of reality" (2002: 168). As such, they are heuristic tools to grasp "the dynamic dimension of social relations, as groups compete in the production, diffusion, and the institutionalization of alternative systems and principles of classification" (*idem*).

Symbolic and territorial boundaries *can*, but *need not*, overlap: some territorial borders are drawn upon symbolic lines while some symbolic boundaries reflect on territorial frontiers. Furthermore, the relations between symbolic and territorial boundaries are mobile and can change according to the historical and social situations. Boundaries can be crossed, blurred or shifted (Zolberg and Woon 1999; Alba 2005; Wimmer 2008a, b). A boundary is crossed when individuals move from on side to another, from one group to another without affecting the characteristics of the demarcation line. A boundary is blurred when its characteristics are not salient anymore, but become fuzzy. As a consequence, individuals are no longer sure on which side of the boundary they can be found. A boundary is shifted when individuals change locations or groups without even moving: "former outsiders become insiders" (Alba 2005: 23).

45.3 The Swiss Juridical Framework

The Swiss juridical system is strongly influenced by its constitutive federalist nature. The Federal Constitution of the Swiss Confederation of April 18, 1999, states: "the principle of subsidiarity must be observed in the allocation and performance of State tasks" (art. 5*a*). This means that the cantons enjoy a large degree of autonomy in the administration of many public sectors, including public education. Public instruction is indeed compulsory but the Federal Constitution provides that: "Education is a cantonal matter" (art. 62). With respect to religion, the Federal Constitution follows a 'double logic,' being simultaneously liberal and communitarian (Ossipow 2003). On the one hand, it guarantees the freedom of religion and philosophy and explicitly asserts that: "no person shall be forced [...] to follow religious teachings" (art. 15). On the other hand, the Constitution declares that: "the regulation of the relationship between church and state is a cantonal matter" (art. 72). This means that the cantons can grant special privileges to specific Churches – including the right to organize RE classes in public schools –, as long as the freedom of religion of the pupils is guaranteed. The law does not forbid dispensing religious education in public schools, as long as the pupils have the right to be excused.

One of the major consequences of this juridical framework is that until very recently Switzerland has known a great diversity of cantonal Church-State systems and an equally large variety of cantonal RE models. As we shall see in the next point, the confessional traditions of the cantons have played a major role in shaping this complex patchwork.

45.4 The Geo-historical Evolution of RE

45.4.1 The First Threshold: The Stabilization of Confessional Tensions

In the sixteenth century, the Protestant Reformation broke the confessional unity of the Swiss Confederacy and, with the exception of few bi-confessional cantons, lead to a political and territorial realignment according to the principle of *cuius regio,*

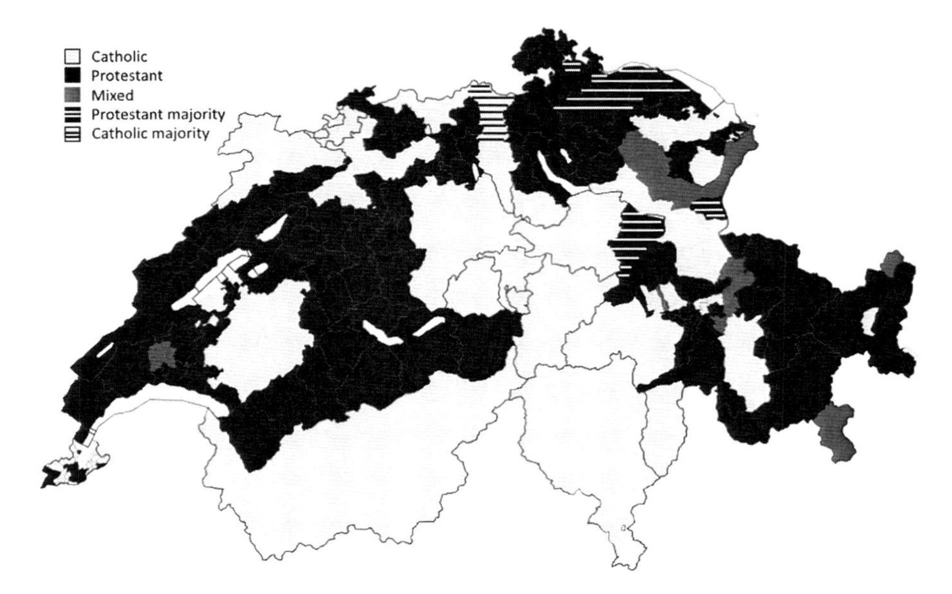

Fig. 45.2 Catholic, Protestant and mixed-confession cantons (and bailiwicks) at the beginning of the eighteenth century, according to Altermatt 1994: 370 (Map by A. Rota with data from Altermatt 1994)

eius religio (Whose realm, his religion). After numerous religious civil wars, Switzerland achieved an instable religious coexistence (Vischer et al. 1995: 99–191; Forclaz 2009) (Fig. 45.2).

Under the Old Regime (*Ancien Régime),* public education was rudimentary and administered by the churches whose teachings were mainly oriented towards the confessional religious education of children. In many Protestant Cantons, for instance, the end of a child's education coincided with the first participation to Eucharist (Caspard 2002).

The attempt to unify and secularize the educational system under the 'Helvetic Republic' (1798–1803) imposed by Bonaparte proved short-lived. It is only after the liberal revolutions in the 1830s that the cantons began to establish secular schools to prepare citizens for their new democratic duties (Bütikofer 2008; Hofstetter et al. 1999). Even though RE became just one subject among many, the confessional orientation of schools was not abolished and the Churches remained strongly influent over public instruction. However the emergence of new political doctrines portended stormy times on the horizon. During the nineteenth century schools became a battleground between the liberal, centralist and (predominantly) Protestant faction and the conservative, federalist and (predominantly) Catholic one.

The Constitution of 1848 ensured the freedom of religion and residence to both Catholics and Protestants throughout the territory of the Confederation for the first time, thus beginning the slow erosion of the cantonal confessional monopolies (Vischer et al. 1995: 207–212). However the new federal government did not dare legislate on the very sensitive question of public instruction, which remained a cantonal matter and thus often subject to confessional segregation (Späni 1997; Criblez 2008).

In the 1870s confessional tensions reached a new high point during the so-called 'culture struggle' (*Kulturkampf*) ignited by the escalation of the Catholic anti-modernism. In spite of the opposition of the Catholic cantons, a new Federal Constitution was voted in 1874 containing several discriminatory articles against the Roman Church. Its article 27 also introduced new basic regulations for public schools, which must be free of charge, under the exclusive supervision of the State, and open to all pupils regardless of their religious affiliation. These new directives caused heated reactions in particular in the Catholic cantons and the right to organize a *confessional* religious education – while allowing the pupils of other confession to be exempted – became the symbol of the demands in favour of the cantonal school autonomy (Criblez and Huber 2008; Späni 1997). The constitutional article of 1874 concerning public education can be interpreted as a first example of the State's intervention to "blur" the confessional boundaries in the public sphere. However, these measures had often the opposite effect of reinforcing the confessional identity against what was considered to be an improper interference especially in the Catholic cantons.

With the reaffirmation of broad cantonal school autonomy and the relaxing of confessional tensions at the end of the century, the debates surrounding RE subsided, and in each canton, the different models of RE crystallized under the various compromises between the liberal political parties and the dominant Church. Figure 45.3 presents an overview of the French-speaking cantons.

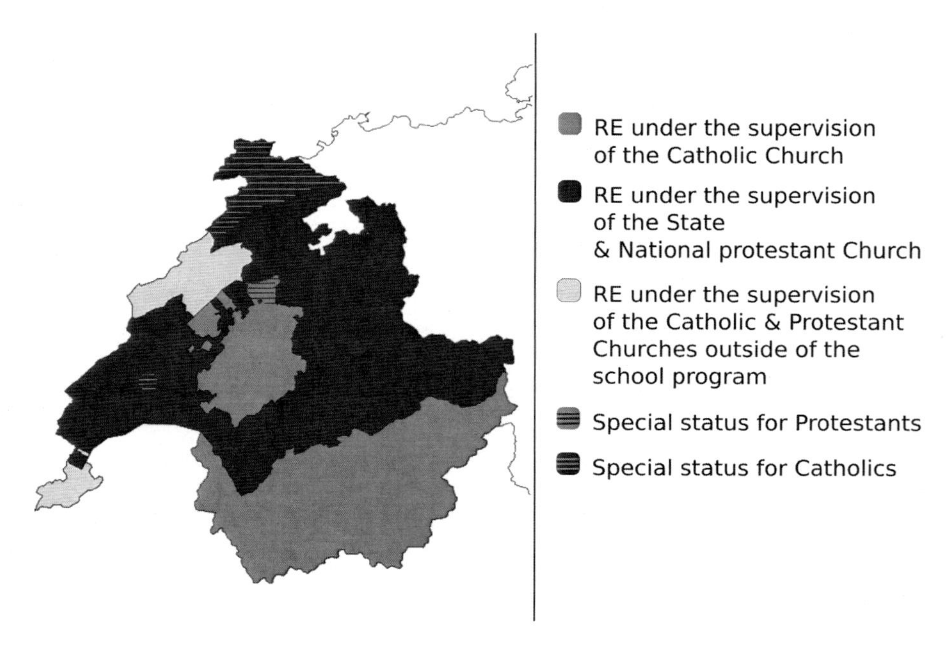

Fig. 45.3 The organization of RE in the French-speaking public schools at the beginning of the twentieth century, according to Rota 2011 (Map by A. Rota 2011)

These configurations were unquestioned for almost a century. However, already after World War II, important changes are afoot.

45.4.2 *The Second Threshold: The Ecumenical Turn of the 1960s*

45.4.2.1 The Rise and Fall of the Catholic 'Parallel Society'

From the middle of the nineteenth century, the Catholic minorities, isolated themselves in a 'Catholic parallel society' with its specific (confessional) institutions: hospitals, universities, journals, recreational facilities, workers associations, and so on (Altermatt 1994). Religious education was an important means to maintain the structures of the parallel society, especially in the Protestant cantons, where the Catholics were a social minority (Späni 1997).[4] Thus, even in the bi-confessional cantons, the confessional boundaries remained unchallenged for a long time.

After World War II, the unity of the Catholic milieu started to crumble. The rising technological and industrial modernization along with the diffusion of mass media favoured the inter-Cantonal mobility and fostered a rising uniformity of the values across the entire Swiss population. In addition, immigration from the Southern European countries significantly increased the number of Catholics in Switzerland in the 1950s and the 1960s, surpassing the Protestants in 1970s. Over these years the juridical differences between the two confessions were appeased and in several Cantons – both Catholic and Protestant – the minority church received the same juridical privileges as the traditionally dominant one. In many cases, these privileges included the possibility of organizing RE in the public schools.

These deep social changes were intermingled with the equally profound theological *aggiornamento* of the second Vatican Council (1962–1965). Among other things, the Council fostered new ecumenical and interreligious reflections and social initiatives across the confessional borders.[5] In Switzerland, since 1965 new official contacts were established between the Protestant and the Roman Catholic Churches and in many cantons, the Churches joined their forces in the field of RE and organized interconfessional-cooperative courses (Bräm 1978; Frank and Jödicke 2009: 286).

The dissolving effects of this ecumenical renewal on the confessional boundaries are particularly visible through the evolutions in the RE schoolbooks. In this respect, the history of the Lausanne-based publishing house *ENBIRO*,[6] which today

[4] In 1917 the reform of the Codex Iuris Canonici offered new doctrinal arguments in favour of the separate religious education of the young Catholics by forbidding them to join non-catholic, neutral or mixed schools opened to non-Catholics as well (Can. 1374).

[5] See the Decree on Ecumenism Unitatis Redintegratio and the Declaration on the Relation of the Church with Non-Christian Religions Nostra Aetate.

[6] ENBIRO is an acronym for French-Speaking biblical and interreligious teaching (ENseignement BIblique et Interreligieux Romand). In 2013 the publishing house was renamed "Editions Agora" to mark its institutional separation from the churches.

prepares and distributes the RE textbooks in five cantons, is very interesting since it allows us to analyze the evolution of the confessional boundaries prior to the intervention of the State in the field of RE. In fact, although this publishing house was originally a Church initiative and is now a non-profit private enterprise, it has become a major partner of the State in reshaping the form and contents of RE in public schools in the French-speaking part of Switzerland.

45.4.2.2 A Short Case Study: The ENBIRO Publishing House

The idea of establishing a publishing house to facilitate the production and distribution of RE textbooks for schools stemmed from an internal reflection of the French-Speaking Protestant Churches. In the early 1960s, these Churches experienced some difficulties in the management of their religious education at school, which was considered to be somewhat redundant with respect to the Sunday's teaching in the parish. The prospected solution was the harmonization of RE programs in all the French-speaking cantons and a clearer distinction between the biblical history classes to be proposed in school with the catechism taught in the parish.

This distinction was not revolutionary in the Protestant theological tradition and was already advocated by liberal pedagogues and theologians in the nineteenth century (Späni 2003). The Catholic Church on the other hand historically opposed it considering that a religious education free of dogmatic references would be "absurd" (Rohn 1879 cit. in : Späni 1999). However the pastoral reforms promoted by the second Vatican Council introduced a genuine paradigm shift in the Catholic understanding of education. The traditional catechism was replaced by an ecumenical life-oriented teaching where missionary attitudes were critically reflected and place was made for dialogue between religions (Klöcker 1996). In particular, the new interest of the Catholic Church in the biblical studies[7] offered a common ground for a fruitful cooperation between Christian denominations. It is on this ground that soon after its foundation the ENBIRO-board was opened to representatives of the Catholic Church as well.

In the successive decades, the guidelines of the ENBIRO textbooks underwent several interesting changes. First the ecumenical concern fostered to the idea of a RE addressed to all the pupils, since a basic understanding of "the Judeo-Christian heritage" of the Western civilization was deemed to be important for everyone in order to learn about their "roots," and the culture where they live. Second, the books integrated an interreligious approach. Although Christianity still occupied the greater part of the program, the textbooks included chapters on other religions as well, among others Islam and Buddhism. Third, the ties between the RE at school and the religious teachings in the parish were progressively severed. RE is now presented as an ordinary school subject in the textbooks with content and approaches that are independent from the specific interests and needs of the churches. Today, the ENBIRO publishing enjoys a great deal of consideration from the public authorities and the original ties to the churches have lost public visibility.

[7] See the encyclical letter Divino afflante Spiritu, 1943.

In short, the history of ENBIRO shows a double process of rapprochement on the one side and differentiation on the other side. Thanks to the social and theological reforms of the 1960s, the Catholic and the Protestant churches found a common ground to teach religion leading to an 'internal *deconfessionalisation*' of RE at school. If, in the private domain of the parishes, the confessional doctrines and practices still play an important role, in the public sphere of the school these distinctions became blurred and lost their *raison d'être*. In parallel, this adjustment lead to a sharper distinction between the RE classes proposed at school and the formative offers in the parishes, the first one becoming increasingly autonomous from the second. Although particularly visible in the field of RE, this kind of process is not limited to this domain. More generally, the differentiation's process causes the secularization (*Verweltlichung*) of those religious services and institutions that are also relevant for other social spheres (Knoblauch 2009: 16). In the case of ENBIRO, this process is intertwined with the (re)emergence of 'religion' as a political factor both at local and international levels and the subsequent transformations of RE at the turn of the twenty-first century. The emergence of the State as an actor in the field of RE is most certainly the central feature of these transformations.

45.4.3 *The Third Threshold: The Intervention of the State*

45.4.3.1 The State as an Actor in the Field of RE

In the last decade, all of the French-speaking cantons (as well as a large number of the other Swiss cantons) have introduced a new RE under the supervision of the State (Jödicke and Rota 2010). These lessons are mandatory for all the pupils regardless of their confessional and religious affiliation. The State's courses replaced those organized by the Churches or are offered as an alternative to them, thus drastically reducing the direct influence of the religious institutions in school.[8] The State's religious education is expected to be 'non-confessional' offering neutral information covering different religions. Indeed, in several cantons the new RE classes are no longer considered to be a 'religious teaching' according to art. 15 of the Federal Constitution and therefore the exemption possibility is no longer granted.[9] Although cantonal specificities still remain, the introduction of these classes is a clear sign toward a standardization of RE policies in Switzerland (Fig. 45.4).

Nevertheless, the path followed by the reforms of RE is only apparently straightforward and its consequences for the redefinition of the confessional and religious boundaries are indeed complex. In order to grasp them we must turn our attention to the evolutions of the Swiss religious landscape and to the debates they have raised.

[8] However, the churches may still have an indirect influence since they are often invited to participate in the deliberative assemblies to discuss the RE reforms.

[9] In this respect one should note that no such decision has yet been referred to the Swiss Federal Tribunal.

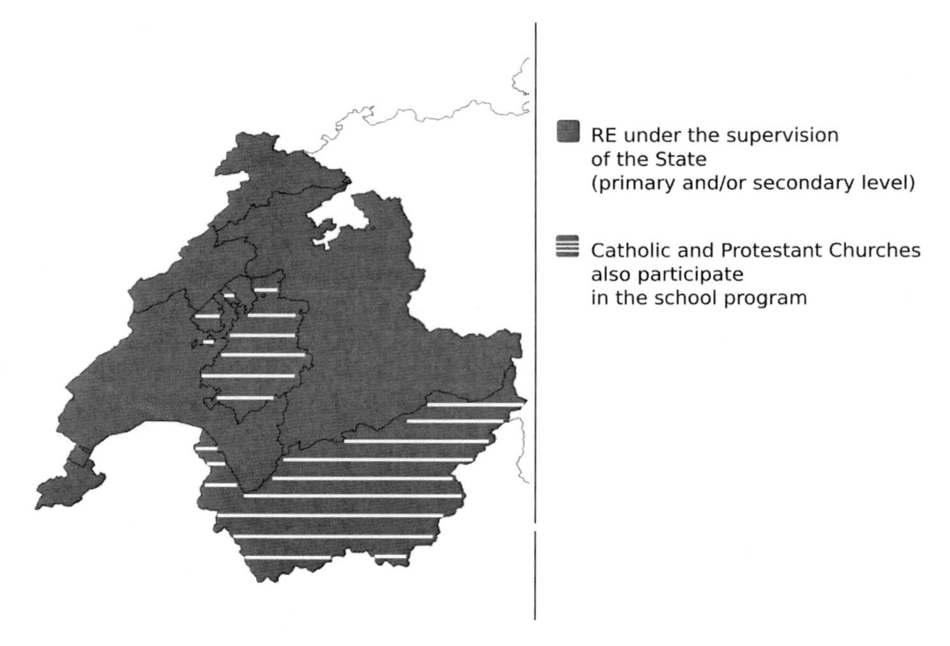

Fig. 45.4 The organization of RE in the French-speaking public schools at the beginning of the twenty-first century, according to Rota 2011 (Map by A. Rota 2011)

45.4.3.2 The Changes of the Religious Landscape as a Call to Action

The increasing uniformity in the organization of RE in the Swiss cantons is not a consequence of the homogenization of the national religious landscape. At the opposite, as we pointed out in the introduction, religion in Switzerland is becoming increasingly plural. This phenomenon is due on the one hand to the immigration of Muslim and Christian orthodox workers and refugees and their families since the beginning of the 90s,[10] and on the other hand to the diversification of the religious beliefs and practices of individuals. Simultaneously, both the Roman Catholic and the Reformed Churches are rapidly loosing members.[11]

These transformations of the religious landscape and the individual religiosity strongly influenced the future of RE in the public schools. It would however be hasty to categorize these reforms as the *automatic* consequence of these religious changes. In fact, these changes are perceived and discussed by social actors sensitive to these issues who then act according to their particular understanding and interpretation of the situation.[12] In this respect, despite the different juridical and

[10] In 2000 4.4 % of the persons living in Switzerland declared to belong to "other Christian denominations" and 4.26 % were Muslims (see Bovay 2004).

[11] In 1970 only 1.14 % of the Swiss population declared no formal religious affiliation; in 2000 this concerned more than 11 % and recent statistical analysis suggest a figure of 25 % in 2009. The same year the Catholics and the Protestants are estimated to represent each about 30 % of the population (Stolz et al. 2011).

[12] We draw here on the methodology proposed by Blumer 1969.

historical background of each canton, it is possible to reveal striking similarities *not only* across the new institutional forms of RE but *also* across the arguments supporting them. Three major intermingled argumentative lines can be identified through the analysis of the public and political debates in the French-speaking cantons between 1985 and 2011 (Rota 2011).

- *Secularization as lost*: In the debates the increasing detachment of a part of the Swiss population from the churches and the resulting crisis of the churches' RE are associated with a loss of precious resources for social and individual life. On the one hand, according to this argument, children are missing important 'knowledge' indispensable to fully comprehend their surrounding cultural environment and their cultural identity. On the other hand, they are lacking ethical references and even losing the possibility to access a 'spiritual dimension of life.'
- *Individualization as danger*: The weakening of the traditional religious identities is depicted in public and political discussions as a risk, in particular for young people. The proponents of this argument stress that the lack of any clear orientation in the religious field leaves children defenseless against the attacks of potentially dangerous 'cults.' Although very present in the 1980s and 1990s when the passions surrounding this question reached a high point in Switzerland, this concern has almost disappeared in recent years (Mayer 1993: 274–354).
- *Pluralization as challenge*: The religious pluralization of Switzerland has become of paramount interest in the public and political arenas in particular in the last decade and the debates are focused on the growth of the Muslim minority in Switzerland (Gianni 2009; Behloul 2010). Pluralization is presented as a challenge to peaceful living in the country. Thus, the promoters of this argumentation underline the need for tolerance and for better reciprocal understanding between members of different religious communities. However, according to a more conservative point of view, they also discuss RE as a means to gathering the young "Swiss" around a shared "Christian culture" and to familiarizing the children from "other cultures" with local norms and values.

At the institutional level these considerations more or less directly question the ability of the churches to provide a religious education adapted to the needs of the contemporary society. Even persons that are themselves close to religious institutions put these concerns ahead. The State is therefore called to intervene.

Behind these conceptions there is however an intricate work of (re)definition of the symbolic boundaries between past, present and future Swiss citizens.

45.5 Conclusion: Past and Present Boundaries

The introduction of a RE class under the supervision of the State has primarily a regulative function. The State is required to take action to prevent the supposed consequences of the new religious configuration in the country. These regulations are legitimized by a certain understanding of symbolic boundaries and at the same time, they contribute to the (re)definition of those boundaries. As the reforms of RE reveal, three main boundaries are being reshaped: the boundaries between traditional Christian confession in Switzerland are being blurred and so are the boundaries between the religious and the secular. The boundaries between in-groups and out-groups have not been blurred, but shifted with the borderline no longer between Catholics and Evangelical Reformed, but between Christians and Muslims.

45.5.1 Blurring the Borders Between Traditional Christian Confession

The symbolic boundary between the Roman Catholic and the Protestant confessions that has conditioned the history of Switzerland for many centuries has lost its clarity today. Although in the theological debates and in the private sphere of religious communities this distinction is still somewhat pertinent, in the public discourses this distinction is no longer capable of producing social or political (and let alone geographical) divisions. This traditional separation has been replaced first by an ecumenical cooperation under the direct action of the Churches and then by an inclusive perspective appropriated by the State for its courses where the specificities of the different religious symbolic systems lose significance in favor of a "cultural" understanding of Christianity and a 'generic religiosity' centered on individual 'existential needs.'

45.5.2 Blurring the Borders Between the Religious and the Secular

Because of its supervision over RE classes, the State handles resources that were previously part of the churches' monopoly. Its purpose is not the religious socialization of children and indeed its lessons are presented as 'opposite' to those previously organized by the churches; however, the public and political debates on the reform of RE in public schools show a peculiar (con)fusion of different conceptions of the role of religion in society In fact, next to a 'cognitive' understanding of the lessons oriented toward the transmission of a certain number of notions a 'moral and existential' conception is also clearly visible. According to this latter perspective, RE should help the children to build their personal religiosity and acquire a particular sensibility for the 'religious dimension' for instance when dealing with their 'existential questions' (see the argumentation in the section above). Because they are considered to be opposite to the 'explicitly confessional' RE of the churches, these courses are not considered to be *religious* teachings. Nevertheless, these conceptions can be considered secularized forms of a liberal Christian theological and pedagogical tradition, though their religious origins are no longer recognized as such in the public arenas (see Rota 2011, 2013).

45.5.3 Drawing the Border Between Christianity and Islam

The introduction of RE classes under the supervision of the State draws attention to an *ambivalent boundary* between Christianity and the other religions. As a *religious category*, Christianity rarely intervenes as a marker in the boundary making processes. When considered as systems of symbols, beliefs and practices, religions are conceived in the public discourses on an egalitarian base, each and every tradition deserving equal respect and consideration, particularly in the RE lessons. However, in the discourse, the religious criterion is often activated as a socio-cultural marker. Christianity is thus presented as the "indigenous" religion, as a (secularized) Western heritage that contributed to shaping local culture and in which the moral values of people are rooted. According to these conceptions, the religious marker can easily be activated in order to draw a clear-cut boundary between groups and to exclude members of other religions from fully participating in the Swiss society. In today's context, Islam is considered as a particularly "foreign religion" with customs and practices that are incompatible with the Swiss way of life. In this respect, the State's RE is stretched with potentially contradictory goals. On the one hand, RE should aim to foster toleration and peaceful cohabitation between religions; on the other hand, it is required to stress the Christian roots at the foundation of the country as an implicit framework to assimilate the children from other cultures and/or religions.

45.5.4 From Clear Confessional Boundaries to Deterritorialized Politics of Religions

In this paper, we discussed the history of RE in public schools as an example to illustrate the (re)formulation of the confessional boundaries in Switzerland. In our analysis we could identify three major moments: the establishing of the confessional boundaries beginning in the sixteenth century and their stabilization at the end of the nineteenth, the ecumenical turn of the twentieth century in the 60s and their definitive blurring in the public sphere at the beginning of the twenty-first century due to secularization, individualization and pluralization.

This new situation does not put into question the territorial boundaries of the cantons. However it addresses new questions concerning the way they will deal with religion(s) in the public sphere in the future. Indeed the cantonal autonomy in regulating the State-Church relations, coupled with the guarantee of freedom of religion at a national level, has been an important instrument to ensure peaceful cohabitation within the Helvetic Confederation. But the fading of confessional identities and the pluralization of religious orientations confront everyone with similar needs and challenges. Although the sovereignty of the cantons in the handling of religion on their territory is not questioned, the convergence of the cantonal politics of RE suggests a trend toward an increasingly deterritorialized regulation of religion in the public sphere. In fact, a common RE curriculum has already been proposed to all French-speaking cantons as a part of a broader harmonization of the school programs. A similar program is to be introduced in the German-speaking cantons starting 2014.

The reform of RE and the new symbolic boundaries that they reveal have practical consequences also for the religious communities. The new symbolic boundaries created to exclude Muslims can result in actual forms of social exclusion as well as new forms of territorial segregation, for instance in the urban space. However, for the traditional Churches the question is how not to loose all visibility in the public sphere. Indeed, the disappearance of the confessional categories as social markers means new possibilities of cooperation, but also the loss of a clear profile in the religious field.

References

Alba, R. (2005). Bright vs. Blurred boundaries: Second-generation assimilation and exclusion in France, Germany, and the United States. *Ethnic and Racial Studies, 28*(1), 20–49.

Alberts, W. (2007). *Integrative religious education in Europe: A study-of-religions approach.* Berlin: W. de Gruyter.

Altermatt, U. (1994). *Le Catholicisme au défi de la modernité*. Lausanne: Payot.

Baumann, M., & Stolz, J. (Eds.). (2009). *La nouvelle Suisse religieuse*. Geneva: Labor et Fides.

Behloul, S. M. (2010). Religion und Religionszugehörigkeit im Spannungsfeld von normativer Exklusion und zivilgesellschaftlichem Bekenntnis. In M. Sökefeld & B. Allenbach (Eds.), *Muslime in der Schweiz* (pp. 43–65). Zurich: Seismo.

Blumer, H. (1969). The methodological position of symbolic interactionism. In *Symbolic interactionism: Perspective and method* (pp. 1–60). Englewood Cliffs: Prentice-Hall.

Bovay, C. (2004). *Le paysage religieux en Suisse*. Neuchâtel: Office fédéral de la statistique.

Bräm, W. K. (1978). *Religionsunterricht als Rechtsproblem im Rahmen der Ordnung von Kirche und Staat*. Zurich: TVZ.

Bütikofer, A. (2008). Das Projekt einer nationalen Schulgesetzgebung in der Helvetischen Republik (1798–1803). In L. Criblez (Ed.), *Bildungsraum Schweiz* (pp. 33–53). Bern: Haupt.

Campiche, R. J., et al. (1992). *Croire en Suisse(s)*. Lausanne: l'Age d'Homme.

Campiche, R. J., et al. (2004). *Les deux visages de la religion. Fascination et désenchantement.* Geneva: Labor et Fides.

Caspard, P. (2002). Examen de soi-même, examen public, examen d'Etat. De l'admission à la Sainte-Cène à aux certificats de fin d'études, XIVe-XIXe siècles. *Histoire de l'éducation, 94*, 17–74.

Criblez, L. (2008). Die Bundesstaatsgründung 1848 und die Anfänge einer nationalen Bildungspolitik. In L. Criblez (Ed.), *Bildungsraum Schweiz* (pp. 57–86). Bern: Haupt.

Criblez, L., & Huber, C. (2008). Der Bildungsartikel der Bundesverfassung von 1874 und die Diskussion über den eidgenössischen "Schulvogt". In L. Criblez (Ed.), *Bildungsraum Schweiz* (pp. 87–122). Bern: Haupt.

Forclaz, B. (2009). La diversité religieuse en Suisse depuis la Réforme. In M. Baumann & J. Stolz (Eds.), *La nouvelle Suisse religieuse* (pp. 95–105). Geneva: Labor et Fides.

Frank, K., & Jödicke, A. (2009). L'école publique et la nouvelle diversité religieuse. In M. Baumann & J. Stolz (Eds.), *La nouvelle Suisse religieuse* (pp. 283–293). Geneva: Labor et Fides.

Gianni, M. (2009). Citoyenneté et intégration des musulmans en Suisse : adaptation aux normes ou participation à leur définition? In M. Schneuwly Purdie, M. Gianni, & M. et Jenny (Eds.), *Musulmans d'aujourd'hui: identités plurielles en Suisse* (pp. 73–94). Geneva: Labor et Fides.

Hofstetter, R., et al. (Eds.). (1999). *Une école pour la démocratie: naissance et développement de l'école primaire publique en Suisse au 19e siècle*. Bern: P. Lang.

Jödicke, A. (2009). Das Verhältnis der römisch-katholischen und der evangelischen Kirche in Deutschland zum Ethik-Unterricht am Beispiel Berlins. In M. Delgado, A. Jödicke, & G. Vergauwen (Eds.), *Religion und Öffentlichkeit* (pp. 221–241). Stuttgart: Kohlhammer.

Jödicke, A., & Rota, A. (2010). *Unterricht zum Thema Religion an der öffentlichen Schule*. Final Report for the Swiss National Program 58 "Religions, the State, and Society." Retrieved: April 4, 2012, from www.nfp58.ch/files/downloads/Joedicke_Schule_Schlussbericht_def.pdf

Klöcker, M. (1996). Der Paradigmawechsel der römisch-katholischen Erziehung und Bildung. In F.-X. Kaufmann & A. Zingerle (Eds.), *Vatikanum II und Modernisierung* (pp. 333–352). Paderborn: Schöningh.

Knoblauch, H. (2009). *Populäre Religion*. Frankfurt am Main: Campus.

Lamont, M., & Molnár, V. (2002). The study of boundaries in the social sciences. *Annual Review of Sociology, 28*, 167–195.

Mayer, J.-F. (1993). *Les nouvelles voies spirituelles. Enquête sur la religiosité parallèle en Suisse*. Lausanne: L'Âge d'Homme.

Ossipow, W. (2003). La double logique des relations Eglise / Etat en Suisse. Une perspective de théorie politique. *Archives de sciences sociales des religions, 121*, 41–56.

Rohn, J. A. (1879). *Das Unding eines confessionslosen Religionsunterrichtes*. Vortrag gehalten den 12. September 1878 an der Generalversammlung des schweiz. Pius-Vereins zu Stans. Solothurn.

Rota, A. (2011). *L'enseignement religieux de l'Etat et des communautés religieuses. Une étude sur la présence publique des religions en Romandie et au Tessin*. PhD thesis, University of Fribourg, Fribourg.

Rota, A. (2013). Religious education between the state and religious communities. In A. Jödicke (Ed.), *Society, the state, and religious education politics* (pp. 105–127). Würzburg: Ergon.

Schneuwly Purdie, M. (2011). *Peut-on intégrer l'islam et les musulmans en Suisse?* Charmey: Editions de L'Hèbe.

Späni, M. (1997). Umstrittene Fächer in der Pädagogik. Zur Geschichte des Religions- und Tunrunterrichtes. In H. Bertscher & H.-U. Grunder (Eds.), *Geschichte der Erziehung und Schule in der Schweiz im 19. und 20. Jahrhundert* (pp. 17–55). Bern: Haupt.

Späni, M. (1999). La laïcisation de l'école populaire en Suisse au 19e siècle. In R. Hofstetter et al. (Eds.), *Une école pour la démocratie: naissance et développement de l'école primaire publique en Suisse au 19e siècle* (pp. 229–251). Bern: P. Lang.

Späni, M. (2003). The organisation of public schools along religious lines and the end of Swiss confessional states. *Archives de sciences sociales des religions, 121*, 101–114.

Stolz, J. (Ed.). (2006). Salvation goods and religious markets. *Social Compass (special issue), 53*(1), 5–123.

Stolz, J., Koenemann J., Schneuwly Purdie, M., Englberger, T., & Krueggeler, M. (2011). *Religiosität in der modernen Welt*. Final report for the Swiss National Program 58 "Religions, the State, and Society." Retrieved April 4, 2012, from www.nfp58.ch/files/downloads/Schlussbericht__Stolz.pdf

Stolz, J., Chaves, M., Monnot, C., & Amiotte-Suchet L. (2012). *Die religiösen Gemeinschaften in der Schweiz : Eigenschaften, Aktivitäten, Entwicklung*. Final report for the Swiss National Program 58 "Religions, the State, and Society." Retrieved April 4, 2012, from www.nfp58.ch/files/downloads/Schlussbericht_Stolz_Chaves.pdf

Stolz, J., Stolz, J., Koenemann, J., Schneuwly Purdie, M., Englberger, T., & Krueggeler, M. (2014). *Religion und Spiritualität in der Ich-Gesellschaft. Vier Gestalten des (Un)Glaubens*. Zurich: Theologischer Verlag Zurich (TVZ).

Vischer, L., Schenker, L., Dellsperger, R., & Fatio, O. (Eds.). (1995). *Histoire du christianisme en Suisse: une perspective œcuménique*. Geneva: Labor et Fides.

Wimmer, A. (2008a). Elementary strategies of ethnic boundary making. *Ethnic and Racial Studies, 31*(6), 1025–1055.

Wimmer, A. (2008b). The making and unmaking of ethnic boundaries: A multilevel process theory. *American Journal of Sociology, 113*(4), 970–1022.

Zolberg, A. R., & Woon, L. L. (1999). Why Islam is like Spanish: Cultural Incorporation in Europe and the United States. *Politics Society, 27*(5), 5–37.

Chapter 46
Missionary Schools for Children of Missionaries: Juxtaposing Mission Ideals with Children's Worldviews

John Benson

46.1 Introduction

For people living in their native land, the local public school is the main schooling option used by most parents, whatever walk of life. For missionary children, living in a land they called their own but their parents did not, often the main option was a private boarding school that reflected some of the values that their parents hoped would become part of their children. These two worldviews often did not mesh because of how the different generations saw this place: for the missionaries, the land where they were living was seen as the "mission field"; whereas for the missionary children, it was just "home" or a country such as "Tanzania." This chapter examines two generations of a Lutheran missionary community who served in Tanzania from the late 1940s through the early 1990s and their Baby Boom children of which the author was a member. The chapter begins by looking at missionary children education through time, delineates why the children may have developed a different worldview than their parents, the role of boarding schools, and then highlights the philosophical basis of three schools attended by this group of children and the response by parents and children to these schools. Differing geographic and religious views intersect in how parents and children saw the roles of these missionary children's educational institutions.

J. Benson (✉)
School of Teaching and Learning, Minnesota State University, Moorhead, MN 56563, USA
e-mail: bensonj@mnstate.edu

S.D. Brunn (ed.), *The Changing World Religion Map*,
DOI 10.1007/978-94-017-9376-6_46

46.2 Historical Background to Missionary Children Education

A call for missionary service is usually felt by an individual. Often missionaries would try to find a spouse who shared this same interest in a call. But for many, the parents of this missionary couple and their children were not seen as really part of the call. In fact when Jesus was exhorting some of his first disciples for missionary services in the villages of Judea in Matthew 10: 37–39, he stated:

> Anyone who loves his father or mother more than me is not worthy of me; anyone who loves his son or daughter more than me is not worthy of me; and anyone who does not take the cross and follow me is not worthy of me. Whoever finds his life will lose it; and whoever loses his life for my sake will find it. (New International Version)

This is how the first Lutheran missionaries, and others from many denominations, often dealt with their children, leaving them back in the United States while they went to fulfill their calls. If children accompanied them, they might be educated through home schooling programs like Calvert School (This was one of the original organizations that delivered home school material around the world from their office in Baltimore, MD) or they might attend schools formed by other mission societies, such as Rift Valley Academy in Kijaba, Kenya. The Lutheran mission tried to set up a school for missionary children for many years, but the first supplies for a new school were lost in the sinking of the S.S. *Zamzam,* by German torpedoes. The Egyptian-owned *S.S. Zamzam* was a neutral ship carrying 201 passengers, of which 144 were missionaries, and supplies to Africa in 1940 (Anderson 2000; Swanson 1941). It was not until 13 years later that the mission was able to set up Augustana School in Kiomboi, Tanzania.

In reading about the schooling choices for the children of the early missionaries, it is heartbreaking to think of some of the options that were done with some of those children about 80 years ago. V. Eugene Johnson spoke about this as their family prepared to return to Africa in 1934 (*Lutheran Companion* 1934: 1194):

> We feel the pain of separation especially in regard to our eleven-year old daughter Doris, whom we are leaving in America because of the lack of educational facilities for our missionaries' children on the mission field; also on account of the severity of the climate and the degradation of the heathen environment…. God has most wonderfully answered our prayers…. He has provided a home for our daughter Doris, where she may receive both the physical and spiritual care she needs. She is to be the "adopted" daughter of Pastor and Mrs. J. Henry Bergren of St. Cloud, Minn…. Another marvelous answer to prayer is that God has made our daughter willing, for Christ's sake and for the sake of the souls in Africa for whom He died, to be separated from her parents and brothers during the long years that we shall be laboring in Africa.

In his book, *Pioneering for Christ in East Africa*, V. Eugene Johnson (1948: 106–7) elaborated on how they found the Bergren family only 2 weeks before leaving for Tanganyika and the struggles they had in considering the possibility of Doris staying back in the United States:

> We were to leave for Africa the first part of September. I came to St. Cloud, Minnesota about the middle of August to speak in the Salem Church there, being the guest of Pastor and Mrs. J. Henry Bergren. During our evening meal together our conversation touched

> upon our children, and I mentioned the fact that my wife and I were still waiting upon God for guidance as to what arrangements we should make for the care of our daughter Doris. I could see that this announcement made a deep impression upon Pastor and Mrs. Bergren. They withdrew for a few minutes to another room and when they came back they asked if my wife and I would consider an offer for them to care for Doris. Something within me whispered: "This is of the Lord." And, like Abraham's servant before Rebekah (Genesis 24: 26), I bowed my head and worshiped…. My wife and I were of one mind in believing that we had been guided of the Lord to leave our beloved daughter Doris with Pastor and Mrs. Bergren. At that time they had no children. When we told Doris of the whole matter, she too believed it was of the Lord and, like Rebekah, consented to his guidance…. Many times after that in Africa I was tempted to believe that in only in a moment of mental aberration could I have done such a thing. But the Lord granted grace both to us and to Doris sufficient to bear the separation, and Pastor and Mrs. Bergren eased both her and our hearts by proving themselves in every respect kind and understanding and generous foster parents. We believed that for Doris as well as for us the Lord has truly fulfilled his promise: "Everyone who has left houses or brethren or sisters or fathers or mothers or children or lands, for my name's sake, will receive a hundred fold" (Matt. 9: 29). And for Pastor and Mrs. Bergren, too, there was a promise fulfilled: "Whoever gives to one of these little ones even a cup of cold water because, he is a disciple, truly, I say to you, he shall not lose his reward" (Matt. 10: 42). They gave infinitely more than a cup of cold water. Is it too much to say that one of their rewards was a son of their own born January 3, 1938?

There was a sense in Rev. Johnson's writing that this was part of fulfilling his Call, but it seems somewhat cruel for the child. Having a school in-country was something desired by most missionaries.

46.3 Development of a Differing World View by the Children

The missionaries' views of Tanganyika (later in 1965, the country became known as Tanzania) centered on the natures of their call and it remained a "mission field" throughout their lives. For the children, being raised on the "mission field," playing regularly with African children, exploration of the landscape, and a different religious development than their parents made for a connection to this East African country different than that of their parents. While the missionaries often developed their sense of a call to mission at a very young age and this was nurtured through their lives in mission-minded churches in the United States; the missionary children did not share the same sense of call. Unlike their parents, their own religious development took a long time in coming, often not being finalized until long into adulthood.

The missionary children did not see themselves as Americans. While regular furloughs every 4 years took the missionary families back to America, these were seen as aberrations in the lives of the missionary children. Often they longed for this year to be over, so they could go back "home." Their closest friendships were formed with the Tanganyikan/Tanzanian families at their mission stations, but often this was severed with the 9 months of each year that was spent at boarding school. The children explored the landscapes both at home and at school and formed a deep attachment to these places because of it. The worldview of the missionary children, then, was not tied to the religious landscape of their parents, but one that was tied to their geographic landscape. The lives of missionary children were tied to the local people, although there was not a shared culture. The missionary children did not view themselves entirely as Americans, nor were they Tanzanians. Their worldview was somewhere in-between.

46.4 Boarding Schools and Their Part in Fulfilling the Call

Eventually, schooling in-country became an option largely composed of boarding schools established by different mission societies. In this country, the Native American boarding schools are seen as a horrific part of Native American cultural history as they tried to assimilate the children into becoming more "American" (McBeth 1983; Coleman 1993). They are also part of the historic framework of most of the overseas lives of children of missionaries. While most of us look on our schools with some fondness, placing their own children in boarding schools is never considered a desirable option. It certainly did not feel right for many of the missionary children.

While much of the time in boarding school was very enjoyable, the bouts of homesickness could make it devastating to young children. Ruth, one of those who attended a missionary boarding school, described what it was like for her:

> I do remember coming home after the first three months and my concern was that I had no facial memory of what my mother looked like, or my father. I just couldn't remember what they looked like and so I was kind of worried we'd get to the place where we met them and I wouldn't know who they were…. I remember being on the bus coming back and the older kids, Ozzie and my older brothers, trying to reassure me that I surely would remember what she looked like and that she had red hair. And I did of course, but I had not seen them for three months. My mother always sent a picture along after that. But I really didn't have a picture that first term.
>
> I did not like going to boarding school. I think I was fine at boarding school but, boarding school is probably the trauma of my childhood. The trauma was because of the separation. The vacations in-between I would be fine for a couple weeks but as I got closer to a week or two before we were going to leave, it became very difficult. When I was little I would cry, probably a week or two before the vacation ended. Once there, I was pretty adjusted to it and had pretty good friends, but I do think you're pushed too far into your peers without any parental alternative in that setting with so few parental figures. I think the separation was hard for me.

Many discussions have gone on between missionary parents and children long into adulthood about the impact of boarding schools on the second generation. Missionaries Dean and Elaine discussed their thoughts on boarding schools and their work after a long discussion with their son Mike:

Dean: In all these decisions, you have to just believe that grace covers your inadequacies. The same is true, I wrote in this journal … we got talking at Mike's table one night and he told how tremendously homesick he had gotten, and it really wrung my heart….

Elaine: We all went through that I think.

Dean: Well, you know our kids still say that it was a good experience and they don't condemn us for doing it, but you know when you look back at that and you see that, that night I thought a long time and the only thing I could say was that we went in those days with an overriding conviction that God had called us to spend our life out there and so you had to juxtapose all other things, your family and everything under that umbrella, but I mentioned in there that maybe it wasn't right all the time, but you still have to trust that grace takes care of this.

Elaine: But we saw so many families come and go home and never could finish their work because they went home for their kids to go to school at home and they never could fulfill any of their work that they started to do in Africa.

It is within this framework of very different views of boarding school that we see another juxtaposition of differing values between the generations.

46.5 Three Schools for Missionary Children

For the missionaries in Tanganyika/Tanzania in this period there were three boarding school options: Augustana School in Kiomboi, Tanzania; Rift Valley Academy (RVA) in Kijabe, Kenya (located about 45 miles north of Nairobi), and International School Moshi (ISM) in Moshi, Tanzania. Augustana School was a small school (maximum attendance of 100 pupils) located in central Tanzania and run by our mission (Fig. 46.1). The school went up to eighth grade when most students would continue to high school at Rift Valley Academy in Kijabe, Kenya. This school was run by the interdenominational Africa Inland Mission (AIM). It had a very different religious flavor to it than Augustana School. When Augustana School closed in 1971, many of the missionary children transferred to ISM which had opened as a day school in 1969. The following section provides the values of these schools as detailed in various documents, the experience of Lutheran missionary children attending these schools, and the reaction of parents to these schools.

46.5.1 Augustana School in Kiomboi, Tanzania

Located on the high and dry Iramba Plateau of central Tanzania, Augustana School was situated about a mile from the main Lutheran mission hospital for this area. The school had two parts, a living area of large dorm rooms for up to ten children, a large eating area, and the house parents living area. This was separated from the classroom area with four large classrooms and an office area by a short walk of several hundred feet. To the east of the school was a soccer field, with a walking path bisecting it diagonally, and the *shamba*, an orchard-like area with guava trees and orange trees surrounding some gardens and a stream. This area was explored by the missionary children on a daily basis. The landscape was dry and open with different

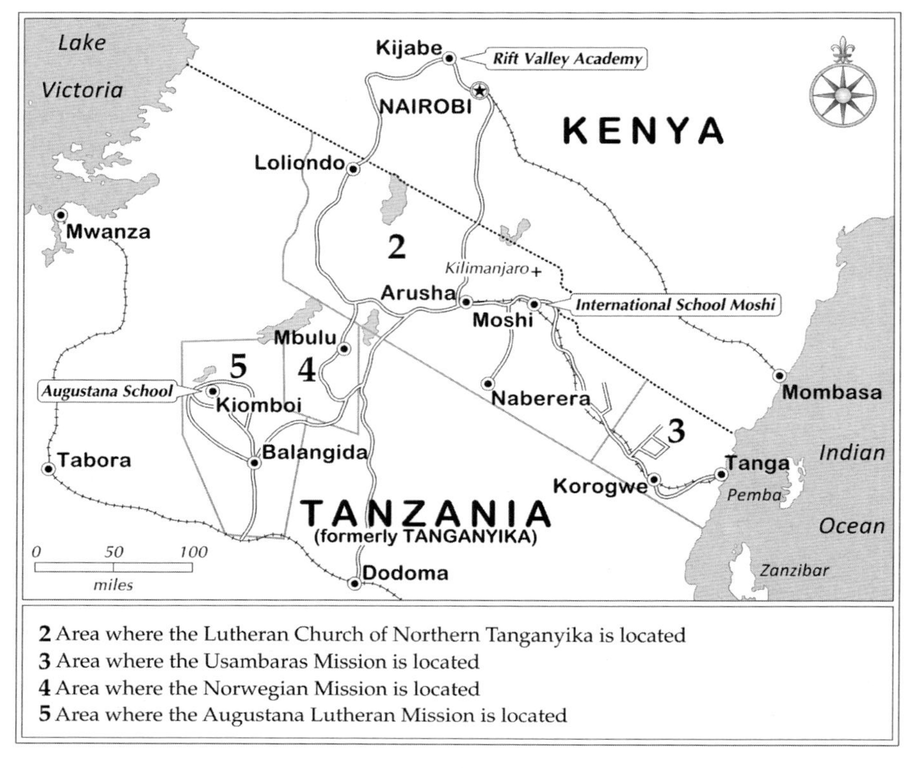

Fig. 46.1 Lutheran mission fields in Tanganyika/Tanzania and locations of boarding schools attended by Lutheran missionary children (Map by Dick Gilbreath, University of Kentucky Gyula Pauer Center for Cartography and GIS; commissioned by the editor) (Johnson 1955)

rock outcroppings and various large trees for climbing, many of which were given names (Big Tree, Police Tree, Jungle Jim, etc.). Behind the living area was an area where roads were created for Dinky and Corgi toys (familiar icons for British children but not American ones) to drive along. At times, shovels and *jembes* (the large African hoes) were used to dig tunnels under ground, some large enough to include several rooms. About a mile below the school to the south, was "The River." Often a dry river bed, water could be found by digging about a foot down in its sandy bottom. About 5 mi. (8 km) from the living area, lay the Blue Hills. On some Saturdays, a group of students might hike there, taking small mirrors with them and when they reached there, they would shine their mirrors back towards the sun and the school to signal their arrival. It was a landscape that was free and open and to the children a somewhat magical place to explore. Apart from the name of Kiomboi village, none of these places could be found on a map. Instead, it was a landscape explored and named by children. Decades later, the children who went there could still identify which tree was which and its approximate location (Fig. 46.2).

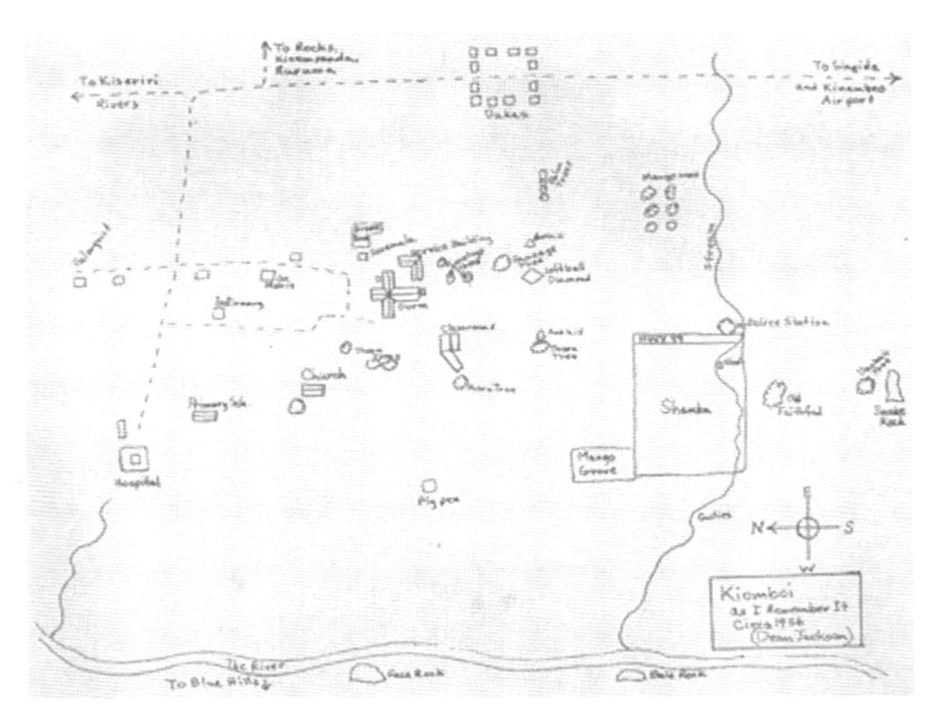

Fig. 46.2 Map of Augustana School and environs made from memory by Dean Jackson, a former student at the school (Source: Missionary Schools for Children of Missionaries, by Chris Nelson 1981; used with permission of the author and the artist)

To residents of Kiomboi, not living at Augustana School, these places might remain un-named or go by completely different names. The children's worldviews continued to be influenced by our strong ties to the landscape home mission stations and at the school.

However, the school remained a fairly culturally isolated island. In setting up the school, some of the main planners and first directors stressed that this would be a school that prepared the children for a life in America. In a letter written in 1940, 13 years before the school opened, pioneer missionary, Elmer Danielson wrote to the Africa School Committee of the Augustana Mission Board, stating that

> The paramount issue in regards to the school is that what is best for the children should be the sole determining factor in the whole problem ....
>
> The children must be guided educationally and socially for life in the United States, and to fit into American life normally and happily. If God guides one here and there to return to Africa to serve Him and his fellow-men, we shall be very happy—but we cannot for a moment, either let them drift educationally and socially, or guide them for life in Africa, but we must guide them in their foundations, for life in the States, in all fairness to their future. We do not want children returning to Africa when they have grown up, because they are unhappy in the States, or because they are misfits in American life—but because God

> called them. So I believe school for the children on the field, run on lines as comparable to school-life in the States as possible, is a debt which is owed to the children and their nearly-as-possible normal development and training .... If the school isn't normal enough along all lines, it can fail just as much as the isolated home in preparing the children for life in America; for I believe the guiding principal of the school should not merely be to give the children an education, but to prepare them for life in the States. (ELCA Archives).

Arthur Anderson, the first director of the school when it opened in 1953 described it to readers of *The Lutheran Companion*, the main periodical of the Augustana Lutheran Church, in a poetic essay; he entitled "What is the American School?" Some key phrases in his essay highlight the importance placed on America and Christianity:

> ... It is the face of the children as they salute the flags, and pledge allegiance to the Savior and His Word, or sing America, or bow their heads in prayer.... Our American school in Tanganyika is all these, but much more; it is a place where we learn to know and love our country, to ride again with Paul Revere, to fight with Sam Houston, to sorrow with Lincoln... It is place where we learn to know our Savior better from day to day and try to live in ways that please him.... It is here we learn to value our freedom and our heritage, to measure the achievements of our nation, to dream of new undertakings and enterprises which will ennoble it, and to pledge ourselves to win victories for our Lord and His Kingdom... This is our American School in Tanganyika.

Despite the American emphasis of the curriculum and its founder, it was a school that welcomed students from many different nationalities. In addition to the American national anthem, the Norwegian anthem, the Tanzanian anthem (when it became independent in 1961), and the Christian Anthem ("Stand Up, Stand Up for Jesus") were heard at school assemblies. Only two Tanzanian children attended the school. It was a small school that never had more than 100 students. It had 56 students in 1965 and only 23 students at its end in 1971. The curriculum was strongly American-based. The study of Tanzania was not included in these early days after its freedom from colonial rule. The students learned American history and geography, celebrated American holidays, and watched Walt Disney movies on weekends.

In 1967, Al Gottneid, an educational missionary who had children at the school, wrote to Rube Pederson of the Board of World Missions about some concerns that the Tanzanian government expressed about Augustana School. He said that the socialist-leaning Tanzanian government was not against private education for foreigners living in the country, but they supported schools that were international in spirit rather than "nationalistic enclaves." Al Gottneid did not think that Augustana School with its American emphasis, or a Swedish school in Nzega, or a German school in Dar-es-Salaam, could fit the international spirit of the law:

> Official and semi-official reaction aside, the very concept of nationalistic enclaves—and that is exactly what these schools are—is incompatible with the spirit and policy of both Church and State in building a multi-racial Church and Society. The opinion expressed by a very few missionaries, that we are already in an international situation by the fact that we are in Tanzania, is a myth as far as our children in these schools are concerned. They live, breathe, read, think, play and work as Americans or Germans or as Swedes. The relationship to Tanzanians working in these schools (and with the odd exception of token acceptance of

> one or two nationals, their only contact is with workers) is that of a member of a household to household servants. Other nationals, always in a very minority, must be Americanized (sic) or Germanized (sic) or Swedishized (sic), to adequately adjust. (Gottneid 1967)

While many formed close friendships with Tanzanian children on mission stations, boarding school often broke these bonds. As Tanzanian children developed, they were responsible for more chores in the homes, and the contrast was especially strong between the missionary daughters and African daughters.

For many years, Augustana School had its own English language worship services, but eventually after much debate in the mission, the students also started attending Kiomboi Lutheran Church, 300 ft (91 m) below the school, behind Big Tree. Religiously, the school fit in with the religious lives the parents were molding for the children at home. The school went up to eighth grade to coincide with the Lutheran rite of passage of Confirmation. No graduation services were held for the 8th graders, only a final Confirmation service.

For most of the children, Augustana School was a cherished place. The house parents were kind, the four teachers that compromised the staff of the school were dedicated teachers and some remain friends of the alumni to this day.

Still, the children were wedded to Tanzania, and not to America. Furlough years in the United States were simply to be endured. The school with its emphasis on an American curriculum and cultural celebrations did not fit the "Third Culture" nature of the student body. The curriculum should have celebrated Tanzania, rather than ignore it, nor did it address the in-between place as children raised by American parents, but tied to the landscape of Tanzania. While reflecting well the religious traditions of the parents, it did not fit the lives of the children who lived and studied there.

46.5.2 Rift Valley Academy in Kijabe, Kenya

Rift Valley Academy (RVA) is a school that serves the children of missionaries from throughout Eastern and Southern Africa. It is located on the edge of the Great Rift Valley about 45 mi (73 km) north of Nairobi in Kijabe, Kenya, which was known as the world's largest mission station. The location is stunning as one has beautiful views of the bottom of the valley and the extinct volcano, Mount Longonot. Winter jackets are often required in the morning as the elevation is above 7,000 ft (2,296 m). Like Kiomboi, it was a wonderful place to explore, perhaps more for the boys as the girls were not given permission to hike off campus when the author attended there in the mid-1970s. Key places around Kijabe included Celebration Point for climbing on rainy days or hikes over to Jubilee Falls on other days. These were the names given by generations of RVA students and not mapmakers. Students might try jump a train and head to Mount Longonot or hitchhike to the plains below. All of this was touched with danger as there were many rules at this school and many were afraid of what might happen to them if they were caught.

Most Lutheran missionary children started attending RVA from the 9th grade. While as adolescents they loved the setting, the legalism and the emphasis on religious ideals of being saved, having a born again experience, and adult baptism were hard for the Lutheran students who had lived in a Lutheran cocoon most of their lives to understand. Some Lutheran missionary families were more evangelical in their outlook and the switch to RVA was not as hard for children from those families, but for some, it was a huge change. These Lutheran adolescents had the feeling that they were looked on as different, because they saw things very differently than many of the other students who grew up in more evangelical missions. Because of these often stark religious differences between the Lutheran mission and the much more dominant evangelical focus of RVA, many of the Lutheran students found it difficult to make sense of their own Christianity until well into adulthood.

One of the key sites at RVA is the cornerstone to the main Kiambogo building laid by former President Teddy Roosevelt in 1909 on his African hunting trip. Still, despite this early American stamp on its character, the early years of the institution reflected the multinational nature of the AIM that ran the institution. In the 1920s, when the Augustana Lutheran Church first began working in Tanganyika, the American Lutheran missionaries often felt more at home working with the AIM missionaries, than they did with the German Lutheran mission, the Leipzig Mission, whom they had worked with jointly in northern Tanganyika (*Lutheran Companion* 1920a, b; Johnson 1930; Smedjabacka 1973). In the early years, no Lutheran children attended the school, and its curriculum was much more British (Dow 2003). By the early 1970s, RVA felt very American. It appeared to be a Southern US high school in the middle of Africa. Teachers were called, "Sir" and "Ma'am." Visiting local soccer, basketball, and rugby teams enjoyed the beautiful cheerleaders. American textbooks were used in almost all classes. The curriculum, like that at Augustana School, consisted of little that connected its student body to its African setting. The student body was somewhat diverse. The 1976 graduating class of 62 students consisted of 46 Americans, 9 Canadians, 5 Kenyans, 1 Nigerian and 1 Norwegian. But apart from the setting, the American influence dominated.

While Augustana School's founders had emphasized its American-ism, the founders of RVA did not articulate a clear philosophy on what their school would do. Phil Dow in his history of the institution felt like the school was more of an afterthought of the mission that was emphasizing evangelism rather than the education of their children. In an article in the AIM newsletter, *Africa Inland*, in 1946, RVA's principal, Herbert Downing (quoted in Dow 2003: 101) then and when the author began at RVA in 1973, articulated what he felt were the two main goals of the school: (1) as a support for the missionaries so they could remain working in Africa, and (2) to train future missionaries. Within the AIM, many generations of missionaries were raised in Africa, who then returned to serve as missionaries as adults. Missionary dynasties remain a part of the AIM, whereas the Lutheran mission believed that they were there to hand over control to the African church. The Lutheran mission peaked at the time of Tanzanian Independence in 1961 with less-and-less missionaries thereafter. Lutheran missionary children growing up in the same East African setting saw little chance of returning as missionaries, in the way an AIM missionary child might.

The strongest part of RVA culture was its evangelical nature. Phil Dow stated this in his book and the Lutherans, coming from a tradition of liturgical worship and infant baptism, felt it most acutely. For his book, Dow interviewed some Lutheran alumni from the 1950s and found that they had been given the message consistently that Lutherans might not be able to be saved because of their closeness to the Catholic faith. Most of the Lutheran "missionary kids" that were interviewed for my own study reiterated this feeling of not being fully accepted into the school because of our parents' faith tradition.

RVA is an American evangelical school located in the middle of Kenya. The setting was African and again the children grew from exploring it. The institution, however, emphasized American culture more than any other. Generally for the American students this was acceptable, but it did not reflect the non-American cultures of over one-quarter of that example class of 1976 not from the US. With RVA's heavy emphasis on evangelical Protestant Christianity, those of us from the Lutheran tradition felt excluded. RVA, like Augustana before it, did not fully reflect the cultural core of the entire student body.

46.5.3 International School Moshi in Moshi, Tanzania

The International School Moshi (ISM) was the third major school option for Lutheran missionary children when Augustana School closed in 1971. When Augustana School opened in 1953, the majority of the American Lutheran missionaries worked in the central area of Tanganyika, fairly close to the school. When the school closed 18 years later, the vast majority of the dwindling group of missionaries lived in the northern area near Arusha and Moshi (see Fig. 46.1). Augustana School only had 23 students in the whole school, not even the size of a regular American classroom. When it closed, the missionary parents had two main options: send their children to RVA early or send them to ISM, a new school in northern Tanzania located much closer to many of our homes.

The choice was often made within families over whether they felt that schooling needed to have a strong Christian influence or if they were open to schooling in a more secular fashion. Many of the Lutheran missionaries with more evangelical leanings chose to send their children to RVA earlier. ISM started in 1969 when an earlier English-medium school in Moshi had closed, and expatriate families who were involved in building the new Kilimanjaro Christian Medical Centre needed a school for their children. For the first year, children went to school in temporary quarters. The Lutheran Church in America (a successor church to the Augustana Lutheran Church) provided some initial funding and its successor church the Evangelical Lutheran Church in America remains part owners of the school. By September 1971, boarding facilities were completed and many of us Augustana School students migrated over to ISM.

ISM was very different from Augustana School and RVA. It is a much more diverse student body, with many non-missionary children as our friends. While there were few Tanzanians, the student body included Asians from many different religions, children of missionaries from many countries and European expatriate children. When the first students came over from Augustana, they found that rules were not really enforced. Many parents were shocked at the lack of rule enforcement. As the school grew and aged, students and parents felt comfortable with the enforcement of rules. While the Christian religion was at the forefront of the experience at the other two schools, it was not emphasized very much at ISM given the truly multi-sectarian student body and philosophy. Its goal was to provide a "quality education to expatriate children living in Tanzania regardless of their nationality, race, or religion" (Heyse 1980, 15). In many ways, it echoed mission statements that might be said for public schools in many countries, except for its emphasis on expatriate children.

Teachers are from many different countries and the curriculum tries to maintain an international focus. The school has adopted the International Baccalaureate (IB) model and it provides a very strong academic model for students at the school. The study of Tanzanian history, language and culture is included in the curriculum in a way that was missing from the other schools. Research by students within the IB program examined issues that were important to the country.

The environs around the school called for exploration, but being at the edge of one of Tanzania's larger towns, it was not as exciting as the surroundings for RVA and Kiomboi. Still, Mount Kilimanjaro looms over the school and many groups regularly explore the mountain.

Of the three schools, ISM made a better fit of the cultural milieu of the Lutheran missionary children. As expatriate children of Tanzania, ISM exposed the children to all the manifestations of this group. In terms of our religious backgrounds, Augustana School was almost the same as the religion experienced at home, but ISM exposed the children to the multitude of religious experiences that were part of Tanzanian life. While its lack of emphasis on an American curriculum may have been seen as a shortcoming to the American Lutheran parents, it did provide a rigorous and demanding education which perhaps prepared the students better for American college than the other two schools. The non-American students felt equally as well-prepared for further studies in their home countries, as well.

It is unclear if the Lutheran missionary parents felt that ISM was exactly what they wanted from a school for their children. Christian ideals dominated their thinking about why they were in Africa and the school did not emphasize any particular religion. Its multi-sectarian emphasis may have not felt right to many missionary parents, who were called to spread the Christian message. Furthermore, the international curriculum was a significant departure from the American education of the Lutheran missionaries and what they hoped to provide for their children. Tanzania was included in the curriculum at ISM which was a welcome departure from the Augustana School and RVA curricula which largely ignored their own environments to prepare their students for life in the United States. ISM was a school that was built for its student body and not for a larger goal that excluded some of its population. In that way, it reflected the lives of the children, rather than those of the parents.

46.6 Conclusion

For the Lutheran missionary children the education they received mainly reflected the worldviews of the missionary parents and not that of the missionary children. Fundamentally, both geography and religion played roles. Geography was important because of the way the two generations viewed the African landscape. The missionaries' viewed this place through the ideological lens of Christianity and through their connections to their home country which led them to develop schools that fit these ideals. The children's connection to their landscape and finding that they belonged to a culture somewhere in-between made these schools places that did not necessarily address their own issues. While the missionaries and their children share a closeness through having lived through their lives together in this place, these differences between the generations show through most profoundly in the ethos of the schools where the Lutheran missionary children were educated. Religious ideals played a fundamental role in the purposes of these schools and the parents' satisfaction with them. Children's understanding of what was involved in religion and the religious directions of the school also had an impact on how they felt about a school. Both religion and geography had a deep and lasting impact on the lives of both generations.

References

Anderson, E. (2000). *Miracle at sea: The sinking of the Zamzam and our family's astounding rescue*. Springfield: Quiet Waters Publications.

Coleman, M. C. (1993). *American Indian children at school, 1850–1930*. Jackson: University of Mississippi Press.

Danielson, E. R. (1940, January 15). *Letter to the Africa School Committee*. Augustana—Tanganyika File, ELCA Archives, Elk Grove Village, IL.

Dow, P. (2003). *School in the clouds: The Rift Valley Academy story*. Pasadena: William Carey Library.

Gottneid, A. (1967, November 6). *Letter to Ruben Pederson*. Found in Ruben Pederson Papers, Lutheran Church in America, ELCA Archives, Elk Grove Village, IL.

Heyse, H. (1980). Learning at the foot of Mount Kilimanjaro. *World Encounter, 17*(Summer), 14–16.

Johnson, V. E. (1920a). Black diamonds. A brief sketch of the African Inland Mission. *Lutheran Companion*, 28, 259 and 261.

Johnson, V. E. (1920b). The Africa Inland mission. *Lutheran Companion, 28*, 419–420.

Johnson, V. E. (1930). Asafari into Nzega district. *Lutheran Companion*, 38, 76 and 108–109.
Johnson, V. E. (1934). Back to Africa. *Lutheran Companion, 42*, 1194–1195.
Johnson, V. E. (1948). *Pioneering for Christ in east Africa*. Rock Island: Augustana Book Concern.
Johnson, D. E. (1955). A growing church in Africa. North Tanganyika Mission is flourishing. *Lutheran Companion, 100*, 8–9.
McBeth, S. J. (1983). *Ethnic identity and the boarding school experience of west-central Oklahoma American Indians*. Washington, DC: University Press of America.
Nelson, C. (1981). *Augustana school, 1953–1957: A 25-year-tribute*. Nevada City: Self-Published.
Smedjebacka, H. (1973). Lutheran church autonomy in northern Tanzania, 1940–1963. *Acta Academiae Aboensis, Ser. A Humaniora, 44*(3), 1–326.
Swanson, S. H. (Ed.), & The Augustana Synod Passengers. (1941). *Zamzam The story of a strange missionary odyssey*. Minneapolis: The Board of Foreign Missions of the Augustana Synod.

Chapter 47
The Role of Place and Ideology in the Career Choices of Missionary Children Who Grew Up in Tanzania

John Benson

47.1 Introduction

Our choice of an occupation can flow from many different directions. It can come from needing to find a job in a particular place and taking anything that is available. It can come from the influence of a family and following in the footsteps of a parent. Or, we can set out in a completely new direction from those people we have known before and what they did. For the children of missionaries, occupational choices grew out of both a connection to the missionary ideals of their parents and a wanting to return to the place where they had grown up, but also an awareness that their religious development was different than their parents and that they could not look on the landscape of East Africa, the example discussed here, in the same way as their parents. Geography and religion had a large influence on occupational choice. Many asked themselves when looking in a particular career direction: Will this type of work allow me to get back to Africa? Some did look at work in the Church, but the deeper question they asked themselves related to their missionary background: How can I be of service to others in this type of work?

Their parents had gone out to work in the "ideological landscape" of a mission field, but this "mission field" was no longer there for the children, nor had it ever been. The landscape of the children who had grown up there was one that they learned to love, but one that they could not see through the same ideological lens with which their parents had seen it. This chapter examines the role this ideological landscape had on the vocational choices of an American Lutheran missionary community that served in Tanganyika/Tanzania during a time of both rapid extension and decline of the mission, and the impact this place and their missionary ideals had

J. Benson (✉)
School of Teaching and Learning, Minnesota State University, Moorhead, MN 56563, USA
e-mail: bensonj@mnstate.edu

S.D. Brunn (ed.), *The Changing World Religion Map*,
DOI 10.1007/978-94-017-9376-6_47

on the occupational choices of the next generation. The author is a member of the second generation of this community and interviewed many members of both generations for a community history that he is writing.

47.2 Influence of First Generation and Place on the Lives of the Second Generation

Missionaries from the Augustana Lutheran Synod first arrived in Tanganyika from the United States in 1922. Tanganyika had been a German colony, but many of the German missionaries had been expelled from the country with the changeover after World War I to a League of Nations trust territory administered by the British. The German Lutheran mission in Tanganyika needed help and asked for help from their fellow American Lutherans of the Augustana Synod. The Augustana Synod (eventually in 1948 becoming the Augustana Evangelical Lutheran Church) was primarily a church of Swedish immigrants and could be said to be the Church of Sweden transported to the United States. Most immigrants from the different Lutheran countries of Scandinavia and Germany formed their own Lutheran bodies in the beginning and often worshipped in the mother tongue into the 1930s. This was true of Augustana, as well (Erling and Granquist 2008). In the beginning, the Augustana missionaries worked closely with the few Leipzig missionaries in the three mountainous areas of Mount Meru, Kilimanjaro, and the Pare Mountains in northern Tanganyika. Augustana missionaries worked mainly on Mount Kilimanjaro and in the Pare Mountains, and overlapped in their work with Leipzig in this area. The missions had different philosophies and it was not easy for the Americans who believed more in individual redemption to work so closely with the Germans who saw their work more in the collective redemption of the different northern ethnic groups (Smedjebacka 1973). In 1927, the Leipzig Mission was allowed to return with its full contingent of missionaries, and Augustana decided to move its mission field to the Iramba area of Central Tanganyika (Kiomboi, at the center of this mission field is approximately 170 mi (274 km) south of Arusha and around 440 mi (708 km) to the northwest of Dar es Salaam). While much more remote and with a harsher climate, the missionaries were grateful to have a "place" of their own, one where they could promote their mission ideals. For about 12 years, this was where Augustana concentrated its work, until once again the Leipzig Mission and all other German mission societies were forced to leave the country in World War II. Because of this, Lutheran missionaries from the originally neutral countries in the conflict were recruited to maintain the mission field. The American Augustana mission and the Swedish mission societies were forced to help these "orphan" fields with very minimally recruited staff (Swanson 1960).

This "ideological landscape" of a mission field needed certain key occupations to be recruited to work on it. The Augustana Lutheran Church believed in a holistic Christian message (Olson 1999). Different mission stations were built and developed that offered the holistic services that the church believed in—evangelism, education, and health care (Fig. 47.1). The mission stations usually were seen as a

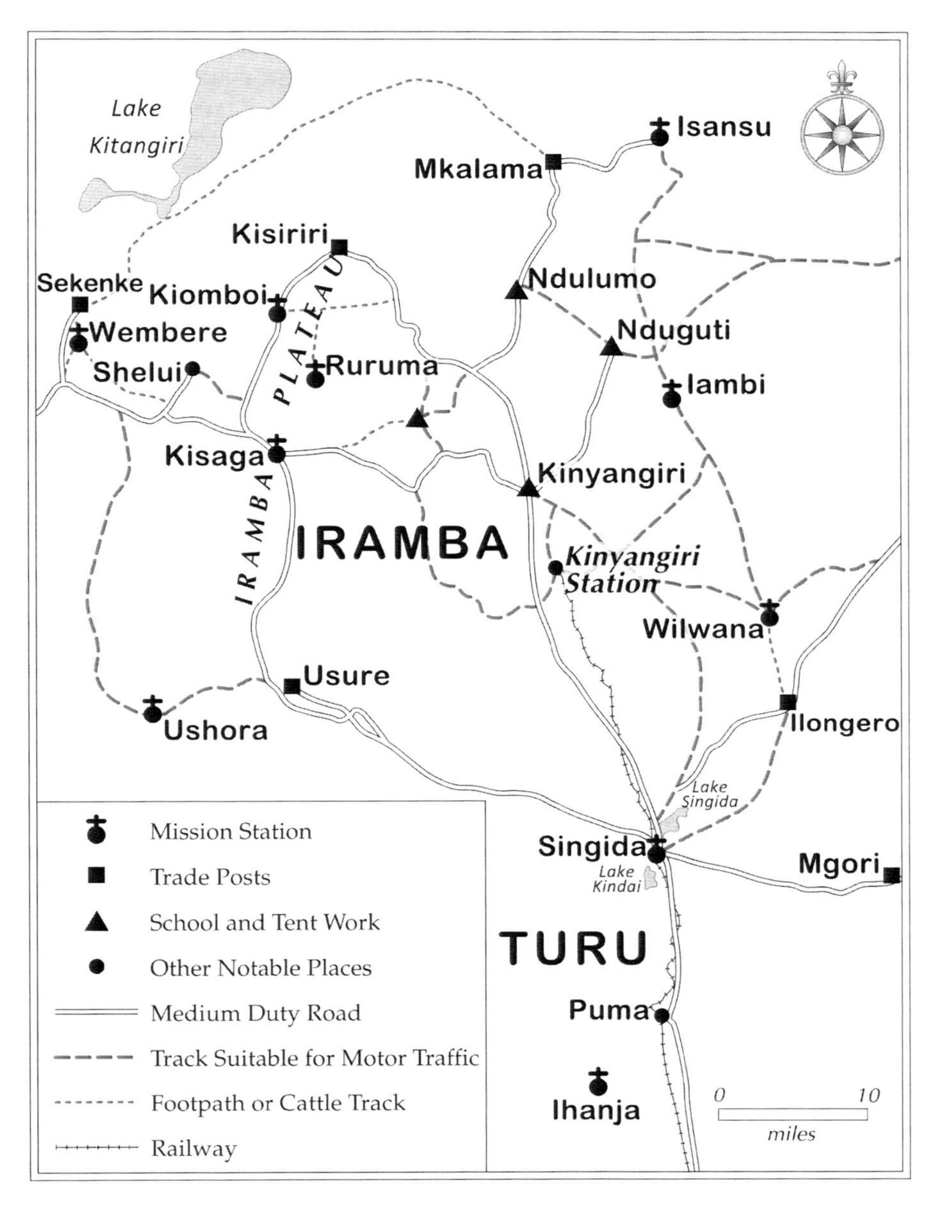

Fig. 47.1 Augustana mission stations in Iramba and Turu (1926–1960) (Map by Dick Gilbreath, University of Kentucky Gyula Pauer Center for Cartography and GIS, based on Swanson 1960; commissioned by the editor)

central service points for many outlying outstations. At each station, a mission house was built, a church, a school, and a clinic. The pastors carried out the training of evangelists and visited the outstations in their area. Single missionaries often helped run the schools. All of this work was intertwined. Edna Miller, one of the early pioneer missionaries in education, described how the missionaries saw the need for each of these areas:

> The evangelistic work dominates the medical and educational. The medical work is a means of contact with the natives to gain his confidence and to bring him to the mission where he can hear the Word of God. The educational work seeks to teach him how to read so that he can read his Bible. It also teaches him health and sanitation—thus linking medical work with educational work. (Miller 1929: 651)

Most missionaries that were recruited fell into these three main areas. Many missionary couples consisted of a husband who was a pastor and a wife who was a nurse. These occupational straitjackets were sometimes hard for the missionaries to accept easily or to know how they would use their talents. One missionary wife, who had been a dental hygienist, switched to being a nurse, as she and her husband felt that there were more needs for nurses in colonial Tanganyika, than dental personnel. Les and Ruth, missionaries for many years in Tanganyika, later Tanzania, related that while Les went into the ministry, he had often wished he could have been an agricultural missionary. Ruth earned a Master's degree in Library Science, but struggled with how to use this background overseas. She had often thought of being a nurse, but felt pulled into this field. Two of her daughters eventually became nurses.

The missionary calls seldom went according to plan. A preacher might end up being more of a builder and a missionary wife could be led in many different directions, as Howard Olson and June expressed in their interviews:

> In Swahili the word for pastor is mchungaji, and the word for builder is mjengaji. Now within one month of taking up my pastorate I was involved in so many building projects that I didn't know if I was a mchungaji or mjengaji. The immediate demands were for two school units, two boys' dormitories for 60 boys in the third and fourth grades, several teachers' houses, a missionary residence, a medical dispensary, a church and a cistern. (Olson 2001: 85)
>
> I don't know if anyone else has ever expressed it in your interviews that [being able to volunteer, rather than be assigned work] to me was always a real plus. It was a blessing that I never had a job and I could choose to do what I wanted to do. I could work at the level that I thought I could. I more or less made my own job. Whereas if you were paid, then I think you look at work a little bit differently. Very thankful for our mission board.

Both ended up doing work on the mission field that were ones of service, but in ways that were unexpected.

The peak year of the Augustana Lutheran Church, later the Lutheran Church in America, and eventually the Evangelical Lutheran Church in America, mission was in 1961, the year of Tanganyika's independence. For this study that was done on this missionary community, most of the families that were interviewed spent their time in Tanganyika/Tanzania between the years from the 1950s through the 1980s and 1990s. The parents arrived in this place during the boom years, but we as children saw the mission declining for most of our lifetime. The African church was booming, but the missionaries were needed less and less. The Lutheran World Federation currently identifies the Evangelical Lutheran Church in Tanzania as the third largest of the Lutheran church bodies in the world (5.82 million) after the Church of Sweden (6.5 million) and the Ethiopian Evangelical Church Mekane Yesus (5.84 million) and somewhat larger than the Evangelical Lutheran Church in America with over 4.2 million members (Lutheran World Federation 2012).

As children growing up in this environment, we did not think of ourselves as growing up on a certain mission field or station. It just felt like home or Tanzania. Here are two quotes that show the strong connection to this place and its people by two missionary sons, about 10 years apart in age, but with similar connections to the place. Joe, the older of the two, spoke of his connection to the Usambara Mountains as a child and David to people of Balangida in central Tanzania:

> At Ngwilu, we were the only European family. It was way out in a valley all by itself and so we had African children who were our friends and I spent time with them. We would go on walks. The paths would go everywhere on the mountain and we would go off on these paths. We had a girl called an ayah who would look after us. Dad was often traveling because he was Superintendent at the time. Mom was at home running the house. We had several servants, one to work the garden and one to chop wood. It was not crowded with people. There were people of the tribe around in villages but it was not densely populated at all. We would often go on these walks on worn down paths. We would cross the valley and cross a stream down in the valley. It was very exciting for a little kid, three or four years old, to go on these extended walks and we grew up speaking the language to some degree so we could talk to the African people and we had a lot of fun. The mountains were quite fertile and there were crops in the valley, so we would do such things as get ears of corn and roast them on the fire and the African life up in the mountains, it was rainforest country, and so we experienced life as the Africans did up in the mountains (Joe).
>
> Safety. I always felt that when I was at Balangida that I was safe. It may sound strange but I felt it among the Barabaig and especially, my friends, the Turu and Iramba. It almost felt like nothing would ever happen to me there. If something came up, I would be protected by the tribe… My name in Barabaig, they used to call me was Gida Balangida, son of Balangida. That is what the Barabaig called me. Steve, Linda, and Mark moved from another mission station. Since I was the only one that appeared there…. Steve is six years older than me. A lot of the other families had siblings that they could hang with and that is not to say that Steve did not look after me, but he was older and he had his own friends. People in Arusha had the other missionary families. My friends were my African friends, who else was I going to play with? Quite frankly, I felt that I was part of them, part African. It's really hard to explain. I knew I was different because I was white, but we did what kids would do. We played our games. It wasn't white or black, or Mang'ati, just kids getting together (David).

The connections the second generation felt to this place were different than that of their parents who saw it as a mission field. It was just home to the children. They formed a connection to the place, and often just endured the 1 year furloughs in the United States (the place their parents called home), just so they could return back "home" to Tanzania. The furloughs ended, however, when their collegiate years began. During their years in college, they struggled with different vocational choices. Were these ones that would allow them to return to Tanzania? Would they offer the same sense of service that they had seen in their parents' occupational choices as missionaries?

A mixture of forces had an impact on the vocational choices of the children of missionaries. They were part of a mission that was down-sizing with fewer chances for the children to return as missionaries, compared to the possibilities for many of their classmates at Rift Valley Academy, who were part of more evangelical missions. They had to think about different ways to fulfill their lives. For some, it was taking one of the vocational models that they had been exposed to within the missionary community—preacher, teacher, doctor, or housewife—and work at these occupations in a new place. Almost all the children of missionaries interviewed were children of pastors, but this was the smallest category of those who chose to go into one of the traditional occupations. Another option was to figure out what was needed in East Africa, because this was the place that they knew, and then to plan to assume a new occupation in the land that they knew. Many chose to be engineers, safari company operators, or researchers because these were still options available for expatriates in East Africa. Finally, there was a third category of those who chose a completely new occupation that they had no exposure to growing up in East Africa, but that still had some values inherent in it similar to the missionary spirit of their parents. This second generation retained many of the missionary ideals of their parents, but primarily in non-religious ways, and their ties to Africa were to a real place and not to the ideological place where their parents had gone several decades earlier.

47.3 Occupational Choices of Second Generation

For many people, occupational choice grows out of what the family does for a living. One can think of many family businesses of "So-and-so and Son." At the beginning of each semester in one of his teacher education classes, he asked the students to give him a sense of their parents' occupations by standing in groups according to what one of their parents does for a living. Usually a group of four or five students in a class of 25 formed a "teacher's kid" clump. While becoming a missionary was something many of children of missionaries considered, many factors made it not as easy as joining the family business or following the family pattern of becoming teachers. Factors such as their own religious ideals, the declining state of the mission, and whether a call could occur to a particular place influenced our decisions about considering missionary service.

David Vikner, in a retrospective look at the Augustana Lutheran Church, one of the earlier predecessor churches for most of the Lutheran missionaries who served in Tanzania, examined how many of the missionaries who had served in this church from about 1860–1960 were second-generation missionaries. He found:

> In its entire history Augustana sent 432 men and women missionaries overseas to ten countries. Of these, 407 served in Augustana's four main fields: India, 32; China (including Hong Kong, Taiwan and North Borneo), 109; Tanganyika, 232; and Japan, 34. There were 25 sent to Argentina, Indonesia, Mexico, Persia, Puerto Rico, and Uruguay…The total missionary family included 25 second generation missionaries: 15 from China, 5 from India, and 5 from Tanganyika. (Vikner 1999: 194)

This worked out to a total percentage of 5.8 % of the children of missionaries returning as missionaries with about 15 % of those from China, 6 % of those from India, and 2 % of the Tanganyikan children of missionaries returning to serve as Augustana missionaries.

Out of the 64 children of missionaries (including the author) that were part of this study, 10 returned overseas at some point as a missionary or a volunteer for a church organization, which is about 16 % of the total, similar to the percentages of the China and India missionaries. However, only four of the members of this second generation returned to work for the Lutheran Church in America (LCA) or the Evangelical Lutheran Church in America (ELCA), the successive churches to the Augustana church. The majority worked for other mission or church organizations. Of these ten from the second generation, four returned to Tanzania, and three returned to the neighboring countries of Kenya and the Democratic Republic of Congo (DRC or Zaire, as it was known then). Only one of them served as a missionary outside of Africa. Of this group, only three are still serving as long term missionaries, three more continue living in Africa, and four are working and living in the United States.

When many of the children of missionaries were asked if they could have become missionaries, several stated that they would have easily taken positions in East Africa but they were not sure that they could have returned as missionaries. They were not sure that their own faith was as strong as that of their parents. At the same time, if they had applied to mission agencies, many agencies would have wanted them to go anywhere in the world and many felt a specific connection to East Africa. Were missionary calls always placeless?

With missionary service not being an option for so many of us, this generation had to make occupational choices based on the occupations that were known, where they wished to live, or what was available in the places they ended up settling. Jill spoke about what she felt were her options when she went off to college, based on what she knew:

> I didn't really know what I wanted to do... I knew I didn't want to do nursing. We had so few options presented to us as high school students of what you could go into: teacher, doctor, pastor, or nurse. We had basically four exposures; that was it. So what did I want to go into? I didn't think I wanted to be a pastor. I knew I didn't want to be a nurse or doctor. I couldn't stand the sight of blood. So that left teacher.

The vast majority of this generation (49 %) chose these traditional missionary occupations that Jill mentioned above, a second group chose international occupations (20 %) that allowed them to work back in East Africa, and finally a third group fell into a category that the author labeled "American" occupations (31 %) because these were occupations that they did not have an exposure to when they grew up in Africa. They might have seen local Tanzanians working in these positions, but as their parents depended on work permits to work in Tanzania, they knew they could not work in these type of positions in Africa. These included later occupations like bankers, food and restaurant managers, service station owners and the like (Table 47.1).

Table 47.1 Categories of occupational choices of MKs in the study population

Category	Number	Percent (%)
Traditional Missionary Occupations	31.0	49.21
International Occupations	12.5	20.84
"American" Occupations	19.5	31.95

Source: John Benson

Table 47.2 Categories of MKs in the study population who chose traditional missionary occupations

Category	Number	Percent (%)
Church Work	6.0	9.53
Education	12.0	19.05
Medicine	9.5	15.08
Homemaking	3.5	5.56

Source: John Benson

The majority chose to go into traditional missionary occupations (Table 47.2). Of the 63 children of missionaries that were interviewed for this study, only four were not pastors' children. Ironically, church work was the smallest category of occupations chosen by this group. Instead, education was the largest group in this category and the largest occupational choice overall. The next largest group overall were medical professionals.

Of the six people who were involved in church work in this generation, only three people did it consistently for most of their career. It was a challenging field to be involved in within the United States. The one that the most fulfilled in this work was someone who came to it from nursing, where she had felt less fulfilled. While there were connections to their parents work, most felt that their work here was extremely different from that led by their minister fathers back in Africa. At this point, there were only male ministers who served in the Lutheran church. The first female pastors ordinations in the American Lutheran churches occurred in 1970 (Robinson 2011).

The strongest connections seemed to come in education. While it seems that this generation could be called the "uncalled generation," quite a few people who chose teaching looked on it as a call and they were in it for idealistic reasons. It did allow some of them a chance to work overseas, but most importantly it gave us a connection to the helping values that were seen in their parents work in Africa. Interestingly, all of the people of this generation in this study who described themselves as having given up on religion all worked as teachers, perpetuating some of their parents' values in a non-religious way. Several other studies of missionary groups had similar results to the authors. Steve Van Rooy found that the majority of his classmates from Woodstock School, the missionary school in India, became missionaries, followed by teachers (Van Rooy 1998). In a survey of West African Third Culture Kids (TCKs) done by Nancy Henderson-James, the majority of her respondents (20 %) had gone into medical or mental health work, followed closely by education (19 %) (Henderson-James 2007).

Medicine drew a number of people into its ranks, as well. Many were exposed to it as children, but surprisingly the majority of the people in this group did not have one parent who was either a nurse or doctor. Several who became doctors did end up returning to Africa for a short while or a career. Many who went into this area were drawn into public health, one even who became the state epidemiologist for several Southern states.

Finally, some of the women followed in their mothers' footsteps to work as homemakers, after having careers of different types. But it was not only women; some of the men took some time to take care of their children while their wives worked. Choosing these fields helped them retain a connection to their parents work even if they could do not do them in Africa. They also allowed the people in these fields to retain the non-religious values of their parents' work (apart from ministry) and continue in work that helped people.

47.4 Discussion

As the Lutheran mission seemed to shrink more and more in their childhoods, most of the second generation realized that they would not be able to return to this land that they loved as children of missionaries. They also had a different relationship with this place, than their parents had towards it. They saw it through the ideological lens of their religious convictions as a mission field; this generation's religious convictions were much weaker, but the connection to this landscape much stronger. Many of them wanted to figure out a way to remain in it. They were children of this land. Some people did choose early on to volunteer with different international groups, like the Peace Corps, but this reflects careers where people ended up in as careers.

For people who wanted to return to Tanzania, different international occupations had to be considered that would work here (Table 47.3). As the mission in Tanzania changed from a focus on central Tanzania to one in northern Tanzania, much of the missionaries' work had been centered on the town of Arusha, Tanzania, the tourist center of the country. Two families became involved with their own safari companies. This type of business fit them well as it drew on their love of the Tanzanian land and people, but also allowed them to draw on the in-between nature of being raised as Americans in the African landscape. As they worked with the Tanzanians,

Table 47.3 Categories of careers within the International Occupations chosen by MKs in the study population

Category	Number	Percent (%)
Safari Operators	6.5	10.32
Engineers	3.0	4.76
International Researchers	2.0	3.17
Pilots	1.0	1.59

Source: John Benson

they could explain the nuances of American/European culture to them and vice-versa with their American/European guests. Mike, one of these that became involved in the tourist trade with his brothers, described the connections within this field to his childhood:

> I think our experiences here, our experiences in Tanzania, our love of the [land] … one of the things that maybe we never even touch on in terms of our childhood was how much we spent time out in the bush camping and hunting and all that. That gave us a love to go out and do things…. We knew a lot of nooks and places that people would like to visit. Our tours were very personalized and well-organized. I think because of that we've always had trips that people have always really liked. And so that word-of-mouth advertising worked well because we had a lot of people coming away really happy and having had such a great experience. This is such a special place that you don't have to work too hard if you do a good job and take people to the right places and give them the right experience. It is very special for most visitors so a combination of all that.

Some of this generation saw the needs in Tanzania and knew that the Church would not be able to address them. It was then that they looked into fields like engineering to help address some of Tanzania's needs. The author's brother worked as a civil engineer for an architectural organization that was building low-cost institutional buildings in the country. Another worked as a civil engineer with a refugee-resettlement group. Finally, the Tanzanian government was helped in its early days as it developed its water resources by another member of this generation. Don, spoke about how he made his decision about using engineering back in East Africa:

> From my Africa days, I was very tuned to going to a vocational missionary aspect and originally was looking on the agricultural side and the engineering side and what I had seen at the time I left Tanzania, I felt very much pulled to the water side, in terms of what I had seen out in Africa and the needs and what my interests were.

Several others returned as researchers. The author has done unpaid research off and on in Tanzania as an adult, but his next younger brother spent many years in different African countries doing research on agricultural and nutritional issues. Another daughter of this generation returned to do research on AIDS and women issues.

Finally, while certain mission organizations, such as Missionary Aviation Fellowship, have always employed pilots, it was not a career focus of anyone in the Lutheran mission. One returned as a pilot to work for a small Catholic flying service, but eventually he moved into commercial piloting.

The children of this generation (a third generation) that returned to Tanzania to work have found that it is now much easier for them to find work in Tanzania. They continue to work in the safari business or local farms. The land drew these people back, but they had to think outside of the normal traditional missionary fields in order to return.

This final category of occupations was labeled "American" occupations (Table 47.4) because they were occupations that we were not exposed to traditionally as the children of missionaries, or that could be done by us in an African setting. (As Americans, the missionaries or their children might not be hired to do them in that setting, but they could easily be hired to do them in this American setting.) In looking at people in these fields, the author initially questioned what it was that

Table 47.4 Categories of careers within the "American" Occupations chosen by MKs in the study population

Category	Number	Percent (%)
Airlines	1.5	2.33
Artist	1.0	1.59
Banking	2.5	3.97
Beverage/Restaurant	3.0	4.76
Computers	2.0	3.17
Construction	0.5	0.80
Factory	1.5	2.33
Home Improvement	0.5	0.80
Oil Industry	0.5	0.80
Organization Specialist	1.0	1.59
Probation Officer	0.5	0.80
Rancher	0.5	0.80
Real Estate	0.5	0.80
Secretary	1.0	1.59
Service Station Owner	1.0	1.59
Student	1.0	1.59
Truck Driver	1.0	1.59

Source: John Benson

would give them fulfillment in these positions. These positions did not seem to have the values in them inherent to the traditional "helping" professions of the missionary occupations, nor were they located in the right place. He found more often than not that these people did find that they had a connection in this work that could be traced back to their childhoods.

Some felt that they began learning the skills of their current profession back in Africa. Dan, a ceramic artist, found his earliest influences had been his observations of the Makonde ebony carvers along the streets of Moshi, Tanzania. Duane, a service station operator, who grew up seeing his parents' generation in white-collar occupations (literally, part of the Sunday morning dress of their Lutheran pastor fathers) found himself drawn to the workshop at our boarding school to work closely with Eliasafi, the school handy man, and be his helper:

> What was really good for me and that would stand me apart, I think from a lot of the other kids, was that I was able to fix things and do some mechanical. I was doing that in 6th, 7th and 8th grade, especially 7th and 8th grade, I remember missing a lot of school because the light plant would quit and there would be no one there. Eliasafi was there. I consider him one of the best mentors I have ever had, during those days, because here is a guy with limited education who was brilliant. In fact, I had so much respect for him that I almost condescended to higher education, because I see a person with his brilliance, in fixing things and how he was picking it up ... he had no formal schooling...I don't know if he had a Standard 5 or what...but he was a testimony to me that we are all equal in one sense while different cultures have their different perspectives on what quote unquote an education is, what it really comes down to is how you take care of yourself and your family and how you help other people take care of themselves and their families and how you interact with each other. I think Eliasafi at Kiomboi was possibly one of the strongest mentors I look up to in my life.... I think my favorite times were going with Eliasafi to go to other mission stations to fix the light plants. I rode with him a lot.

Others found connections to the values and place in very different ways. Beth in her work as a beverage and restaurant manager at a hotel in Death Valley NP found that she often saw her work as akin to that of "Aunt Martha" our house mother at Augustana School back in Tanzania in how she had to manage everything. Mark, after being a pastor, switched careers and became a parole officer. When asked if he saw any connection to his missionary childhood in Africa in his work, he replied:

> Oh God, yes. In many ways, it has all kinds of similarities. It's propounding, working with people about their core values, in unusual settings. It's taking a message of grace and of the love of God to where they live in their language. It really is. It's amazing how closely that is.

Finally, though some found no connection to Africa in their work, they found satisfaction in doing a job well in whatever capacity. Even though this category seemed to have few ties to the missionary values of their parents, part of their satisfaction in the work that they chose was that they were able to find connections within it to a skill learned in Africa, something they had experienced there, or to some value held up as important within the mission family.

47.5 Conclusion

While few of the children of missionaries returned to the "ideological" landscape of their parents, many either captured their parents' ideals in their later occupations or returned to the "real" landscape of East Africa to pursue careers that fit this non-ideological landscape. In the lives of the children of missionaries, geography and religion, intersect very well. There was less a sense of religion influencing our lives, as few went into Christian ministry work, but of the values within our parents' occupations influencing our own occupational choices. Laura, the youngest person interviewed for this study, was still in college when she was interviewed. As she considered her future options, you could hear some of these values coming through:

> In terms of what I would like to do, not business or something, but in terms of what I would like to do, I would like to do something with justice. The world is so unfair and as Christians, I don't think we're supposed to leave it like that.

Now, she works with the New York City Public Health Department. Place was important. To be in that place, some chose career paths that allowed them to work in Africa with these skills that were not asked for by the previous generation, but now were seen as more important. This place was no longer the "ideological place" to which their parents had arrived 50 years previously. It was Tanzania in all of its reality. But the missionary ideals lived on in many of the second generation's choice of occupations and what was wanted from a career.

These careers helped them make sense of a new world but helped them stay connected with their childhood places through the values inherent within the careers or they allowed us to return to their childhood homes in new ways and offer something new.

References

All interview quotations are taken from an unpublished manuscript of the author that he has tentatively titled "An In-Between People: A Community History of Two Generations of a Lutheran Missionary Community in Two Continents."

Erling, M., & Granquist, M. (2008). *The Augustana story: Shaping Lutheran identity in North America*. Minneapolis: Augsburg Fortress.

Henderson-James, N. (Compiler and Ed.). (2007). *Africa lives in my soul: Responses to an African childhood*. Self-Published.

Lutheran World Federation. (2012). *Survey shows 70.5 Million members in LWF—Affiliated churches*. Slight increase globally with growth in Asia and Africa. Retrieved March 31, 2012, from www.lutheranworld.org/lwf/index.php/member-statistics-2011.html

Miller, E. (1929). African Mission Annual Conference. *Lutheran Companion, 37*, 650–651.

Olson, H. S. (1999). The miracle of the mustard seed: Augustana in Tanganyika". In A. Hultgren & V. L. Eckstrom (Eds.), *The Augustana heritage: Recollections, perspectives, and prospects* (pp. 225–229). Chicago: Augustana Heritage Association.

Olson, H. S. (2001). *Footprints*. Self-Published.

Robinson, B. A. 2011. *When some denominations or religious traditions started to ordain women*. Retrieved April 5, 2012, from www.religioustolerance.org/femclrg13.htm

Smedjebacka, H. (1973). Lutheran church autonomy in northern Tanzania, 1940–1963. *Acta Academiae Aboensis, Ser. A Humaniora, 44*(3), 1–326.

Swanson, S. H. (1960). *Foundation for tomorrow: A century of progress in Augustana World Missions*. Minneapolis: Board of World Missions, Augustana Lutheran Church.

Van Rooy, S. (1998). Career developments: Woodstock class of 1968. In J. M. Bowers (Ed.), *Raising resilient MKs: Resources for caregivers, parents, and teachers* (pp. 141–145). Colorado Springs: Association of Christian Schools International.

Vikner, D. (1999). Augustana in world mission, 1861—1962: Introduction. In A. Hultgren & V. L. Eckstrom (Eds.), *The Augustana heritage: Recollections, perspectives, and prospects* (pp. 193–195). Chicago: Augustana Heritage Association.

Chapter 48
Evangelical Short Term Missions: Dancing with the Elephant?

Lisa La George

48.1 Introduction

The expansion of Short-Term Missions (STM) is an unprecedented development in American evangelical outreach. While only 1 % of American college graduates have participated in a study-abroad program, Priest and Priest (2007) record that "in many seminaries and Christian Colleges, 50 times this number have traveled abroad in the context of religiously motivated service" (2007: 2). Called an "industry" (Livermore 2013: 12), a "giant" (Kyle 1996: 5), and a "God-commanded, repetitive deployment of swift, temporary non-professional missionaries" (Peterson et al. 2003: 117), STM is considered by some to be a fabulous opportunity and by others as an elephantine missiological mistake.

This author is aware of secular international humanitarian volunteer experiences and travel for the purpose of doctrinal dissemination undertaken by various religions. However, the specific context of the discussion which follows examines STM primarily in the context of the North American Evangelical Christian community. Additionally, while individuals do participate in solo trips, the discussion below will most frequently address teams functioning in STM.

A simple definition of STM is "The deployment of an individual or team to a cross-cultural context for a pre-determined period of time for the purpose of intentionally engaging in Christian ministries such as evangelism, teaching, and the support of residential missionaries and local churches." STM is further defined in light of the relational and philosophical factors which influence any given trip:

L. La George (✉)
International Studies, The Master's College, 21726 Placerita Canyon Road, Santa Clarita, CA 91321, USA
e-mail: llageorge@masters.edu

S.D. Brunn (ed.), *The Changing World Religion Map*,
DOI 10.1007/978-94-017-9376-6_48

- Participant Demographics: Who is going? How old are they? What skills do they have? Are they financially independent or have their raised support to participate? How many people will be on the team?
- Leadership: Who provides leadership before, during, and after the trip? Who facilitates the planning and execution of trip logistics?
- Sending entity: What organization is sponsoring the trip? Is the team sent by a church, a school, and/or a missionary agency?
- Trip location: Where will the team go? Is the destination domestic or international?
- Receiving entity: Who is hosting the team? A national church? An expatriate missionary? An humanitarian or educational institution?
- Training: What cultural and linguistic preparation will the team receive prior to the trip? When they arrive on the field?
- Length: Is the length of the trip defined in days, weeks, months or years?
- Activity: What is the purpose of sending the team? What task will they be engaged in?
- Follow-through: In what way will the team be prepared to return to their home culture? What engagement will they have with the location and the hosts following their return?

The number of North American STM travelers illustrates the exponential growth of the movement in the past three decades. As Stan Guthrie in his text *Missions in the Third Millennium* reports, 22,000 people participated in STM in 1979. Ten years later, that number had escalated to 120,000, and Guthrie suggested that 1998 saw 450,000 individuals participate (2005: 109). Without considering youth participation, Wuthnow reported that 1.6 million adult U.S. church members traveled aboard in STM trips in 2005 (2009: 170).

48.2 Historical Development of Short-Term Missions

STM is a phenomenon that came into fruition toward the last half of the twentieth century. Prior to 1960, few people traveled under the banner of mission unless they were moving to a location for life.

A few examples of early STM are recorded in individual missionary biographies. In 1911, after serving three decades as a missionary in Algeria, Lilias Trotter wrote in her journal that she had been joined by "short-servicers:" young women who could "come on a self-supporting basis for a time of service in all the countless ways in which such can be rendered with a small knowledge of the language, if hands and hearts are ready" (Rockness 2003: 241). Trotter served another two decades in Algeria and was rejoined in long-term work by some of her "short-servicers."

Although early individual STM examples emanate from British missionaries, the development of STM as it is known today has arisen from primarily North American efforts. Agencies with long-term missionaries in other countries began to facilitate the occasional STM work team in the late 1950s. For example, the Association of Baptists for World Evangelism reported taking small groups of business leaders and builders to view various fields prior to 1960. In the 1960s, involvement in STM began to gain noticeable momentum. Agencies were founded which focused exclusively on the mobilizing and facilitation of young adults to go short-term to the mission field, including YWAM (Youth with a Mission) in 1960, Yugo Ministries in 1964, and Teen Missions International in 1970.

While the earlier STM trips noted above focused on adult participants with personal affluence and technical expertise, one noticeable trend in STM in the past 20 years is the exponential involvement of unskilled teen-agers and young adults. By the first decade of the twenty-first century, STM has become an expected part of almost every church high school youth group. As Priest and Priest observed in their study entitled "'They see everything, and understand nothing:' Short-Term Mission and Service Learning,"

> Short term missions, today, is a core part of youth ministry. The very job description of youth pastors, and indeed of many associate pastors or mission pastors, includes the requirement of organizing and leading short-term mission trips at home or abroad. STM is a core part of the internal discipleship ministry of local churches—participated in because of its spiritual benefits to the sending church and its own members. (2007: 4)

Job description postings for youth pastors frequently include the youth pastor's expected involvement in leading Short-Term Mission trips, and youth ministry resource web sites provide access to books, articles, and conferences that address STM in church youth ministry.

University students are involved as well. At the undergraduate institution where the author teaches, over 65 % of the entering students have already participated in them. The participation of college-age students in missions has become a part of the fabric of mission history, from the Haystack prayer meeting of 1806 to the Student Volunteer Movement founded in 1888 to the triennial Urbana Missions Conference which first met in 1946. But each of those movements was focused on sending long-term workers. STM is a new paradigm.

48.3 Participation in Short-Term Missions

With millions of people from the United States participated--individually, through agencies, and with churchs--in STM annually, one must ask, "What motivates people to participate in STM?" The benefits touted by STM practitioners, missionaries, youth pastors, and marketers include obedience to doctrinal commands, adventure travel, personal development, worldview expansion, and the development of social capital for the host field.

Some STM advocates state that the structure and methodology of STM has extensive Biblical precedent. They reason that the sending of the 12 spies into Canaan, Jesus' commission of the 70 in Luke 10, and the Apostle Paul's missionary journeys recorded in Acts serve as models for how short-term missionaries should be deployed today (Peterson et al. 2003: 197 ff.). On the one hand, the Bible has not set a mandate for the implementation of STM by the church, and on the other hand, the Scriptures do not necessarily preclude the use of STM in missions. Throughout Scripture, God demonstrates a clear desire for men and women to know the character of the Creator. Israel was chosen to display God's goodness to the nations, and repeatedly reminded them of their role as a "kingdom of priests" (Exod. 19:6). Secondly, Jesus left his disciples with a command that has come to be known as the Great Commission:

> And Jesus came and said to them, "All authority in heaven and on earth has been given to me. Go therefore and make disciples of all nations, baptizing them in the name of the Father and of the Son and of the Holy Spirit, teaching them to observe all that I have commanded you. And behold, I am with you always, to the end of the age." (Matt. 28: 18–20, ESV)

The original Greek highlights just two commands from Christ: to make disciples and to look to Jesus. The Great Commission assumes that the listeners are going. God will be known and worshipped by all, according to Philippians 2:10–11 and Revelation 5:9–10, and the followers of Jesus are the instrument of introduction. The precise methodology behind that proclamation is not outlined with detail in the Scriptures—either in long-term or short-term methodologies, leaving room for both.

A thirst for adventure is another major reason that students participate in STM. Livermore states that STM agencies cater to students' desire for adventure:

> …some organizations aren't subtle at all about the role of adventure and fun in motivating people to participate in short-term missions. For example, Teen Mania, a youth organization based in Texas, … [recently ran a] full-page advertisement in a magazine for youth workers [which] featured this huge headline: "Missions Should Be Fun!" (2009: 51)

One popular text marketed to STM leaders acknowledges this motivation among participants in its title: *Equipped for Adventure: A Practical Guide to Short-Term Missions Trip* (Kirby 2006). A 2008 Barna study report notes that people under the age of 25 "are globally aware and cause-oriented. They relish risk, stimulation, and diverse experiences" (para. 10).

Participation in humanitarian work is another reason some travelers participate in STM. Edwin Zehner's assessment of STM among North Americans examines this appeal to STM:

> The growth of STM parallels broader developments in North America, including the growth of international tourism, high school community service, college study abroad and service learning program, and humanitarian volunteer programs… Though most still focus on evangelism or evangelistic support, many evangelicals have adopted a broader definition of mission that includes humanitarian concerns. This shift is amply expressed in STM. (Zehner 2008: 187)

Perhaps the most often publicized reason for sending STM teams is for the personal development of the actual team members. Priest et al. (2006) suggest that the environment created by STM can be compared with rites of passage:

> Like pilgrimages, these trips are rituals of intensification where one temporarily leaves the ordinary, compulsory routine life "at home" and experiences an extraordinary, voluntary, sacred experience "away from home: in a luminal space where sacred goals are pursued, physical and spiritual tests are faced, normal structures are dissolved, communitas is experienced, and personal transformation occurs. (2006: 433–434)

This development may include personal, social, and spiritual growth, worldview expansion, and encouragement to long-term missionary service because of their STM experience.

STM is doubtlessly recognized as a force in the personal and spiritual development of participants. A summary of a recent report by the Barna Group stated, "Most people who embark on service adventure describe the trips as life-changing. In fact, three-quarters of trip goers report that the experience changed their life in some way" (Barna Group 2008, para. 2). This 2008 survey report of over a thousand evangelical Christians continues:

> The most common areas of personal growth that people recall-even years later—included being more aware of other peoples' struggles (25 %), learning more about poverty, justice, or the world (16 %), increasing compassion (11 %), deepening or enriching their faith (9 %), broadening their spiritual understanding (9 %), and boosting their financial generosity (5 %). Others mentioned the experience helped them feel more fulfilled, become more grateful, develop new friends, and pray more. (Barna Group 2008, para. 9)

Worldview expansion is one of the benefits of participants' involvement in STM, claims Paul Borthwick, an STM proponent for the past two decades. One artistic description of this participant worldview expansion has been expressed by recording artist Sara Groves in her 2007 song, "I Saw What I saw":

> I saw what I saw and I can't forget it
> I heard what I heard and I can't go back
> I know what I know and I can't deny it
> Something on the road, cut me to the soul
> Your pain has changed me
> your dream inspires
> your face a memory
> your hope a fire…
> Something on the road, changed my world.

Yohannan assents to the positive impact of STM on the sending church: "Alumni of these programs are helping others in the West understand the real needs of the Third World" (2000: 206–207). Wuthnow (2009) assents that in a church, comparisons of STM participants and non-STM participants suggests that participants are more likely to interact at church with issues of justice, religious persecution, and refugees. Another profit the church may garner is higher overall financial giving by STM participants compared to non-participants (Wuthnow 2009:181). Although Wuthnow states that STM participants' "giving is substantially higher than nonparticipants'," including 5 % more giving to international ministries, he does not indicate how much of that giving might go to STM itself rather than long-term missions efforts (2009: 181). All in all, participants may be able to provide their sending church with a clearer understanding of the spiritual and physical needs of their fellow Christians around the world.

The expansion of the participants' worldview and possible reduction of ethnocentrism is also addressed by Priest and Dischinger in a 2005 study of high school students who participated in a work project in Mexico. While they discovered that the high-schoolers' ethnocentric attitudes did diminish somewhat after their STM experience, they also allude to Kurt Ver Beek's (2008) comparison of STM participants after a 2-week missions trip to a sapling that has been bent and then released: both potentially return to their original shape. Priest and Dischinger urge caution to those who would enthusiastically cite the positive results of their study, concluding that the possible benefit of reduced ethnocentrism is only profitable and sustained when coupled with appropriate training and support:

> [What] our own research lead[s] us to conclude is that short-term missions can lower levels of ethnocentrism but only when accompanied by proper training and field-based culture learning exercises. Also, the proper structures need to be in place when participants return in order for the changes to affect positive, long-term change. If they are not, people most likely will just settle back into their mono-ethnic environments. (2005: 10)

STM may also be a means for individuals from a variety of underrepresented socio-economic situations and ethnic backgrounds to experience other cultures. Priest and Priest surveyed over 5,000 students in Christian colleges and graduate schools, and concluded that when "compared to international tourism expenditures, or to study abroad, short-term missions is relatively inexpensive" (2007: 3). They also suggest that STM may contribute to a student's vocational development and focus, as in this anecdotal account of a medical student named John:

> On his desk is a picture of himself and his dad serving in a Mexican slum with Mexican friends. He and his dad return yearly to serve in this economically poor neighborhood. When asked about the placement of this picture on his desk, he replies that this picture is there to remind him every day while he is in medical school why he is there. Rather than inspire himself to study hard with images of money, prestige, or wealth, John inspires himself with a symbol of human need and a reminder to live his life in service to others. (2007: 4)

For such participants, an STM experience can become a symbol of their pursuit of preparation toward a life of service. While critiquing many STM practices, elder missionary statesman Robertson McQuilkin (1994) suggests that STM may propel some participants to return long-term to the mission field. He states that STM participants get away from the routines of home and may get close enough to hear God's call to the mission field (pp. 259–260) He wrote that "as many as 90 % of career missionaries from North America today have had short-term experience" (pp. 259–260).

The Global Outreach program at The Master's College, an undergraduate liberal arts institution in Santa Clarita, CA, has recently undertaken a study of the alumni participants from its 30-year program (Table 48.1, Fig. 48.1). One of the questions raised was concerned with the students' relationship to long-term cross-cultural missions. Of the 374 students surveyed and interviewed, 1 out of every 7 of the respondents was employed as a missionary, and most of these alumni indicated

Table 48.1 Global Outreach locations of mission projects from the Master's College Program

Year	Global Outreach team locations
2007	Cambodia, Germany, Honduras, India, Kenya, Tanzania, Togo
2008	India, Philippines, Uganda, South Africa, France, Mexico
2009	Cambodia, China, Jacksonville, Kurdistan, South Africa, Tanzania
2010	New York City, Dominican Republic, Latvia, Albania, Uganda, India, Philippines, American Samoa
2011	China, Ireland, Kurdistan, Latvia, Malawi, Philippines, Russia, South Africa, Taiwan
2012	Ecuador, Montgomery, Canada, Togo, Germany, India, Thailand, Cambodia, Indonesia, New Zealand

Source: Lisa La George

Fig. 48.1 Short-term mission group in a classroom in Taiwan (Photo from G.O. files)

that their STM participation was highly significant in their current career choices (La George 2009).

The Global Outreach Philosophy statement includes the following:

God's Glory is the Mission
Partnership—not Parenthood
Vocational Service—not Adventure Vacation
Simplicity—not Luxury
Cultural Awareness is not Optional.

Participants may not be the only benefactors of STM. Host missionaries, churches, and communities also profit from their interaction with STM participants through physical assistance such as building projects or literature distribution, personal encouragement through spiritual fellowship and prayer, professional development, and ministry advancement.

Missionary families may be positively influenced by the presence of STM workers. A panel discussion of STM-hosting missionaries is held annually at The Master's College. In the spring of 2007, the missionaries were asked the question, "How does an STM team impact your family?" One 18-year veteran missionary from Eastern Europe stated unequivocally that visiting STM teams were responsible for "Keeping our family on the field." One 30-year veteran from Southeast Asia wept openly before the class as he spoke of the discipleship of his daughters by short-term missionaries who helped his children "to overcome spiritual hurdles and cynicism."

Visiting STM teams may benefit host congregations as the host community experiences the benefits of building relationships with those from other cultures, being affirmed in their faith and ministry, and receiving financial support and resources from the team. "One function, then, of these short-term missions trips is to create links between Christians with material resources and those with less" (Priest, 2007: 1). In his paper "Peruvian Protestant Churches Find 'Linking' Social Capital through Partnerships with Visiting Short-term Missions Groups," R. Priest suggests that the Peruvian churches who host STM teams experience open doors of ministry—literally:

> One Peruvian pastor explained, "If I knock on another Peruvian's door, they will see me, and turn me away. But if I knock with you, a gringo, standing next to me, they will greet us with a smile, open the door, serve us coffee – and listen attentively to what we say." When Peruvian evangelicos join collaboratively with gringos (or others) from abroad, they often find that high schools, English language schools, University classrooms, jails, and hospitals which normally limit access to evangelicos-open their doors wide. (2007: 2)

STM does seem to have many positive benefits for participants and hosts, including self-reported personal and spiritual growth, worldview expansion, and greater relationships in both the local and global church. Unfortunately, STM cannot guarantee risk-free involvement for participants or hosts, so we must turn our attention to a critique of STM.

48.4 Critique of Short-Term Missions

While admirable motivations and benefits in STM do exist, anyone involved in STM should be able to identify the dangers and difficulties that affect participants, host missionaries, and host communities. Critical appraisals come from proponents of STM, from missionaries, and from sociologists.

One especially potent critique of STM is that the motivation of many participants is solely to derive the personal benefit of the experience with little or no consideration of the impact on the receiving community or the ultimate purpose of missions. STM structures may permit participants to go to the field in spite of egocentric motivations, negligible culture and language skills, and incompetent leadership. Some trips result in little more than souvenir braided hair, photos of participants with dark-skinned children, and stories of strange foods and unfortunate toilets. Word Made Flesh mission agency director, Chris Huertz observed the following with regard to the character and commitment of the young people he works with in mission:

> It's often observed that there is among my generation a crisis in the theology and practice of mission. For many Christians today, mission can seem to be little more than sanctified tourism. Raised as opportunistic individuals, we bounce from one short-term experience to the next. We keep our options open and avoid committing to any one organization or set of relationships.... (2007, para. 10)

STM trips have become an adventurous alternative-vacation, and they are advertized as such. The World Race was mentioned previously with regard to participants' quest for adventure. On this missions adventure, teams of participants race against each other through multiple countries and participate in service projects in between their race days. The World Race website claims,

> What we are offering is not just another missions trip. The World Race taps an ancient human compulsion to take a spiritual pilgrimage. Aussies have their walkabout and Muslims have their Hajj. American college students have an abbreviated bacchanal—a week's trip to Ft. Lauderdale in the spring that barely gives voice to the urge to go somewhere and do something different.
>
> A whole new rite of passage is waiting to be born and a million young people are waiting to respond. They are under-challenged and ultra- coddled.... We owe them not just this experience, but the opportunity it represents to tap their deepest yearning to not sell out, at least not before they've given a wild alternative their best shot. (Barnes 2009, para. 7–11)

Participants may view these STM trips as an opportunity to travel and fill some sort of void in their life. Slimbach concludes that the high volume of STM is due to the rise of a simplified tourism industry, increased wealth in the church, more awareness of the world's pain, and the West's shift from a production to a consumer society. The latter contributes to an expanding craving for "peak experiences." Slimbach writes,

> Travel is a school for life, one that generates fresh insights and unforgettable memories. Nevertheless, it primarily enrolls a class of wandering elites.... Travel will confer a certain social status while satisfying an existential need for life-meaning. Peak experiences are accumulated and rehearsed in conversation or on a resume for years to come. (2008: 158)

"We are," he stated in a presentation, "all tourists, all the time" (Slimbach 2007, personal notes).

Alex Smith (2008: 56–57) provides a summary list of the "concerns, dangers, and warnings for the future" of STM:

1. Short-sighted convenience without long-term commitment – inadequate personal vision.
2. Self-centered individualism without deep altruistic concern for others – false focus; What can I get out of this?
3. Instant gratification without distant responsibility – questionable motivation; just experience the here and now; get the biggest bang for the buck.
4. Intense activity without deliberate purpose – lacking significant eternal goals.

5. Social service participation without evangelistic proclamation – short circuited outcomes in the spiritual duty.
6. Immediate satisfaction without eternal consequences – fuzzy ultimate expectations.
7. Humanistic Self-sufficiency without theological reflection and analysis – unevaluated self-dependency.

Even if the primary motivation of participants is personal growth, in spite of its positive press, some research may actually indicate that under specific circumstances, STM leaves no significant impact on its participants. Through his study of 200 North Americans who participated in a week-long building project in Honduras, sociologist and missionary Kurt Ver Beek has been one of the most vocal STM detractors in recent years, challenging the concept of any participant impact, tangible or intangible (2008). Likewise, David Livermore has suggested that little long-term impact is evident because of the short length of time that participants are in the host culture:

> A great deal of research has been done on culture shock, examining the cycles that typically occur for someone encountering a new culture for the first time. Finding common ground is the coping mechanism most often used in the first several weeks in a new place. After a couple of months of being immersed in a new culture, the emphasis begins to shift toward seeing all the differences rather than the similarities. However, most short term participants never go through the paradigm shift experienced by those who move to another context for an extended time period. Short-termers are back to "life as normal" long before they experience the depths of the differences that become apparent after more extended cross-cultural immersions. (2006: 70)

Perhaps the quick degeneration of the STM impact is a result of poor training, short trips, and inadequate reflection. Adeney cautions that training is essential because many short-term workers may "have a heart for the Lord, but only a sketchy knowledge of Scripture, little experience in evangelism or apologetics, and a lackadaisical practice of spiritual disciplines" (1996: 15). Echoing Adeney's concerns, Priest and Priest (2007: 9) examine a failure to adequately train leaders:

> The short-term missions movement is a populist movement, emergent not out of the strategic vision of leading missiologists or theologians, but out of grass-roots impulses. It is largely a lay movement, and the writings intended to train and orient short-term leaders are missiologically unsophisticated and frequently anti-intellectual. While youth pastors are typically expected to lead such trips, sometimes on an annual basis, nothing in the curriculum of most seminaries is oriented towards instructing youth pastors in what is needed for STM.

Inadequate training may also give way to inadequate reflection on the part of participants. In a 1995 study of teenagers involved in community service, Wuthnow addressed the issue of guided reflection as a hallmark of effective programs, and he points back to that study in his 2009 book *Boundless Faith*: "Like other kinds of voluntary service, mission trips seem to have the strongest effects when they are accompanied by adequate preparation and subsequent time for discussion and reflection" (Wuthnow 2009: 182).

Another censure of STM by some critics involves the extensive financial cost involved in sending teams. Some have stated that STM is guilty of jeopardizing the funding of long-term missions. Doosik Kim reported, "One missionary host said that if participants are not "Properly and thoroughly prepared and trained," STM 'is a waste of time and money and may cause serious harm to all'" (2001: 136). Hunter Farrell notes the critique of a Peruvian pastor with regard to the cost of sending an STM team to Peru: "If they sent us the money they spend on their international travel, we could build more churches, feed more children, train more pastors" (2007: 72).

While the 1.6 billion spent by Americans on STM equals "nearly half of the amount spent on all other U.S. missions programs combined" (Wuthnow 2009: 180), Wuthnow warns that "a focus only on the economics of international service is always too narrow" (Wuthnow 2009: 183). Some of the criticism for financial expenditures presupposes that the money spent on STM would be spent elsewhere in missions if it was not spent on STM, but Wuthnow cautions against such criticism: "The aggregate cost of these trips is huge, but church leaders are probably right in arguing that this is money that people would be unlikely to have contributed to other offerings or programs" (Wuthnow 2009: 183).

STM may have a negative impact on the life and ministry of the host missionary. The STM team will definitely cause a disruption in the normal flow of life and ministry for the missionary and may, in fact, through cultural or political blunders, actually damage the reputation and status of the missionary. Likewise, team complaints and conflicts may discourage the hosts. Without proper planning, team orientation, and adequate leadership, the strength of the host missionary can be depleted by caring for a group of short-term workers.

The receiving community may also find STM unwieldy, and can also create financial dependency within the national church by an influx of foreign money. Farrell addressed the powerful divisions that can be caused by a flood of foreign money brought with the STM team:

> Because of these unspoken power relationships, at its worst, STM can deform the relationship between members of the Body of Christ. Without proper preparation in both the STM group and the host community, a "visit of Christian solidarity" can quickly degenerate into fund-raising frenzy with host community members forced to outdo each other in an attempt to win a hearing for needed financial support from visiting North Americans. Over time, the result can be that local Christian leadership is more sensitive to foreign STM group preferences than the will of the local community. (2007: 73)

Host churches and communities in the developing world may struggle with visitors' paternalistic attitudes. Humanitarian trips such as medical care or construction may even serve to displaced local labor. C. M. Brown recommends that such ecclesiastical squabbles may be related to the power dynamics of wealth transference from the visitors to the less wealthy hosts: "Typically, the congregation with the lesser material resource base has less power in partnership. The leadership of the congregation with the greater material-resource base sometimes is unaware of the power imbalance" (2008: 232).

Adeney illustrates the difficulties encountered by communities who work with Americans in mission:

> The "ugly American" stereotype is alive and well. When I was in West Africa several years ago, the African Director of a Christian NGO [Non Governmental Organization] said, "Would you like to know what it is like to do mission with American? Let me tell you a story."
>
> In this story, Elephant and Mouse were best friends. One day Elephant said, "Mouse, let's have a party!"
>
> Animals gathered from far and near. They ate, drank, and sang, and danced. And nobody celebrated more exuberantly than the Elephant.
>
> After it was over, Elephant exclaimed, "Mouse, did you ever go to a better party? What a celebration!"
>
> But Mouse didn't answer.
>
> "Where are you?" Elephant called. Then he shrank back in horror. There at his feet lay the Mouse, his body ground into the dirt—smashed by the exuberance of his friend, the Elephant.
>
> "Sometimes that is what it is like to do mission with you Westerners," the African storyteller commented. "It is like dancing with an elephant." (2008: 131–132)

48.5 Long-Term Impact of Short-Term Missions

The picture of Short-Term Missions painted above is idealistic and bleak at the same time. When faced against each other, the perceived benefits and dangers of STM for the participants and the hosts seem to negate one another. The host communities, however, demonstrate in their critiques, what they believe is the key to a successful STM team: extensive training prior to the trip (Table 48.2).

Table 48.2 Timeline of Global Outreach team training and trip

Date	Team activity
March	Leaders selected to team in following summer
April	Host contacted by GO Director
September–May	Leaders participate in leadership training
September	Leaders begin communicating with hosts
November	Leaders pursue interested students to apply for team
December	Applications and references collected; leaders conduct interviews
January	Global Outreach class begins 15 weeks of sessions
February	Teams participate in an overnight training retreat
February	Team participates in cultural outing to an ethnic center similar to their host country
March	Team conducts various fundraisers
April	Airline tickets purchased and visas acquired
May	Team leaves for host country
May–July	Team between 4 and 10 weeks in host country
July	Team debriefs in host country
July	Team debriefs at The Master's College

Source: Lisa La George

Faith-based collegiate institutions are brilliantly equipped to provide extensive training in both the character and the competencies necessary to carefully utilize the elephantine STM experience. The author's home institution requires over a year of training for team leaders and an entire semester of training for each team. The majority on-campus residency and common faith confession increase the impact of the team training through formal, non-formal, and informal experiential education. About 10 % of the campus voluntarily participates each year and hosts consistently request repeat teams.

A study of was conducted of 25 years of alumni from The Master's College's STM program. When participants were asked if their STM training and trip continued to influence their lives, over 300 (75 %) of the alumni respondents identified relationships, interpersonal lessons, leadership lessons, church ministries, and an appreciation for missions as evidence of the impact. Whether or not that influence can be quantified, the perceptions of the participants must be taken seriously. Many participants wrote extensive essays which contained multiple themes:

> It was a huge influence.... It changed who I was and continues to shape how I look at this world. It exposed me to culture and made me realize how much I try to fit God into my cultural box. God is bigger than that. I remember being in South Africa and just being utterly amazing that I was so far from home and yet the same God was there with me. It was an incredible moment. It has also given me a love for diversity. I love people from all walks of life. It has given me a passion for missions and a love for missionaries. (Female 202)

Many of the answers were relational in nature. Respondents related that some of their best friends were people them met on their trip, some of them 15 years previously. Respondents mentioned how the trip had prepared them with skills to communicate in a cross-cultural marriage or in jobs which work with immigrant populations in the United States. Others referenced learning about teamwork and learning how to relate with others with patience and openness because of the training they received prior to their trip (Figs. 48.1, 48.2, 48.3, and 48.4).

Other respondents commented on the role of training for STM had on the formation of their worldview. They appreciated, understood, and interacted with different cultures as a part of their regular lifestyle because of the skills they gained on the trip. Other participants referenced specific areas of growth which continue to influence their lifestyle, including a consciousness of materialism, greater understanding of the self, sanctification from specific sins, and increased compassion for other people.

Respondents claimed a continued influence on their ministries, particularly in the area of church ministries, outreach, and STM. The largest number of these answers came from participants who claimed a greater commitment to the local and universal church because of their experience traveling. Participants mentioned that they learned to share their faith and were consistently involved in outreach because of their experience in STM.

Finally, the greatest number of participants commented that their STM experience influences their understanding of missions. Respondents claimed a greater knowledge of missions, theology of missions, and missionary life, and are better able to pray for missionaries and people groups because of their experience. Others addressed

Fig. 48.2 Short-term mission group by a minibus in the Philippines (Photo by Lisa La George)

Fig. 48.3 Short-term missions group in front of a soccer goal in Latvia (Photo from G.O. files)

Fig. 48.4 Short-term mission group having a family meal in Russia (Photo from G.O. files)

the issue of continuing to support missionaries with finances and other physical assistance. While 61 (17 %) of the respondents were already serving cross-culturally as missionaries, another ten alumni said that they were preparing to go as missionaries because of their experience. More than a third of these current and future missionaries stated that their training and STM experience contributed in large part to their decision to become a missionary or chose their agency. One of the latter group wrote,

> Well, I am currently a career missionary who was helped tremendously to make that choice through the experience! God used these trips to shape and mold me and, little did I know at the time, to prepare me for all that He has called and continues to call me to in His world. It helped me grasp a larger view of God and gave me a greater heart for His church and for discipleship. I experienced God's perfect timing and sovereign hand that has all been part of how God continues to grow me. The power of prayer was also greatly seen during these trips and the joy of praying with others on the team and the sweet fellowship experienced with team members – having to lay your own desires and self-wants down for the sake of the greater task at hand: the salvation of souls! (Female 277)

Maybe, as testified by the respondent below, STM can actually serve as the unifying force to assist students actively, thoughtfully, and selflessly integrate their faith, worldview, and vocation:

> How does your [STM]experience affect my life today? In every way! When I go to Church how I view the Church is through an international lens. How I see how to do worship services, what is important, and what can be left out has been primarily formed from these trips.

Seeing how other cultures worship, seeing what was good and what was counter-productive. My faith in God's providence has grown, how He provides for my every need! I see how even tiny things we do can impact many for Christ. As a whole it has made me more aware of the world around me, how I affect it and different ways God is working. (Male 264)

Note: *For the sake of anonymity, alumni responses are identified only by their gender and the order in which each submitted the questionnaire.

References

Adeney, M. (1996). McMissions. *Christianity Today, 40*, 14–15.

Adeney, M. (2008). The myth of the blank slate: A check list for short-term missions. In R. J. Priest (Ed.), *Effective engagement in short-term missions: Doing it right!* (pp. 102–120). Pasadena: William Carey Library.

Barna Group. (2008). *Despite benefits, few Americans have experienced short term missions trips*. Retrieved October 12, 2008, from www.barna.org/FlexPage.aspx?Page=BarnaUpdateNarrowPreview&BarnaUpdateID=318

Barnes, S. (2009). *The world race: Why have a world race?* Retrieved February 2, 2009, from www.theworldrace.org/?tab=about&subtab=why-have-a-world-race

Brown, C. M. (2008). Friendship is forever: Congregation to congregation relationships. In R. J. Priest (Ed.), *Effective engagement in short-term missions: Doing it right!* (pp. 209–238). Pasadena: William Carey Library.

Farrell, H. (2007). Short-term missions: Paratrooper incursion or "Zaccheus encounter"? *Journal of Latin American Theology, 1*(2), 69–83.

Groves, S. (2007). I saw what I saw. On *Tell me what you know* [Audio CD]. Nashville: INO Records.

Guthrie, S. (2005). *Missions in the third millennium* (2nd ed.). Waynesboro: Paternoster Press.

Huertz, C. (2007). *A community of the broken*. Retrieved February 22, 2009, from http://www.christianvisionproject.com/2007/02/a_community_of_the_broken.html

Kim, D. (Paul). (2001). Intercultural short-term missions' influence on participants, local churches, career missionaries, and mission agencies in the Presbyterian Church of Korea (Kosin). Unpublished Ph.D.l dissertation, Jackson: Reformed Theological Seminary.

Kirby, S. (2006). *Equipped for adventure: A practical guide to short term missions trips*. Birmingham: New Hope.

Kyle, J. (1996). A long-term leader looks at short-term missions. In T. Gibson, S. Hawthorne, R. Krekel, & M. Ken (Eds.), *Stepping out: A guide to short term missions*. Seattle: Youth With A Mission.

La George, L. (2009). *Short-term missions at The Master's College: An experiential education*. Ph. D. dissertation, La Mirada, Biola University.

Livermore, D. (2009). *Cultural intelligence: Improving your CQ to engage our multicultural world*. Grand Rapids: Baker Academic.

Livermore, D. (2013). *Serving with eyes wide open: Doing short-term missions with cultural intelligence*. Grand Rapids: Baker Books.

McQuilkin, R. (1994, July). Six inflammatory questions: Part two. *Evangelical Missions Quarterly, 30*, 258–264.

Peterson, R. P., Aeschliman, G., & Sneed, R. W. (2003). *Maximum impact short-term mission: The God-commanded, repetitive deployment of swift, temporary, non-professional missionaries*. Minneapolis: Short Term Evangelical Missions Ministries.

Priest, R. J. (2007, February). *Peruvian Protestant churches find "linking" social capital through partnerships with visiting short-term mission groups*. Paper presented at the meeting of the Evangelical Missiological Society, Deerfield, IL.

Priest, R. J., & Dischinger, T. (2005). Are short-term mission trips the key to bridging inter-ethnic conflict? *A research study on ethnocentrism*. Retrieved October, 2007 from www.tiu.edu/tedsphd/ics_research/040905/TerryDischinger.htm

Priest, R. J., & Priest, J. P. (2007, October). *They see everything and understand nothing: Short-term mission and service learning*. Paper presented at the meeting of the Evangelical Missiological Society, Minneapolis, MN.

Priest, R. J., Dischinger, T., Rassmussen, S., & Brown, C. M. (2006). Researching the short term missionary movement. *Missiology, 34*(4), 431–450.

Rockness, M. H. (2003). *A passion for the impossible: The life of Lilias Trotter*. Grand Rapids: Discovery House.

Slimbach, R. (2007, October). *Short-term missions by many other names*. Paper presented at the meeting of the Evangelical Missiological Society, Minneapolis, MN.

Slimbach, R. (2008). The mindful missioner. In R. J. Priest (Ed.), *Effective engagement in short-term missions: Doing it right!* (pp. 154–184). Pasadena: William Carey Library.

Smith, A. (2008). Evaluating short-term missions: Missiological questions from a long term missionary. In R. J. Priest (Ed.), *Effective engagement in short-term missions: Doing it right!* (pp. 35–62). Pasadena: William Carey Library.

Ver Beek, K. A. (2008). Lessons from a sapling: Review of quantitative research on short-term missions. In R. J. Priest (Ed.), *Effective engagement in short term missions: Doing it right!* (pp. 408–474). Pasadena: William Carey Library.

Wuthnow, R. (2009). *Boundless faith: The global outreach of American churches*. Los Angeles: University of California Press.

Yohannan, K. P. (2000). *Revolution in world missions*. Carrollton: Gospel for Asia.

Zehner, E. (2008). On the rhetoric of short-term missions appeals, with suggestions for team leaders. In R. J. Priest (Ed.), *Effective engagement in short-term missions: Doing it right!* (pp. 185–208). Pasadena: William Carey Library.

Chapter 49
Creating Havens of Westernization in Nigerian Higher Education

Jamaine Abidogun

49.1 Introduction

Nigeria's National Universities Commission (NUC), the governmental oversight group for higher (tertiary) education grants credential authority to Nigeria's federal, state, and private universities. In little more than a decade Nigeria's higher education system has experienced exponential growth primarily due to increased demand for access to these institutions. From 2000 through 2012, 12 new federal universities were established and of these 9 were established in 2011 alone. During this same period 24 new state universities were established, while private universities experienced the most significant growth (Fig. 49.1).

Of Nigeria's 50 private universities 47 of these were established between 2000 and 2012. In fact no private universities were listed from 1983 to 1999. As of June 2012 NUC officially listed a total of 37 federal, 37 state, and 50 private universities as fully or provisionally credentialed universities (NUC 2012) (Fig. 49.2).

This significant increase in Nigerian universities, especially private ones, drew national attention as Nigeria's *Tribune* newspaper (Idoko, March 8, 2011) reported the Minister of Education, Professor Ruqayyatu Ahmed Rufa'i's stern warning, "I must warn that no private university is allowed to run postgraduate programmes without having graduated at least one set of students from mature programme and programmes with full accreditation status … any unwholesome practice or operation outside the provisions of the NUC guidelines is unacceptable and shall attract appropriate sanctions." Her words underscore Nigeria's commitment to stop the tide of unaccredited institutions often referred to as degree mills. The above numbers reflect the massive growth of tertiary or higher education in Nigeria. This growth in the

J. Abidogun (✉)
Department of History, Missouri State University, Springfield, MO 65897, USA
e-mail: jamaineabidogun@missouristate.edu

S.D. Brunn (ed.), *The Changing World Religion Map*,
DOI 10.1007/978-94-017-9376-6_49

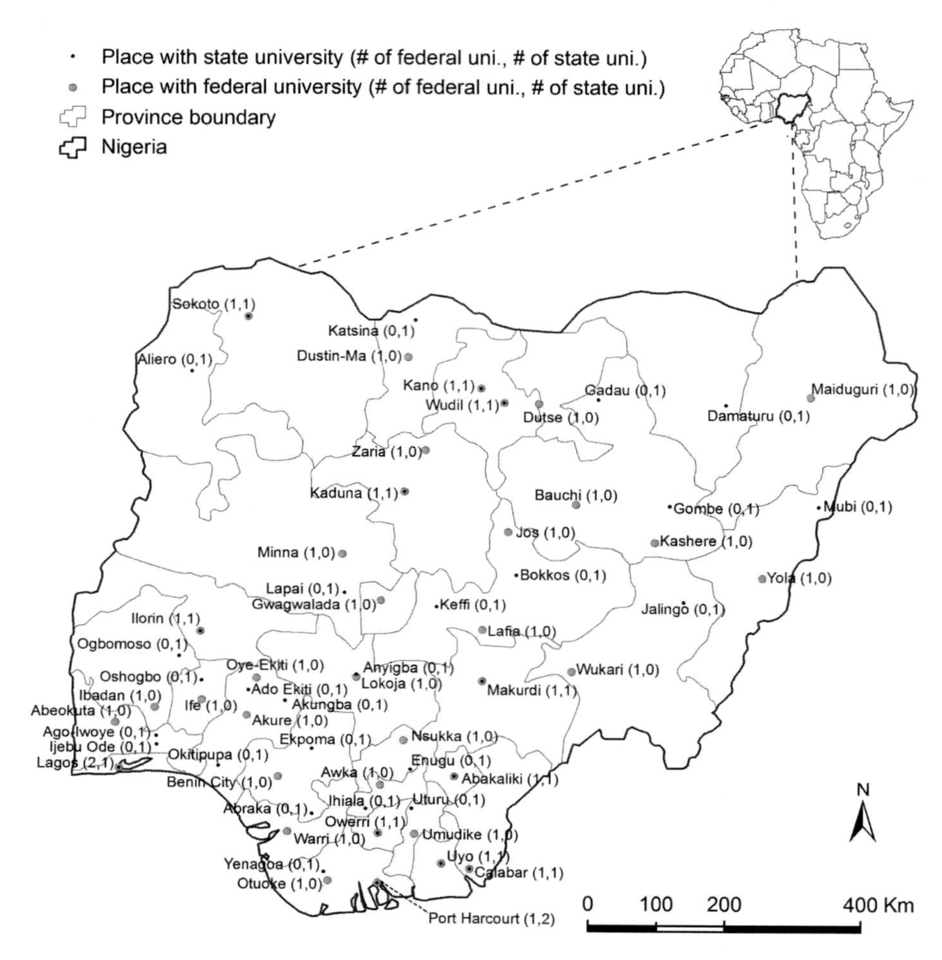

Fig. 49.1 Nigeria's federal and state universities, 2012 (Map by Jamaine Abidogun)

number of universities was accompanied by an equally impressive growth in university student enrollments. Clark and Sedgwick, World Education Services researchers, reported, "In 1998, …the public universities [federal and state] enrolled 411, 347 students" (2004: 4). In October 2010, Dr. Jamila Shu'ara, Director of Tertiary Education, Nigeria Federal Ministry of Education reported the total enrollment in 2010 for all universities (private and public) at 1,330,531 students. This demonstrated more than a tripling of the enrollments that coincided with the expansion of public and private universities.

The recent growth in universities is representative of Nigeria's consistently increasing demand for western education. This demand was initially reflected through a mandate for universal basic (primary) education as part of independent Nigeria in 1960. That developed into federal, regional, and later state expansion of

Fig. 49.2 Nigeria's private universities, 2012 (Map by Jamaine Abidogun)

the secondary education system by the 1970s. In the past few decades this mandate for western education was reflected in the establishment of dozens of small private colleges, academies, and even "universities" throughout the country. These institutions include secondary education preparatory schools, vocational schools and, in some instances, non-credentialed higher education institutions. Such spectacular growth in education within the short span of 50 years is worthy of examination. What is the cultural significance of this increase in private western universities during this latest period of western education growth beginning in the 1990s?

What do many of these private universities have in common? In Nigeria the majority are Christian based institutions. This chapter looks at the historical development, roles and scope of Christian based institutions of higher education within the context of education, culture, and changing worldviews (Table 49.1).

Table 49.1 Denomination affiliation of Nigeria's private Christian universities

Private university name	City/Town location	Christian denomination	Additional church information
McPherson University	Ajebo (Lagos)	4 Square Gospel	
Babcock University	Ilisha-Remo	7-Day Adventist	
Paul University	Awka	Anglican	
Ajayi Crowther University	Ibadan	Anglican	
Joseph Ayo Babalola Univ	Ikeji-Arakeji	Christ Apostolic Church	Pentecostal
Samuel Adegboyega University	Ogwa	Apostolic	Pentecostal
Crawford University	Igbesa (Lagos)	Apostolic Faith Mission	Pentecostal
Evangel University	Eka-Eze (Enugu)	Assemblies of God	
Bowen University	Iwo	Baptist	
Veritas University	Abuja	Catholic	
Caritas University	Enugu	Catholic	
Madonna University	Okija	Catholic	
Godfrey Okoye University	Thinkers Corner – Enugu	Catholic	
Tansian University	Umunya	Catholic	
Gregory University	Uturu	Catholic	
Benson Idahosa University	Benin City	Church of God	Pentecostal
Redeemer's University	Mowe	Church of God	Evangelical
University of Mkar	Gboko	Church of Sudan Among the Tiv	Affiliate of the World Alliance of Reformed Churches
Obong University	Obong Ntak	Churches of Christ	
Bingham University	New Karu	Evangelical	
Salem University	Lokoja	Foundation Faith Church	Evangelical
Landmark University	Omu-Aran	Living Faith Church	Pentecostal
Covenant University	Ota (Lagos)	Living Faith Church	Pentecostal
Rhema University	Aba	Living Word Ministries	Pentecostal
Wesley University of Science & Technology	Ondo	Methodist	
Caleb University	Lagos	Unaffiliated	Website Introduction includes commitment to Christian principles

(continued)

Table 49.1 (continued)

Private university name	City/Town location	Christian denomination	Additional church information
Igbinedion University	Okada	Unaffiliated	University history & founder's background demonstrate Christian orientation.
Wellspring University	Owa	Unaffiliated	University mission statement includes commitment to Christian principles

Data source: National Universities Commission, http://www.nuc.edu.ng/

It seeks answers to several related questions, including: What roles do Christian based universities play in Nigeria? Is it one of maintaining and strengthening Christian religion or do they also act as havens for the promotion of neo-colonial Westernization? Is Nigeria unique in its proliferation of Christian based universities within West Africa, as compared to other Anglophone and Francophone countries?

In 2012, based on their stated university history, mission, philosophy, and/or goals, of Nigeria's 50 private universities 25 were established as Christian affiliated institutions and an additional 3 as Christian unaffiliated institutions. A Christian affiliated university is either an officially sanctioned institution of a Christian denomination or church or it officially endorses and aligns with a specific Christian denomination or church. For example, Bowen University is a Baptist affiliated private university owned by the Nigerian Baptist Convention. A Christian unaffiliated university is not officially sponsored or endorsed by a specific denomination or church, but it identifies itself as a Christian institution as professed through its official documents and institutional curricula. Wellspring University (2012) is a Christian unaffiliated university as it includes in its website's opening statement, "Wellspring University is a private university promoted by Management Science Centre and approved by the Federal Government. It operates on Christian ethics and principles." Yet there is no indication in its history or administrative oversight of a direct endorsement by a particular Christian denomination or church. In contrast only 4 private universities identified as Muslim affiliated or unaffiliated and the remaining 32 indicated no religious affiliation or preference as reflected in their stated history, mission, philosophy, and/or goals. In order to understand this recent proliferation of Christian based universities it is instructive to review the historical development of Nigeria's education system.

49.2 Historical Development

A brief timeline of western education in Nigeria describes to a large extent the corresponding development of westernized enclaves of Nigerians. These enclaves developed during the colonial period and held the majority of national political, economic, and social leadership positions in the post colonial or neo-colonial period. This timeline also represents Christian mission expansion in Nigeria as historically Christian missions were the first to establish formal western education institutions beginning in the mid to late 1800s. The British utilized missionary schools to provide literacy and vocational training. In this way early mission schools were further developed through formal contracts and partial funding from the British government. This pattern of western education development was replicated throughout Anglophone Africa and its diaspora with only slight variations and was eventually formalized in the British government's "Educational Policy in British Tropical Africa" (Colonial Office 1925). This partnership to establish western education is responsible for the basic structure of primary and secondary education found throughout Nigeria today.

While the curriculum changed significantly over the past century, one lasting element that is found in most government, publically funded schools is the official endorsement and practice of religion in primary and secondary school buildings. For example, "Christian" government schools open with Christian prayers, work on the premise that their students are majority Christian, and more often teach Christian Knowledge as the required religious studies course in the primary and secondary curriculum. This same structure is found in Islamic dominated areas with Islamic prayers in school, the expectation that students are majority Islamic, and an Islamic Knowledge course taught as the required religious studies course in primary and secondary curriculum (Nigeria Education of Ministry, UNESCO-IBE 2010; Abidogun 2011).

Today Nigeria's western educated no longer exist as isolated yet powerful enclaves, but rather western education is the status quo education system with Nigeria's vast majority completing at least primary education. According to UNESCO in 2009 Nigeria's primary education enrollment was at 93 % with a 74 % completion rate with approximately a 50 % transition rate to junior secondary schools (JSS). In 2010 secondary education enrollment (JSS and senior secondary schools) was at 44 %. The current national curriculum for primary and JSS continues to require a "religious studies" course which is most often represented as Christian Knowledge, with the exception of Muslim dominated schools (UNESCO-UIS 2011; UNESCO-IBE 2010: 11, 13).[1] While there is a national curriculum that dictates the official content of this course, Religious Studies classroom textbooks are written by regional authors. For example, in Eastern Nigeria, which historically is Catholic dominated, a recommended textbook is *Christian Religious Knowledge for Senior Secondary* (1996) by A.N. Ebizie. While there is nothing about the author in the text, the book is set up like a series of mini-homilies about the Bible beginning with the "Old Testament" on page 3 and working through to the "Mission of the Gentiles" on page 218. The national curriculum, like the colonial curriculum before it, ensures that the

---

[1]The exception in the required Religious Studies course appears in Moslem dominated areas, where Islamic Knowledge is taught instead of Christian Knowledge. These schools represent a minority within Nigeria. Notably Traditional religions are not taught as a Religious Studies course, but is discussed in the JSS social studies textbooks.

majority of Nigerians effectively experience a Christian western education regardless of their personal religious beliefs. The texts and lessons (as observed by this author in 2004–2005, 2012–2013) are structured to provide Christian religious knowledge, as well as to provide moral guidance. Religious affiliations of the students and teachers ranged and many expressed a synthesis of Christian and Traditional beliefs, but all students learned the Christian Religious Knowledge course as a foundation for moral guidance.

Christian religious studies or theology was required along with other Christian practices during the early years at the University of Ibadan, but these required curricular and non-curricular components did not survive intact as required components in the post-independence era. Originally the Christian Council of Nigeria in its 1947 meeting at Onitsha confirmed the following information regarding a Christian framework for the then University College at Ibadan, as follows:

> Theological Courses will be arranged at the University, but the Council agreed that provision must be made for a Chapel which would be a permanent source of help and inspiration to the student, most of whom will have received preliminary education in Christian schools …. Dr. Mellanby, the Principal of the University, has gladly accepted the offer…. (Christian Council of Nigeria 1949)

While religious aspects remain a part of campus life and Christian dominance is still evident at many of Nigeria's federal and state universities by the 1970s; there was no longer a formal endorsement of Christianity or Islam as university policy and students were no longer required to attend church or mosque or take theological courses. For example, there are no religious affiliation requirements for faculty, staff, or students and religious studies courses are no longer required coursework. Today most of these campuses include Christian church(s) and Islamic mosque(s) or center(s) on their campuses and theological course work as academic content from an array of religions is available, but is not required with the exception of students enrolled in Religious Studies programs. Also the informal curriculum has changed as many traditional Christian western practices of prayers and blessings at convocations and graduations often include Christian, Islamic and even Traditional sentiments. Traditional religions per se are comparatively less evident on these campuses, but its practices are often entwined in the daily lives of faculty and students as Traditional culture that is specific to each ethno-national group. African Studies as an established academic area since the late 1960s provided an academic venue that increased Traditional religious content in coursework at many of these universities (University of Nigeria, Nsukka 2012; University of Ibadan 2012).

Today all federal and state universities teach religious content as academic content not as religious doctrine. This shift was largely a product of Pan African and Nationalist movements as Zeleza, a historian, explains, "The postcolonial university was founded to promote the dreams of African nationalism: decolonization and development. Both the political class and the intelligentsia saw these as essential for achieving and sustaining African self-determination" (2009: 114). In Nigeria's case it was also a result of post Civil or Biafran War (1967–1970) politics as Nigeria struggled to reunify itself within a hotbed of religious and regional divisions

(Bray and Cooper 1979). It was not until the recent upsurge in private universities that required religious academic curriculum and practice were once again intertwined with official curricula. So how did the Christian mandate from these historical origins survive to reappear as almost a mass movement in the plethora of recently established private Christian universities?

49.3 The Roots of Christian Universities

By the late 1800s the earliest British holdings in Africa developed secondary schools or "colleges" and one of these, Sierra Leone, produced the first class of African Anglophone Christian educated elite. These elite would help to replicate this early college model across West Africa. Even though their influence on university official curriculum was filtered of most of its required religious doctrine during the 1970s, the Christian western educated elite would prove to have a lasting impact on Nigerian society and particularly on education.

Early on Christian western educated Africans constituted a distinct, elite group. These initial western educated elite, known as *Saros*, were freed slaves from throughout the African Diaspora, "liberated Africans" from slave ships most often repatriated to Sierra Leone, and their descendants. They were mission school educated with the majority culminating their western education at Fourah Bay College in Sierra Leone. Fourah Bay College served as the precursor to western universities in West Africa. Its history is indicative of the steady development of a distinctly Christian western education model. As described in this brief historical summary from its current website:

> Fourah Bay College was founded on the 18th of February, 1827, by the Church Missionary Society [CMS] essentially for the training of teachers and missionaries to serve in the promotion of education and the spread of Christianity in West Africa. In 1876 it became a degree granting institution with an affiliation to the University of Durham in England, and since then has maintained a reputable tradition of higher education in Africa. In 1960 it was granted a Royal Charter to become "FOURAH BAY COLLEGE" The University College of Sierra Leone,"[sic] by the Senate of the University of Durham."[sic] (University of Sierra Leone website, Feb. 15, 2012)

According to Esedebe, a historian, some of the *Saro* educated elite and the later (second) phase of Nigerian Christian educated elite made a major impact on colonial Nigerian society, "Between 1860 and 1940 the Saro showed themselves to be a force in Nigerian affairs providing leadership in practically every walk of life: Bishop Ajayi Crowther and James Johnson in the Church; R. B. Blaise, J. S. Leigh, J. J. Thomas and Henry Robbin in the world of business; J. Otunba Payne and Nash Williams in the Judiciary." Esedebe noted indigenous Nigerian educated elite who followed in the Saros' footsteps included, "H. R. Carr and Herbert Macaulay in the civil service, Isaac Oluwole and Mobjola Agbebi in the church, Kitoyi Ajasa in the legal profession and J. P. Haastrup in business" (Esedebe 1980: 113). There is much historical debate about whether and how these early leaders either colluded with or resisted colonial powers as early nationalists (Ayandele 1974; Afigbo and Falola

2005). Some historians like Ayandele (1974) insisted that these early western educated elite benefitted from British colonialism and therefore promoted support for the colonial system. Other historians, including Afigbo and Falola (2005) and Ezedebe (1980) concluded that while they may have materially benefitted, they used these resources to communicate the need for empowerment and education of the people in either a pro-Nationalist or Pan-African call for independence.

For purposes of this chapter their key contribution is the extent to which their experiences signaled an historical shift toward a Christian western cultural mandate that continues to this day. Though this mandate is often challenged, it constitutes a tangible aspect of Nigerian society. The recent surge in Christian based universities may be viewed as evidence of major cultural shifts from a range of ethno-national groups or perhaps it is evidence of these early Christian western educated leaders' lasting impact on the wider society. This chapter posits the latter, that this elite group had an enduring mandate that stoked by a resurgence of western Evangelical neo-colonialism toward the end of the twentieth century caused the Christian revival in private universities.

The catalysts to this resurgence were many, but the main two catalysts were witnessed in Nigerians' everyday lives as the massive expansion of communication and financial systems. These that allowed Nigerian churches to communicate effectively and efficiently with world churches in their fellowship and development efforts. In return these world churches were all to willing in their proselytizing missions to maintain partnerships that increased their folds. The development of these partnerships on their current scale was not possible prior to internet, mobile phones, and web based financial systems that created world wide exchanges in multiple ways. These multiple ways set the stage for this early mandate to realize its vision for a Christian western educated society.

While there were a number of secondary schools by the early 1900s, it was not until 1948 that the University of Ibadan in Nigeria (Inter-University Council for Higher Education in the Colonies 1948) and the College of the Gold Coast (Ghana) in Legon were established as the first fully accredited western universities in Anglophone West Africa (The University of Ghana 1996: 5). Both of these were initially secondary schools or "colleges" and then later licensed as extension campuses as part of the University of London and finally accorded recognition as fully accredited universities. With independence came an increase in federal then regional and later state universities. Nigeria's private universities were slow to emerge initially due to low secondary matriculation rates and uneven development and then later due to legal barriers.

Secondary schools were few and far between prior to World War II and remained low in numbers until the late 1960s. Many churches, beginning in the 1940s expanded their focus from primary education to include secondary education. A summary of Rev. H.D. Hooper's comments at a CMS Conference on Education in Africa reflected this shift:

> Mr. Hooper thought we were denying the sovereignty of God if we [CMS] handed over one aspect of educational work, any secondary schools .... He considered we should set the standard for Government in all sides of educational work. Because of our shortage of man power the training of Africans was becoming more and more important. (1945: 5)

This conference reflected a new commitment to secondary education as well as an increased recognition of the need for indigenous educational and religious leaders in the field. For the next two decades leading up to independence, the churches partnered with the British government to establish secondary schools that included religious studies as required coursework, as well as informal curricula that further indoctrinated students to Christian western cultural views. This created a third phase or wave of Christian educated elite within Nigerian society.

In practical terms this meant 90 % of Nigerians who entered western schools received some level of Christian indoctrination. Toyin Falola's personal account is insightful as he describes his experiences during the transition to independence in missionary and public schools:

> On getting to [a missionary primary] school, one must also confront the Bible, not just as a discipline called Bible studies, but as songs, sermons, and prayers. School opened and closed with prayers. I never understood many of the prayers, but I said 'Amen' anyway… (2005: 84) [compared to a milder version in the public school]. As Ode Aje was a public school with students from all religious backgrounds, the songs were secular in nature. Rather than Christian prayers, people were asked to engage in 'silent prayer'… We also sang [prayer] songs to introduce lunch ….
> Come with us, Lord
> Let us praise thy name
> Bless the food
> So that we can eat with you in heaven. (143–148)

Falola's experience bears witness to the inevitable ambiguities created by the imposition of a Christian western education on a decidedly multi-religious society. He explains, "Not many people apologized for being Christian or Muslim and still taking part in [Traditional religion] festivals…No one stopped us from participating in the Okebadan festival." He described Okebadan festival as "a huge one-day carnival…to celebrate the spirit force of the venerated Oke (hill), the city's [Ibadan] deity…. The belief was that the hill had offered protection to the earliest occupants…. The carnival turned the hill into a god, worshiped as the Okebadan (the hill of Ibadan)" (Falola 2005: 222–223). Although he qualifies this experience by noting that in later years fundamentalism would drive clear divisions between the Christian and Muslim religions (230). This apparent collusion between colonial education and religion certainly contributed to a moratorium on the establishment of private universities during the early years of independence.

Private universities as legal entities experienced a rocky and scarce existence prior to the passage of Education National Minimum Standards and Establishment of Institutions, Amendment Decree No. 9 of 1993 that allowed private individuals or organizations to establish universities in Nigeria within federal guidelines. Jekayinfa and Akanbi (2011) document this history noting that by the first prohibition on private universities in 1972 there were two federal and four regional public universities and even during the first lifting of the ban (1979–1983) private universities "were poorly planned with neither good infrastructure facilities conducive for learning, nor serious-minded academics in their foundation list" (2011: 298). By 1983 another decree was imposed placing all universities under government control. Even though this decree was lifted in 1985, it was not until the Amendment Decree No. 9 of 1993 that private universities in Nigeria began to develop blueprints that would materialize in accredited institutions of higher education. Which is to say that many non-credentialed "universities" or institutions of higher education were available in Nigeria, but none carried national accreditation, so did not have legal authority to grant degrees.

One factor that initially limited the growth of universities, public or private, was lack of adequate staffing. Higher education scholar, Enahowo, documented a consistent use of ex-patriates from the UK and other nations, "the level of provision of academic staff between 1966 and 1982... in Nigerian universities show that about 30 % of the overall staff for the respective years were expatriates... The figures from 1973 to 1982 varied from 28 % to 21 %" (1985: 313). Shu'ara, Director of Tertiary Education reported a shortfall of 39.1 % in academic staffing in a 2010 report. He attributed this shortfall to "inter- and intra-sectoral brain" (2010: slide 17). These numbers indicate an important historic shift in academic staff, as early colleges, including the University of Ibadan were primarily staffed with UK ex-patriates with a few Nigerian educated elite primarily trained at Fourah Bay and the University of London. Yet only 6 years after independence (1966) Nigerians appear to staff the majority of positions and many of these personnel are in high demand in other professions and in other higher education institutions outside Nigeria. Private universities encounter similar staffing histories, first as academies or colleges in many cases and then in their current university status. The abundant advertisements that appear through Google results for higher education jobs in Nigeria attest to the constant demand for adequate staff. The majority of which are recruited from Nigerian universities.

The Secretariat of the Standing Committee on Private Universities (SCOPU) is the NUC department that oversees the accreditation application process for private universities. Amendment Decree No. 9, Criteria 6.1 speaks directly to equity within higher education, as it stipulates, "(6) (1) A proposed institution shall have an adequate environment base and shall be open to all Nigerians irrespective of ethnic derivation, social status, religious or political persuasion" (Onwuliri 2012). While Criteria 6.1 allows any individual to apply for admission at any private university in Nigeria, it stands to reason that most individuals will not apply to a university that formally endorses religious principles counter to their perspective. So long as each university officially endorses Amendment Decree No. 9, Criteria 6.1 and meets other accreditation criteria, they are free to operate. The result to date is that 28 of Nigeria's 50 private universities self-identify as Christian institutions that were established from a range of denominations as noted in Table 49.1.

49.4 Havens for Westernization

The emergence of private parochial universities in Nigeria, as well as throughout much of Africa, beginning in the late 1990s is not mere happenstance nor is their exponential expansion throughout the next decade. Two factors combined to make it easy, almost inevitable for this historic expansion. One factor was the internal reality of a growing Nigerian Christian presence that seized this opportunity to expand their private school systems to include universities. While a range of denominations is represented, the benefactors for many include a combination of Western and Nigerian sources. For example, those that are Catholic are partially supported by central Catholic offices, as well as the local Nigerian orders that established each university. This type of partnership is common across the affiliated private universities, that is, the University of Mkar is part of the World Alliance of Reformed Churches and its Nigerian based Church of Christ in the Sudan among the TIV, Rhema University is part of the Living Word Ministries International and their Nigerian churches, etc. The other factor was a major shift in western government perspectives regarding hitherto held mandates of separation between church and state in development projects.

This brief history recorded above lays the foundations for understanding the presence of the first factor which involved the reduction of western church control through the introduction of a 4th phase of Christian western educated Nigerians who established, controlled, and developed churches across Nigeria. This group, many of whom used the support of western churches, developed strong congregations with centralized control in Nigeria. Most kept some tie to a specific denomination, but began to ordain their own pastors and maintain direct control of administration and funding. A few, like Joseph Ayo Babalola, initiated independent churches in Nigeria. His church, the Christ Apostolic Church, a independent Pentecostal church, is also the founder of Joseph Ayo Babalola University. These churches followed suit with their colonial predecessors as they quickly established lucrative private primary and secondary schools systems that included many "academies" or "colleges" that fed federal and state university enrollments. These churches, with or without western partners, sat poised to establish the first private Christian universities in Nigeria.

The first three private universities, Igbinedion University at Okada, Babcock University at Ilishan Remo, and Madonna University at Okija, received national accreditation on May 10, 1999. Babcock University (Seventh-day Adventist) and Madonna University (Catholic) are Christian affiliated (NUC 2012). They are each offspring of Nigerian churches with established primary and/or secondary schools. Babcock University originates from "the Adventist College of West Africa (ACWA), established on September 17, 1959" (Babcock University 2012). ACWA's founding was the culmination of combined efforts by a then Methodist Nigerian businessman, Samuel Olatunji Adebawo and the Seventh-day Adventist (SDA) Mission. In this case Mr. Adebawo worked with a local Chief J.S.K. Osibodu at Ibadan and the Pastor Roger Coon of the SDA Mission to broker a deal with Ilisan's Traditional Authority, the *Olofin* of Ilisan, Oba Green Adebo. Due to these negotiations 11 Ilisan families sold land to the SDA Mission and ground was broken for the ACWA.During the mid-1960s Mr. Adebawo attended ACWA and converted to Seventh-day Adventist. Interestingly, Mr. Adebawo became both a district pastor in ACWA and a titled member of Traditional council, *Olori ebi* (community leader). A brief biography by Adesegun, notes, "He relishes this position (*Olori ebi*) and his activities in the church because with these opportunities he is at a vantage point to do good to all (2009)." These activities include involvement in ACWA's transition to Babcock University making it a Nigerian owned and managed Seventh-day Adventist institution of higher education.

Madonna University, recipient of the third certificate, No. 003, was founded by a Nigerian Catholic priest, Rev. Fr. Prof. Emmanuel Edeh. He is the founder of a Catholic Igbo based organization and also founded several primary and secondary schools, including a post-secondary polytechnic school at Enugu. As a testament to a uniquely Nigerian Christian institution, Fr. Prof. Edeh authored several books and other publications on African philosophy, more specifically on Igbo metaphysics which are found on his website promoting Catholicism within the Igbo cultural context (Edeh 2012; Madonna University 2010).

Igbinedion University (IUO) at Okada was the first fully accredited private university. As their official history explains, it was "the realisation of the dream of Sir, Dr. Gabriel Osawaru Igbinedion, the *Esama* or "godfather" of Benin kingdom to bequeath to future generations a University with international standard where academic calendar is faithfully run without interruption" (Igbinedion University 2012). As the Esama who is charged under the *Oba* (high leader) of Benin to assist in the economic wellbeing of Benin people the establishment of Igbinedion University may be seen as a means to ensure the economic wellbeing for its people. This university was the first to receive accreditation as it was the recipient of NUC Certificate No. 001. IUO represents a distinctly Nigerian blend of Christian and Traditional philosophy and practice. Unlike Babcock or Madonna Universities, there is no mention of Christianity or a specific Christian denomination in IUO's philosophy, mission statement, or goals. These Christian principles are rather represented informally in the presence of their Vice Chancellor, Prof. Rev. Eghosa E. Osaghae, Ph.D., J.P. who also serves the Education Advisor to the Lord Bishop, Nigeria's highest ranking Anglican Church official. The university's founder Sir, Dr. Gabriel Osawaru Igbinedion, the *Esama* of Benin kingdom, is a high level official in Benin's Traditional Authority as well as a product of Christian western education. He is founder of Okada Baptist Church and underwriter to at least ten other Christian churches in Edo State. He is truly one of Nigeria's 4th phase Christian educated elite as he rose through the mission school system, attending Eko Boys' High School, one of the oldest schools in Lagos, founded in 1913 by a Methodist Priest (Ohonusi 2012; The Official Website Igbinedion 2012). This university actively combines Benin culture with western education through its founder and Vice Chancellor who are committed Christian benefactors. In this instance, Falola's experience of a multi-religious (Christian and Traditional) philosophy, although not specifically stated, is demonstrated at Igbinedion University. This results in marginal emphasis on religion and full focus on establishing "extensive and rewarding links with industry commerce and a number of other prestigious Universities both at home and worldwide for the benefit of its students" (Igbinedion University 2012).

The fourth certificate was awarded to Bowen University which, like Babcock and Madonna Universities, represents the continued tie between the Nigerian Christian western educated elite and Missionary "Evangelical" Western Churches. Their historical development originates from evangelical efforts of Baptist, Catholic, and Seventh-Day Adventists, not unlike earlier missionary efforts of the Anglican and Methodists societies. Bowen University self-identifies as "a private institution owned by the Nigerian Baptist Convention and named after Reverend Thomas Jefferson Bowen who pioneered Baptist work in Nigeria in 1850" (Bowen University Website 2012). As with Babcock and Madonna, the Nigerian Baptist Convention had established a secondary college and a seminary (Ajayi 2011: 26–27). These private Christian universities exemplify a pattern of development that continues to the present. The original mandate of the early Saros and Nigerian Christian western educated grew exponentially in the twenty-first century in little more than a decade.

These Nigerian founded Christian universities also receive support from endowments and projects sponsored by western churches. For example, Bingham University is a part of the international Soudan Interior Mission (SIM) Evangelical ministry, a Canadian based organization that includes the Evangelical Churches of West Africa (ECWA), founders of Bingham University (Soudan Interior Mission 2012). Several initiatives from the West worked together to create a haven for Christian western education institutions in Nigeria and elsewhere. Amy Stambach notes that the two significant events included the establishment of the World Faiths Development Dialogue (WFDD) in 1998 sponsored by James D. Wolfensohn, former president of the World Bank, Archbishop of Canterbury George Carey and His Highness the Aga Khan, Imam of the Shia Imami Muslims. This organization convened a conference in Nairobi, Kenya held in 2000 that identified education as a priority in Sub-Saharan Africa. The second event was President George W. Bush's signing of the executive orders, Establishment of White House Office of Faith-Based and Community Initiatives and Agency Responsibilities with Respect to Faith-Based and Community Initiatives on January 29, 2001. These orders established the White House Office of Faith-Based and Community Initiatives to function in all federal departments (Office of the Press Secretary 2001). This mandate was realized by 2003 and included the United States Agency for International Development (USAID (2010: 2). These and other events brought an end to long-held practices of separation of church and state in development policy, including education projects (see Stambach 2010).

This shift in Western policy from the maintenance of separation of church and state in development policy to a conscious partnering of church and state was similar in practice, if not ideology, to earlier British policy and practice during the colonial period. Effectively this policy shift resulted by the early 2000s in the development of funding partnerships between churches and the U.S. throughout much of Africa and other parts of the "developing" world. One example is a multi-year program that the Centre for Development and Population Activities (CEDPA) described as:

> With funding from the President's Emergency Plan for AIDS Relief through the U.S. Agency for International Development, … will implement the 4-year, $12.8 million Positive Living project from 2006 to 2010, working hand-in-hand with a consortium of local faith-based organizations … key partners in the effort are leading faith-based organizations including the Anglican Communion, Church of Christ and the National Supreme Council for Islamic Affairs. (CEDPA website 2012)

The major difference was the ideology found within the new practice of church and state partnerships. In comparison to colonial Britain, the U.S. partnership with Christian churches resulted in an increasing "evangelical" environment which more overtly attacked Traditional religions and cultures. The results of which are discussed further in a later section of this chapter. The colonial British missions, based on missionary accounts, took a comparatively more anthropological approach that demonstrated a practiced tolerance for other religions and their associated cultures (Dallimore 1930; Freeman 1884). Much of this tolerance was a byproduct of their arrogant sense of cultural superiority and inevitable sense of victory, but it proved much less intrusive or damaging to African societal structures when compared to newer neo-colonial "evangelical" messages of Christian education. Within little more than a decade of experience these new private Christian western universities are demonstrating both their connection to the past and their impact on the present and future.

49.5 Private Christian Universities in Context

This proliferation of private Christian universities is not unique to Nigeria as current UNESCO and national documents provide substantial evidence of their increased presence as a general trend across Anglophone West Africa. For example, in Ghana with one-sixth the population of Nigeria (Information Please 2012) demonstrates a comparatively similar increase in private universities from 7 founded by the 1970s to 21 currently listed as accredited by Ghana's National Accreditation Board (Embassy of Ghana 2012; National Accreditation Board of Ghana 2010). Of these 21 at least 7 or one- third of the total are self-identified Christian institutions with denominations that are comparatively the same as those indicated in Table 49.1 for Nigeria only with proportionately fewer Pentecostal denominations. The historical timeline for the increase in Ghana parallels Nigeria's, as Asamoah-Gyadu observed at a recent conference on Christian higher education, "Until the influx of neo-Pentecostal churches into the campuses of the country's main universities in the 1990s, the trailblazer in the provision of Christian education and fellowship for students was the Ghana Fellowship of Evangelical Students the local arm of the well-known International Fellowship of Evangelical Students" (2007).

In Francophone West Africa this trend is not evident. The examples of Senegal and Côte d'Ivoire indicate a very restricted environment for private Christian universities. Even though the demand for access to higher education has increased across Francophone Africa, these countries have not experienced the same rapid growth in private universities, especially Christian universities. For example Côte d'Ivoire with a population similar to Ghana's has only three private universities of which none are identified as Christian while Senegal with slightly more than half the population of Côte d'Ivoire also has three private universities (Universities in Cote d'Ivoire 2012; Universities in Senegal 2012). In Côte d'Ivoire and Senegal there are no nationally accredited universities identified as Christian affiliated or non-affiliated universities. A quick look at the history of Francophone African education explains the lack of Christian western universities. Garnier and Schafer describe the Francophone system as the result of:

> A centralized bureaucratic structure, and ministry officials ensured that national educational policy was carried out: …[centralization] is so strong that interstate agreements make it virtually automatic for students … in one Francophone country to enter another country's educational system. (2006: 155)

This transfer agreement includes direct transfer to France's schools. The exclusion of Christian or religious based institutions is found in French colonial education policy as limits on missionary schools were set in 1903 and further restricted by 1924 to place all education directly under government control. In 1903 up to 15 % of schools were mission schools and by 1922 new legislation severely restricted curriculum, so by 1924 they operated as long as they had, "government permission, government-certified teachers, a government curriculum and the exclusive use of French as the language of instruction" (White 1996: 11). As religious studies was not then nor now part of French authorized curriculum, it is easy to see why Francophone West African countries do not share in Anglophone West Africa's rapid increase in private Christian universities.

Across Anglophone West Africa a pattern demonstrating common British colonial education policy as tied to religious institutions contributed substantially to their national education systems and resulted in similar patterns of development in private higher education. The major difference between Anglophone and Francophone Africa in their sharp contrast in the development of Christian western universities is due to France's more centralized and secularized policies, and its continued inclusion of Francophone countries' education system as an extension of their system. It is only with the recent global changes with a demand to increase technology and global communications that Francophone countries are seeing the emergence of a few private, secular western universities. Some would say this is a benefit as it means that Francophone Africa's ethno-national cultures are not as heavily impacted by the latest surge of evangelical Christianity. This holds some truth, but is tempered by the historical realities of France's assimilation policy that did not allow or tolerate indigenous ethno-nations to maintain their social, economic, education, or political institutions. As France was largely successful in marginalizing the majority of these ethno-nations' political and economic systems, the lack of Christian or any other religious institutions does not indicate the maintenance or growth of ethno-national religion or education. While Francophone ethno-nations claim survivals, much of their religious practice consists of either Catholicism or a hybrid of Catholic and Traditional beliefs and practices of which neither are supported through education institutions.

49.6 Conclusion

This chapter demonstrates the mapping of religion in Nigeria and West Africa through the presence or absence of private Christian western universities. The historical evidence provides a foundation to understand the presence or absence of these institutions and demonstrates western education as the status quo education system in West Africa. Nigeria, as with most African countries, has experienced significant increases in public and private universities. An International Institute for Educational Planning report notes several reasons for the increase in private universities. These include:

> De-regulation policies under the structural adjustment programmes, the fiscal incapacity of the state to expand higher education through public universities, and the inability of public universities to respond immediately to household demand for certain market-friendly courses [that is, Information and Communication Technology, Business Administration, etc.]. (Varghese 2004: 12)

In light of the significant number of private Christian western universities located in Anglophone West Africa, the demand for Christian education may also be argued. Many of the founders' stories indicate how they filled a need in civil society and how these churches and their universities continue to fill that need.

Is Nigeria unique in its recent proliferation of Christian based universities in West Africa? No, they are not unique in Anglophone West Africa. As British colonialism left a legacy of church and state partnership in education, these countries have continued to varying degrees with this model. Although challenged and even prohibited in the 1960s and 1970s during the height of Pan African nationalism by the 1990s most Anglophone African nations reopened this historic partnership in order to meet the demands for access to higher education. More importantly this partnership was never inactive as these Christian churches continued in the post-independent era to provide primary and secondary schools. This education base made it easy, almost inevitable for the development of Christian private universities once the legal barriers were removed.

What roles do Christian private universities play in Nigeria and in Anglophone Africa? It can be argued that they maintain a cultural battle to win the minds, if not the souls, of African ethno-nations. The immediate role is one of providing thousands of Nigerians and other Anglophone Africans access to higher education. A much needed commodity as globalization makes it impossible for any country to compete without sufficient numbers of highly western educated citizens to meet the demands of a global infrastructure. These countries cannot successfully compete for wealth and security in a worldwide arena without meeting these educational demands. For addressing these needs Christian private universities may be applauded, but another role they fill complicates the evident benefit of such access and opportunity. The second major role, possibly for Christian organizations their primary role, is the expansion and strengthening of Christianity in Nigeria and throughout Africa. When both roles are examined together, it is this unique version of blending westernization and "modernity" with a curriculum of evangelical doctrine that prompts important ethical and historical questions.

Do these roles only act to maintain and strengthen Christian religion or do they also act as havens for the promotion of neo-colonial Westernization? The goal to advance on the global stage of technology, science, and business, in and of itself, does not automatically translate to westernization, rather it indicates recognition of worldwide knowledge systems and the need to access them. What does indicate a neo-colonial pattern across the private Christian universities is the continued formal practice of the large scale adoption of western Christian religious doctrine in partnership with the maintenance of accreditation systems that adhere to western standards. In effect private Christian universities because they combine western Christian doctrine with western education structures and curricula move their faculty, staff, and students, as well as much of the larger society through social interaction in a decidedly neo-colonial pattern. This continues the colonial "civilizing mission" that insists on western social and cultural norms, only this time it is sold as a "globalization strategy" necessary to compete in the modern world.

The majority of this continued neo-colonial process is found in each university's curriculum content. While the federal universities, like the University of Ibadan or the University of Nigeria, often have an African Studies Center and a fully developed African Studies curriculum, this curriculum is not widely integrated across students' academic experience. It is the exception, rather than the norm, for students to learn about African histories, cosmologies, economics, politics, etc. from a clearly Afrocentric perspective. Instead the majority of these disciplines include African content as it relates to the world, i.e. compares with and to Western and Eastern competitors in the global arena. For example political science, outside of African Studies, maintains a focus on nation-state organization, that is, a Western imposed notion of Nigeria as a nation. This obviously meets a national need to maintain Nigeria as just that, a nation-state, but it also serves to maintain a neo-colonial policy of marginalizing ethno-national political systems.

Much of this is economic driven, but there appears to be few Nigerian universities that reflect a Renascent Africa (Azikiwe 1973) that stand as totally independent of western Christian ideology, knowledge, and infrastructure. There are few isolated efforts, like the Institutes of Africa Studies in Nigeria's federal universities, and Center for American Studies at the University of Nigeria, that encourage ethno-national political participation to widen the scope of democratic representation within Nigeria. Still the general hidden (unofficial) curriculum message with the academic coursework is that Indigenous Knowledge and their associated systems are of non-academic consequence. What is important is a mastery of Western Knowledge and its associated systems.

Even though the general pattern is indicative of further westernization there is some hope of a fully Africanized higher education institution as universities like Igbinedion University build their foundations on combining ethno-national philosophies and authority with western education structures and curricula. The upfront acceptance and endorsement of Traditional Authority and practice within the system ensures the infiltration of Afrocentric practices and the higher likelihood of incorporation of African Indigenous Knowledge.

For instance involvement of the Benin Traditional Authority with Christian churches develops Afrocentric aspects that work to preserve respect for and inclusion of ethno-national and Afrocentric cultural components. This allows a transformational synthesis that makes what was Western Judeao Christianity into an Afrocentric Christianity that aligns more clearly with Traditional belief systems, whether the issue is about women's reproductive issues or sexual orientation. These issues are considered through the histories, norms, and morals of the ethno-national society that indicate layers of understanding and acceptance or resistance based on individual contexts and larger philosophical frameworks. An expectation that seems natural coming from societies who are centuries old. Rather than the adoption of liberation theology, which is not common outside the Catholic church, the adoption of Afrocentric authority and philosophies provides a conduit for increased tolerance within Christian frameworks.

This is true to some degree with all African churches as it is impossible to fully divest ones' collective history and culture from religious practice. Unfortunately, many Christian organizations who own universities in recent years have adopted fundamentalist agendas that actively seek to end Traditional religions and much of the ethno-national and Afrocentric cultural elements identified with those religions (Shah and Toft 2006). One stark and disruptive result is the upsurge in the past decade of "wars on idols" that is an organized attempt to force Traditional believers into Christian churches. Historically, individual Christians did what they saw as "necessary" to persuade others to enter the church, but this varied from nothing to destruction of shrines and haranguing of family and friends. More recently, as this author witnessed in 2004 and 2005, some Christian churches have posted signs along highways declaring "war on idols" to encourage Christians to protest Traditional religious practice and even encourage the destruction of shrines and other Traditional ritual symbols. As might be expected from a country as large as Nigeria, there are as many types of Christianity as there are people is some respects. The one common element is that Christianity definitely provides a specific lens that impacts the adherers interpretations of and therefore interactions with other societal constructs, including education.

In the case of Anglophone West Africa, religion plays a major role in cultural developments to the extent that it infuses itself into other institutions, including higher education. Dermerath, an anthropologist, explains, "religion and ethnicity can at times coalesce and compound each other as in, say, Guatemala, Israel, and Indonesia. But religion can also operate separately from ethnicity as in, for example, Northern Ireland, Egypt, India, and China. The differences between ethnicity and religion with their respective dynamics are many and critical" (2002: 8–9). Based on Demerath's observations (in this case religion), I have discussed here the ways that many private Christian universities are seen as major contributors to westernization. In Francophone West Africa religion also plays a major role in cultural developments. In contrast to an Anglophone country like Nigeria, Francophone countries negotiate the legal reality that religion is not allowed to influence other institutions, in particular higher education. This has meant possibly a more secular academic society that insists on tolerance. It has also meant the establishment of very few private universities in countries that like Nigeria have an increased demand for higher education. The colonial histories and the development or absence of private Christian universities creates a map of religion and colonial policy that documents sustained westernization throughout the post independent period. Religion is just one haven to support neo-colonial Westernization.

References

Abidogun, J. (2011, November 17–19). *Christianity, education, and nation building in Anglophone Africa, 1960s–1990s*. Paper presented at African Studies Association Conference: 50 Years of African Liberation, Washington, DC.

Adesegun, A. A. (2009). *Samuel Olatunji Adebawo b. 1925 Seventh-day Adventist Nigeria*. Retrieved August, 18, 2012 from www.dacb.org/stories/aa-print-stories/nigeria/adebawo_samuel.html

Afigbo, A. E., & Falola, T. (2005). *Nigerian history, politics and affairs: The collected essays of Adiele Afigbo*. Trenton: Africa World Press.

Ajayi, S. A. (2011). The place of Ogbomoso in Baptist missionary enterprise in Nigeria. *Ogirisi: a New Journal of African Studies, 8*, 16–38.

Asamoah-Gyadu, J. K. (2007, October 11). *Christian higher education in Ghana*. Paper presented at the Meeting of the International Association for the Promotion of Christian Higher Education at the Trinity Theological Seminary, Legon, Ghana.

Ayandele, E. A. (1974). *The educated elite in the Nigerian society: University lecture*. Ibadan: Ibadan University Press.

Azikiwe, N. (1973). *Renascent Africa. reprint*. London: Frank Cass.

Babcock University. (2012). *History*. Retrieved July 17, 2012, from: www.babcock.edu.ng/main/index.php?option=com_content&view=article&id=136&Itemid=144

Bowen University. (2012). *About us: Heritage*. Retrieved July 15, 2012, from www.bowenuniversity-edu.org/pages.php?page_id=2

Bray, T. M., & Cooper, G. R. (1979). Education and nation building in Nigeria since the civil war. *Comparative Education: Unity and Diversity in Education, 15*(1), 33–41.

Centre for Development and Population Activities. (2012). *USAID/Positive living: The positive living project (2006–2010)*. Retrieved September, 21, 2012 from www.cedpa.org/section/projects/positive_living.html

Christian Council of Nigeria. (1949). Chapel for University College, Ibadan. Source: School of Oriental and African Studies Archives, University of London. doc:CBMS/03/24/02 Box 307.

Christian Missionary Society. (1945, October 26). *Conference on education in Africa*. Source: Cadbury Archives, University of Birmingham. doc:CMS ACC22 F9.

Clark, N., & Sedgwick, R. (2004). Education in Nigeria. *World Education News and Reviews, 17*(5), 1–8. Retrieved September, 21, 2012, from www.wes.org/ewenr/04sept/Practical.htm

Dallimore, H. (1930). *The religious beliefs of the Ekiti peoples*. WEA. Source: Cadbury Archives, University of Birmingham. doc:CMS ACC 718 F8.

Demerath, N. J., III. (2002, March). A sinner among the saints: Confessions of a sociologist of culture and religion. *Sociological Forum, 17*(1), 1–19.

Ebizie, A. N. (1996). *Christian religious knowledge for senior secondary*. Awka: Meks Publishers.

Embassy of the United States Accra, Ghana. (2012). *The educational system of Ghana*. Retrieved September 4, 2012, from http://ghana.usembassy.gov/education-of-ghana.html

Enaohwo, J. O. (1985, June). Emerging issues in Nigerian education: The case of the level and scope of growth of Nigerian universities. *Higher Education, 14*(3), 307–319.

Esedebe, P. O. (1980, December). The educated elite in Nigeria reconsidered. *Journal of the historical society of Nigeria, 10*(3), 111–130.

Falola, T. (2005). *A mouth sweeter than salt*. Ann Harbor, MI: University of Michigan Press.

Freeman, T. B. (1884). *Reminiscences and incidents of travels and historical and political sketches in and of countries bordering on the Gold and Slave Coasts and in Ashantee, Dahomey, etc., forty six years a resident on the Gold Coast*. Source: School of Oriental and African Studies Archives, University of London. doc:MMS/17/02/03/10/03/02.

Garnier, M., & Schafer, M. (2006). Educational model and expansion of enrollments in Sub-Saharan Africa. *Sociology of Education, 79*(2), 153–175.

Great Britain, Colonial Office, Advisory Committee on Native Education in the British Tropical African Dependencies. (1925). *Educational policy in British Tropical Africa* (Cmd. 2347, Vol. 2). London: H.M.S.O.

Idoko, C. (2011, March 8). *FG grants operation license to 4 new private varsities: Warns against flouting of guidelines*. Tribune, Nigeria.

Igbendion University. *University at a glance*. Retrieved May 8, 2012, from www.iuokada.edu.ng/AboutUS/Default.aspx

Information Please. (2012). Nigeria. *Pearson Education*. Retrieved June 6, 2012, from http://www.infoplease.com/country/nigeria.html.

Inter-University Council for Higher Education in the Colonies. (1948, May). Memorandum: University College, Ibadan, Nigeria. Source: School of Oriental and African Studies Archives, University of London. doc:CBMS /03/24/02 Box 307.

Jekayinfa, A. A., & Akanbi, G. O. (2011). Society's violation of state laws on the establishment of private universities in Nigeria. *International Journal of Humanities and Social Science, 1*(17), 297–302.

Madonna University. (2010). Retrieved June 10, 2012, from www.madonnauniversity.edu.ng/

National Accreditation Board Ghana. (2010, August). *NAB guidelines for affiliation*. Retrieved August 20, 2012, from www.nab.gov.gh/

National Universities Commission. (2012). *List of Nigerian universities and years founded*. Retrieved: June 10, 2012, from www.nuc.edu.ng/pages/universities.asp?ty=2&order=inst_name

Office of the Press Secretary. (2001, January 29). *Executive order: Establishment of White House office of faith-based and community initiatives*. Retrieved May 30, 2012, from http://georgebush-whitehouse.archives.gov/news/releases/2001/01/print/20010129-2.html

Ohonusi, E. (2012, February 26). *State of Eko Boys High School*, Part 1. Retrieved April 20, 2012, from http://ekoboys-hs-oldboysassociation.org/page1.php?SessionID=747f0f4af432518ce07

Onwuliri, C. (2012). *Guidelines for establishing institutions of higher education in Nigeria*. Retrieved June 7, 2012, from www.nuc.edu.ng/pages/pages.asp?id=26

Shah, T. S., & Toft, M. D. (2006, July–August). Why God is winning. *Foreign Policy, 155*, 38–43.

Shu'ara, J. (2010). *Higher education statistics – Nigeria experience in data collection*. Paper presented at UNESCO Institute of Statistics Workshop on Education Statistics in Anglophone Countries, Windhoek, Namibia, October 17–21, 2010. Retrieved September 21, 2012, from www.ibe.unesco.org/en/worldwide/unesco-regions/africa/nigeria/national-reports.html

Soudan Interior Mission. (2012). *Country profile: Nigeria*. Retrieved September 24, 2012, from www.sim.org/index.php/country/ng

Stambach, A. (2010). *Faith in schools: Religion, education, and American evangelicals in East Africa*. Stanford: Stanford University Press.

The official website of Sir (Dr.) Chief G.O. Igbinedion LLD, D. LITT, GCKB, the Esama of Benin kingdom. Retrieved May 28, 2012, from http://igbinedion.net/index.html

UNESCO – IBE. (2010, September). *World data on education 2010/2011: Nigeria* (7th ed.). Geneva: UNESCO-IBE, Publications Unit.

UNESCO – UIS. (2011). *Education: Nigeria*. Retrieved May 20, 2012, from www.unesco.org/new/en/education/resources/unesco-portal-to-recognized-higher-education-institutions/dynamic-single-view/news/nigeria

Universities in Cote d'Ivoire. (2012). Retrieved August 15, 2012, from http://www.classbase.com/Countries/C%C3%B4te-d'Ivoire/Universities

Universities in Senegal. (2012). Retrieved August 15, 2012, from http://www.classbase.com/Countries/Senegal/Universities

University of Ghana. (1996). *University of Ghana Legon: Handbook for the Bachelor's degree*. Accra: The University of Ghana.

University of Ibadan (UI). (2012). *Institute of African Studies*. Retrieved July 16, 2012, from http://ui.edu.ng/africanstudies

University of Nigeria at Nsukka (UNN). (2012). *Institute for African studies*. Retrieved July 16, 2012, from unn.edu.ng/institutes/institute-african-studies.

University of Sierra Leone. (2012). *Historical background*. Retrieved June 13, 2012, from www.tusol.org/historical

Varghese, N. V. (2004). *Private higher education in Africa*. Geneva: UNESCO.

Very Rev. Fr. Prof. Emmanuel M.P. Edeh. (2012). Retrieved July 20, 2012, from http://fatheredeh.com

Wellspring University. (2012). *About*. Retrieved September 21, 2012, from www.wellspringuniversity.edu.ng/about.html

White, B. W. (1996). Talk about school: Education and the colonial project in French and British Africa (1860–1960). *Comparative Education, 32*(1), 9–25.

Zeleza, P. T. (2009). African studies and universities since independence. *Transition: Looking Ahead, 101*, 110–135.

Chapter 50
Religious Influence on Education and Development in Twentieth Century Tanzania

Orville Nyblade

50.1 Introduction

Religions often have ambiguous roles in a society. They may be a disruptive element that challenges that which is dominant in the culture, or they may serve to support and maintain the status quo, or some of both. They may serve to largely meet the emotional and intellectual needs of individuals or largely provide for building community and social cohesion, or some of both.

The introduction of Islam and Christianity into the non-literate societies of Tanganyika was disruptive to the traditional religions and culture and was also supportive of individual needs and the building of new communities. The disruption was especially true of most early Christian efforts, for example, as missionaries required monogamy and sought to replace some traditional rites of passage. However, some things such as Islamic dietary practices were also disruptive. As we trace the development of the Church and Mosque in the twentieth century, these functions of religion will be kept in mind, seeking to show how they functioned both to support and disrupt the status quo.

When German colonial efforts began the last two decades of the nineteenth century, Islam was already well established in Tanganyika, dating back to the rule of the Kilwa Sultanate in the fourteenth century. Coastal areas as well as towns such as Tabora and Ujiji along the central trade route had large Muslim communities. Christian missionaries made their first contacts in the middle of the nineteenth century, but widespread missionary activity did not begin until the beginning of German colonization.

O. Nyblade (✉)
Makumira University College, P.O. Box 55, Usa River, Tanzania
e-mail: owjsn@embarqmail.com

S.D. Brunn (ed.), *The Changing World Religion Map*,
DOI 10.1007/978-94-017-9376-6_50

50.2 German Colonization

By the turn of the twentieth century, German colonial rule had been established in many parts of Tanganyika. The mission activity before German rule was primarily by Anglicans and Roman Catholics, but with German colonization German mission societies, Lutheran and Moravian became active in both evangelism and education.

By 1900 the German colonial government had established 60 three-grade primary schools, 9 two-year vocational schools and one high school of 500 students. These were primarily for the purpose of meeting the clerical and industrial needs of the government and for providing teachers for the schools. This was education for boys as there was little interest in the African population for the education of girls.

By way of contrast the missions had established about 600 schools with about 50,000 students, mostly up-country. The missions' purpose was basic literacy that would serve their educational and evangelistic ends. For this reason the vernacular languages were often used while the government schools were conducted in Swahili. The missions were concerned with teacher training as was the government at their secondary school in Tanga. The Lutheran Mission in Northern Tanganyika opened a teacher training college and Marangu near Moshi, but given the large numbers of pupils reported by the missions, no doubt many of those reported as pupils were taught by men with minimal literacy in two-grade bush schools. Bush schools were structures built like the indigenous homes with poles for the children to sit on. Writing material (Fig. 50.1) for the pupils was usually slates.

The pace of educational development prior to World War I can be seen in that by 1914 there were 99 government schools with about 6,100 students while missions

Fig. 50.1 A pupil practicing numbers in sand outside a bush school, 1956 (Photo by Orville Nyblade)

reported 1,853 schools with 155,287 students ("Education" 2012). With a population estimated at 5.8 million in 1913, of which probably well over 5.7 million were indigenous Africans, this represents a small proportion of school-age children which, if they represented 40 % of the indigenous population, would be about 2.2 million. Nonetheless, a pattern of a western-style education with a very large Christian influence had been established during German colonial rule. Both colonial and mission education were westernizing agents and disruptive elements in traditional African society.

50.3 British Colonial Rule to 1944

Following World War I, the British established colonial administration under a mandate from the League of Nations. They continued the educational policy of German rule with both government and mission schools. In 1920 the British administration stated as their overall purpose "to develop the people, as far as possible on their own lines and in accordance with their own values and customs" ("Education" 2012).

The Phelps-Stokes Commission was established in the early twenties to bring recommendations on the way forward in education. In 1924 they presented their report, recommending government-mission co-operation in advancing education in Tanganyika. As a result the government established their authority to exercise supervision over and establish guidelines for education, and began to subsidize mission schools ("Education" 2012). This provided an element of control of some of the education provided by the missions, but missions continued to operate village schools with unqualified teachers that barely provided for basic literacy and numeracy and were also centers of evangelism. While the missions provided some supervision of these bush schools, it was rudimentary from an educational point of view. The continued influence of the missions under British colonial rule can be seen by the educational statistics. In 1931 7,505 students were reported in government schools, while 159,959 were reported in mission schools ("Education" 2012). Even if a majority of the mission schools were sub-standard, it indicates a continued and growing Christian presence in Tanganyikan society. However, it represents a relatively small advance over the number of students reported in schools in 1914.

50.4 Education in Tanganyika: 1944–1961

50.4.1 Provision for Rapid Expansion

Great Britain administered Tanganyika during the years between the two world wars largely out of economic resources developed within the country. The effects of this can be seen in that the 1935 education budget was only $290,000 ("Wikipedia" 2012). During World War II Great Britain decided to increase aid to the colonies

Fig. 50.2 Building of a two-room four-grade elementary to upgrade a bush school (Photo by Orville Nyblade)

through the Colonial Development and Welfare Acts of 1945 ("Nature" 1945). This enabled the Tanganyika colonial administration to develop an ambitious 10-year development plan in 1947 for education and general development of 8,000,000 lb (approximately 1947 US$ 24 million) of which 7,000,000 lb were to be grants from the British government.

In order to leverage this for maximum development, the colonial government increased their aid to non-governmental organizations. The Christian missions in Tanganyika took advantage of this to increase their efforts in both the educational and medical fields. This initiative provided a large opportunity for the missions in that they were able to up-grade many of their 2-year bush schools to 4-year elementary schools (Fig. 50.2) and to develop teacher training colleges for the training of teachers. The curriculum allowed for religious instruction and for many years the salaried teachers provided this instruction. As the teachers were at least nominally Christian, this provided a captive audience of children for Christian teaching. While the law provided for the excuse of children from Christian education if the parents requested it, and for the schools to provide facilities for any religious community that wanted to provide for the instruction of their children during religious education periods, in actual practice it resulted in almost universal Christian education in mission schools. The result of this was to put the Muslim communities in rural areas at a disadvantage which was the cause of considerable resentment. The schools in towns and cities were government schools that often had Muslim teachers, or schools of the Aga Khan Muslim community.

The one major educational undertaking by a Muslim community was the formation of the East African Muslim Welfare Society (EAMWS) founded in 1945 by the Aga Khan with headquarters in Kenya at Mombasa. This was primarily financed by

Fig. 50.3 A modern mosque in Moshi, Tanzania (Photo by Orville Nyblade)

the Ismaili community but it had pan-Islamic ambitions. It started schools, hospitals and dispensaries in population centers and was active in building mosques (Fig. 50.3). When Tanganyika became independent in 1961 they moved their headquarters to Dar es Salaam in Tanganyika and it was then headed by a prominent African leader, Chief Abdallah Fundikira, a political opponent of President Nyerere (Lodhi and Westerlund 1997).

The level of literacy and general education in the African population was significantly increased during this period. While the large expansion of 4-year primary schools did not always result in permanent literacy for all of its students, it provided literate students for an increasing number of 4-year middle schools, which in turn provided students for teacher training colleges and the development of some elite secondary schools. It also provided opportunities for intelligent and ambitious children of poor families that were not influential in their communities to gain a westernized education. This in turn gave them access to the skills and employment that enabled them to challenge traditional authority and become an active part of the independence movement.

It also provided a base for a more advanced religious education program. The Protestant missions started seminaries and Bible Schools to increase the number of African leaders as well as enhance their educational competence. The Roman Catholic Church had developed seminaries for the training of African priests dating back to the first one in 1904, so they were ahead of the Protestants in providing for an educated African clergy.

The large part that the missions played in the educational development also had an impact on the political development. While the colonial government exercised oversight of the whole educational enterprise, and provided the funds to pay the teachers, the teachers were regarded as employees of the missions. The teachers in government schools were civil servants and were not allowed by law to engage in political activity. Since mission teachers were not civil servants, they became major sources for recruitment for the developing political movement for independence, the Tanganyika African National Union (TANU). In order to control this, the government put pressure on the missions to forbid their teachers to engage in political activity or to join political organizations. Many missions complied and any that did not were subject to subtle restriction of any expansion of activity in which they needed government approval.

In the development of TANU in the mid-1950s the emerging African churches did not take any active role, although individual African church leaders were active in the political party. Institutionally, policy was still largely under the control of the missions.

The influence of religion on political development is illustrated by the life of Julius Nyerere, Tanganyika's first president. In the elite Tabora Government Secondary School, some of his teachers were Roman Catholic priests. It is probably through their influence that he was baptized in 1943 and remained a lifelong Catholic. Following a diploma from Makerere University College in Uganda, he taught for 3 years in a Roman Catholic secondary school. After 3 years he went to the University of Edinburgh and then returned to teach again at a Roman Catholic school. While there he became head of TANU.

The influence of Christianity on the independence movement and the subsequent political developments are evident in the liberal social teachings of some elements of the Roman Catholic Church which were influential in Nyerere's life, but also in the number of Christian leaders in the newly independent government, a result of the educational advantage that Christianity had developed over Islam. However, Muslims were highly influential in the leadership of TANU, and after the merger with the Afro-Shirazi Party of Zanzibar in 1977 to form the *Chama cha Mapinduzi* (revolutionary party) they increased their influence in the party (Lodhi and Westerlund 1997).

The rapid growth of Christianity through much of the twentieth century was in the rural areas. As urbanization gained momentum in the later part of the century, the influx of Christians into the cities challenged what had been the dominance of Islam in many of the population centers on the coast such as Dar es Salaam and Tanga.

It may be argued that the Christian educational activities served the purposes of the colonial government in maintaining political and economic control and in developing an educational system that thwarted full African development, but the situation was much more ambiguous. Certainly many in the colonial administration were aware that the large increase in educated Africans would result in leaders that would seek independence. And while it is true that the missions were necessarily supportive of government educational policy, a large number of them were not British or related to British interests and had no direct interest in furthering British control. Most of them did come from Europe and North America, however.

Another criticism expressed by some was the slowness in the education of girls. This had many facets and was not just due to colonial education policy although this policy did favor the education of boys to meet economic needs. In many traditional African rural societies girls were seen as largely homemakers and sources of bridewealth in cultures in which wealth was transferred from the family of the groom to the family of the bride. Educating the girls often disrupted the early marriages of girls. In order to fill the classrooms in rural areas the local native leader often intimidated parents so they enrolled their girls in school. The education of girls eventually led both to women becoming active in politics and government and to their contribution to the work force in an independent Tanzania. After independence women held responsible positions in the National Bank and became part of the police force. Anecdotally, they were regarded as less likely to be corrupt. Titi Mohamed was a founding member of TANU and was deputy minister of health from 1962 to 1969. Lucy Lameck was a prominent member of Parliament from the time if independence, active in promoting community development and health and social welfare. Julie Manning was minister of Justice from 1975 to 1979.

50.4.2 Protestant Training of Religious Leaders

The Protestant missions had always recognized the need for educated clergy, but the institutionalization of this training did not take place until after World War II at the beginning of the rapid expansion of general education. The earliest unified effort of the Protestant missions was by the Lutherans with the start of training of clergy for the Lutheran Church at one location for all of Tanganyika at Lushoto in the Usambara mountains in 1947, eventually moving to Makumira near Arusha in 1954. Originally this was an upgrading of African leaders who had established their abilities. With the expansion of the educational system and the large number of men trained in teacher training colleges, elementary school teachers became a major source of recruitment for the clergy. This also set a standard for educational qualifications for clergy (Fig. 50.4).

Fig. 50.4 Makumira Lutheran Theological College chapel, 1982 (Photo courtesy of Archives of the Evangelical Lutheran Church in America, used with permission)

With an increase in educational standards and the number of educated African clergy, the process of the development of the authority of the African churches and a decline in the control of the Western missions was accelerated. The Lutheran Church in Northern Tanzania, the largest of the Lutheran Churches, elected its first African leader, Stefano Moshi, in 1958, over 3 years before Tanganyika's independence. The first African Roman Catholic Bishop preceded Bishop Moshi by six years when Laurean Rugamba became the first African Roman Catholic Bishop in 1952.

The pattern of the development of education was similar in all of the British colonies following the British system. Teachers and students could move from one country to another and recognize the system. The three colonies of East Africa, Uganda, Kenya and Tanganyika can be viewed as a unit with Makerere University in Kampala, Uganda serving as the only university for all three colonies until 1963. The development of theological education was different since the theological colleges represented traditions from many countries as well as the denominational differences. However in the early 1960s the faculties of many theological colleges and the faculty of religious studies at Makerere began yearly seminars and formed an Association of Theological Institutions. Out of these meetings developed a common diploma in theology awarded by Makerere University on the basis of external examinations. Eventually many Roman Catholic seminaries joined the Association, forming a broad-based theological forum. Later Mekane Yesus Seminary in Addis Ababa, Ethiopia also joined.

50.5 Early Years of Independence

50.5.1 Period of Transition: 1961–1967

With the independence of Tanganyika from Great Britain in December of 1961, the time of ambivalence in the life of the missions and churches had changed. They no longer had the problem of pressure from the colonial government to support their policies of restraining the independence movement and the pressures of the independence movement to support their cause. They now had to relate to one entity, the independent state. Tanganyika was established as a secular country with a guarantee of religious freedom. This has continued throughout its history, although after the formation of Tanzania with the union of Tanganyika and Zanzibar in 1964 there have been occasional calls for Islamic *shariah,* at least on Zanzibar and Pemba.

Initially there was a smooth transition, with no changes in those areas of subsidies and support of the educational and medical work. The most notable effect was the increased pressure from the African churches for control of the policy held by the missions. An example of the result of this was the formation in 1963 of the Evangelical Lutheran Church in Tanzania (ELCT). The ELCT, along with its constituent jurisdictions, replaced the Lutheran missions in the making of policy for the national church. The executive council made up of the clerical head and a lay delegate from each jurisdiction became the main policy body. Since the churches were heavily dependent on the partner churches in the West that provided missionaries for the support of their medical and educational institutions, the overseas agencies continued to exercise considerable control, but they did this in dialogue with Tanzanian rather than missionary counterparts, or with missionaries who held their position in the African church because they had been selected by the African Church for leadership.

The churches supported the independent government, as its members had generally supported the independence movement. There was little criticism of the establishment of a one-party state under the leadership of the TANU, or the formation of Tanzania in 1964 with the union of the mainland with the islands of Zanzibar and Pemba. Since the islands were largely Muslim, with some estimates as high as 95 % of the population, this increased the influence of Islam in the government of the country. While Christians were over-represented in the government of Tanganyika, the Muslim leaders of Zanzibar now became a part of the government of Tanzania and partially redressed this imbalance.

50.5.2 The Arusha Declaration: 1967

While Nyerere had enunciated a policy of *ujamaa* (a Tanganyikan version of African Socialism) from the time of independence, a formal declaration of this policy took place with the Arusha Declaration of 1967. The Declaration called for state control of the major means of production and exchange. An immediate action was the nationalization of the banks, large scale industry, part of the trade sector and 60 % of the sisal industry (Ibhanawoh and Dibua 2003). With this formal declaration of *ujamaa,* there was political pressure for all parts of society to give active support to the actions of the state. This the churches did, certainly because of the promise of improving the life of all of the citizens, especially the rural poor, which was congruent with church teaching, but also because any opposition could be met with strong community disapproval and possibly fear of negative political action against the church. The Muslim community was also supportive of the declaration.

A program that more directly affected the life of the religious communities was the villagization *(vijiji vya ujamaa)* program that sought to bring farmers into planned villages from their scattered homesteads. The goals of the program were to both increase rural productivity as they worked together and enable a more efficient provision of services to the community. This was initially to be a voluntary program encouraged by incentives and the promise of both economic advance and improved services. While there was enthusiasm for the program by some resulting in successful villages, the actual progress was slow. In 1970, in response to this slowness it was decided to force villagization in areas where the need for improvement of rural life was particularly acute. This proved to be disastrous in terms of poor implementation and opposition on the part of the farmers. In 1975 the program was abandoned (Ibhanawoh and Dibua 2003).

In spite of the obvious abuses in the forced villagization, there was no public criticism either by religious leaders or others. This does not mean religious leaders condoned the abuses, but indicates the level of control of the media that took place after the Arusha Declaration. Very likely, religious leaders had access to policy makers, including Nyerere himself, and expressed their concerns over the abuses but there was no possibility of public pressure.

By 1968 the EAMWS had established 70 schools in Tanzania. However, in that year the government banned EAMWS and some other Islamic organizations. With the support of TANU an organization was formed to represent all Muslims in Tanzania with the acronym of BAKWAKTA thus aligning Islam closely to the political party (Lodhi and Westerlund 1997).

50.5.3 Nationalization of the Schools

Of more direct impact on the churches was the nationalization in 1970 of all of the schools that were managed by the churches with government support. This meant that a major source of the influence of the church in the community was removed. While the government provided for the possibility of religious instruction in the schools, it was now the responsibility of churches and mosques to provide instructors from their own resources. While teachers with a strong religious commitment were not prohibited from volunteering to provide this instruction, there could be no external pressure for them to do so. The Christian presence in public education was greatly reduced, not only by the loss of management functions, but because even in those cases where Christian education was continued, it was often carried out by poorly trained catechists. However, there was no interference with the Christian secondary schools that were independent of government support and in the training of leadership for the churches.

The Muslim reaction to the nationalization of the schools was positive. This addressed many of their long-standing grievances and they felt that this would begin to address the educational gap between Muslims and Christians. The actual effects are difficult to assess, especially in the light of government policy since 1967 not to include religion in census data. At what may be a small indication of the narrowing of the gap is the report of an increase in enrollment of Muslims in the University of Dar es Salaam from 15 % in 1985 to 18 % in 1989 ("Islamtanzania" 2012). However, these are not official university numbers.

50.5.4 Economic Failure: 1978–1990

The rapid expansion of the state sector of the economy resulted in management problems in the implementation of nationalization, but a number of other issues also resulted in the collapse of the Tanzanian economy, including the world oil crisis, a weak world commodities market for Tanzanian products, and the conflict with Uganda in 1979. All of these drained Tanzanian resources in the 1970s. At the same time there was an increased demand for education and medical care from many parts of the country, and the population continued to grow at a rapid pace.

During this period the membership of the Church and Mosque grew. The seminaries continued to train men and women who provided leadership in the villages. With increasing demands in the social sector, and continued management problems, religious leaders were often called upon to provide help. Thus the influence of the Church and Mosque continued in the educational and medical areas.

50.6 The Impact of Modern Medicine

While the major impact of the Christian religion on Tanzanian culture and society has been through education and evangelism, note needs to be paid to the effect of modern medicine and the part the missions and churches have played in it. As colonialism and the Christian missions entered a Tanganyika that was largely non-literate, it also entered an African society and culture in the latter part of the nineteenth century in which the medical practitioners were herbalist or sorcerers. The introduction of modern medicine challenged the Africans' belief in these practitioners.

The early colonial intrusions brought modern medicine with it for the care of the colonialists. Many endemic diseases, like malaria, were a health risk for the colonists. The medicine the colonists used for themselves was made available to Africans in contact with the colonialists. For those working for the colonialists it was of economic value to keep them healthy. While early missionaries often did not have formal medical training, they shared with the Africans the remedies that they had, for example aspirin for pain and fever and quinine for malaria. The missions' development of medical work was largely humanitarian, although its medical centers did provide occasion for Christian witness to those who came for care.

As with education, the mission medical work initially was local with no regional or territorial planning and with no coordinated approach to government medical policy. This changed with the formation of the Tanganyika Missions Council in 1934 that was open to all non-Roman Catholic missions. The Council formed a Missions Medical Committee in 1936. The Roman Catholics had their own organization through the regular meetings of the Bishops and the organizations they developed. Medical work developed an educational thrust early with the need to train paramedics (dressers) to provide basic health services in village clinics and nurses for the hospitals.

50.7 Post Arusha Declaration Tanzania

50.7.1 Political Change

In 1985 Nyerere stepped down as president of Tanzania, although he continued as chair of the ruling party. He had succeeded in developing a degree of social unity and of maintaining political stability, but had failed to bring economic development to the country. His successor, Ali Hassan Mwinyi, took steps to reverse the country's economic policies and during his second term multi-party democracy was introduced with the active support of Nyerere.

50.7.2 Educational Developments

With these political changes the religious organizations were again able to become major players in the educational field. While local communities retained control of the elementary schools, religious organizations, Christian and Muslim, expanded their involvement in secondary and tertiary education well beyond that of training leaders for the Church and the Mosque. Funds from Saudi Arabia and other Arab countries through the World Council of Mosques with headquarters in Jeddah became available for the development of secondary schools and an Islamic University (Lodhi and Westerlund 1997).

This development was aided in 1992 when the government agreed to a "Memorandum of Understanding" in which they agreed not to nationalize independent educational institutions and also to share grants for education and medical work from foreign governments with non-governmental organizations ("Kilaini"

1998). The government also authorized the formation of the Christian Social Service Commission, with two Executive Boards, one for medical work and one for education. These included both Roman Catholic and non-Catholic Christian groups and provided one Christian agency to relate to the government on medical and educational matters. BAKWATA continued as the umbrella Muslim organization recognized by the government to speak for Islamic interests.

The churches had the organization and history of working together in relating to and influencing government through the non-Catholic Christian Council of Tanganyika (CCT) dating to the formation of the Tanganyika Mission Council in 1934 and the Roman Catholic Bishop's Tanganyika Episcopal Conference (TEC). Both the CCT and the TEC had ongoing education and medical committees from the 1940s. The co-operation of the churches at the national level in the educational and medical field was not always evident at the local level. The colonial government sought to minimize this through comity agreements that assigned a rural area to a particular church. However, with increased urbanization, this was less effective. Also in the later decades of the twentieth century, Pentecostal groups became very active throughout Tanzania and were outside the ecumenical framework of the Christian Council and the comity arrangements.

50.8 Contemporary Tanzania

50.8.1 Demographics

In 1900, the estimated population of Tanzania was 3,938,000. While Muslims were numerous in the coastal areas and dominated the off-shore islands, the majority of the population followed traditional African religions. Christians represented a small portion of this population. Missionaries had been in Tanganyika since the Roman Catholic White Fathers came in 1878. The population grew to 5.2 million by 1930 and to 10.3 million by 1960. In 2000 it was 35.3 million and is estimated to grow to 48.1 million by 2015 ("Population" 2003).

Since the turn of the century, both Islam and Christianity have grown and traditional religious practice has declined. There has been no religious census data collected by the government since the Arusha Declaration in 1967, the adherents of various religions can only be estimated. Since some major Christian denominations keep records of their members, a somewhat clear picture of the number of Christians is available. The following list shows the latest reported members along with estimates of those groups for which no reports were available. These unreported groups included what are probably sizable churches, including some Moravians, the Africa Inland Church and the Pentecostal churches. With approximately 18,000,000 Christians out of a population of about 42,000,000, Christians would represent about 42 % of the population (Table 50.1).

Table 50.1 Religions in Tanzania (Source: Orville Nyblade)

Religion	Followers
Roman Catholic	8,500,000
Lutheran	5,601,271
Anglican	2,500,000
Moravian (Western Tanzania)	80,000
Mennonite	60,000
Other (estimated)	1,500,000
Total	18,141,271

Fig. 50.5 A small modern rural church (Photo by Orville Nyblade)

The estimates of the Muslim population range from 35 to 40 % of the population while estimates of the Christian population range from 30 to 62 %. Probably the Christian and Muslim populations are about the same size with the followers of traditional religions representing about 20 % of the African population. Small numbers of other religions are also present, such as Hinduism, but the members would be non-African. With the rapid growth of the population the last 30 years some of the growth of the Islamic and Christian communities can be accounted for by natural increase, and less by proselytizing.

50.8.2 Congregations and Mosques

With the growth of Islam and Christianity, the result was the establishment of houses of worship throughout the country in which large numbers of people gathered every week (Fig. 50.5). The religious communities are an efficient means for the dissemination of information. With the advent of modern communication in Tanzania and the growing ubiquity of the cell phone, this is not as important as it previously was. However, they remain an important influence in the establishing of values and building community beyond clan and family loyalties.

In a society in which there is little independent institutional development, the Church and the Mosque become important as non-state centers of influence. While the educational and medical institutions often depend on subsidies from outside Tanzania, or from the national government, the local church or mosque is more likely to depend on its own resources. It is also a resource that the local government can draw upon. Leadership is developed in the local fellowship that then can be called upon to give leadership in the larger community. This can be important for local educational, medical and other social services and for political leadership.

The institutions that train the local leadership of the congregations and mosques are important centers of influence. As the theological institutions of the churches are led by Tanzanians and with a majority of Tanzanian staff, they are more likely to be more sensitive to local and national interests than a generation ago when the staff was largely expatriate, and thus have more cultural, social and political impact on the students.

50.8.3 Social Services

When the government took over the management of the schools in 1970, there were some demands for the nationalization of medical services. However, most medical services of non-governmental organizations remained under their management. A number of factors contributed to making this nationalization of social services difficult for the government. Perhaps the major factor was the failure of economic development in the 1970s and 1980s so the government could not finance its ambitious plans. Another was the increasing demand for secondary and tertiary education as the country moved toward universal primary education. A third factor was a lack of adequate corruption-free oversight. A fourth was the ability of NGOs to tap financial resources and personnel beyond what the government was able to do. This led the government to adopt the colonial government pattern of state-voluntary organization partnership in the providing of social services (Fig. 50.6).

The result was an explosion of the church's involvement in education as well as a larger presence of the Islamic communities in education. The Roman Catholic Church reported more secondary schools in 1991 (82) than they had at the time of nationalization in 1970 (44). The difference is that they did not report any elementary schools and two teacher training colleges rather than six (Kilaini 1998). The Lutheran Church also reported 47 secondary schools by the end of the century.

These schools followed the curriculum established by the Department of Education. This included a core curriculum in religion and optional courses in Bible Knowledge and Islamic Knowledge examinable at the end of 4 years of education by examinations set by the Ministry of Education. This curricular program is carried

Fig. 50.6 Rural hospital at Haydom, Mbulu 40th year anniversary, 1980s (Photo by Orville Nyblade)

out by a parastatal organization, the Tanzania Institute of Education, for all pre-primary, primary, secondary, and teacher training schools ("Tanedu" 2012).

The involvement of the churches in tertiary education is even more dramatic. In 1996 the Lutheran Church opened Tumaini University with three constituent university colleges. In 1998 the Roman Catholic Church opened St. Augustine University with three constituent university colleges. In 2006, the Muslim University of Morogoro was founded. The Anglicans opened St. Johns University in 2007 (Fig. 50.7).

One of the perennial problems in the development of a country like Tanzania is improving and maintaining the quality of the education institutions. These problems have been met in two ways, by the recruitment of qualified staff from other countries, and by the upgrading of able local staff. Often this means early staffing with under-qualified staff in the early stages of institutional development. With the rapid expansion of secondary and tertiary education in Tanzania in the 1990s and the first decade of the twenty-first century, this was likely the situation. While the Muslim University of Morogoro is the only one to post faculty members information online, it may well represent the case in other institutions. Of 30 faculty members, 6 have doctorates and 17 have masters. While they do not specify the nationality of faculty members, the first degrees of ten members are from institutions outside of East Africa –six from Saudi Arabia and one each from Pakistan, Sudan, India and Canada ("Mum" 2012).

Christian schools have traditionally had both extensive expatriate faculty support and graduate training of national faculty from North America, Europe and the Commonwealth countries. While the churches have prepared a large number of nationals for faculty positions, with the rapid expansion of secondary and tertiary institutions they will continue to need expatriate assistance.

Fig. 50.7 Entrance to Makumira University College of Tumaini University, 2006

As the country became covered with houses of worship and schools it also became covered by dispensaries and hospitals offering various levels of modern medicine. Although they have not reached into the villages in the same way that houses of worship and schools have, they do offer major medical help in population centers, with many dispensaries in the rural areas. The churches continue to play a major role in providing medical services, tapping resources outside of Tanzania as well as receiving subsidies from the government. In 1998 the Roman Catholic Church reported managing one consultant hospital, 35 other hospitals and 225 health centers and dispensaries. The Lutheran Church reported one consultant hospital and 21 other hospitals and numerous clinics. Other Christian churches were active in medical work, especially the Anglicans. The Aga Khan medical work continued and other Muslim organizations were active in the establishing of hospitals and dispensaries. The churches coordinate policy and provide a single approach to government through the Medical Board of the Christian Social Service Commission. A major ecumenical venture in medicine was the establishment of the Good Samaritan Foundation in Tanzania by the Anglican, Lutheran and Moravian churches to build and manage a 450 bed consultant hospital in Moshi ("KCMC" 2012). This resulted in the opening of the Kilimanjaro Christian Medical Center KCMC in 1971. KCMC has now become a teaching institution for doctors and medical specialties.

50.8.4 1900–2000

The twentieth century was a time of transition for Tanganyikan society. It started from a largely non-literate society composed of over a hundred ethnic groups each practicing their own traditional religions and with minimal contact with one another. It ended the century with a stable democracy with a highly literate population, (estimated at over 70 % of adults) with approximately equally strong competing Christian and Muslim communities. Traditional religionists were marginalized except in so far as they influenced Christianity and Islam. In this development religion has played a major role, with the adherents of Islam particularly using the commercial sector to spread its message and Christianity using education and medicine with both establishing worship centers. Christianity used publishing to provide a literature in Swahili and the radio to spread its message in addition to using public preaching. Islam, while late in using modern communication methods, has become adept and aggressive in these methods. The use of the cell phone and the internet will increase in importance as the infrastructure supporting them is developed. Christianity provided management and personnel for a large part of the social services that made development possible (estimates as high as 50 %) and both religions provide intellectual and ethical stimulation that contributes to political ferment.

Christians and Muslims co-existed in Tanzanian society with a minimum of tension, although the more fundamentalist and radical of both religions have on occasion engaged in more serious conflict. The U.S. Department of State International Religious Freedom reports on Tanzania contains information of occasions of violence that has a basis in Christian-Muslim tension ("State" 2012). There are no easy solutions to relieving this tension.

The complaint of the Muslims that they are disadvantaged educationally because of the colonial development has been partially addressed at the elementary level by the nationalization of the schools. But at the secondary and tertiary levels with the proliferation of Christian managed institutions, it is still true.

Religions continue as dynamic elements in Tanzanian culture and society. Christianity and Islam have a major role in Tanzania's future. The peaceful development of the two religions and their continued contribution to the social and political life of Tanzania depends on their ability to respectfully work together and to resolve the tensions that arise.

References

"Education." (2012). www.education.stateuniversity.com/pages/1516/Tanzania-EDUCATIONAL-SYSTEM-OVERVIEW. Accessed 8 June 2012.

Ibhanawoh, B., & Dibua J. I. (2003). Deconstructing Ujamaa: The legacy of Julius Nyerere in the quest for social and economic development in Africa. *African Journal of Political Science, 8* (1), 65–66. Archive.lib.msu.edu/DMC/Africa9620Journals/pdfs/political9620science. Accessed 11 June 2012.

"Islamtanzania." (2012). www.islamtanzania.org/nyaraka/Elimu2.html. Accessed 6 June 2012.

"KCMC." (2012). www.kcmc.ac.tz/index.html. Accessed 19 July 2012.

"Kilaini." (1998). www.rc.net/Tanzania/tec/tzchurch.htm, Fr. Method M.P. Kilaini, PhD. Accessed 11 June 2012.

Lodhi, A. Y., & Westerlund D. (1997). (A chapter in the revised English edition of their Swedish book Majoritetens Islam, Stockholm, 1994)) at Curran Press, London/New York. www.islamfortoday.com/Tanzania.htm. Accessed 31 May 2012.

"Mum." (2012). www.mum.ac.tz/academic.htm. Accessed 19 July 2012.

"Nature." (1945). www.nature.com/nature/journal/v155/n3934/abs155358d0000.html. Accessed 8 June 2012. Colonial development and welfare bill. *Nature, 155*, 358–359.

"Population." (2003). www.popul.stat.info/Africa/Tanzania.htm. Accessed 20 Apr 2012.

"State." (2012). www.state.gov/j/drl/rls/irf/2006/7138.htm. Accessed 19 July 2012.

"Tanedu." (2012). www.tanedu.org/index.php?=com_content&task. Accessed 18 July 2012.

"Wikipedia." (2012). en.Wikipedia.org/wiki/history_of_Tanzania. Accessed 8 June 2012.

Chapter 51
Kansas Versus the Creationists: Religious Conflict and Scientific Controversy in America's Heartland

Alexander Thomas T. Smith

51.1 Introduction

On Sunday, 31 May 2009, the controversial abortion clinic doctor George Tiller was serving as an usher at his local church in Wichita, Kansas. At 10.00 am, Scott Roeder, a 51-year old man who had once allegedly been linked to an anti-government militia group in Montana called the Freemen, entered the foyer of the Reformation Lutheran Church just as the service was about to begin. He approached Dr. Tiller and shot him. A few hours later, he was arrested on Interstate 35 in the outer suburbs of Kansas City, a large Midwestern city of some two million residents that straddles the Kansas-Missouri border in the state's northeast corner. By then, Dr. Tiller was dead; Mr. Roeder would later be found guilty of first-degree murder on 29 January 2010 after a short trial in Wichita, Kansas.

Almost immediately, the impact of Dr. Tiller's murder was felt across Kansas and the rest of the United States. The then president of Planned Parenthood of Kansas and Mid-Missouri, Peter Brownlie, stated that his death was "an enormous loss for our movement and for women and their families across America." (*Topeka Capital-Journal*, 06.01.2009) Both Dr. Tiller's supporters and his opponents were quick to condemn his murder. A Wichita member of Catholics for Choice said that it was 'absolutely unbelievable [...] people that call themselves pro-life could do this.' (*Topeka Capital-Journal*, 06.01.2009) Troy Newman, then president of the pro-life organization Operation Rescue, denounced vigilantism and described the killing as a "cowardly act" while Mary Kay Culp, executive director of Kansans for Life, said that members of her organization "value life, completely deplore violence and are shocked and very upset by what happened." (*Topeka Capital-Journal*, 06.01.2009) Later that day during an address to graduates at the University of Notre

A.T.T. Smith (✉)
Department of Sociology, University of Warwick, Coventry CV4 7AL, UK
e-mail: alexander.smith@warwick.ac.uk

S.D. Brunn (ed.), *The Changing World Religion Map*,
DOI 10.1007/978-94-017-9376-6_51

Dame – a private, Catholic liberal arts college in northern Indiana – President Barack Obama said: "However profound our differences as Americans over difficult issues such as abortion, they cannot be resolved by heinous acts of violence" (*Topeka Capital-Journal*, 06.01.2009).

Nevertheless, many others welcomed the news that Dr. Tiller had been killed. "George Tiller was a mass murderer," explained one of the founders of Operation Rescue to local newspaper reporters. "We grieve for him that he did not have time to properly prepare his soul to face God." (*Topeka Capital-Journal*, 06.01.2009) Some of his opponents drove past his church honking their car horns in celebration of his death. At a vigil held in Wichita only a few hours after his murder, members of Topeka's notorious Westboro Baptist Church turned up with placards reading "Baby killer in hell." (*Topeka Capital-Journal*, 06.01.2009) Infamous throughout the United States for opposing abortion and gay rights and picketing the funerals of US military personnel killed in the wars in Afghanistan and Iraq, the Westboro Baptist Church is led by Fred Phelps. Members of his church would be regularly seen picketing meetings organized by pro-choice groups in the greater Kansas City metropolitan area in the aftermath of Dr. Tiller's murder, with signs declaring "God hates fags" and "Your preacher is a whore."

Dr. Tiller's killing had two important consequences for pro-choice campaigners in Kansas. Firstly, his medical clinic closed within a fortnight of his death. As one of just a handful of doctors in the United States who had been willing to provide controversial late-term abortions, which is an abortion performed in later pregnancy when the foetus is considered more "viable" (that is, able to survive outside the womb), this was a major setback in the provision of the full range of family planning services across the American Midwest. Secondly, the pro-choice political action committee (PAC) Pro-Kan-Do folded, its chairwoman leaving for the Missouri Progressive Vote Coalition in St Louis. While the closure of this PAC had been planned in advance by some months, the timing of its closure so soon after Dr. Tiller's murder demoralized pro-choice activists who had relied heavily on Dr. Tiller for leadership and financial support. Without Pro-Kan-Do, pro-choice activists now lacked a dedicated PAC focused on Kansas and some pro-life supporters appeared to be emboldened, sensing a new opportunity in the aftermath of Dr. Tiller's death. Downtown a couple of months later in the liberal college town of Lawrence, Kansas, a pro-life activist stopped my wife in the street and asked her to sign her petition. "Now Tiller is dead we want to get the rest of the abortionists out of Kansas," she explained.

51.2 Narratives of Creation: An Ethnographic Approach

Dr. Tiller's murder cast a long shadow over the 6 months of ethnographic fieldwork I conducted from July 2009 on the interface between politics, religion and science in the greater Kansas City metropolitan area. In my previous research on the understudied Scottish Tories, I had explored how a group of marginalized political elites and their supporters had sought to re-empower themselves after being cast to the geographical and institutional margins of Scotland in the 1997 general election (cf. Smith 2011). I had come to Kansas in July 2009 to develop a second field site in which to gain a fresh perspective on my long-standing anthropological interests in activism, electioneering, knowledge-making and statecraft (cf. Coles 2007; Paley 2001; Riles 2001) by studying a similarly dispossessed political elite in the US: the moderate Republican.

In socio-cultural anthropology, "moderate" political movements have attracted little empirical attention, despite posing interesting theoretical challenges and possibilities. In my view, this is an important omission that might tell us more about the political sympathies and predilections of professional anthropologists and other social scientists than it does the relative importance of moderates in a variety of cultural, political, religious and social contexts. My interest in moderation, therefore, seeks to address a neglected topic in the ethnographic record of contemporary politics and society.

The focus of my fieldwork in Kansas has therefore been on the activism of moderate-secular activist organizations seeking to counter-mobilize against the Christian Right in grassroots Republican Party politics. In theoretical terms, my ethnographic approach draws inspiration from the philosophical tradition of American pragmatism, employing in particular John Dewey's conception of the public (1954) to consider politics as a process of *creating a public*, rather than as a reflection of public opinion. The central question with which I am concerned is how one creates a moderate (but activist) public – particularly in socio-cultural contexts characterized by division, electoral polarization and militancy – capable of successfully challenging the worst excesses of political and religious extremism. I study those political and social practices that might be considered constitutive in the making of a moderate politics, displacing the usual representation of moderation as a default position defined by the absence of strong ideological or other convictions in favor of understanding it as *a disciplined engagement with divided publics*. In taking this approach, I engage with recent developments in pragmatist thinking in the philosophy of science and education (for example, Rorty 1998, 2000) that highlight those qualities of contemporary publics that are more properly understood as contingent, negotiated and temporal.

As religion has re-entered the public sphere in Western liberal democracies like the U.S., it has come to play an increasing role in the framing of political disputes around science funding, research and teaching. In my fieldwork, which is ongoing in the greater Kansas City metropolitan area, I have interviewed and met with activists, citizens and political and religious leaders engaged with the inter-related and highly controversial issues of abortion, embryonic stem cell research and the teaching of Creationism (or Intelligent Design) in the high school science curriculum. I suggest these issues are inter-related because I approach them, primarily, as narratives of (and about) Creation. Each of these issues is capable of generating political, religious and scientific controversy. But they also afford new opportunities for once-marginalized and varied religious claims and traditions to re-engage and remake politics – abortion particularly so, as I will argue. How moderate-secular groups respond to, and seek to counter-mobilize against, such creative re-workings of politics, religion and society in contemporary America is the question that underpins and drives my ethnographic research.

51.3 What's the Matter with Kansas?

Over the last 50 years, single-issue interest groups have grown in electoral significance, transforming the two main political parties in the U.S. and marginalizing moderates (Aldrich 1995; Crotty 1984). For Republicans, this period was marked by the rise of the Christian Right (Andrew 1997; Berlet and Lyons 2000; Bjerre-Poulsen 2002; Diamond 1989, 1995; Hardisty 1999; Klatch 1999; Micklethwait and Wooldridge 2004; Viguerie and Frank 2004), particularly in the South – which was targeted explicitly under President Richard Nixon, with his so-called "Southern Strategy" – the Southwest and the West (cf. Phillips 1969). With time, these regions came to command greater economic and political clout than the old moderate "strongholds" in New England, the Mid Atlantic, the Upper Midwest and the Pacific Northwest (Rae 1989). As a result, the Party's once-dominant "liberal" or moderate wing, which was closely associated with political figures like New York State Governor Nelson Rockefeller, disintegrated. By the time of the 1994 Midterm Congressional elections, the religiously and socially conservative "capture" (to paraphrase Brennan 1995) of Republican politics seemed to be complete when the GOP seized control of Congress for the first time since 1954 and elected Newt Gingrich Speaker of the House of Representatives.

In the two decades since, Kansas has emerged as a primary battleground in the culture wars that have swept the United States over the last two decades (see Sharp 1999, 2005). As Thomas Frank has argued, "[Americans] are accustomed to thinking of the backlash as a phenomenon of the seventies (the busing riots, the tax revolt) or the eighties (the Reagan revolution); in Kansas the great move to the right was a story of the nineties …" (2005: 91). Since the early 1990s, the Christian Right has mobilized on a wide range of issues in Kansas. These have included abortion, embryonic stem cell research, gay marriage and the teaching of abstinence-only sex education in public schools. In addition, the conservative Christian majority on the Kansas State Board of Education drew newspaper headlines from around the world when they sought to revise science standards in the early 2000s to allow the teaching of Creationism in the high school curriculum.

Largely as a result of these controversies, Kansas is often presented as a crazy place, a state in America's heartland where passions run strong and political disputes too often escalate into the kind of violence that claimed Dr. Tiller's life. However, while never quite the quiet backwater depicted in *The Wonderful Wizard of Oz*, Kansas has demanded more attention historically as a hotbed of progressive political and religious activism. In the 1850s, for instance, guerrilla warfare raged on the Kansas-Missouri border between abolitionist militia and pro-slavery "Border Ruffians," in an episode from US history known as "Bleeding Kansas." Entering the

Union as a free state in 1860, just a few months before the American Civil War commenced, Kansans overwhelmingly identified with Abraham Lincoln's new Republican Party.

The legacy of this progressive Republican politics survived the emergence of the Populist movement as well as Prohibition and Temperance in Kansas during the nineteenth and early twentieth centuries. During the Great Depression, Kansas Governor Alf Landon – a standard-bearer for the Republican Party's progressive wing nationally – challenged Franklin Delano Roosevelt unsuccessfully for the Presidency in 1936. After the Second World War, Kansas continued to break new ground for progressive politics. In 1951, a class action filed against the School Board District in the state capital Topeka eventually became the landmark 1954 case *Brown v. Board of Education*, in which the US Supreme Court declared that state laws permitting racial segregation in public schools were unconstitutional. Kansas was also one of the few states to liberalize its abortion laws prior to the *Roe v. Wade* ruling in 1973.

In the 1980s, the Kansas legislature was dominated by Republicans who identified with their party's Lincolnian heritage and supported civil rights and a pro-choice consensus across the state. In the early 1990s, however, the ascendancy of such "traditional" Republicans faced a new challenge when the national pro-life organization Operation Rescue decided to target Dr. Tiller's medical practice in Wichita. Known for adopting aggressive tactics against abortion providers, the leaders of Operation Rescue sought to draw anti-abortion activists to the city to commit acts of civil disobedience in July 1991.

The campaign that Operation Rescue mounted during their "Summer of Mercy," as they called it, succeeded in closing medical clinics for a week, including that run by Dr. Tiller. The demonstrations climaxed with a pro-life rally at the Wichita State University football stadium. Twenty-five thousand people turned up, a significant number hailing from Kansas. According to Thomas Frank, whose book *What's the matter with Kansas?* became a *New York Times* bestseller in 2005, this was the moment when the Kansas conservative movement "got an idea of its own strength; this was where it achieved critical mass" (2005: 93). The campaign that Operation Rescue built that summer in opposition to Dr. Tiller's practice would have a profound impact on the state's political culture in general – and the Kansas Republican Party in particular – as well as politics nationally.

51.4 The Making of Moderates

Dr. Tiller was no stranger to anti-abortion violence. His medical practice had been fire-bombed in June 1986, and in 1993 a pro-life militant had shot him five times in his car. Since then, he had always travelled with a bodyguard – except when he was at church. However, it is unusual for most Americans engaged with the abortion conflict to experience such violence. More commonly, the militant wing of the pro-life movement has adopted tactics of political protest that, when faced with stubborn opposition, have involved harassment in the hope of exhausting their foes (cf. Doan 2007). In the months that followed the "Summer of Mercy," for instance, anti-abortion activists and other conservatives targeted pro-choice Republicans in primary races across the state:

> They had lain beneath cars to stop abortion [during the Summer of Mercy], and now they were putting their bodies on the line for the right wing of the Republican Party. Most important of all, the conservative cadre were dedicated enough to show up in force for primary elections, which in Kansas are held in the distinctly unpleasant month of August. And in 1992 this populist conservative movement conquered the Kansas Republican Party from the ground up: in Johnson County [...] and in all the other heavily populated parts of the state, they swamped the GOP organizations with enthusiastic new activists and unceremoniously brushed the traditional Kansas moderates aside. (Frank 2004: 95)

Many of my research subjects recounted stories of political intimidation during these fraught years. For instance, Nancy Brown was a pro-choice Republican state representative from affluent Johnson County, which borders Missouri and forms part of the greater Kansas City metropolitan area. At the time, pro-life protestors were engaged in tactics of harassment against her family that will be familiar to anthropologists, sociologists and other social scientists who have studied local struggles over abortion in other parts of the US (for example, Ginsburg 1989; Ragone 1994). These included picketing her house, following members of her family when they were out in public, making obscene telephone calls and depositing the carcasses of dead animals on her front porch during the night.

Responding to these provocations, a group of her supporters formed the interfaith, non-partisan Mainstream Coalition[1] in 1993. As one of the founders recounted, many perceived a need for an organization that provided a home for people of "a more moderate mindset:"

> I remember us saying: 'The religious right [has] organizations. They've got the Christian Coalition. They've got all these organizations. [Yet] there are lots of us out there. We don't have any place to come together. What we need is a place for us all to come together.[2]

One of the organization's most prominent leaders was Dr. Bob Meneilly, a pastor at Village Presbyterian Church. Later that year, he delivered a sermon denouncing the impact of the Christian Right on American democracy, which was then reprinted in the *New York Times* as an opinion piece. Dr. Bob, as his supporters affectionately call him, soon became a target of the Christian Right, which bombarded him with telephone calls and, in later years, email. "Dr. Bob really took the flak for most of [our work]," one of the founders of Mainstream explained. "[He] was seen as the flashpoint, so

[1] The name of the organization is credited to Nancy Brown, who drew out the first four letters (M.A.I.N.) as an acronym for Moderate Alliance of Informed Neighbors.

[2] This quote and the one below are drawn from an interview that Angie McGaw conducted with the founders of the Coalition. A summary of the interview by Dr. Lore Messenger, entitled "Mainstream beginnings: An interview with our founders," is available via the Coalition's website: www.mainstreamcoalition.org/. All other quotes in this section are drawn from the author's ethnographic field notes.

[suddenly, it's] "we're unhappy with that sermon, we're unhappy with this organization.'"

Most of the founding members of the Mainstream Coalition were registered Republican Party voters. The experience of being targeted by the Christian Right was formative in reshaping their political identities as moderates. As one former Republican explained to me, she and others would never have described themselves as "moderate" before the Christian Right mobilized in Johnson County. "We were all just Republicans," she said. Another explained that while she might have considered herself a "liberal" or "Rockefeller" Republican during the 1970s and 1980s, "moderate" was a label that just would not have made sense to her. Meanwhile, Dr. Bob was one of several Mainstream supporters who preferred to describe their politics to me as "progressive." Pro-choice Kansas Republicans acquired their identities as political moderates partly because that was the label with which the Christian Right branded them and their politics during the early 1990s.

In the years that followed, Mainstream Coalition grew into an organization of some 3000 members, mainly drawn from Johnson County. Several other grassroots organizations also emerged to oppose the agenda of the Christian Right. Some of these groups, such as the GOP Club – which was later re-launched as the Kansas Traditional Republican Majority in the early 2000s – have worked explicitly on trying to wrest control of the Kansas Republican Party from the pro-life movement. Others have focused on specific issues and sought to form bipartisan coalitions across the party political divide. Two examples include Kansas Citizens for Science, which has engaged with debates over the teaching of Creationism in the high school science curriculum, and the Kansas Coalition for Lifesaving Cures, which campaigned to protect state funding and investment in embryonic stem cell research.

Before I consider these respective groups and issues, I want to note here that the most important political dividing line in a "red" Republican state like Kansas is not between "Republican" and "Democrat" or even "conservative" and "liberal;" it is between "conservative" and "moderate" Kansas.[3] During my fieldwork in 2009, for example, I attended the annual Torch dinner organized by the Greater Kansas City Women's Political Caucus. An occasion to celebrate the political achievements of women in public life, the dinner provides Caucus members with an opportunity to recognize the community activism of women in the greater Kansas City metropolitan area as well as advocates of women's rights more generally. Their special guest that evening was Dr. Tiller's widow, who was collecting a posthumous award on behalf of her late husband. At the beginning of the evening, around 100 elected officials in attendance formed a line so that they could introduce themselves individually to the audience. Many officials repeatedly called upon Kansas voters to elect "more Democrats and moderate Republicans" in their brief speeches. My research subjects told me numerous similar stories of Democrats and moderate

[3]This potentially reinforces the argument that U.S. politics is "off centre," so to speak, suggesting that conservative interests exert a greater influence on national policy debates than those that might be described "liberal" because institutions of the left have become much weakened in contemporary American society (see Hacker and Pierson 2005).

Republicans "reaching out" to each other to make common cause in the struggle against the Christian Right.

51.5 Amendment 2: The Missouri Stem Cell and Cures Initiative

One issue that emerged as particularly divisive for the Republicans in the early 2000s was embryonic stem cell research (Smith 2010b). From the late 1990s onwards, a series of dramatic breakthroughs in the biomedical sciences demanded an ethical and regulatory response from governments around the world. The first of these occurred in February 1997 when scientists at the University of Edinburgh cloned Dolly the Sheep in Scotland. A year later in the American Midwest, a group of doctors isolated stem cells from 14 surplus embryos to an IVF program in Wisconsin. Despite public optimism that these developments might lead to the discovery of cures for an array of human diseases and disabilities, President George W. Bush introduced a ban on Federal funding for biomedical research that created new lines of embryonic stem cells from disused fetuses shortly after his inauguration. Drawing on his conservative Christian faith, President Bush consistently argued subsequently that he remained strongly opposed to "the use of Federal money – taxpayers' money – to promote science which destroys life in order to save life" (Washington Times, 21/05/05). Pro-life groups, most conservative Republicans, the Roman Catholic Church and the US Conference of Bishops all backed the President's stance.

This discussion had important policy implications across the U.S. for an issue of enormous medical and social significance. In the absence of a Federal regulatory framework, states like California and New Jersey introduced legislation to invest in embryonic stem cell research while anti-science religious conservatives (cf. Mooney 2005) in state legislatures elsewhere sought to curtail or even criminalize those engaged in such work. During this time, the United States came to resemble a patchwork quilt of legislative responses to the ethical and policy issues raised by this issue, which similarly to abortion and Creationism served to highlight deep, moral divisions amongst American conservatives. In 2006, President Bush acted again to restrict such research when he vetoed the Stem Cell Research Enhancement Act on 19 July, the middle of the primary season in Kansas and many other states for crucial Midterm elections being held later that year. The vetoed legislation had passed finely balanced Houses of Congress with the support of Democrats and a small group of "centrists" from the Republican Main Street Partnership (RMSP), which represents "disenfranchised Republicans" alienated from the Party's religious Right Wing.[4] In domestic politics, the Christian Right and pro-life movement greeted his decision with enthusiasm.

[4] For more on the history and mission of the Republican Main Street Partnership, go to the RMSP website: www.republicanmainstreet.org.

In few states was embryonic stem cell research more politically contentious than in Kansas and neighboring Missouri, largely because of the major concentrations of biomedical industries, infrastructure and research expertise in both Kansas City and St Louis. The former is home to the Stowers Institute – a major investor in embryonic stem cell research that was founded with a multi-billion dollar endowment from the Stowers family – and the Kansas University Medical Center (KU Med) while the latter hosts the internationally prestigious Washington University Medical School. Like Kansas, the Christian Right and the pro-life movement have come to dominate Republican Party politics in Missouri throughout much of the last two decades. In St Louis, the centrist Republican Leadership Council (RLC) maintained its largest and, arguably, best-organized state chapter. In February 2008, I interviewed the RLC state captain for Missouri, Dena Ladd, a woman who had previously worked for the Missouri Coalition for Lifesaving Cures, an organization co-led at the time by national RLC co-chair and former Missouri state Governor, Senator John Danforth. In both states, in fact, there was significant crossover between moderate GOP activist groups and those campaigning in favor of embryonic stem cell research. At the time of my visit to Kansas City in early February 2008, the Pro-Vice Chancellor responsible for embryonic stem cell research at KU Med was David Adkins, a man who had previously been a leading GOP moderate in the Kansas state senate. Prior to taking up his appointment at KU Med, Adkins had run for Kansas Attorney General and been defeated by a candidate backed by the Christian Right in the state's Republican Party primary.

That moderate Republicans in Kansas and Missouri should embrace biomedical research should not be surprising. For supporters of the biomedical and life sciences in both Kansas City and St Louis, much was at stake economically and politically. In Missouri particularly, embryonic stem cell research emerged as a pivotal issue during the 2006 midterm Congressional elections. Emboldened by polling that suggested public opinion was solidly behind them, the Missouri Coalition for Lifesaving Cures proposed an amendment to the state constitution to guarantee access for Missouri patients to cures and treatments derived from embryonic stem cell research. In response, the Christian Right mobilized to try to defeat the Missouri Stem Cell and Cures Initiative, known as Amendment 2. As summer approached and the primary season got underway, the struggle attracted national and international media interest when Hollywood celebrity and Parkinson's sufferer, Michael J. Fox, appeared in a 30-s television commercial supporting Amendment 2. When the amendment passed later that year with only 51 % of the vote, some activists I interviewed from the Missouri Coalition for Lifesaving Cures were shocked at how close they had come to defeat.

According to the executive director of the Kansas Coalition for Lifesaving Cures and other supporters of embryonic stem cell research in greater Kansas City, proposing Amendment 2 had been a major tactical error for their Missouri counterparts. In contrast, the Kansas Coalition sought to be less provocative towards the better-resourced Christian Right. It directed its focus instead on the state legislature to make sure that legislation introduced by opponents of embryonic stem cell research was quietly defeated, usually in committees. With a coalition of Democrats and moderate Republicans in control of the state senate and the popular Democrat, Kathleen Sebelius, in the Governor's mansion, this approach appeared to prove effective in quietly killing off anti-science legislation.

51.6 Kansas vs. The Creationists

Another very good example of how moderate Kansans united across the party political divide to challenge the Christian Right occurred during the 2005 State Board of Education hearings on the teaching of Creationism and evolution in the high school science curriculum. Despite generating interest from media organizations around the world[5] Kansas was not the only state debating whether to teach Creationism (or Intelligent Design) alongside evolution in the classroom during the early 2000s.[6] Furthermore, these debates have a long pedigree in the US, dating at least as far back as the landmark legal case *The State of Tennessee vs. John Thomas Scopes* (more usually known as the "Scopes Monkey Trial") in 1925 (cf. Laats 2010). In the 1970s, over 20 states debated whether to teach Creationism alongside evolution in the classroom, with both Arkansas and Louisiana passing laws to do so. The status of Creation "science" eventually became the subject of a 1987 US Supreme Court ruling, which declared that Creation science is religion and therefore cannot be taught in American public schools. In recent years, these debates have raged in states as far afield as New Hampshire, Tennessee and Texas (Madren 2012). And with science standards in Kansas schools set to be reviewed again by the State Board of Education before 2015, there was also a good chance that this issue would surface again in Kansas.

The 2005 hearings in Kansas were the culmination of a struggle over high school science standards stretching back to the 1990s, the first skirmishes of which had led to the founding of Kansas Citizens for Science in 1999. From then until 2007, Kansas adopted five different sets of standards for the high school science curriculum depending on whether conservative Republicans or moderates were able to win a majority on the State Board of Education. The issue appeared to be settled politically in 2006 with the bipartisan, grassroots "Take Back Kansas" campaign, in which Mainstream Coalition took a leading role along with like-minded groups like Kansas Citizens for Science and Kansas Families for Education. For these groups, maintaining good science standards in the high school curriculum was partly about protecting the quality of Kansas public education and reaffirming the importance of the constitutional separation of church and state. But according to some activists I interviewed from Kansas Citizens of Science, achieving victory was also about ensuring that Kansas could not be misrepresented as an "anti-science" state. This was an important consideration given the opposition outlined above to the biomedical industries in Kansas City, supporters of which have had to maintain vigilance in

[5] See, for example, Jeff Tamblyn's 2007 documentary *Kansas vs. Darwin*.

[6] In *Kitzmiller v. Dover Area School District*, Judge John E. Jones III ruled against Creationism in the United States District Court for the Middle District of Pennsylvania in December 2005.

order to discourage funding of embryonic stem cell research from leaving Kansas or Missouri for other states where the biosciences might enjoy greater support from local policymakers.

The 2006 "Take Back Kansas" campaign succeeded in overturning the conservative religious majority on the State Board of Education, electing an anti-Creationism coalition of Democrats and moderate Republicans who then agreed to adopt the science standards that are being used in Kansas today. Many of my ethnographic subjects, therefore, described this achievement as a victory for moderate Kansas. This verdict was confirmed by at least one pro-life activist I interviewed, who was not a supporter of Creation "science" himself. In his view, the hearings had been a mistake and had set the wider agenda of the Christian Right back by a generation in Kansas.

51.7 The 2010 Elections and Beyond

It is clear that debates over embryonic stem cell research and the teaching of Creationism/Intelligent Design in the high school science curriculum provided moderate-secular activist groups with important opportunities to counter-mobilize and re-empower themselves against the Christian Right. However, more important than both these issues remains the enduring legacy of abortion politics, a powerful resource for the pro-life movement that has helped turn the tide in favor of religious conservatives in Kansas.

In the 20 years that followed 1991s "Summer of Mercy," the Christian Right has grown to dominate state politics in Kansas. It would seem, then, that the adoption of tactics of political harassment, intimidation and even violence (though rare) has yielded political results for the pro-life movement. Today, even in their traditional stronghold of Johnson County, pro-choice Republicans often find themselves struggling to win primary races they were once able to take for granted. This experience can be very unsettling for moderates, who are sometimes left questioning their own assumptions about the political complexion of the communities within which they live. A young, pro-choice woman who ran for an open primary for Kansas House District 20 – a safe Republican seat in leafy Overland Park, Johnson County – in 2010 told me of her disappointment when she lost to a candidate endorsed by Kansans for Life. "Is District 20 as moderate as we always used to think it was?" she asked herself afterwards when she realized that the political assumptions on which she had built her campaign had been found wanting.

Until 2010, however, the Christian Right had never succeeded in winning a gubernatorial election for a conservative Republican and, as a result, they had been unsuccessful in introducing state legislation to curtail abortion services further in Kansas. Throughout most of the 1990s, Bill Graves, a moderate Republican, occupied the governor's mansion. Kathleen Sebelius, a Democrat who resigned to join the Obama administration as Secretary of State for Health and Human Services in April 2009, followed him. However, her Democratic replacement Governor Mark Parkinson, a moderate who had previously served as Chairman of the Kansas

Republican Party in the early 2000s, did not seek re-election in 2010. Without a strong Democrat or moderate Republican candidate to oppose him, the deeply conservative Senator Sam Brownback – a national figure on the Christian Right who has voted against abortion and embryonic stem cell research and supports the teaching of Creationism in high school science classes – would go on to win the gubernatorial election comfortably. With moderate Kansas in disarray, in part as a result of Dr. Tiller's murder and the closing down of Pro-Kan-Do, the Christian Right and the pro-life movement may well have drawn the conclusion that they were confronted with an opportunity to reshape Kansas politics again.

In late 2009, Mainstream Coalition organized a series of meetings to discuss how to support pro-choice candidates in Johnson County during the 2010 primary and general elections. This issue has remained close to the hearts of Mainstream members since the organization began. In the absence of a dedicated PAC to fight for abortion rights in Kansas, several office-holders felt that it was important to discuss a potential strategy. At one meeting I attended, one activist argued passionately that the only way to respond to the Christian Right was to appropriate their tactics and harass elected officials if they did not oppose pro-life legislation. However, others at this meeting remained unconvinced. The organization's executive director explained to me afterward that she thought most Mainstream members would reject such tactics as "it is not in the nature of moderates" to want to emulate them. After all, she said, "they are moderates, not militants." But this was the problem, as far as the activist who had argued for a more aggressive approach was concerned. "The problem with moderates is that they won't get off their backsides and do anything until there is blood on their patio," she was fond of telling me.

How to mobilize moderates in the face of political and religious extremism remains one of the most complex of challenges for the Mainstream Coalition and other similar activist organizations in strong Republican states like Kansas. To put this problem another way, how might Republican "moderates" render an identity that many of them feel their pro-life opponents imposed upon them operational in grassroots politics, especially if they remain ill at ease with such a label? This question has vexed leading national GOP moderates like former governor of New Jersey Christie Whitman, who has called for "radical moderates" to come forward and "reclaim Lincoln's legacy" by retaking the Republican Party from the religious Right (cf. Whitman 2005).

In Kansas, however, where moderates and secularists square up to a well-resourced Christian Right and face a pro-life movement in the ascendant, the political stakes could not be higher. This became clear during the 2012 Republican primary season (Figs. 51.1 and 51.2). Backed by the Christian Right as well as the billionaire, Wichita-based Koch brothers – long-time supporters of libertarian politics and Right-Wing causes in the United States – and other supporters of Governor Brownback, conservative Republicans challenged incumbent moderates in their respective primary races. Their efforts focused on the state senate, in which moderate Republicans had been able to hold leadership positions with the support of Democrats to block legislation moved by religious conservatives in control of the House. The moderates were swept aside in a string of dramatic defeats; only a

Fig. 51.1 Many Kansas churches opposed to abortion urge members of their congregations to vote in the 2012 Republican Party primary elections (Photo by A. T. T. Smith, Tonganoxie, Kansas, July 2012)

Fig. 51.2 A mainstream yard sign urged passing motorists to "make a difference" and "vote moderate" in the 2012 election (Photo by A. T. T. Smith, Shawnee, Kansas, July 2012)

handful survived. For the first time in the state's political history, both chambers of the Kansas legislature as well as the Governor's mansion were in the hands of political and religious conservatives. Whether moderate Kansans can rebuild the state's proud traditions of supporting civil rights and tolerance for minorities, public education and other socially progressive causes, remains to be seen.

Acknowledgements The title of this chapter was inspired by a 2007 documentary, directed by Jeff Tamblyn, called *Kansas vs. Darwin*, which explored the debates over the teaching of evolution in Kansas public schools in the mid-2000s. For an earlier version of this chapter, see Smith 2010a. For a more detailed account of religion, science and, specifically, the politics of embryonic stem cell research under President George W. Bush, see Smith 2010b.

References

Aldrich, J. (1995). *Why parties? The origin and transformation of political parties in America*. Chicago: University of Chicago Press.

Andrew, J., III. (1997). *The other side of the sixties: Young Americans for Freedom and the rise of conservative politics*. New Brunswick: Rutgers University Press.

Berlet, C., & Lyons, M. (2000). *Right-wing populism in America: Too close for comfort*. London: Guildford Press.

Bjerre-Poulsen, N. (2002). *Right face: Organizing the American conservative movement 1945–65*. Copenhagen: Museum Tusculanum Press/University of Copenhagen.

Brennan, M. (1995). *Turning right in the sixties: The conservative capture of the GOP*. Chapel Hill: University of North Carolina Press.

Coles, K. (2007). *Democratic designs: International intervention and electoral practices in post-war Bosnia Herzegovina*. Ann Arbor: University of Michigan Press.

Crotty, W. (1984). *American parties in decline* (2nd ed.). Boston: Little, Brown and Company.

Dewey, J. (1954). [1927] *The public and its problems*. Athens: Ohio University Press.

Diamond, S. (1989). *Spiritual warfare: The politics of the Christian right*. Boston: Pluto Press.

Diamond, S. (1995). *Roads to dominion: Right-wing movements and political power in the United States*. New York: Guildford Press.

Doan, A. (2007). *Opposition and intimidation: The abortion wars and strategies of political harassment*. Ann Arbor: University of Michigan Press.

Frank, T. (2004). *What's the matter with Kansas? How conservatives won the heart of America*. New York: Metropolitan/Owl Book.

Ginsburg, F. (1989). *Contested lives: The abortion debate in an American community*. Berkeley: University of California Press.

Hacker, J. S., & Pierson, P. (2005). *Off center: The Republican revolution and the erosion of American democracy*. New Haven: Yale University Press.

Hardisty, J. (1999). *Mobilizing resentment: Conservative resurgence from the John Birch Society to the Promise Keepers*. Boston: Beacon.

Klatch, R. (1999). *A generation divided: The new left, the new right and the 1960s*. Berkeley: University of California Press.

Laats, A. (2010). *Fundamentalism and education in the Scopes era: God, Darwin and the roots of America's culture wars*. New York: Palgrave Macmillan.

Madren, C. (2012). *Classroom clashes (Part 1): Teaching evolution*. http://membercentral.aaas.org/blogs/stemedu/classroom-clashes-pt-1-teaching-evolution. Accessed 29 May 2012.

Micklethwait, J., & Woolridge, A. (2004). *The right nation: Conservative power in America*. New York: Penguin.

Mooney, C. (2005). *The Republican war on science*. New York: Basic Books.
Paley, J. (2001). *Marketing democracy: Power and social movements in post-dictatorship Chile*. Berkeley: University of California Press.
Phillips, K. (1969). *The emerging Republican majority*. New York: Arlington House.
Rae, N. (1989). *The decline and fall of the liberal Republicans: From 1952 to the present*. Oxford: Oxford University Press.
Ragone, H. (1994). *Surrogate motherhood: Conception in the heart*. Boulder: Westview Press.
Riles, A. (2001). *The network inside out*. Ann Arbor: University of Michigan Press.
Rorty, R. (1998). *Achieving our country: Leftist thought in twentieth century America*. Cambridge, MA: Harvard University Press.
Rorty, R. (2000). *Philosophy and social hope*. New York: Penguin.
Sharp, E. B. (Ed.). (1999). *Culture wars and local politics*. Lawrence: University Press of Kansas.
Sharp, E. B. (2005). *Morality politics in American cities*. Lawrence: University Press of Kansas.
Smith, A. T. T. (2010a). Fear and loathing in Kansas City: Political harassment and the making of moderates in America's abortion wars. *Anthropology Today, 26*(4), 4–7.
Smith, A. T. T. (2010b). Faith, science and the political imagination: Moderate Republicans and the politics of embryonic stem cell research. *Sociological Review, 58*(4), 623–637.
Smith, A. T. T. (2011). *Devolution and the Scottish conservatives: Banal activism, electioneering and the politics of irrelevance*. Manchester: Manchester University Press.
Viguerie, R., & Franke, D. (2004). *American's right turn: How conservatives used new and alternative media to take power*. Chicago: Bonus Books.
Whitman, C. (2005). *It's my party too: The battle for the heart of the GOP and the future of America*. New York: Penguin.

Chapter 52
Religious and Territorial Identities in a Cosmopolitan and Secular City: Youth in Amsterdam

Virginie Mamadouh and Inge van der Welle

52.1 Introduction

Over the past decade, religion has been a much contested issue in Dutch politics in general and more specifically in Amsterdam. The Dutch capital city can be characterized as a cosmopolitan and secular city. The majority of the population has no religious affiliation, but new groups of immigrants are more religious and often see their identity largely defined by their religious background, distinct from the religions traditionally present in the city. The largest of the new religions brought to Amsterdam by immigrants, is Islam which has been widely framed as a social and political problem in public debates about immigration and integration. This chapter discusses religion in Amsterdam in this highly politicized context. More specifically, it will examine the attitudes of Amsterdam youth from different ethnic backgrounds toward religious and territorial identities.

Whereas in the Netherlands religion has mainly been portrayed as an obstacle for the integration of second generation immigrants, studies in other countries have proven religion an avenue to integration, for example for second generation immigrants in New York. Those who are involved in organized religion attend churches and temples where they are likely to come into contact with other ethnic groups (Kasinitz et al. 2008). By focusing on the identification strategies of the second generation immigrants, we examine the contribution of religion to integration and a sense of belonging in Amsterdam and the Netherlands.

The chapter is structured as follows. The first section provides a brief introduction of the role of religion in Dutch society in the formation of the nation state and in the twentieth century, with special attention to the situation in the city of Amsterdam.

V. Mamadouh (✉) • I. van der Welle
Department of Geography, Planning and International Development Studies, University of Amsterdam, Postbus 15629, 1001 NC Amsterdam, The Netherlands
e-mail: v.d.mamadouh@uva.nl; i.c.vanderwelle@uva.nl

S.D. Brunn (ed.), *The Changing World Religion Map*,
DOI 10.1007/978-94-017-9376-6_52

The second section describes the main changes of the postwar (World War II) period in the religious landscape of the country and the city: secularization, depillarization and immigration. The third section introduces the contemporary public and political debates about immigration, integration, religion and national identity. In the final section, we present some results of a research done among Amsterdam youth of different ethnic origins to explore their religious and territorial identities.

52.2 Religion Nation Building and National Identity in the Netherlands

Present debates on religion in the Netherlands should be viewed in historical perspective. Religious diversity has been a characteristic of the Netherlands during modern history (Van Rooden 2002). The establishment of the Dutch Republic of the United Provinces and their resistance to the King of Spain were directly linked to the Reformation and the religion wars that devastated Europe in the sixteenth and the seventeenth centuries. Protestantism was predominant, but religious tolerance was a major trait of the new Republic. The Union of Utrecht (1579) acknowledged the freedom of religion and the freedom from persecution for religious reasons. There were Catholic majorities in certain cities and regions. The Republic welcomed religious minorities displaced by religious persecution elsewhere: Sephardic Jews from Portugal, Ashkenazy Jews from Easter Europe, Huguenots from France and Southern Netherlands under Spanish rule, Lutherans from Salzburg, Quakers from England and Mennonites from Switzerland. They found a safe haven in many Dutch cities (especially in Amsterdam) and largely contributed to the economic growth and cultural development of the new Republic.

With the Batavian Revolution and the French occupation (1795–1815) and the establishment of the Batavian Republic later the Kingdom of the Netherlands in the French Empire, all citizens became equal and the Calvinist Church lost its privileged position. After the Napoleonic Wars, nation-building in the new Kingdom was marked by the idea of the Protestant nation (Knippenberg 1997, 2002, 2006). Although freedom of religion was a key tenet of the 1815 Constitution, Catholics and Jews were second class citizens. Catholics were often suspected of being more loyal to Rome than to the national state. Religion rather than language motivated the Belgian revolt and its separation in 1830. In 1848 a new constitution was adopted and nation building shifted from the "Protestant nation" to the "Dutch nation" and in 1853 the Episcopal hierarchy was restored. Meanwhile religious diversity increased due to several schisms in the Reformed Church.

In the nineteenth century, the struggles about the enlargement of the electoral franchise gave birth to modern political parties: liberal, socialist, Catholic and Orthodox and Protestant. In 1888 a coalition of Orthodox Protestants and Catholics took office in the national government, and the main cleavage in Dutch politics became the division between believers and unbelievers. In 1917 a Pacification was achieved through a package deal introducing both universal franchise (a demand of the socialists) and equal financing of confessional and public education (a demand of the confessional parties) (Knippenberg and De Pater 1990).

The period following pacification is characterised by pillarization, or consociational democracy, as the Dutch American political scientist Lijphart 1968 would name it (see Wintle 2007 for a recent appraisal), the organization of Dutch society in separated pillars. This vertical organization of social life meant that anyone would live in pillarized institutions, with little interaction with people belonging to other pillars. A Catholic would not only go to a Catholic Church, he would go to a Catholic school, vote for Catholic party, read a Catholic newspaper, be the member of a Catholic trade union, visit a Catholic hospital, play sports in Catholic club, rent a dwelling owned by a Catholic housing association, etc. and last but not least, have friends and marry in the same pillar. Cross-pillar cooperation was the work of the elites.

The Catholic and the Protestant pillars (the latter divided religiously among several Protestant denominations, of which the Dutch Reformed was the largest) were complemented by a socialist and a liberal pillar, which were organized more loosely (especially the liberal one). As a result of this pillarisation system, the number of inhabitants declaring no denominations was high as this was recorded in the municipal administration and consequential for the financing of different institutions. This accurate and detailed registration of religious denomination greatly facilitated the deportation of Dutch Jews during the German occupation. Pillarized institutions, most notably political parties, were re-established after the Liberation in 1945, but since then the Dutch religious landscape drastically changed.

52.3 Post-war Changes in the Dutch Religions Landscape

Two main processes characterize the relations between religion and national identity in the postwar period: the first is driven by individualization and secularization, the second pertains to the religious consequences of international immigration.

The 1960s were a particularly important turn in Dutch society with a strong movement against the hierarchical and vertical organization of the society. Individualisation and secularization eroded the pillarised institutions (Sengers 2005). Religion became marginalized as a sociopolitical issue through a wide array of emancipatory movements in the 1960s and 1970s: youth movements (Provo, Kabouters, squatters), feminist movements, gay and lesbian movements. They undermined the religious and patriarchal forms of authority on which the pillarised society was grounded (Mamadouh 1992; Kennedy 1995; Van der Donk et al. 2006). Religion became less important: the proportion of mixed marriages increased and confessional parties lost their taken for granted electorate. In addition, an ever growing number of people reported not to belong to any religious denomination. While in the first half of the twentieth century all Protestant churches, but the orthodox

Table 52.1 Religious composition of the Dutch population age 18 years and older

	1899 (%)	1920 (%)	1930 (%)	1947 (%)	1960 (%)	1971 (%)	1985[a] (%)	2000[a] (%)	2005[b] (%)	2009[b] (%)
Dutch/Reformed churches[c]	56	50	44	40	37	33	28	21	21	18
Roman Catholic	35	36	36	38	40	40	37	32	30	28
Other	6	6	5	5	4	2	5	8	9	10
No affiliation	2	8	14	17	18	23	31	40	41	44

Data source: Statistics Netherlands 2012

[a]Since the last census in 1971, data are based on surveys

[b]From 2005 onward the weighing method has been changed to better represent ethnic minorities in surveys. As a result their religious affiliations are better represented than in previous surveys. This has caused an increase of the category 'other denominations'

[c]Since 2005 this category also features the Protestant Church in the Netherlands (PKN), a merger of the Netherlands Reformed Church, Reformed Churches in the Netherlands and the Evangelical Lutheran Church

Calvinist, lost members, secularisation affected the Roman Catholics and Orthodox Calvinists from the 1960s onwards. As a result, the percentage of the population not linked to a denomination continued to grow (Table 52.1).

One of the main outcomes of this process of depillarization was the erosion of the religious determination of voting behavior, hollowing out the electorate of the confessional parties. In 1980 the main Catholic and Protestant political parties merged into a single Christian Democratic Party to conserve their dominant position in national politics. The Labour Party also lost its privileged ties with the red pillar and the electorate became more volatile. This – in combination with an extreme proportional voting system – opened the Parliament to successive protest parties.

In the postwar period, while traditional religions became less and less important in Dutch society, the Netherlands, previously a country of emigration, became a country of immigration (Penninx et al. 1994; Lucassen and Penninx 1997; Özüekren and Van Kempen 1997; Vermeulen and Penninx 2000; Fennema et al. 2000; Lucassen et al. 2006; Doomernik and Knippenberg 2003; Knippenberg 2005; Bernts et al. 2007). Repatriates and migrants from the former colonies (Indonesia and later Suriname), from guest worker recruitment areas (Spain, Portugal, Italy, Greece, Yugoslavia, Turkey, Morocco) and from conflict areas (with asylum seekers from Asia, Southeastern Europe and Africa), settled in the Netherlands. Many of these immigrants and refugees were Muslims. Based on the country of origins of immigrants and their descendants the number of Muslims in the Netherlands is estimated to be about 6 % of the population (Table 52.2). The Muslim population is ethnically diverse, including Dutch converts, but Turks and Moroccans form by far the two biggest national groups. Most of them live in the larger cities. In 2010 about 13 % of the population in the Amsterdam agglomeration was Muslim, about twice as much as the national average.

Table 52.2 Muslim and Hindu percentage of the population in the Netherlands, 1971–2004

	1971	1975	1980	1985	1990	1995	2000	2004
Muslim	0.5	0.8	1.7	2.3	3.1	4.1	5.1	5.8
Hindu	0.0	0.1	0.2	0.3	0.4	0.5	0.6	0.6

Data source: Statistics Netherlands 2012

Table 52.3 Religious denominations in Amsterdam, 1900–2010

Religious denomination	1900 (%)	1930 (%)	1947 (%)	1960 (%)	1971 (%)	1984 (%)	1992 (%)	2000 (%)	2005 (%)	2010 (%)
Dutch Reformed Church/Reformed Churches in The Netherlands/ Protestant Church in the Netherlands	46	25	24	22	18	13	9	4	4	3
Roman Catholic	23	22	23	24	23	24	20	11	9	9
Judaism	12	9	1	1	1	1	1	1	1	2
Islam	0	0	0	0	0	7	9	15	11	13
Other	14	9	7	5	5	7	4			
No denomination / affiliation	6	35	45	48	54	48	58			

Data source: O+S Amsterdam 2012
1900–1992: Amsterdam city register (based on registered membership in municipal population register)
2000–2010: Burgermonitor O+S Survey (based on reported affiliation in local survey Burgermonitor)

Religious diversity and tolerance were historically always greater in the Dutch cities than in the countryside and this is particularly true for Amsterdam (Mak 2001). The city has a particularly rich history of religious pluralism (most notably German and Portuguese Jews and Huguenots) and of secularism. In Amsterdam 35 % of the inhabitants had no religion in 1930 (Kaal 2011). The Dutch Reformed and the Catholics were groups of equal size, and Jews formed an important minority until their deportation and extermination during the Holocaust (Table 52.3). The city has attracted a large share of immigrants over the past decades, guestworkers, asylum seekers, artists, students and expats. Since the 1960s it also has a solid world reputation for alternative lifestyles and sexual emancipation (Mamadouh 1992; Kennedy 1995; Deben et al. 2000; Musterd and Salet 2003).

In Amsterdam the national trend towards secularization is sharper than in the rest of the country and a larger share of the population is not affiliated with any religion (see Table 52.3). Immigration has brought religious diversity to the city, where Islam is now the largest religion. In 1900 one Muslim was listed in the register of the municipality of Amsterdam, in 2000 this had increased to over 88,000 Islamic adults (Van der Steenhoven 2001). Since 2000 the share of Amsterdammers reporting some affinity with a religion or a spiritual group has hardly changed: about 4 out of 10. Islam and Christianity are by far the largest religions in town (Booi et al. 2011).

Table 52.4 Religious and spiritual affiliation in Amsterdam and Rotterdam, 2008 and 2010, based on the survey question: Do you feel affiliated with a religion or philosophy of life?

	Amsterdam		Rotterdam	
Ethnic group[a]	2008 (%)	2010 (%)	2008 (%)	2010 (%)
Surinamese	50	64	86	84
Turkish	78	82	93	97
Moroccan	76	88	96	98
Other Non-Western[a]	60	62	75	71
Western[a]	37	35	55	53
Dutch	24	21	35	36
Total	39	39	51	49

Data source: O+S (Amsterdam) and COS (Rotterdam) in Gemeente Rotterdam & Gemeente Amsterdam (2012: 128)

[a]Dutch statistics use the place of birth of parents to categorize individuals and distinguish between Western and Non-Western foreign places of birth

There is a difference the importance of religious affiliations between ethnic groups. Moroccans and Turks are the most religious groups (90 and 82 %), inhabitants of Dutch origin the least (21 %). In addition, young people (until 25 years old) are more religious than other age groups. This is related to the population structure. There is a larger share of inhabitants of Moroccan and Turkish origin (the two groups with the largest share of believers) in this age group (Phalet and Ter Wal 2004; Crul and Doomernik 2003; Crul and Heering 2008; Van der Steenhoven 2001).

Table 52.4 shows that Amsterdam is a more secular city, even compared to the second Dutch city Rotterdam. Overall respondents are (10 percentage points) less often prone to see themselves as belonging to a religious denomination than in Rotterdam. This is true for all ethnic groups. However, while the figure seems rather stable for most groups, in Amsterdam there is a relevant raise among Surinamese and Moroccan respondents.

52.4 The Public Debate About Religion and National Identity

Due to immigration, the place of religion in society and its relations with the state are back as important objects of public debate (Kennedy 2010: 155, but also Bader 2007). In that debate, religion is often discussed in relation to immigration and integration problems. The Dutch approach to immigration and integration has often been called "multicultural" (Vink 2007; Maussen and Bogers 2010). More specifically, similarities with the eroded but still visible pillarisation system have been noted (Landman 1992). It seems indeed that the past pacification arrangements inspired the Dutch authorities to approach migrant communities as potential new pillars and to give religious leaders and religious organizations much leverage in the organization of social and cultural activities. Institutions created to sustain Catholic and Protestant schools, for example, made it easy to set up Muslim schools.

By contrast, the presence of immigrants and their place in society was much disputed. In the 1980s a small extreme right party gathered some electoral support and – thanks to a nation-wide proportional voting system – a seat in the Second Chamber of the national parliament. In the 1990s mainstream conservative politicians, such as the leader of the Conservative party VVD, Frits Bolkestein, who became later European Commissioner, voiced doubts about the place of Islam in Dutch society (Bolkestein and Arkoun 1994; Bolkestein and Von der Fuhr 1997). The political debate further broadened after 2000, when the Labour Party also came to problematize the multicultural society (Scheffer 2000, 2007). By 2001 Pim Fortuyn entered the political debate with a new anti-Islam political platform (Fortuijn 1997, 2001; Pels 2003; Mamadouh and Van der Wusten 2004). The attack of 9-11 2001 and the assassination of Fortuyn by an animal rights activist in May 2002 brought the divisions in the consensus oriented political culture of the Netherlands to the fore. The polarization continued with the campaign of the Conservative MP Ayaan Hirsi Ali (a former Muslim of Somali origin) against Islam and for the emancipation of Muslim women (Hirsi Ali 2005; Hirsi Ali and Wilders 2003). In November 2004 a young Dutch Muslim of Moroccan origin assassinated Theo van Gogh in Amsterdam in November 2004, after he made a movie called Submission for Hirsi Ali (Buruma 2006; Mamadouh 2008; Uitermark and Hajer 2008). Another politician Geert Wilders dissented from the conservative party VVD in September 2004 because he refused to accept his party's decision to support the opening of the accession negotiations with Turkey as he opposed the accession of this Muslim country to the European Union (Wilders 2005). He turned out to become a major political factor with his Party for Freedom PVV with electoral success in the 2006 national elections (Lucardie 2007, 2008; Vossen 2008, 2009, 2010; Brinks 2006, 2010; Fennema 2010; Kuitenbrouwer 2010; Van der Waal et al. 2011; Mamadouh and Van der Wusten 2012; Van Gent and Musterd. 2012, see also Van der Valk 2003; Van der Veer. 2006).

During the past decade the political discourse against the anticipated Islamization of the Netherlands has been harsh, and it sharply contrasted with the previous immigration and integration policies that were often described as multiculturalist. Many issues were discussed including domestic violence, delinquency among Moroccan youths, gender relations in the public sphere, veils and burqas, radicalization and Islamic fundamentalism. In these debates, the visibility of religion in public space was at stake. Islam was often framed as the root of many societal problems and more bluntly as a religion incompatible with Dutch society. There was also a fear of intervention from abroad from undemocratic states (Dijkink and Van der Welle 2009) and fundamentalist movements through foreign imams in the mosques. For this reason some efforts have been deployed to establish imam training tracks at Dutch universities, but this is also much disputed as it affects views of the relations between state and religion. Moreover the relation of the second generation youth to the national society has been problematized (Mamadouh 2001, 2003) as were national identity and identification (Meurs 2007).

In Amsterdam, the combination of the presence of a large share of Muslims in the city population and a specific political culture characterized by left-wing orientation and tolerance to diverse lifestyles, the national debates had a slightly different impact, although contentious too (Uitermark and Gielen 2010). The Mayor of Amsterdam Job Cohen, affiliated with the Labour Party, was both acclaimed and criticized for his handling of the situation after the murder of Theo van Gogh. Some saw it as a great achievement that clashes between ethnic groups did not take place (he was listed by *Time Magazine* as one of the "European Heroes" 2005); others were outraged by his consensual approach of marginal and marginalized groups. He was mocked for his appeasing approach to "keep everyone together" through tea-drinking and talking. This was criticized as soft and ineffective.

Meanwhile, the PVV gained ascendance and many were worried about the increasing influence of Wilders on the public debate. The party obtained 6 % of the votes at the 2006 general elections, but this score was lower in Amsterdam (4.5 %) than nationally, and much lower than in other cities (especially Rotterdam) (De Jong et al. 2011). The PVV advocated a no tolerance policy towards young delinquents and drop outs that were framed as potential fanatics and terrorists. For the PVV transnational terrorism (that is, Al Qaeda) and loitering teens in the streets were part of the same Muslim threat of Western civilization. In this context being Muslim, and religion itself, became framed as key societal and political problems. How has it affected the perception of young Amsterdammers (many of them with a Muslim background) of their religion, their city and the Netherlands?

52.5 Youth and Religion in Amsterdam: A Survey

The remainder of this chapter is based on a study of young Amsterdammers of different ethnic origins, representing the three main immigrant groups and those with a Dutch background.

Between April and July 2007 we conducted a survey of over 1,000 young adults living in the city of Amsterdam. The sample for this research was generated from the register of personal data of the municipality of Amsterdam (GBA) in which each (legal) resident is registered. This register includes data on ethno-national origins, such as the place of birth of parents and religion. The sample criteria were: (1) living in Amsterdam for a continuous period of 5 years, (2) born in the Netherlands, (3) aged 18–30 years old, (4) with both parents born in Morocco, in Turkey, in Suriname (they are known as second generation migrants) or in The Netherlands (known as native Dutch). The survey was mainly conducted through the internet using unique log-in codes, and partly face-to-face. The response rate was 25 %. Females and 25–30 year olds were slightly over represented. For the analysis, we weighted the data to take into account possible gender or age differences.

In addition to the survey, open-ended face-to-face interviews with 50 respondents were conducted. The interviews lasted from 1 to 3 h. Most took place in the respondents' homes, although some were conducted in public spaces (cafes, restaurants, workplace) or at the university. All the interviews were recorded, transcribed and coded. The quotes in this chapter come from these in-depth interviews. The translation from Dutch into English is ours, and names are fictional.

Considering the political context sketched in the previous section, a context marked by the national debate on the position of Islam in the Netherlands and poor integration of recent immigrants and their descendants, what were the attitudes of Amsterdam youngsters with diverse ethnic backgrounds towards religious and territorial identities? We are particularly interested in territorial identifications with the city, the country, and the country of birth of their parents (if they were born abroad). We first look at religious attachment and religious practices, then consider the importance of religious identification in the light of other identification strategies. Finally we discuss their attitudes to religion and the public sphere.

52.5.1 Religious Attachment and Practices

Our second generation respondents were more religious than the young adults of Dutch origin. Over 90 % of the youth of Moroccan origin, 83 % of those of Turkish origin and 48 % of those of Surinamese origin considered themselves religious, compared to 7 % of those of Dutch origin. The young Surinamese and native Dutch were less religious than their parents. For the young Moroccan and Turks, there was no difference with their parents.

Almost all religious Moroccans and Turks were Muslim. The religious affiliation of the religious Surinamese was more diverse (in line with the great ethnic and religious diversity of the Surinamese population): 46 % Christian, 33 % Hindu and 17 % Muslim.

Being religious for most young adults did not necessarily mean that they were practicing it. For some of them, it was even hard to make a distinction whether it is culture or religion. The 28-year-old Farida of Moroccan background noted:

> [Muslim?] Well it is of course what you have been brought up with. Is it culture or does it belong to your faith? These are two totally different things that are linked together.

Organized religion played an important role in the lives of only a minority of the second generation respondents. Amongst the religious respondents, about a quarter of the Moroccans and Turks and about 10 % of the Surinamese attended a house of worship once a week or more often. The Surinamese Muslims attended the mosque a bit more often than the Surinamese Christians or Hindus attended the church or the temple.

In Amsterdam participating in organized religion is not necessarily a site of integration. Young Muslims especially were not likely to come into contact with other ethnic groups in the mosque, because mosques are, besides being attached to specific religious persuasions, often organised along ethnic (and linguistic) lines. The young Muslims did shop between mosques, but while doing so, they generally did not cross ethnic divisions. The 28-year-old Erdem, of Turkish origin, for example, shopped between Turkish mosques depending on the quality of the imam:

> I go to different mosques. But only to Turkish mosques. It just happens. Most of the time in Amsterdam West [where he lives], but we used to go out of town to a mosque along the highway .... The better the imam, the more you prefer to go to that mosque. Every once in a while there is an imam who is really good in getting his messages, what

he knows comes across. But there are also imams who know a lot, but cannot deliver the message. So, if there is a talented imam, than I will go there. For a year or two, I have been attending a mosque linked to the Grey Wolves [a Turkish nationalistic political movement]. I went there because the imam was very good and not because I feel connected to the Grey Wolves.

An exception was 19-year-old Mohammed, of Moroccan origin. For him, going to the mosque is also a social event. It is a place where he met new people, and, therefore, he visited mosques linked to various ethnic groups. From his experiences, it also becomes clear that language is a barrier when it comes to "mosque shopping:"

There are various types of mosques, for example a Turkish mosque, an Indonesian and a Surinamese mosque. I try to attend a lot of different mosques, because I would like to broaden my social circle, meet new people of different backgrounds. For example, I go to the Turkish mosque where almost only Turks are going. I hardly ever see a Moroccan … The sermon is in Turkish, so Moroccans cannot understand it… But my Turkish friends translate a bit for me. In the Moroccan mosque, sermons are in Arabic, not in Moroccan. There I see many more cultures. It is "pure Arabic," the Arabic they speak in Saudi Arabia and such. There they speak as in the Koran, so everybody understands at least a little bit. And there is one mosque in my neighbourhood, where they translate everything into Dutch, so everybody understands.

Most young adults stressed their individual relationship with faith. Private prayer, and searching for information on the internet or in books have replaced church or mosque attendance. The 29-year-old Fidan did not attend the mosque, and stressed her individual relationship with faith:

I hardly ever attend the mosque. It is very individual the way I experience my faith. It is very personal. And I only talk about it with few people. Of course going to the mosque is something beautiful, the communal prayer. But faith also is sometimes very individual and faith says that you practice meditation secretly. You don't have to show it off or brag about how faithful you are. It has to remain your [individual] relationship with God. And that is very private for me.… And besides, going to the mosque does not fit into my way of life. I do not have time for that.

For about half of the religious young adults, religion gained importance over the years before the survey (Fig. 52.1). For most of the others, religion remained equally important. Only a few religious Surinamese felt that religion lost importance for them over the last years. All of them were Christian or Hindu. When the young adults talked about their religion gaining importance, coming of age was an important reason for that change. The 21-year-old Sarah of Surinamese origin explained that for her Christianity has gained importance because of the decisions she had to take, growing up:

Religion gains importance, especially with decisions I have to take. Then I ask for example for a sign, or help me to make things all right, that kind of stuff. Yes, that happens more when you grow older.

Just a few of the young Muslims specifically referred in the interviews to the 9/11 2001 attacks as a reason for their religion gaining importance. Many of them were still young when the attacks happened. The 23-year-old Çelik, was 17 years

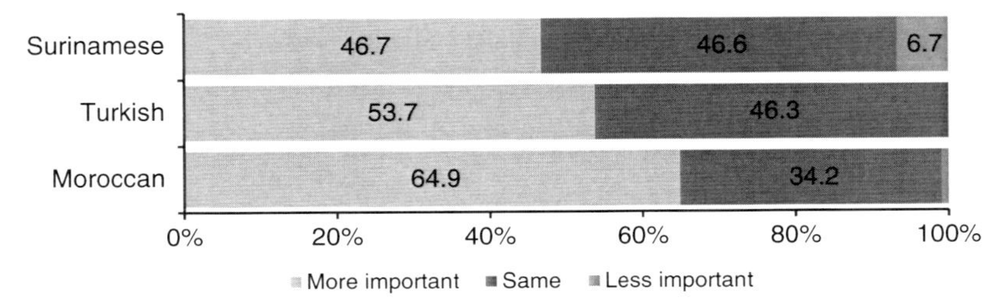

Fig. 52.1 Reported change in importance of religion among Amsterdam youth, by ethnic background, according to the survey question: Over the past years, have your religious beliefs become a lot more important, more important, more unimportant, a lot more unimportant or have they remained of equal importance to you? (Source: Author's survey 2007)

old in 2001; he stressed the influence of 9/11, but for him it was also a process of thinking about what he wants in life, of growing to maturity:

> I think that especially after 9/11 many young people started reflecting. What is happening here? What is this? [...] I had a confusing period, not such a long time ago. I was tending towards atheism. But, and this sounds as a cliché, you start questioning yourself, what do I believe in? You learn more about other religions and ideologies. And for me, personally it led to the Islam.

52.5.2 Importance of Belonging to Religious Group and Territorial Identity Strategies

All young adults strongly identified with Amsterdam. They felt they belonged in this city, they felt at home and they felt "Amsterdammer" (Van der Welle and Mamadouh 2009; Van der Welle 2011). In fact, most of them primarily identified themselves as "Amsterdammer" but not as Dutch. There was no difference between religious affiliations regarding their identification with the city, but being Muslim did affect the identification with Netherlands. A considerable share of the Muslim youth felt discriminated against, based on their religious beliefs on a regular basis. This was one of the reasons they felt excluded from identifying as Dutch. Sometimes, they had the feeling they were forced to choose between feeling Dutch and feeling Muslim. For them, these identities were not incompatible, but they felt that the society around them assumed they were, as the 22-year old Meliha of Turkish origin explained:

> I feel Dutch. So I am that. But I am of Turkish origin and I am a Muslim. I just am. You know, forget about Turkish origin, I am Muslim, but also Dutch (Hollander). I cannot choose, because I love my faith and I love Holland. Easy as that. I hope that it will not get so far, that you really have to choose. [...] But it already begun. I experience it already among my acquaintances. They feel they can no longer stay here, you are not a human being anymore, You are not respected anymore, because you are Muslim.

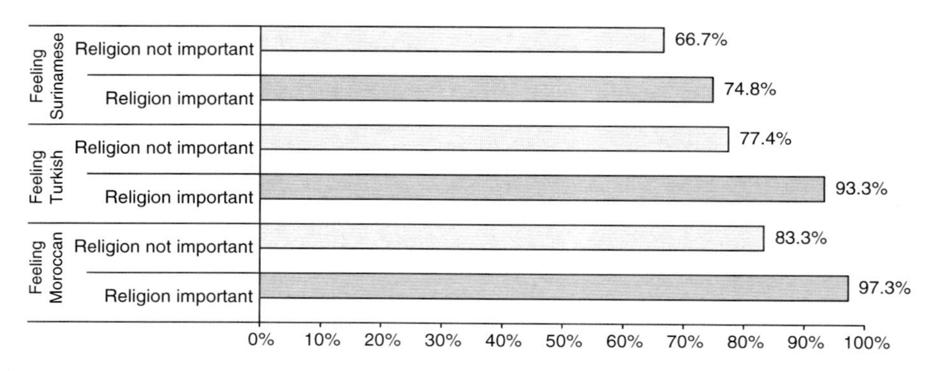

Fig. 52.2 Ethno-national identification by religiosity (Source: Author's survey 2007)

Religion did play a part when it came to ethno-national identification, but not for all second generation respondents. Surinamese Christians more often identified themselves as Surinamese (94 %), than non-Christians (76 %). For example, the Hindu Surinamese more strongly identified with India and as Hindustani. They did not describe themselves as Surinamese, as, for example, the 30-year-old Anja, Hindu and of Surinamese origin. For her there is a huge difference between Creoles and Hindustani, and she explained:

> Yes, what is my relationship with Suriname? To be honest, I feel no affinity for Suriname. I feel more affinity for India.
>
> Interviewer: Because of Hinduism?
>
> Yes, that is my culture, my identity even. I don't feel Surinamese. This is mainly because Surinamese are always seen as one of a kind and that is annoying. That really annoys me.

Almost all Moroccan and Turks considered themselves to be Muslim. There was, however, a difference between them in the importance they attached to belonging to a religious group. Respondents who attached great importance to belonging to a religious group, more often identified as Moroccan or Turk (Fig. 52.2). Besides, they more often felt more Moroccan or Turk than Dutch or only Moroccan or Turk. The 23-year-old Kadir, of Turkish origin had been brought up as Muslim but was not practicing anymore and felt he had been "Hollandized". This caused problems with his parents. He did not feel Turkish and explained:

> I always was the black sheep of the family. I am of Turkish origin, but it does not really fit. I am too "Hollandized" so to speak. The whole Dutch culture has been drilled into me, and that clashed with Turkish culture at home.
>
> Interviewer: Hollandized?
>
> Yes, I call myself the worst Muslim of the Netherlands. [...] Fasting? You will never catch me doing that. I am not attending the mosque; I do not believe in God, Allah or whatever you will call it.

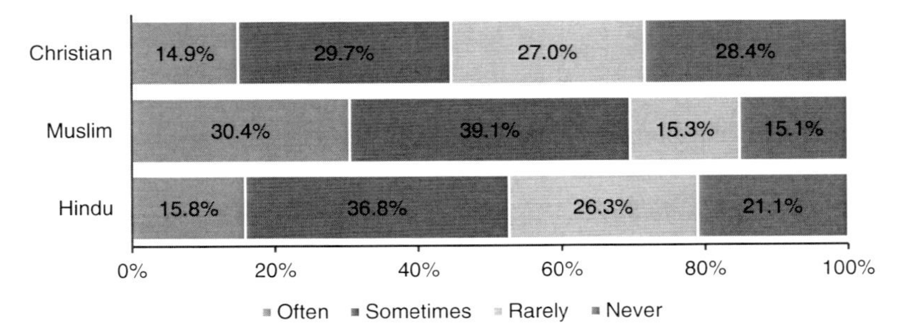

Fig. 52.3 Religious identity in the public sphere, by religious affiliation, according to response to the question: How often were you called to account for your religion over the past years? (Source: Author's survey 2007)

52.5.3 Attention for Religion, Religion and the Public Sphere

Because of the enormous amount of attention paid to the Islam in Dutch public debates over the last 10 years, the Muslim youth often felt they had to explain or defend their religious practices. The young Muslims were more often called to account for their religious affiliation, than the young Christians or Hindu (Fig. 52.3). This increased their awareness of their religious identity, or of being labeled Muslim by others.

The majority of the religious young adults reacted positively to being asked about their religion by others. Especially the Muslim youth appreciated it when people asked them about their faith. The 29-year-old woman Fidan, Muslim of Turkish background, expressed her feelings about this as follows:

> I think it has become more difficult to talk about Islam, because it is associated with extremism. [..] People think, "oh yeah it is wrong, it is scary." Then I prefer to be asked directly.
>
> Interviewer: And, do they?
>
> Sometimes they do. [...] And I always show that I appreciate it, even so when questions are harsh or confronting. [...] I feel it is the only way to break down prejudices. Because often they will tell you: "hey, you don't match the image I had in mind about Muslims or Turks." And you only have to do tiny things to accomplish this. Often people lead such isolated lives. They only read the paper and watch television. In their daily lives they do not meet Turks or Muslims.

Although there was a lot of discussion about radicalization and religious conservatism within the Islamic population in the Netherlands, when asked about tensions between groups in the city, our respondents did not frame these primarily in religious terms. A majority of them agreed there are tensions between groups and many felt that tensions had increased over the last years. However, they mainly mentioned tensions between native Dutch and other ethnic groups. Tensions between religious groups, or between Muslims and non-Muslims were mentioned a lot less.

52.6 Conclusion

Even though Amsterdam could be characterized as a secular city, that does not mean religion is invisible in the everyday life of its inhabitants. People dressed up going to church on Sundays, ladies wearing colorful headscarves, men in djellabas going to the mosque on Fridays, are common in the Amsterdam scene. A large share of young inhabitants in cities consider themselves religious. There is however a difference in religiosity between ethnic groups, for example between the young adults of Moroccan, Turkish, Surinamese and Dutch origin in our study. Almost all the young adults of Moroccan and Turkish origin feel affiliated with religion, compared to just a small share of the young adults of Dutch. The young adults of Surinamese origin take a middle position.

In any event, organized religion is not a site of integration in Amsterdam, because those of Dutch origin are in great numbers not religious at all, and the main new religion Islam is mainly organised along ethno-national (and linguistic) lines. And besides, affiliation with religion for many young adults does not necessarily mean that they are attending the church, mosque or temple on a regular basis. Many of them stress their individual relationship with faith.

In the Dutch context, Muslim is possibly more an ethnic identity than a religious one, like Moroccan is more an ethnic identity than a territorial one (Van der Welle and Mamadouh 2009; Van der Welle 2011). For the young Moroccans and Turks, religion and the Turkish or Moroccan identity are strongly interlinked. Those who attach more importance to belonging to a religious group, are more likely to more strongly or exclusively identify as Turk or Moroccan. Besides, for a part of them it is difficult to make a distinction between culture and religion. The Dutch public debate has increased their awareness of their religious as well as ethnic identity. There is also a link between religious identity and Surinamese identity. The Surinamese Christians more often identified themselves as Surinamese, than non-Christians. Part of the Hindu Surinamese more strongly identified with India and as Hindustani, than with Surinam and the Surinamese.

All of the young adults strongly identify with the city of Amsterdam, but not all necessarily feel Dutch. A considerable share of the Muslim youth felt discriminated against on a regular basis, based on their religious beliefs. This was one of the reasons they felt excluded from identifying as Dutch. The Muslim youth also were more often called to account for their religious affiliation. The young adults with other religious affiliations hardly felt excluded based on their religious beliefs.

Five years later the public debate about the position of Islam and Muslims in the Netherlands is still on-going. The prevalence of the PVV is limited in Amsterdam (9.4 % in 2010, against 15.5 % of the national average, but still much higher than in 2006). The geopolitical context changed dramatically with the election of Barack Obama at the American presidency, which was perceived by many immigrants in western Europe as a sign of hope for non-white citizens in the rest of the western World. Job Cohen, the former major was called to lead the list of the Labour Party for the 2010 national elections but failed to obtain a sufficient plurality to conduct a coalition government. He left national politics in 2012. Wilders's PVV, by contrast enlarged its share of votes, became the third party in size, and with his conditional support became the necessary partner of a minority government led by Mark Rutte. As a result he was in the position to continue his campaign against immigrants, but he since diversified his targets with the immigrants from the new EU Member States (Poland and Bulgaria). More surely, however, religion was partially displaced from the spotlights by the U.S. mortgage crisis and the subsequent banking, economic and sovereign debt crises. Social economic issues regained prominence upon identity politics in the public debate. It would warrant a new survey to assess whether and how this has affected the religious and territorial identification of Amsterdam youth, but organized religion is unlikely to provide an avenue to integration in this secular city.

References

Bader, V. M. (2007). *Secularism or democracy? Associational governance of religious diversity*. Amsterdam: Amsterdam University Press.

Bernts, T., Dekker, G., & Hart, J. D. (2007). *God in Nederland 1996–2006*. Kampen: Uitgeverij Ten Have.

Bolkestein, B., & Arkoun, M. (1994). *Islam en de democratie: Een ontmoeting*. Amsterdam: Contact.

Bolkestein, B., & Von der Fuhr, G. (1997). *Moslim in de polder: Frits Bolkestein in gesprek met Nederlandse moslims*. Amsterdam: Contact.

Booi, H., Laan, N., Lindeman, E., & Slot, J. (2011). *Diversiteits- en integratiemonitor 2010*. Amsterdam: Gemeente Amsterdam/Dienst Onderzoek en Statistiek.

Brinks, J. H. (2006). Les Pays-Bas entre islam et populisme. *Politique étrangère, 70*(3), 587–598.

Brinks, J. H. (2010). Fragilités du pacte démocratique aux Pays-Bas: Geert Wilders et l'islam. *Politique étrangère, 4*, 899–911.

Buruma, I. (2006). *Murder in Amsterdam; The death of Theo van Gogh and the limits of tolerance*. New York: Penguin Press.

Crul, M., & Doomernik, J. (2003). The Turkish and Moroccan second generation in the Netherlands: Divergent trends between and polarization within the two groups. *International Migration Review, 37*(4), 1039–1064.

Crul, M., & Heering, L. (2008). *The position of the Turkish and Moroccan second generation in Amsterdam and Rotterdam. The TIES study in the Netherlands*. Amsterdam: Amsterdam University Press.

De Jong, R., Van der Kolk, H., & Voerman, G. (2011). *Verkiezingen op de kaart 1848–2010: Tweede Kamerverkiezingen vanuit geografisch perspectief*. Utrecht: Uitgeverij Matrijs.

Deben, L., Heinemeijer, W., & Van der Vaart, D. (Eds.). (2000). *Understanding Amsterdam; Essays on economic vitality, city life and urban form*. Amsterdam: Het Spinhuis.

Dijkink, G., & Van der Welle, I. (2009). Diaspora and sovereignty: Three cases of public alarm in the Netherlands. *Tijdschrift voor Economische en Sociale Geografie, 100*(5), 623–634.

Doomernik, J., & Knippenberg, H. (Eds.). (2003). *Migration and immigrants: Between policy and reality; A volume in honor of Hans van Amersfoort*. Amsterdam: Aksant.

Fennema, M. (2010). *Geert Wilders. Tovenaarsleerling*. Amsterdam: Bakker.

Fennema, M., Tillie, J., Van Heelsum, A., Berger, M., & Wolff, R. (2000). *Sociaal kapitaal en politieke participatie van etnische minderheden*. Amsterdam: IMES.

Fortuijn, P. (1997). *Tegen de islamisering van onze cultuur: Nederlandse identiteit als fundament*. Utrecht: Bruna.

Fortuijn, P. (2001). *De islamisering van onze cultuur: Nederlandse identiteit als fundament (met een kritische reaktie van Abdullah R.F. Haselhoef)*. Uithoorn: Karakter Uitgevers.
Gemeente Rotterdam, & Gemeente Amsterdam. (2012). *De staat van integratie Rotterdam Amsterdam*. Rotterdam/Amsterdam: Gemeente Rotterdam/Gemeente Amsterdam.
Hirsi Ali, A. (2005). *Insoumise*. Paris: Laffont.
Hirsi Ali, A., & Wilders, G. (2003). Liberale Jihad NRC Handelsbald, 12 Apr 2003.
Kaal, H. (2011). Religion, politics, and modern culture in interwar Amsterdam. *Journal of Urban History, 37*(6), 897–910.
Kasinitz, P., Mollenkopf, J. H., Waters, M. C., & Holdaway, J. (2008). *Inheriting the city. The children of immigrants come of age*. New York: Russell Sage Foundation.
Kennedy, J. C. (1995). *Nieuw Babylon in aanbouw: Nederland in de jaren zestig*. Amsterdam: Boom.
Kennedy, J. (2010). Globalization, the nation-state and religious newcomers. Reflections on two countries. In E. Sengers & T. Sunier (Eds.), *Religious newcomers and the nation state. Political culture and organized religion in France and the Netherlands* (pp. 155–166). Delft: Eburon.
Knippenberg, H. (1997). Dutch nation-building: A struggle against the water? *GeoJournal, 43*(1), 27–40.
Knippenberg, H. (2002). Assimilating Jews in Dutch nation-building: The missing 'pillar'. *Tijdschrift voor Economische en Sociale Geografie, 93*, 191–207.
Knippenberg, H. (2005). The Netherlands, selling churches and building mosques. In H. Knippenberg (Ed.), *The changing religious landscape of Europe* (pp. 88–106). Amsterdam: Het Spinhuis.
Knippenberg, H. (2006). The changing relationship between state and church/religion in the Netherlands. *GeoJournal, 67*(4), 317–330.
Knippenberg, H., & De Pater, B. (1990). *De eenwording van Nederland* (2nd ed.). Nijmegen: SUN.
Kuitenbrouwer, J. (2010). *De woorden van Wilders & hoe ze werken*. Amsterdam: De Bezige Bij.
Landman, N. (1992). *Van mat tot minaret. De institutionalisering van de islam in Nederland*. Amsterdam: VU Uitgeverij.
Lijphart, A. (1968). *The politics of accommodation: Pluralism and democracy in the Netherlands*. Berkeley: University of California Press.
Lucardie, P. (2007). Rechts-extreem, populistisch of democratisch patriottisme. Opmerkingen over de politieke plaatsbepaling van de Partij voor de Vrijheid en Trots op Nederland. Retrieved May 22, 2012, from http://dnpp.eldoc.ub.rug.nl/FILES/root/jb-dnpp/jb07/JB07LucardieDEF1.pdf
Lucardie, P. (2008). The Netherlands: Populism versus pillarisation. In D. Albertazzi & D. McDonnell (Eds.), *Twenty-first century populism: The spectre of Western European democracy* (pp. 151–165). Basingstoke: Palgrave Macmillan.
Lucassen, J., & Penninx, R. (1997). *Newcomers, immigrants and their descendants in the Netherlands, 1550–1995*. Amsterdam: Het Spinhuis.
Lucassen, L., Feldman, D., & Oltmer, J. (Eds.). (2006). *Paths of integration: Migrants in western Europe (1880–2004)*. Amsterdam: Amsterdam University Press.
Mak, G. (2001). *Amsterdam: A brief life of the city*. London: Harvill Press.
Mamadouh, V. (1992). *De stad in eigen hand, Provo's, kabouters en krakers als stedelijke sociale bewegingen*. Amsterdam: Sua.
Mamadouh, V. (2001). Constructing a Dutch Moroccan identity through the World Wide Web. *The Arab World Geographer, 4*(4), 258–274.
Mamadouh, V. (2003). 11 September and popular geopolitics: A study of websites run for and by Dutch Moroccans. *Geopolitics, 8*(3), 191–216.
Mamadouh, V. (2008). After Van Gogh: The geopolitics of the tsunami relief effort in the Netherlands. *Geopolitics, 13*(2), 205–231.
Mamadouh, V. D., & Van der Wusten, H. (2004). Eindstand van een diffusieproces: het geografisch patroon van de steun voor de LPF. In G. Voerman (Ed.), *Jaarboek 2002, Documentatiecentrum Nederlandse Politieke Partije* (pp. 181–205). Groningen: DNPP.

Mamadouh, V., & Van der Wusten, H. (2012). "Ceci n'est pas un parti": Le véhicule fantôme de l'anti-islamisme de Geert Wilders. *Hérodote, 144*, 113–121.
Maussen, M., & Bogers, T. (2010). Tolerance and cultural diversity discourses in the Netherlands (ACCEPT PLURALISM Working Paper 11/2010). Florence: European University Institute.
Meurs, P. (2007). *Identificatie met Nederland*. Amsterdam: Amsterdam University Press.
Musterd, S., & Salet, W. (Eds.). (2003). *Amsterdam human capital*. Amsterdam: Amsterdam University Press.
Özüekren, S., & Van Kempen, R. (Eds.). (1997). *Turks in European cities: Housing and urban segregation*. Utrecht: ERCOMER (European Research Centre on Migration and Ethnic Relations, Utrecht University.
Pels, D. (2003). *De geest van Pim. Het gedachtegoed van een politieke dandy*. Amsterdam: Anthos.
Penninx, R., Schoorl, J., & Van Praag, C. (1994). *The impact of international migration on receiving countries: The case of The Netherlands*. Den Haag: NIDI.
Phalet, K., & Ter Wal, J. (Eds.). (2004). *Moslim in Nederland, Een onderzoek naar de religieuze betrokkenheid van Turken en Marokkanen*. Den Haag: Sociaal Cultureel Planbureau.
Scheffer, P. (2000). Het multiculturele drama. *NRC Handelsblad*, 29 Jan 2001.
Scheffer, P. (2007). *Het land van aankomst*. Amsterdam: De Bezige Bij.
Sengers, E. (Ed.). (2005). *The Dutch and their gods: Secularization and transformation of religion in the Netherlands since 1950*. Hilversum: Verloren.
Uitermark, J., & Gielen, A. J. (2010). Islam in the spotlight: The mediatisation of politics in an Amsterdam neighbourhood. *Urban Studies, 47*(6), 1325–1342.
Uitermark, J., & Hajer, M. (2008). Performing authority: Discursive politics after the assassination of Theo van Gogh. *Public Administration, 86*(1), 5–20.
Van der Donk, W. B. H. J., Jonkers, A. P., Kronjee, G. J., & Plum, R. J. J. M. (Eds.). (2006). *Geloven in het publiek domein; Verkenningen van een dubbele transformatie*. Amsterdam: Amsterdam University Press.
Van der Steenhoven, P. (2001). *Geloven in Amsterdam. Factsheet, 5, June 2001*. Amsterdam: O+S.
Van der Valk, I. (2003). Political discourse on ethnic minority issues; A comparison of the right and the extreme right in the Netherlands and France (1990–97). *Ethnicities, 3*(2), 183–213.
Van der Veer, P. (2006). Pim Fortuyn, Theo van Gogh, and the politics of tolerance in the Netherlands. *Public Culture, 18*(1), 111–124.
Van der Waal, J., Koster, W. D., & Achterberg, P. (2011). Stedelijke context en steun voor de PVV; Interetnische nabijheid, economische kansen en cultureel klimaat in 50 Nederlandse steden. *Res Publica, 53*(2), 189–207.
Van der Welle, I. (2011). *Flexibele Burgers? Amsterdamse jongvolwassenen over lokale en nationale identiteiten*. Amsterdam: Off Page.
Van der Welle, I., & Mamadouh, V. (2009). Territoriale identiteiten en de identificatiestrategieën van Amsterdamse jongvolwassenen van buitenlandse afkomst: over evenwichtkunstenaars en kleurbekenners. *Migrantenstudies, 25*(1), 24–41.
Van Gent, W., & Musterd, S. (2012). Les transformations urbaines et l'émergence des partis populistes de la droite radicale en Europe. *Hérodote, 144*, 99–112.
Van Rooden, P. (2002). Long-term religious developments in the Netherlands, 1750–2000. In H. McLeod & W. Ustorf (Eds.), *The decline of Christendom in Western Europe, 1750–2000* (pp. 113–129). Cambridge: Cambridge University Press.
Vermeulen, H., & Penninx, R. (Eds.). (2000). *Immigrant integration. The Dutch case*. Amsterdam: Het Spinhuis.
Vink, M. P. (2007). Dutch 'multiculturalism' beyond the pillarisation myth. *Political Studies Review, 5*, 337–350.
Vossen, K. (2008). Van Bolkestein via Bush naar Bat Ye'Or. De ideologische ontwikkeling van Geert Wilders. Retrieved May 22, 2012, from http://dnpp.eldoc.ub.rug.nl/FILES/root/jb-dnpp/jb08/KoenVossenartikelJB2008.pdf

Vossen, K. (2009). Hoe populistisch zijn Geert Wilders en Rita Verdonk? Verschillen en overeenkomsten in optreden en discours van twee politici. *Res Publica, 51*(4), 437–466.
Vossen, K. (2010). Populism in the Netherlands after Fortuyn: Rita Verdonk and Geert Wilders compared. *Perspectives on European Politics and Society, 11*(1), 22–38.
Wilders, G. (2005). *Onafhankelijksverklaring*. La Haye: Groep Wilders.
Wintle, M. (2007). Pillarisation, consociation and vertical pluralism in the Netherlands revisited: A European view. *West European Politics, 23*(3), 139–152.

Chapter 53
Religiosity in Slovakia After the Social Change in 1989

René Matlovič, Viera Vlčková, and Kvetoslava Matlovičová

53.1 Introduction

The Slovak Republic is a relatively young country in Central Europe, which was established on 1 January 1993 after the peaceful division of Czechoslovakia. The development of religiosity and religion in Slovakia is closely related to the country's changing geopolitical, social and cultural context. Socio-political and economic changes that began in 1989 created a new situation for religious development. This fact was reflected in increased interests of geographers and other social scientists about religiosity from various perspectives (see Matlovič 2011). In our contribution we outline the religious situation in contemporary Slovak society. We will also briefly evaluate religious developments of the population in the last two decades and their spatial differentiation in the self-governing regions (SGR).[1] The final section includes results from a survey of university students about their religiosity and attitudes toward contemporary religion.

[1] There are eight self-governing regions in Slovakia: Bratislava, Trnava, Trenčín, Nitra, Žilina, Banská Bystrica, Prešov and Košice.

R. Matlovič (✉) • K. Matlovičová
Department of Geography and Applied Geoinformatics, Faculty of Humanities and Natural Sciences, University of Prešov, Ul. 17. Novembra 1, 080 01 Prešov, Slovakia
e-mail: rene.matlovic@unipo.sk; kveta.matlovicova@gmail.com

V. Vlčková
Department of Public Administration and Regional Development, Faculty of National Economy, University of Economics in Bratislava, Dolnozemská cesta 1, 852 35 Bratislava, Slovakia
e-mail: viera.vlckova@euba.sk

S.D. Brunn (ed.), *The Changing World Religion Map*,
DOI 10.1007/978-94-017-9376-6_53

53.2 The Conceptual Framework

In previous periods, especially since the second half of the nineteenth century, we witnessed a growing conviction of the intellectual elite that religion was less important in daily life because of economic and social modernization. On the contrary with the pre-modern era, religion should have become a less important social factor, whose original functions should have been replaced by other social subsystems. K. Marx, É. Durkheim and M. Weber were concerned about the gradual decline of religion (Giddens 1999: 419). Religion would have more likely been considered a private matter of individual psychology or free-time activities (Lužný 1999: 137). This stream of thought was represented by B. R. Wilson, T. F. O'Dea, T. Luckmann et al. (Lužný 1999). To some extent, these views are also included in the Fukuyama's theory about the end of history and the triumph of liberal democracy over religion in Europe (Fukuyama 2002: 260). The author of this thesis admits that topics which keep religion on the European political scene still exist (for example, problems of induced abortion and euthanasia), mainly due to the action of strong Christian Democratic Parties (Matlovič 2005). Development in the last quarter of the twentieth century significantly disrupted these concepts. The criticism of the secularization paradigm, which was blamed by critics (R. Stark, R. Finke, L. Iannaccone) for its Eurocentric and Christian-centric focus, was more intensive (Podolinská 2008: 54). It was related to the manifestation of religious revival and the role of religion in society in different parts of the world. New situations in Europe led to a review of the previous opinions of several influential theorists who supported the secularization paradigm, for example, P. L. Berger, (pointed out by T. Podolinská 2008: 54). G. Davie (2007) in this context limits the validity of the secularization paradigm in the period of industrial modernity (Podolinská 2008: 56). Desecularization and revival of the religion role in the world are reflected in a broad range of theoretical concepts. G. Kepel (1994) names it *God's vengeance* (*la Revanche de Dieu*) and J. Casanova (1994) call it *religion deprivation*. According to D. Lužný (1999: 114), religions are adapted to the conditions of modern society. Religions respect and use their basic principles (the right to privacy and freedom of religion) and enter the public domain. This process take many forms, for example, religions act to protect traditional values, distrust the absolute legal and moral autonomy of the secular sphere and emphasize the principles of the common good opposite to liberalism which is preferred over individual needs (Lužný 1999: 112–113). S. Huntington (2001) also explained the factors of the global phenomenon of religion revival. He identified not only specific factors but also general factors, where people need to obtain a new foundation for their identity and for new moral laws, which would provide them a meaning, purpose and goal of life, were included. Paradoxically, the reason for religious revival was the process of modernization and secularization, during which the traditional sources of identity and authority were disputed. People migrating from rural to urban areas, were uprooted from their roots. They came into contact with many representatives of other cultural traditions and were exposed to new types of interpersonal relationships. According to R. Stark and W. S. Bainbridge

(1987), secularization is self-limiting process that provokes two opposing tendencies. On the one hand, it is the revival of the existing religious traditions and on the other hand, it stimulates experimentation and the formation of new religious traditions (Matlovič 2005: 317).

A special case of this development was a revival of religion and religiosity in some post-communist countries after the collapse of communist regimes at the turn of 1980s and 1990s. This change was noticed by several authors, for example, M. Tomka (2006), T. Podolinská (2008). Religion and religious institutions in several countries were directly involved in society's transition from the authoritarian and totalitarian regimes to democracy (for example, Poland, Romania and to some extent also in Slovakia[2]). It was highly connected with a realized atheistic policy during the communist regime (1948–1989), for example, persecution of the clergy, property confiscation, the abolition of religious schools, destruction of the Greek Catholic Church, etc. The revival of religious life in post-communist countries has, however, a complex and differentiated character that has been affected by a number of factors, including the pre-communist past, the way how to deal with the communist past (for example, persistent distrust of the Church, lower level of emerging and new religious movements, lower rate of stability of post-transformation conversions) (Podolinská 2008: 60). It is being shown that in studying religion, it is necessary to pay attention to generational changes and differences between urban and rural environment in regards to education, social strata and economic status (Tomka 2000 in Podolinská 2008: 62).

In the context of the above mentioned theoretical concepts, religious development and its forms in Slovakia after 1989 can be interpreted as the result of interference of often contradictory factors. From general factors, there is a process of secularization, demonstrating diversion from religion on the one hand and a process of deprivatization of religion, demonstrating by revival of indigenous religious traditions as well as experimentation and the formation of new forms of religious life. From the specific factors, it is necessary to mention the revitalization of religious life after the restoration of religious freedoms in post-communist countries, which is demonstrated by an increased in the number of priests, the restoration of religious education, the sacralisation of the area (construction of churches, tabernacles and other religious infrastructure), the creation of new parishes, the reforms of church territorial organization (creation of new parishes, dioceses and ecclesiastical provinces), the revival of traditional and formation of new pilgrimage places and a new adjustment of relationship between church and state (Matlovič 2000, 2001). In the next section, we will seek to determine whether this wave of religious revival continues and how it is evident in the current generation of young Slovak university students.

[2] On 25 March 1988, in the Slovak capital city Bratislava, the demonstration for religious freedom and human rights took place, organized by a group of activists from the secret Catholic Communities which were persecuted by communist power. They required to appoint the Catholic bishops for vacant Slovak dioceses and to respect human rights including religious freedom.

53.3 Methodology and Data

Religion development can be measured using a number of indicators. In Slovakia, widely used indicators are the religiosity rate (the percentage of the religious people in the entire population) and the religious structure of the population, which are available in population censuses in 1991, 2001 and 2011. Additional information is provided in various sociological surveys carried out in inter-census periods. The third source of quantitative data is internal data of churches and religious societies themselves. Another way of examining religiosity development is through the development of religious institutions, the production of sacred space, the development of religious education and pilgrimage places and also the relations between church and state.

At this point, it is necessary to note that interpretation of data from the census becomes quite problematic. This is particularly the case using census data from 1991 to 2011. In the census in 1991, a significant part of the population (17.4 %) did not declare any religious affiliation. This was probably because only a short period that had elapsed since the changes in society in 1989; many people were probably reluctant to express their religion for fear of possible re-persecution by state authorities. In the census in 2001, the situation changed dramatically. Only 3 % of population did not declare their religious affiliation. The latest census was held in 2011. Again, however, there were problems declaring religiosity and religious affiliation. 10.6 % of population did not declare any religious affiliation. In this case, the census was distrustful because of conflict between the two state institutions.[3] These facts complicated the interpretation of the development trends in the religious structure of the Slovak population between 1991 and 2011.

Another source of data about religiosity are sociological surveys that were carried out by different institutions. It can be concluded that using this kind of data often results in a discrepancy between the statistical data about religiosity from censuses and actual religiosity that is declared through an active participation of people in religious life.[4] One of the first such studies after 1989 was work of T. Houston, R. Worthing-Davies and R. Russell (1992), which revealed that sincere faith was admitted by only 50 % of Slovak citizens, although in the census in 1991, 72.8 % of the population declared a particular religion. Complementary data sources are the internal database of churches and religious societies. In our study, we implemented own data collection. It was a survey conducted by questionnaire (25 questions) of Slovak students at the universities in Bratislava, Nitra and Prešov. In total, we contacted 600 respondents from ages 19–25 years. It was a survey, where research sample was obtained by accidental sampling. In the final analysis, 509 fully completed questionnaires were used in the analysis below. At this point, it should be

[3]The Office for Personal Data Protection distrusted the anonymity of the census with regard to the application of bar code identifiers. The Slovak Statistical Office rejected this criticism. This bizarre conflict discouraged many people from mandatory participation in the census or unfair manipulation with identifiers, which greatly slowed the processing of the census.

[4]At this point, it should be noted that the results are often influenced by the methodology of data collection and the way of asking questions.

noted that previous views about religion among university students in Bratislava were studied by M. Kováč and T. Krúpova (1993).

53.4 The Development of Religiosity and the Religious Structure of the Population of Slovakia After 1989

Based on data the aforementioned censuses and sociological surveys, it is possible to follow the development of religiosity and religious structure of the population of Slovakia after 1989. M. Tížik (2010) applied a more detailed overview of the results of the religiosity examination using various sociological surveys (Table 53.1). They were ISSP (International Social Survey Programme), CPS (Citizenship and Participation in Slovakia), EVS (European Values Study), PCE (Political Culture in the Central Europe), WVS (World Values Survey) and the ESS (European Social Survey).[5]

The overview (see Table 53.1) shows that the religiosity rate in the last two decades is oscillating between 71.6 and 86.8 %. Due to the different kind of methodologies

Table 53.1 The religiosity rate in Slovakia, 1990–2011 according to censuses and sociological surveys

Survey	Year	Religiosity rate (%)
WVS	1990	79.2
EVS	1991	71.6
Census	1991	72.8
ISSP	1992	86.2
ISSP	1996	79.4
ISSP	1998	83.5
WVS	1998	86.8
EVS	1999	76.8
PCE	2000	71.9
Census	2001	84.1
ISSP	2004	82.1
ESS	2004	74.4
ISSP	2005	85.5
ESS	2006	75.4
EVS	2008	75.3
ISSP	2008	80.0
ESS	2008	72.9
CPS	2008	72.7
ISSP	2009	77.2
Census	2011	76.0

Data source: Tížik (2010), www.infostat.sk/vdc/pdf/DDP68.pdf

[5] For more information about these surveys can be found on the Slovak Archive of Social Data (www.sasd.sav.sk).

Table 53.2 Religious structure of the Slovakia population, 1991–2011

Church and religious community	Number of confessors (1991)	Number of confessors (2001)	Number of confessors (2011)
Roman Catholic Church	3,187,383	3,708,120	3,347,277
Greek Catholic Church	178,733	219,831	206,871
Old Catholic Church	882	1,733	1,687
Orthodox Church	34,376	50,363	49,133
Lutheran Church in Slovakia	326,397	372,858	316,250
Reformed Christian Church/Calvinists	82,545	109,735	98,797
Evangelical Methodist Church	4,359	7,347	10,328
Baptist Union	2,465	3,562	3,486
Brethren Church	1,861	3,217	3,396
Seventh-day Adventist Church	1,721	3,429	2,915
Apostolic Church in Slovakia	1,116	3,905	5,831
Catholic corps in Slovakia	700	6,519	7,720
Czechoslovak Hussite Church	625	1,696	1,782
Religious Society of Jehovah's Witnesses	10,501	20,603	17,222
Central Union of Jewish Religious Communities	912	2,310	1,999
Baha'i community	–	–	1,065
Church of Jesus Christ of Latter-day Saints	91	58	972
New Apostolic Church	188	22	166
Other Churches	6,094	6,214	23,340
Undetected	917,835	160,598	571,437
Non-believers	515,551	697,308	725,362
Total	**5,274,335**	**5,379,455**	**5,397,036**

Data source: Statistical Office of the Slovak Republic, http://portal.statistics.sk/files/tab-15.pdf, 1.5.2012

of the surveys, from these data is not possible to formulate firm conclusions about the trends of religiosity in Slovakia.

Based on population and housing census result in 1991, 2001 and 2011, it is possible to identify several basic trends in the religiosity development of the Slovak population (Tables 53.2 and 53.3):

(A) A decrease in the number and proportion of people who did not express their religious orientation in 1991–2001 and re-growth in 2001–2011. While in 1991, there were 917,865 who did not answer the questions about religiosity and confessional affiliation in the census; this amounted to 17.40 % of the total

Table 53.3 Main religious groups in Slovakia, 1991–2011

Church and religious community	Percent of confessors (1991) (%)	Percent of confessors (2001) (%)	Percent of confessors (2011) (%)
Roman Catholics	60.4	68.9	62.0
Greek Catholics	3.4	4.1	3.8
Orthodoxes	0.7	0.9	0.9
Lutherans	6.2	6.9	5.9
Calvinists	1.6	2.0	1.8
Others	0.5	1.2	1.6
Undetected	17.4	3.0	10.6
Non-believers	9.8	13.0	13.4
Total	**100.0**	**100.0**	**100.0**

Data source: http://portal.statistics.sk/files/tab-15.pdf, 1.5.2012

number of population while in 2001 it was only 160,598 inhabitants or 2.98 % of the Slovak population, but in 2011 it was 571,437 inhabitants or 10.6 % of the total population. It is shown that just after the social changes in 1989, in the first census in 1991 people were reluctant to comment on the issue of religion (perhaps for fear of the communist regime return, which had persecuted religious people). In 2001, those concerns had disappeared. From the growth in the number of people who did not express their religious orientation in 2011, it is possible to relate those numbers to the negative impact of conflicts in state institutions regarding the organization of the census as well as increasing public sensitivity to privacy issues supported by wide media coverage of this topic.

(B) The increase in the number and proportion of the religious people of the total number of population in Slovakia in 1991–2001 and a subsequent decrease from 2001 to 2011. In 1991, 3.8 million registered as religious people (72.8 %), in 2001 it was 4.5 million (84.1 % of the total number of inhabitants) and in 2011 it was 4.1 million (76.0 %). The number was less because of growing number of people expressing no religious preference. It is of interest that the decrease in the absolute number of religious members was recorded in all traditional and larger churches.

(C) A slight increase in the number and proportion of people without religion of the total number of population in Slovakia. In 1991, 515,551 (9.77 %) were identified without religion, in 2001 it was 697,308 people (12.96 % of the total number of population) and in 2011, 725,362 (13.4 % of the total number of population).

We can gain also another picture through the analysis of "pure" confessional structure of the population, that is, when we exclude people without religion and those undetected in the census. Only this approach allows us to identify real changes in the proportions between the individual religions in the last decade. We can identify declines in the two most significant and traditional religions (Table 53.4).

The proportion of the Roman Catholics in the total number of believers was decreased from 82.98 % in 1991 to 82.01 % in 2001 and 81.64 % in 2011, that is, by 1.34 % decline from 1991 to 2001. A smaller decrease was noticed in Lutheran

Table 53.4 Confessional structure of the religious people in Slovakia, 1991–2011

Church and religious community	Percent of confessors (1991) (%)	Percent of confes-sors (2001) (%)	Percent of confessors (2011) (%)
Roman Catholics	82.98	82.01	81.64
Greek Catholics	4.65	4.86	5.05
Orthodoxs	0.89	1.11	1.20
Lutherans	8.50	8.25	7.71
Calvinists	2.15	2.43	2.41
Other Churches	0.82	1.34	2.00
Total	**100.00**	**100.00**	**100.00**

Data source: http://portal.statistics.sk/files/tab-15.pdf

Church (from 8.5 to 7.71 %), that is, by 0.79 % in the two decades. At first, the position of Reformed Christians/Calvinists improved (from 2.15 to 2.43 %), by 0.28 %, but in 2001–2011 it was slightly weaker (from 2.43 to 2.41 %). On the other hand, the position in the confessional structure of the population of Slovakia was stronger in the Greek Catholics (an increase of 0.4 %), Orthodox (0.31 %) and members of other smaller churches and religious communities (by 1.18 %) (see Table 53.4). These results outline the confirmation of the trend of increased experimentation and the formation of new religious traditions, observations mentioned by R. Stark and W. S. Bainbridge (1987). It is also clearly demonstrated by increases in the number of numbers of other churches, which do not have their foundation in the historical traditions of Christianity. These processes in Slovakia have been closely examined by M. Tížik (2006).

53.5 Spatial Differentiation of Religiosity and the Confessional Structure of the Population in Slovakia in 2011

Religiosity is spatially differentiated in Slovakia (Table 53.5, Fig. 53.1). The highest level of religiosity is in the northern part of Slovakia, in the Prešov (84.9 %) and the Žilina self-governing region (80.5 %). The lowest level of religiosity is in the Bratislava (65.5 %) and the Banská Bystrica self-governing region (69.7 %).

In terms of spatial variations among various religions, in recent years there have been no significant changes. The Roman Catholic Church dominates all Slovak regions. The most significant dominance is in the Trnava (90.7 % of all religious people), Nitra (88.8 %) and the Žilina (87.1 %) self-governing region. The lowest percentage of Roman Catholics is in the Kosice (69.6 %) and Prešov (72.4 %) self-governing region. In these two regions of the eastern Slovakia, there is a significant number of the Greek Catholics and Orthodox. In the Košice self-governing region we find the Reformed Christians/Calvinists. Lutherans have a significant representation in the Banská Bystrica, Zilina and Trencin self-governing region.

Table 53.5 Religiosity rate and confessional structure of the religious people in Slovakia, 2011

Region	Religiosity rate (%)	Roman Catholics (%)	Greek Catholics (%)	Lutherans (%)	Calvinists (%)	Orthodoxs (%)	Others (%)
Bratislava	65.5	85.9	1.2	7.8	0.7	0.6	3.8
Trnava	77.3	90.7	0.3	4.8	2.5	0.2	1.5
Trenčín	73.8	86.8	0.3	11.0	0.1	0.1	1.7
Nitria	79.3	88.8	0.3	3.5	5.4	0.1	1.9
Žilina	80.5	87.1	0.4	11.1	0.1	0.1	1.2
Banská Bystrica	69.7	78.7	1.2	15.2	2.3	0.3	2.3
Prešov	84.9	72.4	16.6	5.4	0.1	4.3	1.2
Košice	73.7	69.6	12.9	5.0	7.5	2.3	2.7

Data source: Statistical Office of the Slovak Republic

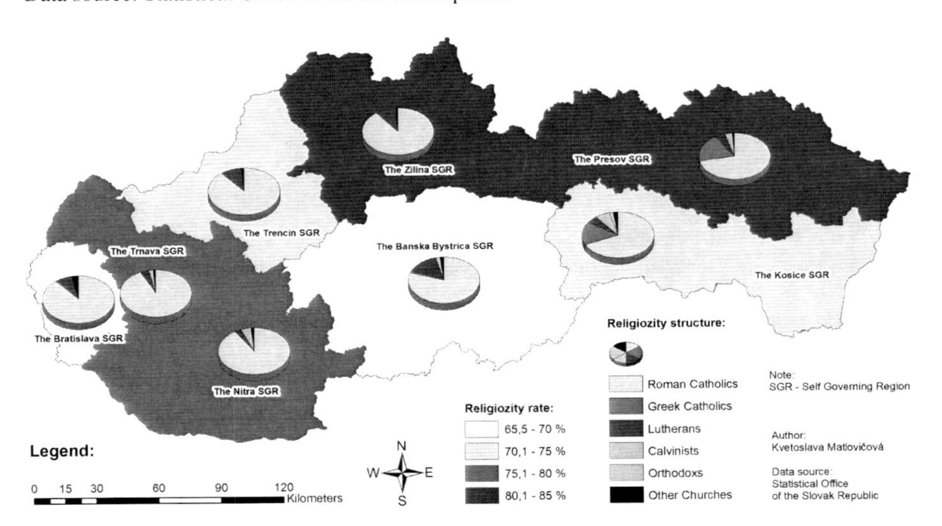

Fig. 53.1 Religiosity rate and confessional structure of religious people in Slovakia in 2011 (Map by Kvetoslava Matlovičová with data from the Statistical Office of the Slovak Republic)

Reformed Christians live in the southern parts of Slovakia, especially in the Nitra, Trnava, Banská Bystrica and the Košice self-governing region. Most of them belong to the Hungarian minority.

From the administrative point of view, the main centers of the Roman Catholic Church are in two archdioceses – Bratislava and Košice. Other centres include these dioceses – Trnava (historical archdiocese), Nitra, Žilina, Banská Bystrica, the Spiš and Rožňava. The main center of the Lutheran Church is in Bratislava and in the districts of Zvolen (western) and Prešov (eastern). The main center of the Greek Catholic Church is located in Prešov and centers of other eparchies (dioceses of the Greek Church) are in Košice and Bratislava. The main centre of the Reformed Christian Church is Komárno and the Orthodox Church is Prešov.

53.6 The Religious Thinking of Slovakia Young People

In today's secularized society, young people are confronted with a lot of inconsistent information. Life attitudes of young people are often determined by the education (especially in the family, but also at school) and a considerable impact of the society they live in.

To determine opinions and attitudes of university students in Slovakia on issues of religious life and faith, we prepared a questionnaire with 25 questions. Research was conducted using 600 questionnaires which were distributed among students aged 19 to more than 25 years (from the 1st to the 5th academic year). The selection was made according to regional representation of the most numerous religions in Slovakia. According to 2011 home census 62 % of population professed Roman Catholicism and since the majority of Roman Catholic population (70.5 %) was reported in the Nitra region, we have decided for the University of Constantine the Philosopher (Faculty of Natural Sciences) in Nitra. The second largest denomination in Slovakia is Evangelical Church of Augsburg Confession (5.9 %). Its strong presence can be felt in Bratislava. Therefore we have chosen three faculties of the University of Economics in Bratislava. Greek- Catholic Church is the third most populous, with the highest concentration in the Prešov region (14.1 %). The University of Prešov with three faculties – the Humanities and Natural Science, Greek-Catholic and Greek-Orthodox Theology – was the third to be chosen. Belonging to other churches and religious communities was declared in a much lesser extent. The proportion of people without religion, in comparison with the census 10 years ago, has increased slightly. While in 2001 there were 13.0 %, in 2011 the Statistical Office reported 13.4 % with the largest presence in the region of Bratislava (26.8 %). The distribution of questionnaires among students has been made at random. Out of 600 questionnaires distributed, we could use 509, because the others were incomplete or provided confusing information. Majority of respondents were women (363), most respondents were aged 20–25 years (399). There were 22 respondents older than 25 years because at this age less than 8 % students were studying. The greater part of students was single (496), 428 lived with their parents.

Nitra has 83,400 inhabitants. In this town there are two universities: the Slovak Agricultural University and the University of Constantine the Philosopher. The questionnaires were used in the latter. We could process only 58 questionnaires. The majority of respondents were Roman Catholic (63.8 %), followed by atheists (17.2 %) and nearly the same proportion was between Protestants of the Augsburg Confession and the Greek Catholics, (8.6 % and 6.9 %) respectively. Respondents were mostly women, and the atheists, compared with men reported only one third. Up to 70 % of students surveyed were believers although practicing faith, prayers, worshipping and regular attendance of the Mass, was reported only by less than half of the respondents. From Christian families, where at least one parent regularly attends the Mass were 62 % students, and it was up to 76 % of their grandparents observing religious services. 70 % of respondents consider religious holidays (Christmas, Easter) to be important religious festivals even though half of them confirmed that those were also important secular holidays, connected with free time and gifts. More than 70 % of respondents would like to get married and christen their children in the church. A negligible number of respondents read religious literature or listen to radio religious programmes or watch religious television channel.

Bratislava was chosen as the capital of Slovakia whose population of 432,800 embrace all religions. The inhabitants of Bratislava have the second largest allegiance to the Evangelical Church of Augsburg Confession. The city has four universities, and public, state and private colleges. Here we have chosen the University of Economics in Bratislava, where 219 questionnaires with a predominance of women were correctly completed. The majority were Roman Catholics (62.8 %), 10.2 % were Protestants, 3.4 % Greek Catholics and 22.6 % atheists. Only 46 % of respondents come from families where religion is practiced at least by one parent, but as many as 80 % of their grandparents regularly attend religious services. As at the University of Nitra, also at the University of Economics in Bratislava almost 70 % of respondents considered themselves to be believers. 51 % pray regularly and 30 % attend religious services. 60 % of students at this university believe in miracles, while in Nitra it was 52 %. 41 % of respondents say they try to practice the doctrine of their religion in their lives. Less than 5 % read religious books, or listen to religious broadcasts on the radio. Television programmes have double ratings. For almost 70 % of respondents Christmas is an important religious holiday. Up to 73 % would like to have a wedding in the church. 82 % of respondents would christen their children in the church.

Prešov has 91, 800 inhabitants. There is the University of Prešov and the College of International Business ISM Slovakia. The questionnaires were distributed at the University of Prešov and 232 were processed. The questionnaires were completed by 166 women and 66 men. Nearly 59 % of respondents were members of the Roman Catholic Church, 21.6 % belonged to the Greek Catholic, 12.1 % to Orthodox, 4.3 % were Lutherans and less than 3 % were agnostics. 93 % of respondents consider themselves to be believers. Prayers are important for 83 % and 73 % practice the doctrine of their religion in everyday life. 83 % of respondents believe in miracles. In Christian families where at least one parent regularly attends religious services grew up 64 % of respondents. When asked if they remember the church attendance of their grandparents 91 % responded positively. 22 % respondents read religious literature, 33 % listen to religious radio broadcasts and 54 % of the students watch TV programmes with the religious topic. Christmas and Easter are important religious holidays for 90 % of respondents, the same proportion would like to have a wedding in the church and even a higher percentage would like to baptize their children in a church. This University has the highest number of students practicing religious beliefs. Certainly this reflects the fact that the questionnaires were filled out by students from two theological faculties.

53.6.1 Questions in the Survey:

1. Are you coming from a family, where religion was practiced by both parents (both visiting church)?
2. At least one parent?
3. Are you coming from a family where at least one parent is religious, but does not practice it?
4. Did your grandparents go to church regularly?
5. Do you consider yourself as a believer?
6. Do you believe in the principles that religions teach?
7. Do you believe in a celestial being?
8. Do you believe in miracles?
9. Do you practice your faith in life?
10. Do you go to church regularly?
11. Do you consider it important that your current or future partner shares similar views about religion?
12. Are you planning to raise your children in a religious atmosphere?
13. Do you follow the religious broadcasting?
14. On TV?
15. Do you read religious literature?
16. Do you consider religion as important for society operation?
17. Is Christmas an important religious holiday?
18. Is Christmas an important worldly holiday for presents, etc.?
19. Do you want your children to be baptized once you have them?
20. Do you want to have your wedding at church?
21. Is praying important for you?
22. Would you encourage religious education at primary schools?
23. Do you ever attend religious services in a church different from the one you attend regularly?
24. Do you believe in God, but do not attend religious services?
25. Do you prefer secular society or a religious society?

53.7 Discussion

Our survey among the university graduates supports and confirms the attitudes of young people towards religion which have been generally known in Slovakia. Based on the survey young people can be divided into several groups. The first group, (34 %) are deep believers, then there is a group of young people of emotional believers (58 %) who, nevertheless do not feel the need to be confession- bound. The rest are atheists, who do not consider religion to be important for society and who know practically nothing about God. Spirituality and belief in miracles have been demonstrated as important for the youth. 76 % of women believe in miracles while it is only 58 % of men. Today there is much more religious literature available than it was before the year 1989, wide selection of radio religious programmes and a special television channel broadcasting the religious issues, but young people take advantage of those opportunities provided by the literature and the media very rarely.

The insights received from 25 questions in the questionnaire were summarized in tables and charts to illustrate the views of undergraduates at selected universities in Slovakia (Tables 53.6 and 53.7; Figs. 53.2 and 53.3). Their current attitudes to religion reflect the information from their study.

The outcome, however, has been affected by the answers of the respondents at Faculties of Theology. The university students' attitudes to religion and its practice

Table 53.6 Religious orientation of students from selected universities

	Female						Male					
University	RK (%)	E (%)	G (%)	PR (%)	I (%)	A (%)	RK (%)	E (%)	G (%)	PR (%)	I (%)	A (%)
Bratislava	69	9	3	–	4	15	43	14	4	–	2	38
Nitra	74	6	6	–	6	9	48	13	9	–	–	30
Prešov	60	5	22	7	4	3	41	3	20	26	8	3

Source: R. Matlovič, V. Vlčková and K. Matlovičová

Table 53.7 Students' answers to selected questions from the survey

	Female							Male						
University	1 (%)	2 (%)	4 (%)	5 (%)	6 (%)	9 (%)	10 (%)	1 (%)	2 (%)	4 (%)	5 (%)	6 (%)	9 (%)	10 (%)
Bratislava	40	49	85	79	60	49	35	29	41	73	52	46	25	16
Nitra	51	54	74	86	63	57	49	43	74	78	48	13	30	13
Prešov	67	65	91	94	81	77	61	52	59	91	94	68	71	55

Source: R. Matlovič, V. Vlčková and K. Matlovičová

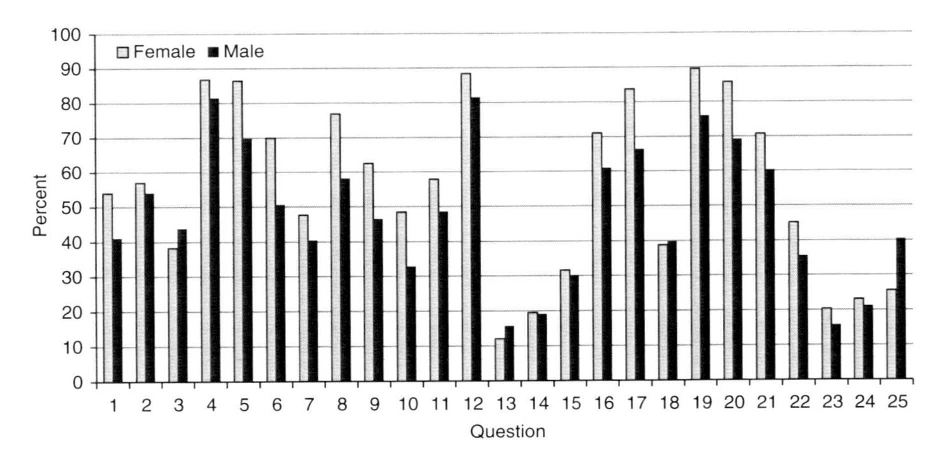

Fig. 53.2 Survey answers, by gender (Source: Viera Vlčková)

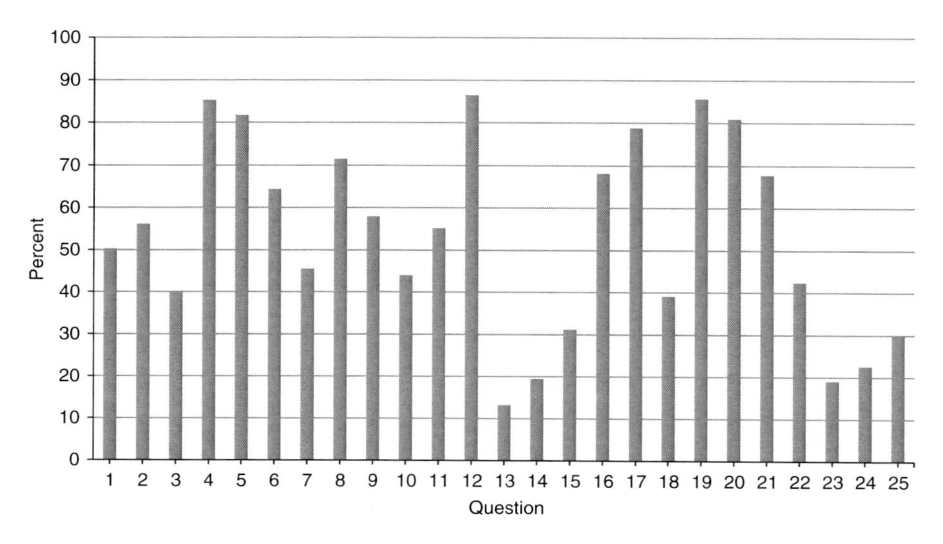

Fig. 53.3 Survey answers (Source: Viera Vlčková)

include not only cognition but also emotions which are associated with their various mental needs and it seems that they are often confused with faith. We can hardly say that they try to acquire deeper knowledge of religion whether from literature, or the media, unless it is a study of theology. Students choose from religion what they need. The prime attitudes are secularism and retreat from the traditional forms of religion.

References

Casanova, J. (1994). *Public religions in the modern world*. Chicago/London: University of Chicago Press.

Davie, G. (2007). *The sociology of religion*. London: Sage.

Fukuyama, F. (2002). *Konec dějin a poslední člověk*. Praha: Ryka Publishers.

Giddens, A. (1999). *Sociologie*. Praha: Argo.

Houston, T., Worthing-Davies, R., & Russell, R. (1992). *Evangelizace v České a Slovenské republice po odstranění marxistické filozofie*. Swindon: Britská a zahraničná biblická společnost.

Huntington, S. P. (2001). *Střet civilizací. Boj kultur a proměna světového řádu*. Praha: Rybka Publishers.

Kepel, G. (1994). *Revenge of God: The resurgence of Islam, Christianity and Judaism in the modern world*. University Park: Pennsylvania State University Press.

Kováč, M., & Krupová, T. (1993). *Správa o výsledkoch prieskumu religiozity študentov bratislavských vysokých škôl*. Bratislava: Chronos.

Lužný, D. (1999). *Náboženství a moderní společnost*. Brno: Masarykova univerzita.

Matlovič, R. (2000). Sakralizacja – cząstkowy proces transformacji przestrzeni miejskiej miasta postkomunistycznego na przykładzie Preszowa. In I. Jażdżewska (Ed.), Miasto postsocjalistyczne – organizacja przestrzeni miejskiej i jej przemiany (pp. 121–128). UŁ a ŁTN Łódź.

Matlovič, R. (2001). *Geografia relígií. Náčrt problematiky*. Prešov: FHPV PU.

Matlovič, R. (2005). The end of geography of religion? Towards the issue of the relevance of religions research in the contemporary human geography. In B. Domański, & S. Skiba (Eds.), *Geografia i sacrum* (2 Vols., pp. 315–327). Krakow: IGiGP UJ.

Matlovič, R. (2011). Geography of religion in Slovakia since 1990. *Peregrinus Cracoviensis, 22*, 49–60.

Podolinská, T. (2008). Religiozita v dobe neskorej modernity. *Prípad Slovensko. Sociální studia, 3–4*, 53–86.

Stark, R., & Bainbridge, W. S. (1987). *The theory of religion*. New York/Bern: Peter Lang.

Tížik, M. (2006). *K sociológii novej religiozity*. Bratislava: Univerzita Komenského.

Tížik, M. (2010). *Metodologické súvislosti merania religiozity na Slovensku*. Demografické popoludnie Infostatu, 25. Marca 2010. Dostupné: www.infostat.sk/vdc/pdf/DDP68.pdf, stiahnuté 26.5.2012.

Tomka, M. (2000). Náboženská situace v zemích východní a střední Evropy. In H. Renöckl & M. Blanckstein (Eds.), *Nová religiozita fascinuje a zneklidňuje* (pp. 22–32). Kostelní Vydří: Karmelitánské nakladatelství.

Tomka, M. (2006). Is conventional sociology of religion able to deal with differences between Eastern and Western European developments? *Social Compass, 2*, 251–265.

Chapter 54
Milwaukee Catholicism Intersects with Deindustrialization and White Flight, 1950–1990

Steven M. Avella and Thomas Jablonsky

54.1 Introduction

For more than a century now, from the late nineteenth to the early twenty-first, Milwaukee and Wisconsin, each in their own context, have been leading centers of American manufacturing. In 1910, Milwaukee, then America's 12th largest city, had the second largest percentage of its work force engaged in manufacturing, next to Detroit. One hundred years later, Wisconsin has the second highest percentage of its statewide work force still engaged in manufacturing at 15.8 %. Explanations for these accomplishments are familiar: locational advantages, proximity to critical transportation networks, entrepreneurial risk-takers, and deep but inexpensive labor pools. In other sectors of the U.S. Rust Belt, this last component routinely included, at some level, African American migrants from the South. The rise of Milwaukee as an industrial center, however, was largely built upon the sweat and muscle of immigrants from Germany, Ireland, and Poland with only tiny cohort of black co-workers until World War II and after. Even the postwar Second Great Migration of African Americans into Northern cities took place a bit later in Milwaukee than most large cities in the East and Midwest. As a result, the central city neighborhoods of Milwaukee's North Side underwent dramatic population transformations at the same time (1970s and 1980s) that deindustrialization stole thousands of local jobs. In turn, the inevitable reshaping of social institutions within these newly established African American neighborhoods – including the composition and mission of Roman Catholic parishes – took place amidst a triangulation of demography, economy, and religion.

S.M. Avella (✉) • T. Jablonsky
Professor of History, Marquette University, Milwaukee, WI 53233, USA
e-mail: steven.avella@marquette.edu; thomas.jablonsky@marquette.edu

S.D. Brunn (ed.), *The Changing World Religion Map*,
DOI 10.1007/978-94-017-9376-6_54

54.2 Triangulation of African American Migration, Deindustrialization and White Flight

Positioning on the southwestern edge of America's Great Lakes system provided Milwaukee with regional advantages in terms of contacts with Eastern cities during the middle of the nineteenth century through the inland water route that connected New York City (and thereby the world) with the Hudson River, the Erie Canal, and all of the Great Lakes cities. In turn, Milwaukee did not have Chicago's additional benefit of being not only along the western shores of Lake Michigan but also at the far southern edge of the lake, thereby providing easier access to the eventual westward extension of East Coast railroads. Nonetheless, the lakefront settlement that would become Wisconsin's largest city still grew into a significant regional entrepôt during the course of the second half of the nineteenth century, first by its domination of wheat exports from the Midwest. Later an improbable accumulation of gifted inventors, engineers, and managers would shape Milwaukee into the metal-bending capital of America. Some steel was produced locally; but most was shipped in from other centers such as Pittsburgh or Chicago or later Gary. What laborers in the Cream City (a nickname that derives from the distinctive off-white color of locally fired bricks) molded were not the raw materials of the industrial era so much as the devices that drove this age: steam turbines and dynamos for power production, earth moving equipment for large scale mining, gasoline engines and electric motors, electronic controls for heating and cooling systems, tractors, hot water heaters, and even motorcycles. In addition, Milwaukee developed a national reputation through brewing empires associated with Pabst, Schlitz, Blatz, and Miller. Lastly, in association with Wisconsin's extensive cattle and dairy herds, both meatpacking and leather manufacturing were dynamic sectors of the local economy, beginning in the post-Civil War era and extending throughout much of the following century (Fig. 54.1).

Fig. 54.1 Factories, warehouses and docks along the Milwaukee River, on south edge of the central business district. Photograph is looking north (Courtesy of the Milwaukee County Historical Society, used with permission)

Later, in the aftermath of the economic devastation of the Great Depression, Detroit was reborn during World War II as the Arsenal of Democracy. Milwaukee was equally reinvigorated during wartime by the nation's desperate need for military hardware. Torpedoes, tanks, walkie-talkies, rifle straps, parachutes, ignition systems for fighters and bombers, bomb casings (80 % of America's needs by one estimate), and even camouflaged beer cans flowed out of the city, carried by lake steamers and southbound railroads heading toward America's coastlines for distribution across the Atlantic and Pacific oceans (Gurda 1999: 308). In Milwaukee's case, this renaissance of its industrial core carried deep into the following decades, blunting meaningful signs of postwar deindustrialization until the 1970s. In the early 1950s, 79 % of workers at Milwaukee largest manufacturers – nearly 83,000 individuals – still labored in metal bending, mostly in heavy machinery: (16,674 for Allis-Chalmers; 9,300 for A.O. Smith; and 6,000 for International Harvester) Almost 10,000 worked for brewing, yeast, and affiliated companies, just under 6,000 in leather works, and another 2,500 in textiles (R. Cutler 2001: 241–248). Whereas shrinkage in industrial employment revealed itself in New England textile centers, Pittsburgh and Gary steel complexes, and Detroit auto manufacturers by the 1960s, noticeable decreases in the Milwaukee metal fabrication sector did not become apparent until a decade later, creating the illusion of exceptionalism: late-onset deindustrialization (Fig. 54.2).

Fig. 54.2 Factories, warehouses and docks along the Milwaukee River (*middle* of photo) as it curves eastward toward Lake Michigan (*far right* of image). Railyard and factories at *lower right* are part of Walker's Point industrial district. The collection of tall buildings to *left* and *center* are the central business district. Photograph is looking northeast toward Lake Michigan (Courtesy of the Milwaukee County Historical Society, used with permission)

Like metropolitan centers whose industrial zones are fractured by mighty rivers (such as St. Louis in relation to Alton and East St. Louis, Illinois), Milwaukee is cleaved into two, not by a vast flowing river but by a nearly mile-wide geological riverbed through which today runs an unimpressive stream. While the contemporary Menomonee River is more of a liquid companion alongside which joggers and bikers carry out their business, the ancient Menomonee River Valley (which runs westward for about three miles from the lakefront) divides the original center of Milwaukee with its industrial, financial, and governmental districts along the northern edges of the Valley from the largely residential South Side. The Menomonee Valley and Walker's Point (a promontory along its eastern edge) in combination with the banks of the Milwaukee River (which flows north/south through Milwaukee's North Side) evolved into the loci for many of the city's industrial employers. Within these zones were located internationally recognized firms such as Allen-Bradley, Pawling and Harnischfeger, Pfister and Vogel Leather, International Harvester, Falk Corporation, Miller Brewing, and the Chicago, Milwaukee and St. Paul Railroad. And in the eastern corner of this labyrinth of factory centers arose the city's central business district (Fig. 54.3).

For decades, as these manufacturing zones developed, the largest ethnic presence in the Brew City, the Germans, favored residential areas north and northwest of the CBD. Here too Yankee-Yorkers from the East Coast, often the original developers

Fig. 54.3 Menomonee River Valley looking east toward the central business district, Lake Michigan, and the St. Paul Railroad. Viaducts cutting across the Valley are the 35th Street (*nearest*) and 27th Street Bridges (Courtesy of the Milwaukee County Historical Society, used with permission)

and financiers, preferred to locate their families and businesses. In contrast, on the South Side, separated from the city's principal employment centers by the vast Menomonee Valley and ill-served by municipal systems, were late- arriving newcomers, most notably thousands of Polish immigrants. Even with this chasm separating major sectors of the city, by the end of the twentieth century's first decade Milwaukee was America's third most densely populated city behind Boston and Baltimore (Simon 1996: 26). Electric streetcars lines steered residential growth north and west of downtown along very precise corridors. In contrast, the absence of viaducts across the Menominee River Valley outside of downtown delayed development within the southern half of the city. As the twentieth century progressed, Polish and German settlements eventually appeared along the south rim of the Valley where working class laborers and middle-level managers filled an exciting mixture of Polish flats (the simplest of frame homes originally built at ground-level and later elevated upon wood or cinder block foundations to permit basement rental units) and brick bungalows. Nonetheless, the city's North Side retained residential and financial dominance. During the post-World War II period, with the local economy still humming along with decent-paying jobs in the manufacturing sector, former G.I.s drew upon their veterans' benefits as well as their wartime savings to invest in suburban lifestyles in contiguous cities such as West Allis and Greenfield in addition to newly annexed, former farmland in the far northwestern edges of Milwaukee County. Suburbanization in Milwaukee proceeded at a modest pace in the immediate postwar era, unshaped by any dramatic changes in the city's demography (Fig. 54.4).

As noted, in contrast to most large, industrial cities of the East and Midwest Milwaukee did not experience the First Great Migration of African Americans from the South in the opening decades of the twentieth century. As with these other cities, jobs were certainly available in Milwaukee – theoretically, if one ignored the racist policies of local manufacturers. Chicago, Gary, Detroit, and other cities south of Milwaukee acted as migration screens, retaining northbound blacks. In 1910, only 980 African Americans lived in Milwaukee, leaving the city 99.5 % white. This number of African American residents doubled during the following decade and then tripled over the course of 1920s, reaching 7,501 by 1930. Still that was a remarkably small number compared to Milwaukee's Midwestern competitors. Chicago, for example, with a more complex economy and a more vigorous Bronzeville community (on an array of scales) hosted 234,000 African Americans (representing 6.9 % of the population) by the outbreak of the Great Depression (Trotter 2007: 41; Levine and Zipp 1993: 42, 47; I. Cutler 2006: 159).

Moreover, the racism that black Milwaukeeans encountered in the most significant sector of the local economy – manufacturing – was unrelenting, making the city less inviting than others. African Americans males were excluded from semi-skilled and skilled labor positions in the metal fabrication industry. Black men were fortunate if they could find employment as common laborers, the only kind of work typically available to members of their race other than domestic and personal service. Only a small handful of companies allowed blacks to do custodial work in the shops and foundries over night or during late-night third shifts. Working alongside unskilled

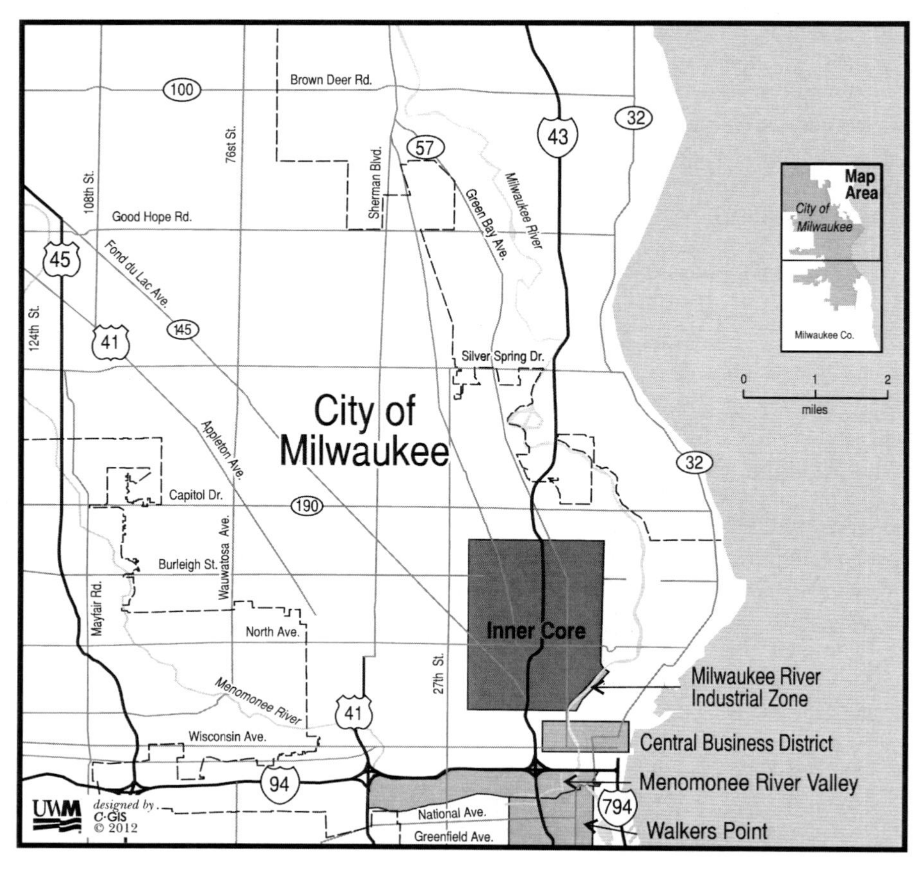

Fig. 54.4 Milwaukee's industrial zones (Map by University of Wisconsin – Madison, C-GIS for the authors, 2012)

whites was almost unheard of in Milwaukee's fabrication factories. For instance, the 1900 census enumerated a *single* unskilled African American male as an "Iron and Steel Worker" and *zero* listed as either skilled or semi-skilled workers in similar categories of employment. By 1930, those numbers had crept up to 123 unskilled workers, 72 semi-skilled workers, and 64 skilled workers in iron and steel work (including machinists and mechanics) (Trotter 2007: 255–256).

With the relatively small population of African Americans, white Milwaukee maintained a segregated society without trouble or fanfare: separate seating in movie theaters, refusal to serve black customers at downtown restaurants, exclusion from churches, and a black-only ward at the county facility that cared for tuberculosis patients. African Americans were constrained to live in a dilapidated, mixed-use district along the northwest edge of downtown near several breweries and leather works. Only late in the 1930s and during wartime did a modicum of unskilled jobs open up in the manufacturing sector, largely due to pressures from the Congress of

Industrial Organizations and to a lesser degree the federal government. The smaller and late-developing community of African Americans in Milwaukee led to less developed commercial, educational, recreational, and political systems compared to Cleveland or Detroit or Chicago. And Milwaukee's economic and governmental power brokers were most content with this state of affairs.

During the late 1950s, as race relations in Milwaukee slowly seeped into the public consciousness, the black areas of Milwaukee (where upwards of 90 % of African American lived) became identified by municipal authorities and academics as the "Inner Core," 26 contiguous census tracts located west of the Milwaukee River from Holton Street on the east to 20th Street on the west and from Juneau Street (the original nexus of black settlement) on the south to Keefe Street on the north. In 1940, three of these tracts adjacent to Juneau Street housed populations that were more than 50 % black; one additional tract (also along Juneau) was between 25 and 49 % African American. Those four areas all became blacker during the 1950s at the same time that two additional tracts to their north also passed the 25 % threshold. What was noteworthy during wartime and immediate postwar period was an increase in nine adjacent tracts whose populations rose from 0 % black to between 1 and 25 %. By 1960, the transformation overall was astonishing. Whereas 20 years earlier, only three of Inner Core's 26 census tracts had black populations more than 50 %, now 21 of those tracts had reached the 50 % threshold.

The pre-1960 changes were soon dwarfed by the arrival of significant numbers of African Americans from other Northern cities such as Chicago and Detroit as well as directly from the South during the second and third decades following the end of World War II. With a black population just over 10,000 in 1945, Milwaukee became home to 62,000 African Americans within 15 years and then doubled during the 1960s, marking a delayed Second Migration compared to other Rust Belt cities. From a level of less than 2 % of the city's population during wartime, black Milwaukee constituted ten times that number by the mid-1970s and 40 % by the opening of the twenty-first century (Trotter 2007: 285).

In an untimely manner, this migration coincided with the onset of deindustrialization in Milwaukee. As noted earlier, Wisconsin's largest city did not feel the full effects of job losses in manufacturing until the 1970s, although unappreciated evidence suggested as early as the Great Depression that manufacturing practices were changing. Holeproof Hosiery, for instance, subcontracted with Southern mills during the 1930s and at the same time two Milwaukee shoe companies moved their production facilities to rural parts of Wisconsin. By the mid-1950s, companies such as National Biscuit Company (Nabisco) and Phoenix Hosiery Company had discontinued manufacturing in Milwaukee altogether. Other enterprises such as a major local food distributor, Roundy, and the internationally recognized small engine manufacturer, Briggs and Stratton, had relocated facilities outside the city limits. Even a major manufacturer such as Chain Belt whose origins rest in the Walker's Point area south of downtown had only a single foundry left in Milwaukee by 1956.

This drama only deepened during the 1970s and 1980s. In 1965, Milwaukee's largest industrial employer, Allis-Chalmers, which had established the nearby city of West Allis in the early twentieth century to handle its portfolio of 1,600 products, employed 11,000 workers in the fabrication of everything from turbines to tractors. Within 15 years, by 1980, that number was cut in half. In that year, annual sales reached $2 billion, with profits approaching $50 million. However, during 1981, Allis-Chalmeers lost $29 million, followed by $207 million the next year, and $261 million in 1984. The company's stock, worth $38.63 per share in 1979, slipped to $3.13 six years later – 2 years before Allis-Chalmers, Milwaukee's largest employer in 1956, filed for bankruptcy on the way to dissolution (Gurda 1999: 415,417).

Whereas 56 % of the city's workforce held manufacturing-related jobs at its heyday in 1910 and still provided 41 % of these jobs as late as 1960, as soon as 1980 that figure had dropped to 32 %. Milwaukee lost 40 % of its manufacturing jobs during the quarter of a century after 1960, introducing a radical remake of its industrial landscape (Levine and Zipp 1993: 43, 48). The South Side Caterpillar Tractor plant was redeveloped into a Pic n Sav grocery store, the Pfister and Vogel Tannery along the Milwaukee River became a business park as did the nearby Schlitz Brewing complex, and key buildings in the Blatz Brewery provided upscale apartments by the end of the twentieth century (Gurda 1999: 378). In 1980, metropolitan Milwaukee was home to 11 Fortune 500 companies. Nine years later than was cut in half to six as the region lost the headquarters for Allis-Chalmers, Clark Oil, Rexnord, Pabst, Schlitz, Bucyrus-Erie, and Koehring (while adding two newcomers to the list) (Gurda 1999: 418). The economic slide, first evident in the 1960s, in time became an avalanche of job losses and factory closings.

Amidst opening phase of this employment collapse, thousands of job-seeking African Americans arrived with promises of fair-paying industrial jobs. Just west of 27th Street and south of Capitol Drive, at the heart of the increasing black community of Lincoln Creek, ran a series of rail tracks that supplied nearly four dozen manufacturing companies their raw materials, before subsequently removing their finished products for international distribution. A. O. Smith, the nation's largest welder of underbodies for the American automotive industry, provided 9,000 jobs into the 1970s at a sprawling complex on the northern rim of this industrial district. A block away was Koehring Machine Company with 1,000 employees and half of a mile further south was the small engine giant, Briggs and Stratton, with 700 workers scattered within six-building campus. Dozens of other metal parts, metal processing, metal finishing, tool and die, leather, and shoe companies lined this railroad corridor. A.O. Smith – which had not hired an African American for the 20 years prior to World War II – subsequently became during the postwar period a regular employer for blacks as well as whites, providing middle-class livelihoods with $8 to $10 per hour wages. This very brief window of opportunity created a well-spring of fond memories among African American workers, even as tens of thousands of jobs vanished throughout the 1970s and 1980s (Milwaukee Journal Sentinel 2011:1, 20–22).

Payroll cuts and permanent closures of Milwaukee's manufacturing shops coincided with the city's greatest influx of African Americans. The arrival of these newcomers timed with crushing factory shutdowns caused white Milwaukeeans from the city's North Side to flee into surrounding municipalities and ultimately over the county line into Waukesha County where a portion of the city's white and blue collar jobs relocated (Terrell 1995: Appendix B). Suburbanization, largely a lifestyle choice up to that point, now became unadulterated white flight. Milwaukee had reached its all-time high in population at 741,000 in 1960, a full decade later than its Rust Belt counterparts. As that number steadily dwindled to below 600,000 at one point, the city's suburban neighbors grew in terms of numbers and influence. From 234,000 residents in 1950, suburbs outnumbered Milwaukee resident handily by 1980 with 761,000 residents. That number rose to 867,000 by the end the twentieth century (Trotter 2007: 284–285).

This out-migration left economic footprints across the local landscape. The per capita income of city dwellers – white and black – fell from 83.6 % of the suburban per capita income in 1970 to 63.4 % 20 years later whereas the black median income dropped from 65 % of the white median to 39.5 % during the same period. Among the "nation's one hundred largest metropolitan areas" Milwaukee had the highest concentration of poverty by the last decade of the twentieth century when the number of high poverty census tracts grew from 11 in 1970 to 59 by 1990 (Squires and O'Connor 2001: 17).

This emptying out of the Inner Core and adjacent neighborhoods by white residents and their steadier incomes remade the retail economy along major shopping stretches such as North Third Street (to be renamed Martin Luther King Drive) and North Avenue. Venerable department stores and specialty shops closed, sometimes for good and sometimes until they relocated to the suburbs. The Milwaukee Public School system, long a skillful manager of racist policies, completely mishandled the transition of elementary and high schools from white to mixed to black student bodies in rapid succession (Dahlk 2010). And also swept up in this demographic transformation were the Inner Core's Catholic parishes and schools. Unlike small shop owners and private residents, the morphological footprint of the Catholic Church demanded a different set of strategies to address these changes among their neighbors from across the street and around the corner.

54.3 Changing Milwaukee Parishes: An Urban Presence Recedes: 1950–1995

For many years now, historians such as Kathleen Neils Conzen have explored the significance of religious institutions on the urban landscape. In addition to their sectarian purposes, churches and their ancillary institutions (for example, schools and institutions of social provision, that is, hospitals, nursing homes and child care agencies) are also major urban agents (Conzen 1996: 108–114). They occupy important urban space, provide significant social and cultural services, enhance neighborhood viability, and provide a common ground for diverse elements in neighborhoods. Churches exist primarily to offer worship services, but they are also repositories of sacred art and venues for cultural events. They teach school children, employ thousands of urban workers whose salaries contribute to the local economy, and offer a limited safety net to those who cannot manage in a changing economy. Church communities are intentional bodies – established generally in the wake of stable economic conditions, that is, the presence of decent paying jobs, home ownership, and some measure of disposable income which is voluntarily given to the church. Either planned by a central agency or the collective will of a congregation, churches overspread the urban landscape, united to their "parent" church by common bonds of belief, cultic practice, and a moral code. Churches can sometimes anticipate the patterns of urban growth, but generally they follow the people. After their creation, they often add an additional incentive to the settlement patterns of neighborhoods. The sometimes informal association of religious leaders with real estate developers in many American cities has worked to the mutual benefit of both. Real estate agents are often pleased to acknowledge the presence of religious bodies in areas they wish to market pointing to the church as an emblem of community stability and respectability. Religious leaders often rely on real estate agents to plan for expansion and garner centrally located or affordable parcels of land on which to plant a church and sometimes a school and residences for clergy.

The Roman Catholic church in the United States has always had a significant presence in the urban landscape of the nation. Following the nineteenth and twentieth century patterns of migration, particularly of immigrants from the Catholic parts of Western Europe, Catholic leaders in Rome and America took special care to provide for their widely scattered flocks and to prevent them from joining other religious bodies or falling away from religion altogether. Missionary priests, religious sisters, and dedicated lay persons helped to create a viable Catholic presence in America's cities from one end of the country to the other. In many places, Catholics put a high premium on large and beautiful church structures and over time huge and often ornate churches thrust their steeples over emerging cities in the East, the Industrial Midwest, and even the Great Plains and American West. Catholic schools were also a priority for most (but not all) Catholic ethnic groups as a shield against the perceived Protestant bias of the public schools and also as protective institutions to maintain language, cultural customs, and above all socialization into a Catholic way of life. A typical Catholic parish in an American city often occupied an entire city block on which sat a large church, a multi-story school, and residences for the clergy and the religious sisters who taught in these schools. These investments in urban property alone were worth millions of dollars. The services of the priests and sisters in religious and other social necessities (that is, caring for the poor and sick, providing marriage and youth counseling, running athletic and social programs such as dances, credit unions, recreation halls) is difficult to categorize. Their remuneration governed both by the vow of poverty (for the sisters and for some priests) and also by the low wages of junior priests who often did most of the work is difficult to reconstruct. Even though offset by having no family to support and by low costs for housing and food – was likely far below the wages of people in the private sector – even in the days of low industrial or white collar wages. Overall, the parish never lost its ecological connection to the specific geographical parcel in which it was located. Parishes were significantly affected by the changing fortunes of the local economy and the rearrangements of urban demography over the course of the years. Many factors could affect the membership and viability of an urban parish: the stability of the job base of the parishioners, the realities of racial succession, the decisions of urban planners especially those connected with the every-shifting needs of freeway expansion and urban renewal. A once thriving parish could be fatally compromised if a freeway went through it or by it, wiping out the base of homes and incomes on which the parish depended for membership, school enrollments, and financial contributions. Fears of racial succession also prompted abandonment of certain neighborhoods. Even if none of these factors were operative, the lure of suburban living with bigger yards, more housing space, and the impression of upward social mobility that attended getting out of the old neighborhood affected the calculus of urban life. Catholic institutions which seemed almost like impregnable fortresses and which bore many significant memories of life-cycles – baptisms, first communions, school days, weddings and funerals- found themselves caught in the cycle of change. Some reduced themselves and adapted – others faded from sight – their buildings either torn down or sold to other entities or just abandoned.

Milwaukee was a typical industrial city to which the Catholic institutional presence contributed significantly to the complex process of urbanization (Avella 2002). What is written of Milwaukee bears a strong resemblance to developments in other Midwestern communities: Detroit, Cleveland, Toledo, Gary, and Chicago to name but a few. In each of these cities the Catholic church was "present at creation" and often provided some of the city's first public services. In Milwaukee, the first hospital, St. Mary's, located near the city lakefront, was operated by the Daughters of Charity. Other large and complex Catholic institutions took their place on the urban landscape. At least five Motherhouses, that is, central administrative headquarters for scores of religious women (nuns), were located in Milwaukee. These large and imposing structures provided basic education, spiritual formation, and required the women who ran them to be financiers and personnel managers as they administrated the lives and destinies of literally thousands of young women who came to join their respective communities. When many of these communities were established in the nineteenth century, these women enjoyed a scope of leadership and responsibility that most American women would not enjoy until the twentieth century. Catholic academies which later evolved into free-standing high schools were often attached to these communities of women. Once high school education was delineated more clearly according to the years 14–18, Catholic secondary schools emerged which also hosted thousands of Milwaukee youth – for education, spiritual formation, athletic competition and these schools (as well as parishes) served as marriage brokers, bringing together young men and women who would marry. Analogs to all of these functions existed in wider society, but Catholics created their own sub-culture in Milwaukee which provided for members from cradle to grave and attempted to preserve religious homogeneity even as they welcomed the ethnic heterogeneity of the wider Milwaukee community.

But the central institution in Catholic life is the parish. This prayer and worship center is the heart and soul of Catholic life and identity. The city of Milwaukee provides an interesting case study for the impact of urban parishes and also of their subjection to the forces described above. The aforementioned changes took place all over the city of Milwaukee, but were particularly felt on the city's north side – an area bounded by the Milwaukee River on the East, Forty-third street on the West,

North Avenue to the north and State Street to the South. Within these boundaries, once a part of an original division of the city called Kilbourntown (later joined to the other two areas founded by different land speculators to form the city of Milwaukee) were an array of Milwaukeeans of different ethnicities and socioeconomic backgrounds. Roman Catholics who lived here were mostly, but not exclusively German-speakers – who were among the most dominant and industrious members of the growing Milwaukee Catholic community. Over the years, 14 parishes would be developed in this part of the city: two for English speakers, one for Slovaks, another for Bohemians and ten for Germans. As German-speaking Catholicism waned over the years in Milwaukee, these large and often ornate churches, all with substantial schools, convents, and rectories continued to welcome the sons and daughters of the original immigrants and thrived in the sturdy neighborhoods and business districts around them (Fig. 54.5). In this mix were the 14 parishes listed in Table 54.1 with their founding dates.

In 1950, these parishes served approximately 50,000 Catholics and schools served nearly 8500 grade school pupils. By 1990, the North Side area now served about 3700 Catholics and 670 children. A once vibrant and heavily Catholic area

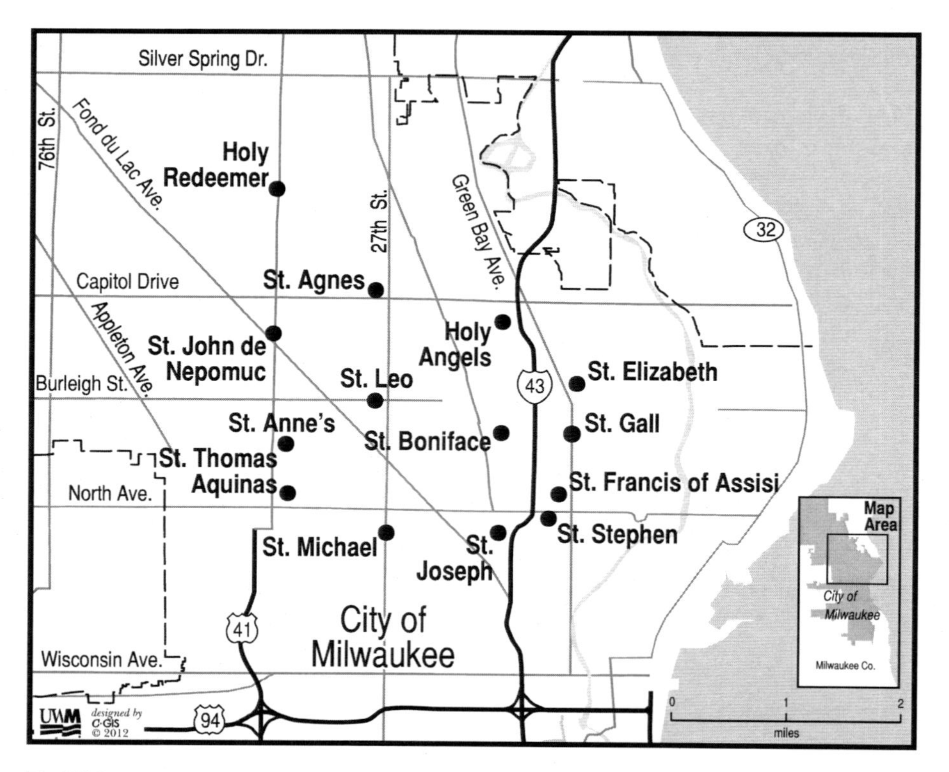

Fig. 54.5 Early dominant Catholic parishes and schools in the 1950s. Most had closed or were merged by the mid-1990s (Map by University of Wisconsin – Madison, C-GIS for the authors, 2012)

Table 54.1 Parishes with location and founding dates

Parish	Location	Founding date
St. Agnes	N. 25th and W. Capitol	1925
St. Anne's	N. 36th and W. Wright	1894
St. Boniface	N. 11th and Clarke	1888
St. Elizabeth	N. 2nd and W. Burleigh	1902
St. Francis of Assisi	4th and Brown Street	1871
St. Gall	N. 3rd and Clarke Street	1906
Holy Angels	N. 11th and Atkinson	1914
Holy Redeemer	N. 38th and North Hopkins	1897
St. John de Nepomuc	38th and Keefe	1863/1928
St. Joseph	N. 11th and W. Cherry	1855
St. Leo	N. 25th and W. Locust	1908
St. Michael	N. 24th and W. Cherry	1883
St. Stephen	N. 5th and W. Walnut (later moved to N. 51st)	1907/1952
St. Thomas Aquinas	N. 36th and W. Brown	1901

Source: S. M. Avella and T. Jablonsky

had virtually emptied out within a generation. By 1995 all of the following had been either closed or merged: St. Joseph's, St. Agnes, St. Anne's, St. Boniface, St. Elizabeth, St. Gall, Holy Angels, St. Leo, St. Thomas Aquinas, and Holy Redeemer. Although some still retained the property, e.g. St. Agnes and St. Elizabeth, their status as free-standing parishes was done-and they were officially "suppressed" and re-opened under new names with members of other parishes. Likewise the schools of these parishes either ceased operations altogether or merged with other parishes to form hybrid schools. This would include St. Agnes, St. Anne, St. Boniface, St. Elizabeth, St. Francis, St. Gall, Holy Redeemer, St. John Nepomuc, St. Joseph., St. Leo, St. Michael, St. Stephen and St. Thomas Aquinas. There had been an effort to reconfigure St. Leo's school in an Urban Day school in the late 1960s, but by the early 1970s, that plan was abandoned and a joint Central Cities school was created which combined Holy Angels, St. Thomas Aquinas, St. Leo, and a parish from the near south side, St. Rose of Lima.

What had occurred? Some of the change was attributable to proposed freeway building. In 1960, the Archdiocese of Milwaukee agreed to relocate the old St. Joseph Parish on 11th and Cherry to accommodate plans to build a portion of the interstate system through it. The old church, one of Milwaukee's founding trio of German parishes established by the first bishop, John Martin Henni (1843–1881), was torn down and its remaining parishioners relocated to a site in the suburb of Wauwatosa where it still exists today. Another smaller mission for African Americans, Blessed Martin DePorres, not really a parish, but a locale for the city's growing African American community, was also taken out by the construction of the Hillside Housing development. In 1952, the struggling St. Stephen's Slovak parish on 5th and Walnut relocated to a parcel of land on 51st.

Between 1950 and 1960 the most precipitous drop had been at the once thriving St. Boniface parish at 11th and Clarke. From a parish membership of 5113 in 1950, the numbers plunged to 1872, a decline of 63 %. A similar drop-off had been noted in St. Gall at Third and Clarke. Parish histories attribute this decline to the changing racial composition of the parishes. The St. Boniface area had already begun to see African American movement in its direction as early as the 1930s. As noted earlier in Sect. 54.1, Milwaukee's African American community had always been small until World War II. African American workers began to appear in larger numbers during the War, but most expansive growth of the black community had taken place during the 1950s as the small Bronzeville area bordered around Sixth and Walnut Streets and served by a number of churches, including the African American Catholic St. Benedict the Moor on 10th and State, began to push north and west for housing. St. Boniface school dipped from 777 to 461 students although St. Gall rose slightly from 305 to 309.

The decade between 1960 and 1970 was even more precarious for these once thriving parishes and schools (Figs. 54.6, 54.7, 54.8). St. Anne went from 4,478 to 2,250 members; St. Boniface shrunk further from 1,872 to 500 members, St. Elizabeth went from 4,450 to 1,100, St. Francis from around 2,000 to 750, St. Gall from 1,388 to 300, Holy Angels from 3,500 to 1,150 and St. Leo from 5,656 to 1,582. St. John de Nepomuc declined only slightly from 2,670 to 2,296 and St. Thomas Aquinas enjoyed a burst of what turned out to be temporary growth from 5,199 to 6,500. But the toll was even heavier on the schools. By 1970 the following schools ceased to exist, St. Boniface, St. Elizabeth, St. Francis, and St. Gall, representing a major collapse of Catholic education in the central city.

The decline between 1970 and 1980 was dramatic. St. Anne shrunk dramatically from 2,250 to 890, St. Boniface from 500 to 300, St. Elizabeth from 1,100 to 437, St. Gall dipped to 194 from 300 10 years earlier; Holy Angels from 1,150 to 690, St. John de Nepomuc began its steady descent from 2,296 to 1,800 and St. Leo plummeted from 1,582 to 750; St. Thomas dropped from 6,500 to 1,100. Further school closing included St. Michael's. Every other remaining school in the area registered drops in enrollment and higher costs.

The final piece came in the decade of the 1980s as parish numbers dropped further. St. Agnes had a mere 415 members by 1990, St. Boniface, long moved out of its old church into smaller quarters, had only 258 on its books; St. Elizabeth reported only 345 members; St. Francis, 702, St. Gall, 228; Holy Angels, 452, St. John de Nepomuc crashed down to 275 members, St. Leo to 423; St. Thomas Aquinas to 345. School closures ensued. St. Agnes, Holy Redeemer, and St. John de Nepomuc all closed their doors.

Affected by the spreading dynamic of neighborhood transition the Catholic institutional presence in this region began to shrink as well. In 1992, one of several collaboration projects undertaken by north side pastors bore fruit in the first major parish consolidation. Facing shrinking financial resources, in the 1980s St. Nicholas and Holy Redeemer parishes began working together on religious education and sacramental preparation. In 1987, both parishes St. Nicholas and Holy Redeemer began sharing staff and priests. In 1992, St. Albert, a parish north of the district, Holy Redeemer and St. Nicholas – all experiencing declines in parish membership-agreed to come together to form Blessed Trinity parish.

Fig. 54.6 St. Anne's Catholic church (Photo courtesy of Milwaukee Archdiocesan Archives, used with permission)

By 1994, a major consolidation of parishes took place under the direction of Archbishop Rembert Weakland (1977–2002). After an extensive study, it was determined to shutter some of the largest and most venerable churches of this area. Put on the block was St. Anne's and St. Boniface (which had not made it even in smaller quarters), so also with St. Gall church which had earlier closed the doors of its original church and tried to remain viable in a tiny center; St. Leo's closed its doors as did St. Thomas Aquinas. Many of these parishes were reconfigured and two new sites absorbed the old parishes, St. Agnes, renamed All Saints welcomed parishioners willing to come to its church: these included, St. Agnes, St. Anne, Holy

Fig. 54.7 St. Leo's Catholic church (Photo courtesy of Milwaukee Archdiocesan Archives, used with permission)

Angels, St. John de Nepomuc and St. Thomas Aquinas. St. Elizabeth's, renamed St. Martin de Porres, welcomed former parishioners from St. Boniface, St. Elizabeth, and St. Gall. A Central Catholic School system picked up the remnants of Holy Angels, St. Rose, St. Thomas Aquinas, and St. Leo Schools.

54.3.1 Reasons for the Change

Explaining the reasons for this decline inevitably connects to the twin realities of racial succession and de-industrialization that hit this part of Milwaukee hard. Most histories of Milwaukee, religious and secular, focus a great deal on the huge influx of African Americans into Milwaukee after World War II as discussed in Sect. 54.1

Fig. 54.8 St. Boniface Catholic church (Photo courtesy of Milwaukee Archdiocesan Archives, used with permission)

of this chapter. The inevitable expansion of the black section of Milwaukee was the petri dish for block-busting and other "scare" tactics that led white Milwaukeeans out of these neighborhoods and into other parts of the cities or the expanding suburbs. Racial tensions were felt in every neighborhood institution – but especially in parishes when African Americans sought admission for their children in Catholic schools and wished to join in Sunday worship and parish groups, such as the Holy Name (a men's society), the Catholic Youth Organization (an athletic group) and the Christian Mothers. A moment of uncertainty followed as efforts were made to reinforce the standing segregation of white and black Catholics by providing separate black Catholic parishes and chapels. However, at the insistence of Archbishop William E. Cousins (1959–1977), white parishes were forbidden to discriminate against African American children and their parents and all of the parishes admitted black parishioners and school children. By the early 1960s, integration went forward but not without some difficulty. As tensions increased between white and blacks in Milwaukee in the 1950s and especially during the 1960s, one of the parishes of this cohort, St. Boniface, would transition to a major center for Catholic-sponsored civil rights activism, spearheaded by white priest James Groppi. But even as parishes welcomed African Americans who moved into white neighborhoods, whites left – withdrawing their children from the school and transferring their financial support to new Catholic parishes. As the aforementioned statistics indicate, declining numbers meant a diminution of religious services (more below) and a withdrawal of priests and sisters from these shrinking communities.

White flight was also amplified by the withdrawal of industrial work from the area. One parish in particular that was hit by this was St. Leo's on 25th and Locust. One veteran priest who had been assigned to the parish after his 1945 ordination related how the parish Mass schedule for Sunday was configured to serve the religious needs of workers on the various shifts from the A.O. Smith Plant on 24th and Hopkins. Many St. Leo's parishioners worked for this industrial giant which made among other things body frames for cars. St. Leo's parishioners at one time enjoyed good wages and benefits from the plant and scores of young families settled around the parish. St. Leo's boasted of one of the largest Catholic Youth Organization groups in the entire ten county Archdiocese of Milwaukee. The parish's later decline can no doubt be directly traced to the reduction of the plant work force and its sale of the automotive division in 1997 after 90 years.

54.3.2 The Human Cost

What were the practical implications of this shrinking presence? Since Catholicism puts much stock in its institutional visibility, the decline meant the end of parish church buildings – some of them of great beauty and relative antiquity. It meant a withdrawal of religious services, the shrinking and disappearance of once large Catholic social groups, and of course a decline in income and expenditures – the latter contributing to the economic decline of the area.

Church buildings were the most powerful symbol of the change. Two parishes were literally torn down: St. Joseph's, a fixture in the city since the 1850s, gave way to freeway expansion. St. Boniface, a huge church which seated 1,400 in its heyday, was torn down in the 1970s to make way for a new North Division High School. Three other large and architecturally and artistically elegant churches were shuttered and sold: St. Leo's, church replete with magnificent statuary, icons, and a large altar had been built in the 1920s, St. Thomas Aquinas, which had the look of an English country church and St. Anne's a majestic church for German speakers, replete with exquisite wood-work, a Tiffany-glass dome and luminous stained glass windows. All closed their doors and some became worship sites for other religious communities. The subtraction of these churches, all of which held precious memories of sacramental moments, school days, and the untold tales of hard work and sacrifice by priests, sisters, and lay people created an unresolved sadness that was difficult to acknowledge at the time they closed since the decisions to end them were so tied to finances. As lovely as they were, few people worshiped them any longer. Some of these churches were sold to Protestant denominations. St. Leo's was purchased by the Little Hill Church of God in Christ, an African American evangelical community. St. Anne's became the Capital Christian Center, another locale for African American Protestants. However, once the ecological "golden cord" between residential parish and the Catholic community had been severed many of the churches, schools, convents and rectories became like shuttered factories and stores: hulking reminders of a once thriving urban civilization.

Worship was of course the primary reason these institutions existed in the first place. Sampling the number of Masses offered – the key worship experience of the Roman Catholic faith – is an important indicator of its vitality. In 1950–1951, St. Anne's had seven Masses on Sunday: St. Boniface, St. Leo and St. Thomas Aquinas each had six. A little more than 10 years later, change was already evident. Perhaps tracking the westward movement of the transition, St. Anne's actually added an additional Mass (they were large enough to have Masses in capacious upstairs and downstairs churches); St. Leo's too added one Mass and St. Thomas held at six. But St. Boniface, one of the first churches to experience the wave of racial succession had already pared its Mass schedule down to four by 1963. By 1973, important reductions in religious services were well underway. St. Anne dropped to six Masses, St. Leo to four, St. Thomas to five and St. Boniface, now on the brink of destruction, was down to two. By 1983, St. Anne had four Masses, St. Boniface (now in new storefront quarters), St. Leo, and St. Thomas all held three Masses each weekend. By 1993 these parishes were virtually extinct. St. Anne had 2 Masses wherein only 100 parishioners attended, St. Boniface, had 2 and St. Leo and St. Thomas were down to 1 Mass. Many of these parishes shared just one priest who traveled to each site.

Parish organizations also reflected these changes. Catholicism's social clubs, sorted out for age and gender, were an important and vibrant part of the Catholic presence in the city. Large groups like the Holy Name or the Sodality held annual marches and meetings that literally turned out thousands for parades, devotional celebrations in the capacious Milwaukee Auditorium and field Masses or Marian devotions that turned out thousands of people – even in inclement weather. St. Anne's Holy Name claimed 840 men in 1950, St. Leo's two CYO groups had nearly 1,000 in the same year. St. Leo still had 650 men in 1963 while another popular parish group, the Christian Mothers, claimed 600 members. These groups held popular fish-fries, smokers, card parties, and dances – all of which made the parish more than a worship and educational site. By the 1970s, these groups had all but disappeared – some of them casualties of a new regimen of lay spirituality ushered in by the liturgical changes of Vatican II (1962–1965) – but also reflecting the decline in parish membership. Lost forever was a component of the social and cultural contribution of Catholics to their neighborhoods, the purchase of goods from vendors, and the occasion to form bonds of unity with other people in the surrounding area. In its place, a new type of engagement with the community took place – albeit on a less visible and less overtly Catholic manner. Empty convent and school facilities were often rented or purchased by secular entities. The large school built by the parishioners of St. John de Nepomuc in the 1950s was acquired by the Milwaukee public school system in the 1980s and re-christened Frederick Douglass School. The empty convent of St. Leo's parish had a number of different uses, but at one point it was a pre-release facility for prisoners from the local jail. Other closed convents and schools as well as parish rectories (residences for parish priests) were also rented out or sold to day care facilities, half-way houses, and homeless shelters.

In summary Catholic institutions provide an important window into the contours of post-war de-industrialization and racial change in many American cities.

Milwaukee's central city parishes were at one time proud emblems of a large and vibrant Catholic population. They were also important contributors to the city's commonweal and provided important services and common ground for urban residents. Although other institutions and groups would take its place, perhaps providing an equivalent and more cultural sensitive set of urban services and presence, the passing of an active Catholic presence in this part of the city of Milwaukee is worth noting and observing as part of the larger evolution of American urban life in the period after World War II.

References

Avella, S. (2002). *In the richness of the earth*. Milwaukee: Marquette University Press.

Conzen, K. (1996). The place of religion in urban and community studies. *Religion and American Culture, 6*, 108–114.

Cutler, R. (2001). *Greater Milwaukee's growing pains*. Milwaukee: Milwaukee County Historical Society.

Cutler, I. (2006). *Chicago*. Carbondale: Southern Illinois University Press.

Dahlk, B. (2010). *Against the wind*. Milwaukee: Marquette University Press.

Gurda, J. (1999). *The making of Milwaukee*. Milwaukee: Milwaukee County Historical Society.

Levine, M., & Zipp, J. (1993). A city at risk. In J. Rury & F. Cassell (Eds.), *Seeds of crisis* (pp. 42–72). Madison: University of Wisconsin Press.

Milwaukee Journal Sentinel, 1, 20–22. November 13, 2011.

Simon, R. (1996). *The city-building process*. Philadelphia: American Philosophical Society.

Squires, G., & O'Connor, S. (2001). *Color and money*. Albany: State University of New York.

Terrell, J. (1995). *The socio-spatial expansion of Milwaukee's African American community*. Milwaukee: Marquette University. McNair Scholar Project.

Trotter, J. (2007). *Black Milwaukee*. Urbana: University of Illinois Press.

Chapter 55
The View from Seminary: Using Library Holdings to Measure Christian Seminary Worldviews

Katherine Donohue

55.1 Introduction

The concept of worldview is loosely defined for the purposes of this chapter. One could easily replace "worldview" with "perspective." The materials found in any given library have the ability to influence the mind of the student, and at the same time they reflect the worldview held by the faculty and staff who put together the library collection. The primary purpose of a Catholic or Protestant seminary library is to educate students on Christianity, and the library materials will reflect this goal. However, Christianity is not a monolith, nor does it exist in a vacuum. Therefore, any given library will reflect the denomination of the seminary, along with the seminary's views on the relationship of religion to international relations and social controversies.

This chapter looks at those specific lenses that form the worldview of seminaries across the globe. The holdings of the library provide commentary on the school's perspective regarding monotheistic religions, Christian denominations, international relations, scientific speculations, and secular debates. This chapter will also examine the expanding role of the Internet in shaping the worldview of seminary libraries. First, however, it will be helpful to examine the background and development of the seminary library.

55.1.1 The History of Seminary Libraries

Few studies document the evolution of the seminary library. A couple of studies have examined the development of individual seminary libraries, like Thomas Slavens' article on the Union Theological Seminary (Slavens 1976). Other studies

K. Donohue (✉)
M.A. Diplomacy and International Commerce, Patterson School of Diplomacy and International Commerce, University of Kentucky, Lexington, KY 40506, USA
e-mail: kjanaed89@gmail.com

S.D. Brunn (ed.), *The Changing World Religion Map*,
DOI 10.1007/978-94-017-9376-6_55

have examined trends in seminary libraries, like Ron Jordahl's study on the development of seminary librarianship (Jordahl 1990). Jordahl theorized that the concept of seminary librarianship evolved during the twentieth century. He analyzed the history of Protestant seminaries in the United States and came up with four distinct phases of librarianship within seminary library development: The *professor-librarian*, the *professor-librarian with an assistant*, the *educator-librarian*, and the *specialist-librarian*. The characteristics of each phase of librarianship provide insight into the development of the seminary library.

During the *professor-librarian* phase, the role of the library in seminary education was minimal. A member of the faculty managed the library collection and frequently selected books based on faculty interests and research. At this point, the library was not a key resource for students. The students were expected to learn primarily through the lectures given by professors. As library collections grew, the professor-librarians realized that a better cataloguing system was needed to make the resources accessible. At this point, many professor-librarians chose to hire a full-time assistant (Jordahl 1990). Assistants with various degrees of library training worked alongside the professors to organize the library. At Union Theological Seminary, Julia Pettee was hired to redesign the cataloguing system for the library. Pettee considered the Dewey decimal system to be insufficient for cataloguing a seminary collection. She constructed a cataloguing system that divided the collection into three main sections: historical sciences (for bibliographies and histories), experimental sciences dealing with the material universe and mental phenomena (including philosophy and ethics), and practical sciences (including education, fine arts, and special collections). With this cataloguing system in place, seminaries like Union Theological could begin to develop strategies for expanding their library collection (Slavens 1976).

The expanding library collections coincided with a dramatic rise in seminary enrollment during the 1950s. During this time period, seminary libraries entered the phase that Jordahl terms the *educator-librarian*. While the teaching continued to place emphasis on education through rote learning, the libraries began to view their mission as an educational supplement for the students (Jordahl 1990). This mode of thinking was expanded in the fourth phase described by Jordahl, the *specialist-librarian*. During this phase, librarians became more proactive in their development of the seminary library. Libraries underwent automation and specialization. Some of these *specialist-librarians* possessed library science degrees; however, more frequently the librarians chose to obtain a degree in theology and took some outside courses on librarianship as a supplement. Through the leadership of these librarians, seminaries worked to expand their special collections and to increase the amount of information and resources available to their students (Jordahl 1990).

Union Theological Seminary provides a good example of the four phases of development in the seminary library. As the largest theological library in the Western hemisphere, with more than 700,000 items according to their website, scholars like Thomas Slavens sought to document the development of this collection. During the late 1800s, William Rockwell, a member of the faculty, managed the library on his own until he brought in an assistant, Julia Pettee, in 1909. Pettee re-catalogued the books in the library and paved the way for further library and collection development (Slavens 1976). During the 1940s, the library's mission shifted to focus on providing educational resources for the students. The mission of the library today is to support the seminary's curriculum and student research, while maintaining a collection that reflects the existing scholarship on theology across the world. As with many seminaries in the Internet age, the library website now connects students to electronic resources that further expand the breadth of information available to the seminary student.

Protestant seminaries outside of the United States appear to have followed similar patterns, although the majority of seminaries with an online presence are in phases one through three. Religious groups from the United States founded many of these Protestant seminaries, and a large percentage of the library materials come from donations made by the sponsoring denomination. The Asia Pacific Theological Seminary was conceived at the 1960 Far East Conference of the Assemblies of God in Hong Kong and then brought to fruition under the leadership of Maynard Ketcham, Field Secretary for the U.S. Assemblies of God. The Faculté de Théologie Evangélique de Boma (FACTEB) in the Democratic Republic of the Congo (DRC) came about through the vision of Dr. Willys Braun, a missionary for the CMA (Christian Missionary Alliance). The seminary website credits the establishment of the Boma Seminary to the support of the CMA in the U.S. and Canada.

In South America and sub-Saharan Africa, many libraries are managed by a member of the faculty, and the library size is often dependent on the quality and quantity of the donations, as very few websites made mention of a separate library stipend. At FACTEB in the DRC, the assistant professor of biblical languages doubles as the director of library services, and the Seminario Sudamericano in Quito, Ecudaor received a large portion of its library resources from donors James and Virginia Beaty and the Carcelén family.

Seminary libraries in Europe, Asia, and the Middle East, by contrast, showed a very strong educational component, placing them in phases three and four of Jordahl's model. The library at the South Asia Institute of Advanced Christian Studies (SAIACS), for example, offers a 3-month theological librarianship course and is a member of the Joint Library Committee in Bangalore for theological studies.

While there were certain national trends for librarianship in seminaries, the development and expansion of collections appear to be largely organic and tailored to each individual seminary. In the nineteenth century, seminary libraries were largely dependent on donors, and the library collections often reflected the views and preferences of the donors. When libraries began to receive stipends for purchasing additional materials for the library, the librarians had to prioritize which materials would make the greatest contribution to the seminary. Most websites for seminary libraries include a mission statement, often pledging the library's commitment to supporting and supplementing the curriculum of its seminary (Table 55.1). As such, many libraries have acquired special collections that reflect their denomination and geographic location. Princeton Theological Seminary, for example, has the Samuel Agnew Baptism Collection and The Grosart Library of Puritan and Nonconformist Theology, reflecting both the seminary's Presbyterian background and its location in the American northeast.

Table 55.1 Seminary library mission statements

Seminary	Library mission statement/objective
Union Theological Seminary	The mission of The Burke Library is to identify, acquire, organize, provide access to, interpret, and preserve for the future information in the field of theology and contextually related areas of study. Furthermore, the Library supports the specific instructional and research needs of Union Theological Seminary and Columbia University and provides resources for the scholarly community in theology and related areas of the humanities. The Library reflects in its collections the pluralistic and ecumenical concerns of the Seminary while maintaining its role as a comprehensive resource within the limits of its collecting policies. In carrying out its mission, the Library cooperates with other institutions both regionally and nationwide.
Arab Baptist Theological Seminary	Education is at the heart of Christian mission, "… teaching them to obey all that I have commanded you" (Matthew 28:20). Educational ministries abound at every level – from discipleship of new believers to higher level theological study. Unfortunately these ministries are too often inadequately resourced and/or poorly thought through. The purpose of the ABTS Resource Center is to provide human and material resources for facilitating excellence in strategic educational ministry in the Middle East and beyond.
Alliance Theological Seminary	The Division of Library Services enriches the learning experience by providing quality resources and responsive service to encourage users to develop life-long critical thinking skills within a Christian framework.
University of St. Mary of the Lake	The mission of the library is to support the education and formation of men for pastoral ministry as diocesan priests. This mission has been expanded to provide educational resources for separate diaconate and lay ministry programs, insuring competent graduates for the parishes and agencies of the Archdiocese of Chicago.
Princeton Theological Seminary	The libraries' major objective is to acquire a comprehensive collection of the basic works of world theology

Source: Katherine Donohue

The subjects of seminary librarianship and collection development continue to evolve. In 1946, the American Theological Library Association (ATLA, www.atla.com) was founded to serve and support the development of theological studies in the United States. ATLA is a professional association comprised of approximately 1,000 individual, institutional, and affiliate members. ATLA also publishes a journal, *Theological Librarianship*, providing a valuable national forum for discussing the past and future direction of trends in the development of seminary libraries. Recently, the journal has shown particular interest in incorporating and utilizing Internet resources within the seminary library system (Table 55.2).

55.1.2 The Geography of Seminary Libraries

When examining the contents of a seminary library, there are certain variables that have a strong impact on the worldview of the seminary. These variables include the size and age of the library, the denomination of the seminary, the geographic location, and the use of Internet databases.

Table 55.2 Articles from theological librarianship

Issue	Author	Article title
Vol. 5, No. 1 (2012)	Andrew J Keck	Mobile Devices and Libraries
	David R. Stewart	Introduction: The Reshaping of Libraries
Vol. 4 No. 1 (2011)	Beth M. Sheppard	A Good Look at the Nook
	Paul Stuehrenberg	Theological Libraries and International Collaboration in Southeast Asia

Source: Katherine Donohue

When a library is fairly young, or the resources are limited, the books that the library chooses to acquire give a strong indication of the worldview of the seminary. When the collection is small, the worldview is, by necessity, limited. The larger the library, the more perspectives it can accommodate. From the data gathered in this study, it was noted that when a library contained more than 250,000 items, the ratio or percentage of a given subject matter provided more insight into the seminary library than the individual books that were acquired. Nearly all of the libraries with more than 250,000 items contained at least the basic materials for the subjects examined in this study, so the percentage of the library devoted to a particular topic proved more indicative of the library's perspective than any set of individual books in its collection. Over half of the seminaries surveyed in this study possess more than 250,000 books. For that reason, the library contents are examined according to the percentage of the library filled by a given subject matter, such as the percentage of library materials devoted to Catholicism or evolution.

In addition to the library age and size, the denomination of a seminary can also shape the worldview of the seminary. The Unitarian seminaries and non-denominational seminaries are more likely to contain a balanced perspective regarding denominations and religions. In the case of certain non-denominational seminaries, like Union Theological Seminary, the theology degree requires a variety of investigative material to be available for the students' research.

The geographic location of the seminary also influences their worldview. The seminary libraries presented in this chapter contain a strong bias toward seminaries in developed countries. While an effort was made to include seminaries from various countries and continents, the distribution of seminaries worldwide is far from uniform. The United States holds the greatest concentration of seminaries (Fig. 55.1). For this reason, the majority of the seminaries in this chapter are in the United States. In addition, it is not uncommon for a seminary in Asia or Africa to be closely affiliated with (and often founded by) a seminary in the United States or Great Britain, as with the Asia Pacific School of Theology in the Philippines and FACTEB in the DRC.

The distribution of Protestant seminaries reflects the impact of these affiliations. The majority of Protestant seminaries are concentrated in the U.S. and the small pockets of Protestant seminaries in other countries began as American or European missions. Figure 55.2 shows the distribution of Protestant seminaries, with the United States, India, Australia, Germany, and Britain holding the greatest concentrations.

The Catholic seminaries are more evenly spread across the globe, radiating from the central governing structure in Rome (Fig. 55.3). Once again, the United States

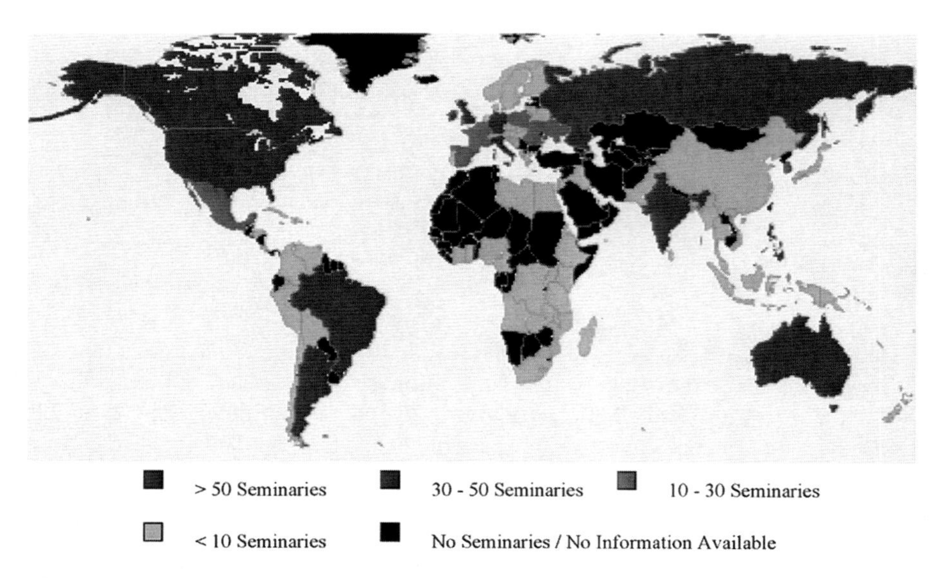

Fig. 55.1 Distribution of seminaries worldwide (Map by Katherine Donohue)

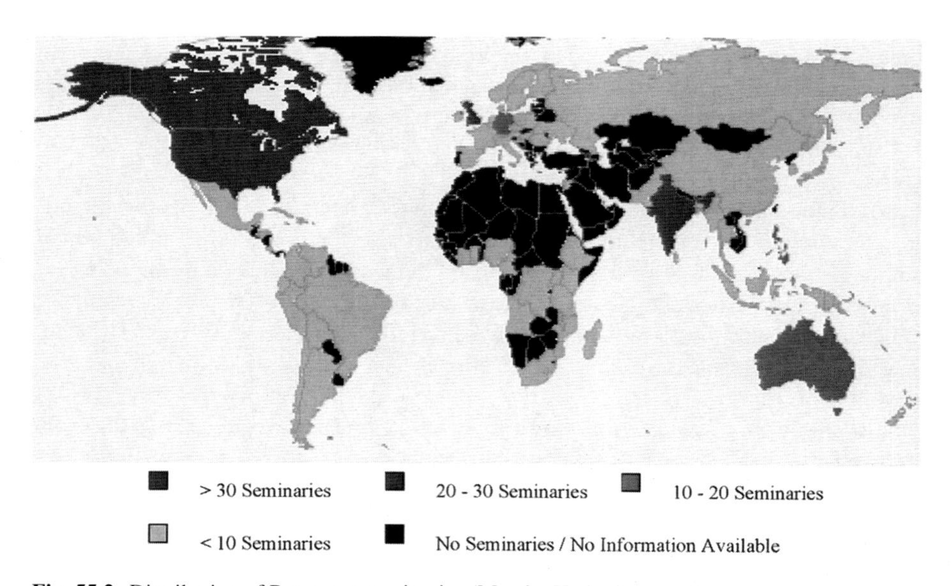

Fig. 55.2 Distribution of Protestant seminaries (Map by Katherine Donohue)

holds the greatest number of Catholic seminaries; however, there are seventeen other countries containing significant concentrations, and these appear in both developed and developing countries. Still, one can see from all three maps that developed nations host the majority of seminaries, and the majority of the data used

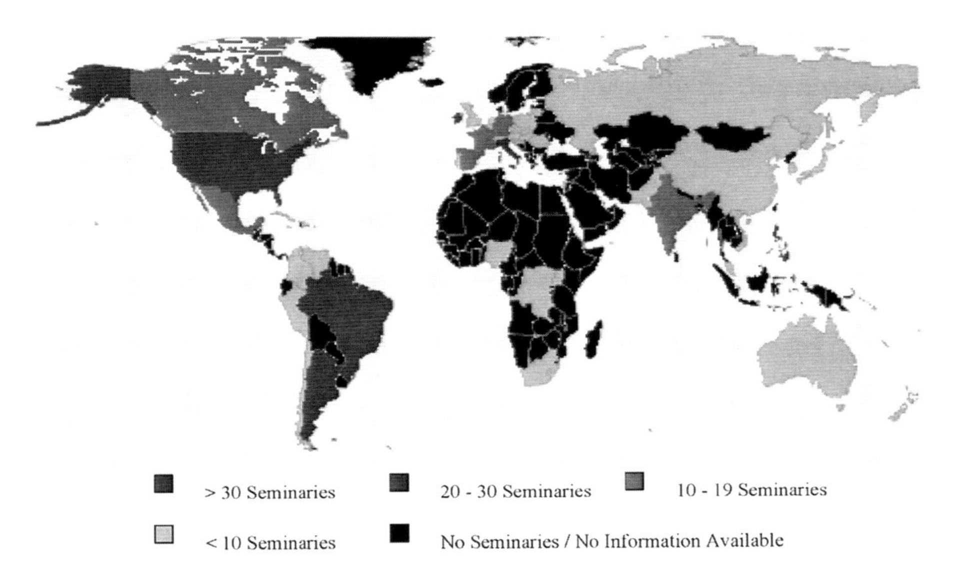

Fig. 55.3 Distribution of Catholic seminaries (Map by Katherine Donohue)

in this chapter comes from these seminaries, giving many of the data sets a similar bent, that is, biased towards Western issues and points of contention.

This study examined 51 seminary libraries (Table 55.3), in which 33 seminaries came from North America, three seminaries came from South America, six seminaries came from Europe, three came from Oceania, four came from Asia, one came from the Middle East, and one came from Africa. The seminaries available for the study were limited to those that had online library catalogues. Many of the seminaries in Africa and the Middle East lacked online catalogues, so only their library mission statements were compared to other seminaries from the geographic region. At least one seminary from each region was analyzed for library content.

55.1.3 Analyzing Seminary Library Content

There are several standard methods for analyzing library content. This article makes use of the conspectus method of analysis. The conspectus method was developed by the Research Libraries Group (RLG) as a way to evaluate the quality of a library collection. The RLG, a U.S.-based library consortium, was founded by a group of research libraries, including Columbia, Harvard, and Yale. The RLG developed the Eureka inter-library search engine and worked to compile various databases for documenting library collections. In 2006, RLG merged with the Online Computer Library Center (OCLC), the group that developed the WorldCat online catalogue. Despite the merger, the RLG Conspectus method is still used by the Library of Congress to assess the extent of the Library's collection on a particular subject.

Table 55.3 Seminary libraries surveyed

Denomination/Movement affiliation	Seminary (located in the US unless otherwise indicated)
Catholic	The Catholic University of America
	University of St. Mary of the Lake
	Mount St. Mary's Seminary
	Universita Cattolica del Sacro Cuore (Italy)
	Seminario Missionario Arquidiocesano (Brazil)
	Universidad Catolica de Salta (Argentina)
	Université Catholique De L'Ouest (France)
Orthodox	St. Vladimir's Seminary
	Holy Cross School of Theology
	University of Balamand (Lebanon)
Mennonite	Eastern Mennonite University
Lutheran	Concordia Seminary
	Evangelisches Stift (Tubingen, Germany)
Anglican/Episcopalian	Church Divinity School of the Pacific
	Virginia Theological Seminary
	Episcopal Divinity School
Presbyterian	McCormick Theological Seminary
	Princeton Theological Seminary
Unitarian	Meadville/Lombard Theological School
United Free Church of Scotland (1900)	University of Edinburgh Divinity School (Scotland)
Baptist	Southern Baptist Theological Seminary
	Golden Gate Baptist Theological Seminary
Methodist	Asbury Theological Seminary
	Claremont School of Theology
African Methodist Episcopal (AME)	Payne Theological School
Mormon	Brigham Young University
Seventh Day Adventist	Seventh Day Adventist Theological Seminary
Christian Missionary Alliance (CMA)	Alliance Theological Seminary
Assemblies of God	Assemblies of God Theological Seminary
	Hansei University (South Korea)
	Alphacrucis College (Australia)
	Harvest Bible College (Australia)
	Asia-Pacific Theological Seminary (Philippines)
Evangelical	Bethel Seminary
	Trinity Evangelical Divinity School
	Dallas Theological Seminary
	Gordon-Conwell Theological Seminary
	South Asia Institute for Advanced Christian Studies, SAIACS (India)
	Faculdades EST [Escola Superior de Teologia] (Brazil)
	School of Mission and Theology (Norway)
	China Graduate School of Theology (Hong Kong)

(continued)

Table 55.3 (continued)

Denomination/Movement affiliation	Seminary (located in the US unless otherwise indicated)
Non-denominational	Harvard Divinity School
	Chicago Divinity School
	Union Theological Seminary
	San Francisco Theological Seminary
	Duke Divinity School
	Seabury-Western Theological Seminary
	Western Theological Seminary
	NUI [National University Ireland] Maynooth (Ireland)
	University of Otago: Theology Department (New Zealand)
	University of Ghana: Religion Department (Ghana)

Source: Katherine Donohue

In the conspectus method, the researcher examines the amount of materials collected for a specific subject. That amount is then placed in one of six levels:

0. Out-of-Scope: No materials acquired on the subject
1. Minimal Level: The library contains the most basic works pertaining to the subject
2. Basic Information Level: The library has up-to-date material to provide an introduction to the subject.
3. Instructional Support Level: A collection sufficient for most undergraduate and graduate education on the subject.
4. Research Level: The collection includes a number of published source materials that would be sufficient for most undergraduate and graduate research needs.
5. Comprehensive Level: A collection that includes most of the significant published works on the subject, often including a significant archive of historical works on the subject matter.

The conspectus method has the advantage of quickly differentiating a library's strong subjects from the weaker ones. Once the strong subjects have been identified, the observer has a good framework for analyzing the seminary's worldview. The disadvantage to the conspectus method is the subjective nature of the category divisions and a bias toward the larger North American libraries (Ekmekcioglu and Nicholson 2002). Although this bias is clearly present in the analysis, the vast majority of the seminaries are in the U.S. and this method is useful to help distinguish between the worldviews of various American seminaries.

Using this method, the seminaries were analyzed for their worldview based on their library collections. First, the religious materials in the libraries were examined. The religious subjects evaluated on the conspectus scale include monotheistic religions, Christian denominations, and Eastern philosophy and religion (Table 55.4). Next, the secular content in the library was analyzed for collections on key countries and contemporary secular social debates.

Table 55.4 Seminary library collections on secularism and the Evangelical movement evaluated using the conspectus method

Conspectus ranking	Materials on secularism	Materials on the evangelical movement
1. Minimal Level	Universita Cattolica del Sacro Cuore	Universita Cattolica del Sacro Cuore
	Concordia Seminary	McCormick Theological Seminary
	NUI Maynoot	Seventh Day Adventist Theological Seminary
	Gordon-Conwell Theological Seminary	Gordon-Conwell Theological Seminary
	San Francisco Theological Seminary	
	Southern Baptist Theological Seminary	
2. Basic Information Level	Chicago Divinity School	Dallas Theological Seminary
	McCormick Theological Seminary	Concordia Seminary
	Asbury Theological Seminary	Princeton Theological Seminary
		Chicago Divinity School
		NUI Maynooth
3. Instructional Support Level	Harvard Divinity School	The Catholic University of America
	Seventh Day Adventist Theological Seminary	Harvard Divinity School
	Dallas Theological Seminary	San Francisco Theological Seminary
4. Research Level	The Catholic University of America	Asbury Theological Seminary
	Alliance Theological Seminary	Southern Baptist Theological Seminary
	Princeton Theological Seminary	
5. Comprehensive Level		

Source: Katherine Donohue

55.2 Perspectives on the Monotheistic Religions

Given that all the seminaries in the study are Christian, one expects to find that among the three monotheistic religions, Christianity will dominate the library holdings. The data supports this hypothesis, showing a large concentration of Christian literature (61 % in Catholic seminaries versus 76 % in Protestant seminaries), with significantly less literature on Judaism (26 % versus 15 %, respectively), and even less on Islam (13 % versus 9 %, respectively). While these figures reflect a fairly consistent order of distribution, the Catholic seminaries selected revealed a slightly more even distribution among the three religions than the Protestant seminaries selected.

55.2.1 *Judaism*

The majority of the seminaries reflected the Instructional Support Level of the conspectus scale. As mentioned in the introduction, the conspectus method has a bias toward North American libraries. The size of North American libraries often dwarfs their counterparts in other continents. Because of the abundance of resources available to many North American seminaries, they can develop collections on many different subjects. Within the North American seminaries, the largest libraries, often connected to larger research and Ivy League universities, attained the Comprehensive Level for Judaic materials. This includes the non-denominational seminaries, like Harvard Divinity School and Union Theological Seminary, as well as the long-established seminaries like Princeton Theological Seminary. Six seminaries fell into the Minimal Level, and that directly corresponded to the seminaries that had less than 50,000 books: Universidad Cattolica, Payne Theological School, Asia-Pacific Theological Seminary, SAIACS, Alphacrucis College, and Seminario Missionario Arquidiocesano. With the exception of Payne Theological School, the seminaries are not located in the United States.

Given the common roots of Christianity and Judaism, one would expect most of the seminaries to have at least a Basic Information Level of Judaic material. It is interesting to note, however, that the amount of Judaic material almost always surpasses the amount of Islamic material. There is a much larger Muslim population (about 1.6 billion people) than Jewish population (about 13.4 million), yet based on the library content, the seminaries appear to place a stronger emphasis on the Judaic religion.

55.2.2 *Islam*

While the Judaic material falls largely within the Instructional Level, the Islamic materials in seminary libraries are evenly distributed between the Basic Information and Instructional Levels. Once again, the largest seminary libraries in North America, like Union Theological Seminary and Seabury-Western Theological Seminary, had a Comprehensive Level of material, and the smallest libraries, like the China Graduate School of Theology and Payne Theological Seminary, had a Minimal Level. The smaller libraries also showed the greatest disparity between the amount of material on Christianity and the other two monotheistic religions. The one exception to this trend was the University of Balamand in Lebanon, which had more material on Islam than Judaism *and* Christianity put together. In this case, the geographic location appeared to have a greater impact on the seminary worldview than the size of the library.

55.3 Denominations and Regional Relations

To get a sense of each seminary's doctrinal worldview, three key denominations were analyzed to help place the worldviews of seminaries within a global perspective: (a) the split between Catholic and Protestant, (b) the growing movement of

Evangelical churches that has dominated much of the Christian dialogue this past decade, and (c) regional relations, viz., interfaith and interdenominational groups, viz., the National and World Council of Churches.

55.3.1 Catholics and Protestants

Among all of the seminaries, published materials about Catholicism outstripped published materials on Protestantism. The Catholic seminaries have almost the same average distribution as the Protestant seminaries (70 % Catholic material and 30 % Protestant material), which suggests that both Catholic and Protestant doctrinal perspectives are present at most seminaries. The prevalence of Catholic materials in the individual Protestant seminaries, however, is widely varied. Within the Protestant seminaries, the largest collections of Catholic materials are found in the larger libraries of the high church denominations, like the Virginia Theological Seminary and the Episcopal Divinity School in Massachusetts. These seminaries did not have collections large enough to be placed in the Comprehensive Level, but they were large enough to rank in the Research Level. Among the Catholic seminaries, the larger North American libraries, like the Catholic University of America, have a more balanced distribution of materials on Catholicism and Protestantism than the ones outside of North America, such as Seminario Missionario Arquidiocesano in Brazil.

55.3.2 Rise of the Evangelicals

In recent years, the Evangelical movement has spread rapidly across the United States and into South Korea and Australia. The Evangelical movement closely mirrored the Charismatic movement, blending with other Protestant denominations instead of developing a distinct denomination. The churches frequently associated with the Evangelical movement are Baptist and Assemblies of God. In addition, many Non-denominational churches have sprung up across America, some of them leaning towards religious Fundamentalism, like the Grace Bible Church of Fresno, and others bridging the gap between religion and contemporary pop culture, like Willow Creek Community Church. Unlike the materials on Protestantism and Catholicism, most of the materials on the Evangelical movement are concentrated in the Evangelical seminaries. Figure 55.4 shows the distribution of materials about the Evangelical movement in Protestant seminaries compared to the amount of library materials on Protestantism, Catholicism, and the Reformation.

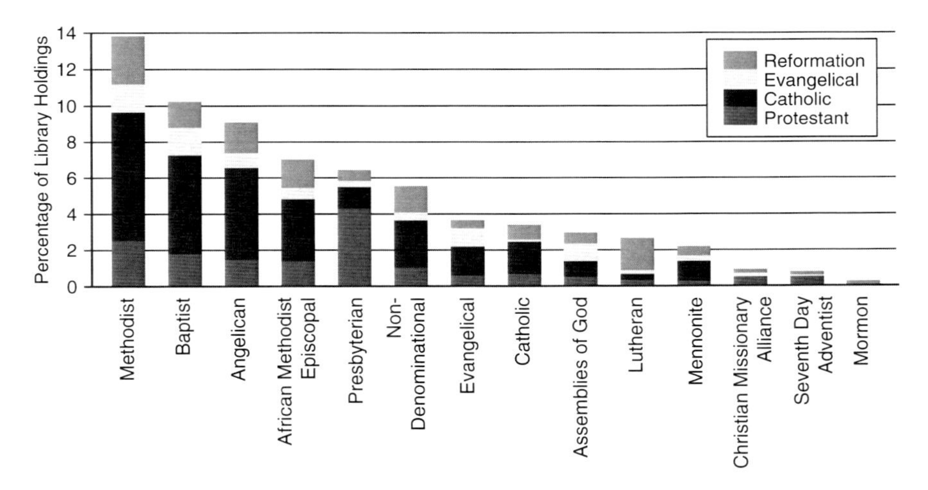

Fig. 55.4 Materials on church denominations in seminary libraries (Source: Katherine Donohue)

55.3.3 Regional Relations

As globalization becomes an increasingly popular catchword in the academic world, there are several religious movements that worked toward increasing the dialogue between faiths and denominations. The Ecumenical movement that began after World War II led to the establishment of two large church councils—the National Council of Churches in the U.S. headquartered in New York City and the World Council of Churches based in Geneva, Switzerland. The interfaith movement gained popularity in the 1960s, after changes brought about by Vatican II called for further dialogue and cooperation between different faiths.

Figure 55.5 shows the average distribution of *interfaith* literature among the various seminary denominations. The widespread appeal of the interfaith movement shows itself in the even distribution in seminaries of multiple denominations. The Unitarian seminary, Meadville/Lombard Theological School, contained a significant collection of materials on the interfaith movement in keeping with their doctrinal views (in the figure the Unitarian seminary was averaged in with the non-denominational seminaries). In addition to the Unitarian seminary, Presbyterian, Anglican, non-denominational, and Baptist seminaries maintained a Research Level collection of interfaith material. Princeton Theological Seminary contained a Research Level of interfaith material, along with Asbury Theological Seminary and Golden Gate Baptist Theological Seminary.

Figure 55.5 also shows the distribution of literature in seminary libraries from the *ecumenical movement*, which includes writings about the National and World

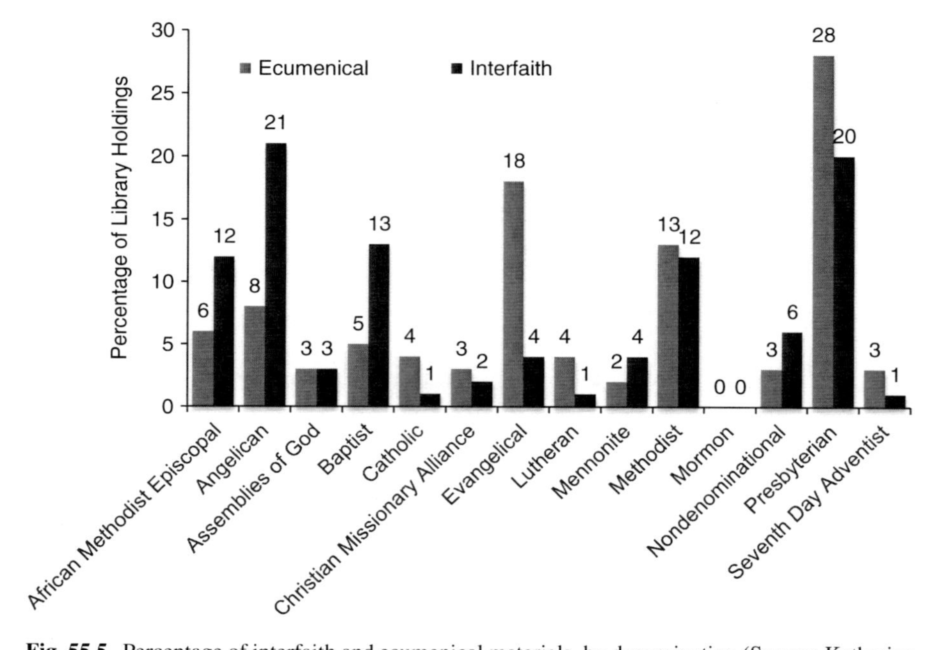

Fig. 55.5 Percentage of interfaith and ecumenical materials, by denomination (Source: Katherine Donohue)

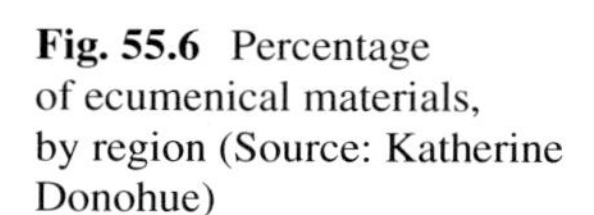

Fig. 55.6 Percentage of ecumenical materials, by region (Source: Katherine Donohue)

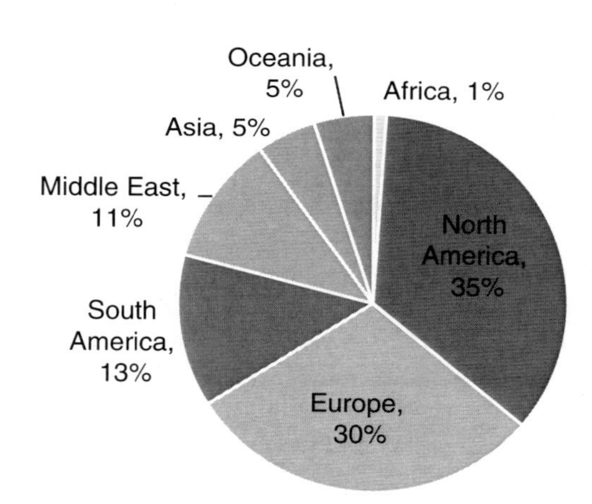

Council of Churches. These councils have a particularly Western bent, as can be seen from the regional distribution (Fig. 55.6), where 65 % of the literature comes from seminaries in North America and Europe. Seminaries in the U.S. and England contain significant amounts of ecumenical material. The Catholic seminaries are notably absent, as the ecumenical movement until recently did not encompass the Catholic Church (note, the Catholic seminaries only make up 1 % of the ecumenical materials). Given that the Catholic Church already had a centralized structure organizationally, scholars observe that Mainline Protestant churches created the World Council of Churches in an effort to parallel the central unity of the Catholic Church.

55.4 Seminary Libraries and the Disparity Between Eastern and Western Religions

In the process of analyzing seminary library holdings, one trend consistently emerged: the seminaries focused the majority of literature on Western monotheistic religions. The religions and philosophies of the East (Hinduism, Buddhism, Taoism, and Shintoism) appeared to have little significance to the seminary libraries. Even the seminaries located in the East (viz., Asia) reflected this disparity. Figure 55.7 shows the proportion of Western religious literature compared to Eastern religious literature. The disproportion exists in both Western and non-Western seminaries. Within the existing seminary literature on Eastern philosophy and religion, Hinduism and Buddhism dominate the literature (Fig. 55.8). Several of the larger libraries, like the Duke Divinity School and Union Theological Seminary, possessed a Basic

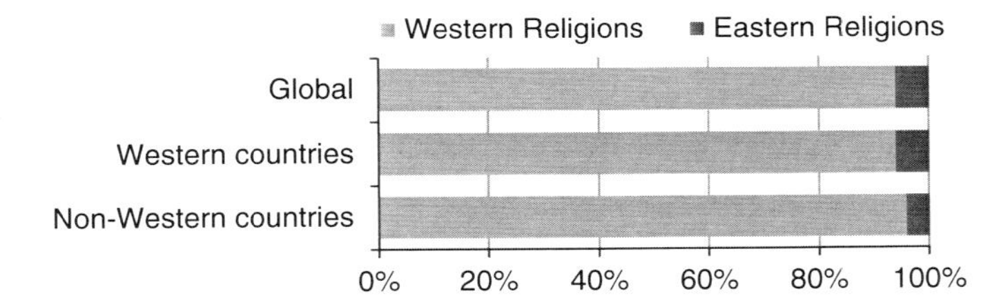

Fig. 55.7 Distribution of western versus eastern religious materials (Source: Katherine Donohue)

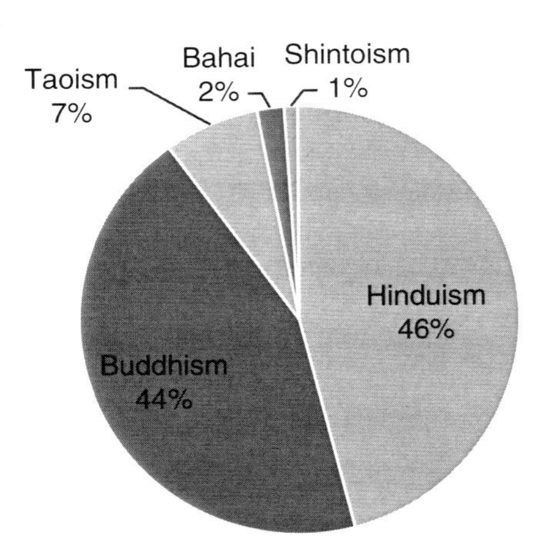

Fig. 55.8 Average distribution of eastern religious materials (Source: Katherine Donohue)

Level of materials on Hinduism and Buddhism. However, few seminaries outside of the largest North American seminary libraries exhibited more than a Minimal Level of materials on any Eastern religion.

On one level this discrepancy is unsurprising, because the population of monotheists greatly outnumbers the adherents to the Eastern religions. On a deeper level, however, the discrepancy is surprising because it reflects a lack of understanding of the culture in that area. The one exception to this trend was a seminary in India, the South Asia Institute of Advanced Christian Studies (SAIACS), where they housed an Instructional Level of Hindu material. The amount of monotheistic material, however, still dwarfed the amount of material on Hinduism.

55.5 Perspectives on International Relations

This study now turns from analyzing religious content to examining the relationship of the seminary to political and social cultures. Two political narratives were selected to examine the worldview of the seminaries. The first narrative is the *conflict between Israel and Palestine*. This conflict, lasting for more than half a century, has generated a vast amount of international interest. This study hoped to discover if certain seminaries contained biased opinions on the conflict based on their library content. The second narrative looks at *international foreign relations*, and the struggle for influence between China and the United States. In some ways, the struggle for influence mirrors the Cold War. In this case, developing countries are aligning with either the U.S. or China to gain economic support. Just as the Soviet Union repressed religion, the Chinese government maintains a similar secular society. Therefore, the China-U.S. narrative proves interesting to seminaries on two levels: on a political level, the library content can suggest which country has more influence in eyes of the seminary, and on a religious level, the library content adds to the narrative of secularism versus spirituality.

55.5.1 Israel and Palestine

The Israel-Palestine narrative revealed interesting trends among the denominations and geographic locations of the seminaries. The denomination of the seminary proved significant when examining the distribution of Israeli-Palestinian materials. Figure 55.9 shows the materials on Israel and Palestine in different seminary denominations. The Catholic seminaries present a much more balanced view of the Israeli-Palestinian conflict, perhaps a measure of the official stance of the Vatican in Rome. Among the Protestant seminaries, the University of Balamand in Lebanon is notable for its fairly even balance between Israeli and Palestinian literature. In this scenario, the proximity of Lebanon to Israel helps to account for the even

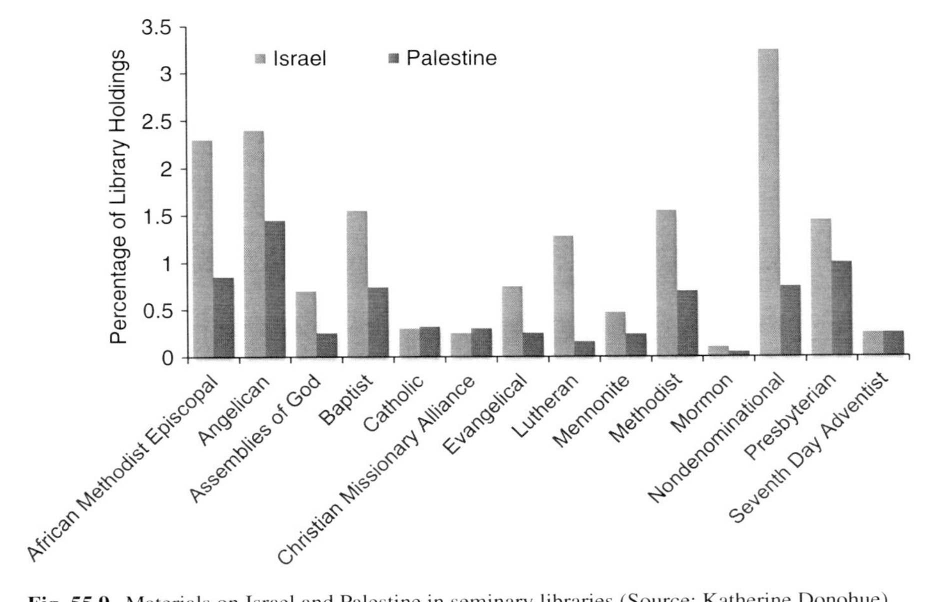

Fig. 55.9 Materials on Israel and Palestine in seminary libraries (Source: Katherine Donohue)

distribution of literature between the two sides. Every seminary housed at least a Basic Information Level of material on Palestine and an Instructional Level of material on Israel. Cleary, the seminaries maintain an interest in the political future of the Holy Land.

55.5.2 China and the United States

The library holdings on China and the United States reflected the struggle for influence mentioned in the introduction of this section. Figure 55.10 illustrates the seminary library holdings for the U.S. and China in various geographic regions. The two are fairly evenly balanced in most of the developing world, with the U.S. occasionally coming out on top and China winning out on others. The geographic location appears to have a measurable effect on the content available on each country. The China Graduate School of Theology in Hong Kong understandably has more literature on China, while the American seminaries consistently house more literature on the United States. However, it is interesting to note that all of the seminaries contained fairly even distributions between China and the United States. All of the non-denominational seminaries had at least a Research Level of information on both countries. If there are any who doubt the growing influence of China, this is certainly an indicator of the rise of Chinese influence. The emergence of China as a key economic powerhouse has greatly enhanced the country's importance in the

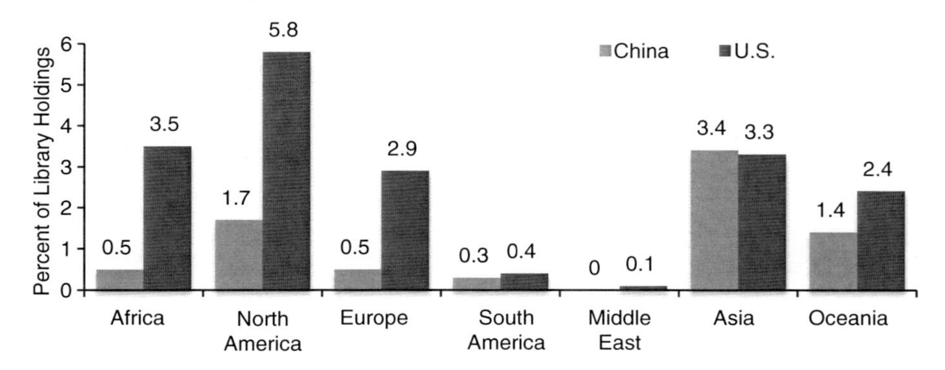

Fig. 55.10 Materials on the United States and China in seminary libraries (Source: Katherine Donohue)

eyes of the seminaries. The subject material on both countries fell into at least the Instructional Level for seminaries worldwide.

55.6 Library Holdings on Science and Secularism

Having examined some of the political elements of the seminary worldview, this study now turns to the growth of contemporary popular debates and social discussions in the religious world. One pattern that clearly emerged was a divergence between Western seminaries and those outside of the developed world. Many of the key issues examined in this section are regional debates, like secularism and evolution, serving as subjects of discussion primarily in the Western social front. This section will examine three key social fronts that color the worldview of seminary libraries. The first front is the blending of religious doctrine and *scientific discoveries*. The second front looks at the growing movement of *secularism*, with the growing platform for atheistic and agnostic speakers. The third and final front examines issues of *sexuality* that have come to the forefront of social debates in the past few decades.

55.6.1 Scientific Discoveries: Evolution and Creationism

This section examines one particular debate between the scientific and religious world that continues to dominate the dialogue in academia—*evolution*. Some denominations teach the compatibility of evolution and religion while others adamantly oppose the theory, offering up the doctrine of Creationism as a viable

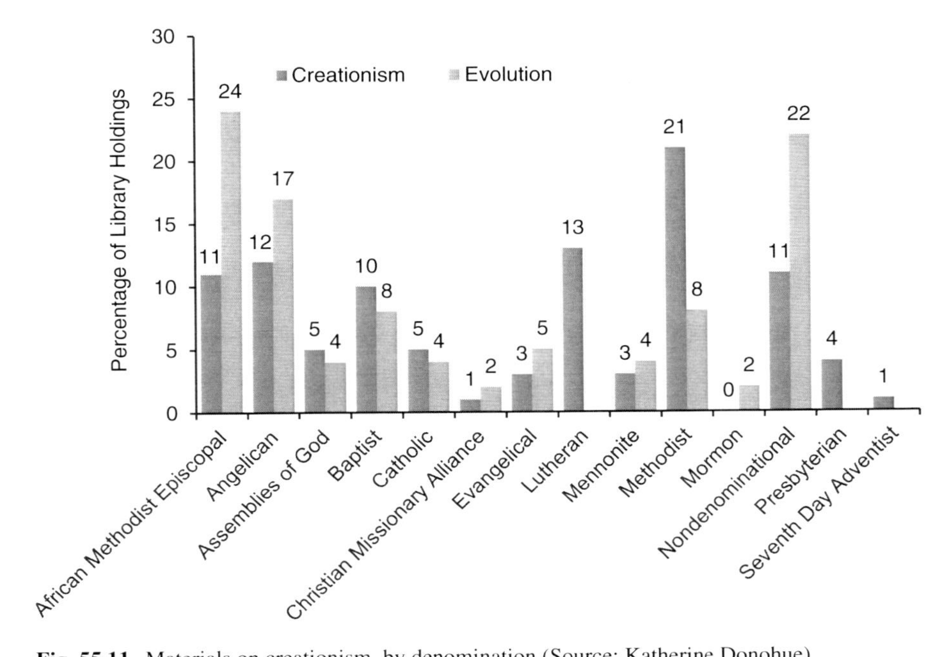

Fig. 55.11 Materials on creationism, by denomination (Source: Katherine Donohue)

scientific alternative. As expected, the library holdings on Creationism break down into denominational components (Fig. 55.11). However, it is surprising to note that the denominations and movements frequently associated with Creationism, like Assemblies of God and the Evangelical movement, do not dominate the holdings on Creationism materials. When compared to holdings on evolution, the amount of literature on that subject in these two groups is also limited. It is possible that because some seminaries consider Creationism to be a doctrinal matter, they do not see the need to acquire many materials on the topic, but more research would need to be completed to make a definitive conclusion.

Figure 55.12 shows the distribution of literature on evolution. All things considered, the distribution of literature on evolution is fairly even with most seminaries containing at least an Instructional Level of material, although the viewpoints taken within the holdings on evolution are not equally distributed. Again, the holdings of the seminaries largely reflect the stance of the denomination or movement that they represent. Materials on evolution from the Assemblies of God Theological Seminary library frequently portray evolution in opposition to Creationism, while library materials in Mt. Saint Mary's Seminary discuss the compatibility of Christianity and evolution, reflecting the official stance of the Catholic Church. The strong presence of material on evolution in Western seminaries, however, signals its continued importance as a social issue that the seminaries must address.

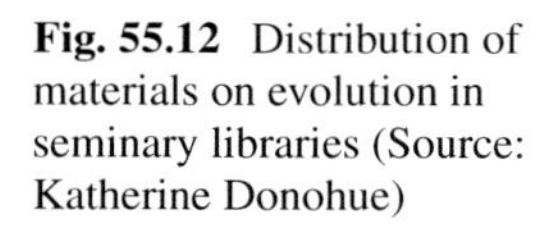

Fig. 55.12 Distribution of materials on evolution in seminary libraries (Source: Katherine Donohue)

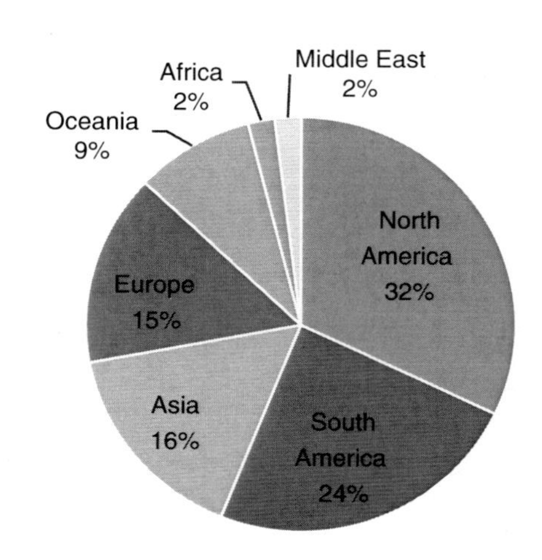

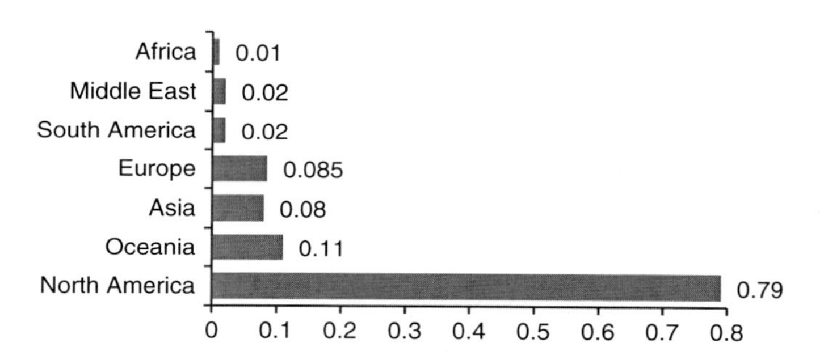

Fig. 55.13 Materials on secularism in seminary libraries (Source: Katherine Donohue)

55.6.2 Secularism: Atheists and Agnostics

The topic of secularism shows a particularly stark contrast between Western and non-Western seminaries. In the developing countries, secularism does not hold the same significance that it does in the developed world. This supports theories that suggest that secularism is a movement that has grown out of the wealthier nations. Figure 55.13 shows the distribution of literature on secularism in seminaries in select geographic regions.

It is interesting to note that the greatest collection of literature on secularism can be found in Evangelical and non-denominational seminaries. Those seminaries typically have an Instructional Level of material, while the others have only a Minimal or Basic Level. Upon closer examination of the materials in these libraries, the contents in the Evangelical libraries, like Bethel Seminary and Dallas Theological Seminary, contain materials that place secularism in opposition to Christianity, a force that must be combated. In contrast, the non-denominational seminaries, like

Harvard Divinity School, contain a great deal of analytical literature, speculating on the reasons for the recent growth in the secular movement, leaning more toward the social science and anthropological analyses.

55.6.3 Issues of Sexuality: Homosexuality and Abortion

Breaking news stories of sexual abuse by priests flooded newspapers in 2008 and 2009, bringing issues of sexuality back to the forefront of religious debates. Churches and seminaries had to grapple with their stances on homosexuality. The resurgence of the debate on sexuality led to the rise of other hot topics, like family planning and abortion. The pro-life and pro-choice movements both experienced surges of popularity.

Given the history of sex scandals in the Catholic Church, it is not surprising that the majority of literature on homosexuality can be found in Catholic seminaries. Figure 55.14 shows the spread of materials on this topic across the various denominations. Given the strong stance that certain Protestant denominations have taken against homosexuality, one might expect them to have a greater proportion of materials on homosexuality. However, a number of Protestant seminaries have only a Minimal Level of books on the subject, like Golden Gate Baptist Seminary and Trinity Evangelical Seminary. One reason for this might be because their official stance on homosexuality leaves little room for investigating the subject, while the scandals in the Catholic world drove many people to study the topic.

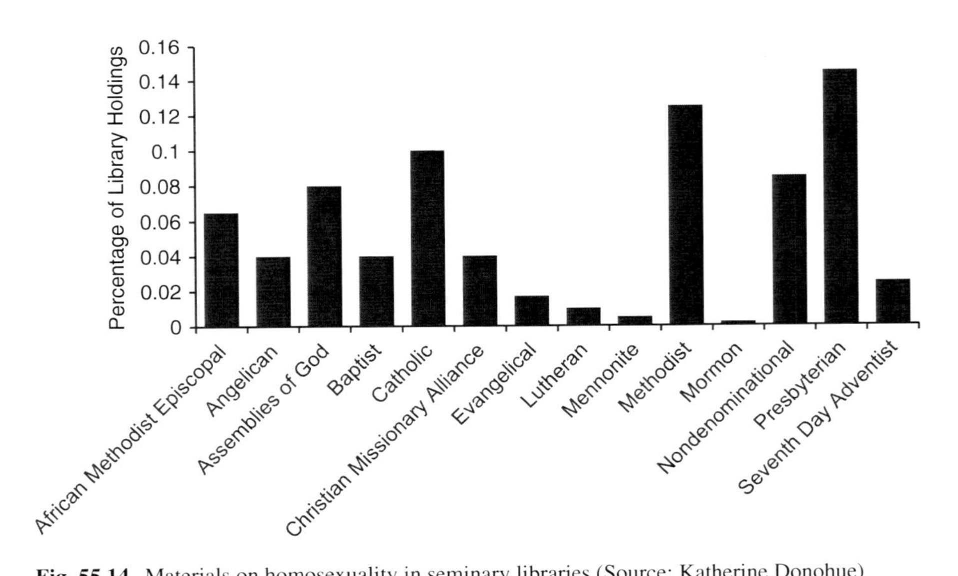

Fig. 55.14 Materials on homosexuality in seminary libraries (Source: Katherine Donohue)

The distribution of materials on abortion and family planning are fairly evenly spread among the seminaries. As with evolution, the stance portrayed in the literature varies (more so with family planning than with abortion); however, the topic is significant enough that seminaries have been forced to collect at least a Minimal or Basic Level of material on the subject for their students. With these secular issues, a seminary's geographic proximity to a culture involved in the social debate, like the maternal health care debate in the United States, appears to play a role in determining the level of materials collected on the subject.

55.7 Seminary Libraries and Internet Databases

One issue that has been largely ignored up to this point is the growing use of the Internet in seminary education. Many of the seminary websites reflect this trend, by placing links to larger databases and combining their libraries with others in the region through online catalogues and inter-library loans. This dramatically expands the potential worldview of the seminary, leaving much more choice for the student to fashion his/her own worldview through the materials available online. In determining the worldview of a particular seminary, the physical library holdings are a good place to start, as many of the books are requested by faculty members and used in course instruction. However, even that practice is beginning to change with the growing use of Internet resources to supplement academic learning.

Many of the seminaries in Asia, Africa, and the Middle East have made excellent use of the resources available through Internet databases. There is a divergent trend in the seminaries in those regions, in which a seminary either makes strong use of the Internet or ignores it to the point of not even having a webpage. The seminaries in those regions that *do* make use of the Internet often have limited resources to develop a large library. To make up for that lack of resources, a number of seminary websites provide links to Internet databases and online resources, such as Biblegateway.com.

In North America and Europe, a growing number of seminaries are using the Internet to link with other seminaries and facilitate inter-library loans. In the United States, seminaries in close proximity to each other have often combined their collections into one online catalogue to give their students greater access to printed materials. For example, seminaries in the San Francisco Bay Area, including the Church Divinity School of the Pacific and San Francisco Theological School, established the Graduate Theological Union Common Library to allow the seminaries to combine their library resources. As this trend grows, more research will need to be conducted to develop methods of analyzing the impact of Internet databases and resources on seminary worldviews.

55.8 Concluding Remarks

This chapter clearly illustrates the different factors that influence the seminary's worldview based on library holdings. While the seminaries share certain trends, like the distribution of monotheistic materials and the shortage of material on Eastern religions, they also have a number of collections that make their perspectives unique. The goals and missions of different denominations help to determine the subject material that the seminary library collects, and the geographic location and library size often influence the collections on international relations and secular debates. As research continues to expand on seminary libraries, the American Theological Library Association has identified specific areas that need further research: literature written by the "global south," the accessibility of seminary library websites, and the use of e-journals and Internet resources. The ATLA website, www.atla.org, provides a number of resources and forums for discussing these issues.

Philip Jenkins, in his book, *The Next Christendom: The Coming of Global Christianity* (2007), called for seminary libraries to increase their holdings on literature published by the "global south," or the developing world. This subject requires further research, because even when library holdings reflect a variety of topics, many libraries rely almost exclusively on materials published by sources in the "global north," or the developed world (Weaver 2008). Studies should be conducted to analyze and encourage seminary libraries to expand their collection of materials from the "global south."

Kate Ganski, in an article for *Theological Librarianship*, designed a survey to analyze the accessibility of seminary library websites. The survey examines the use of links to navigate the webpage, the visibility of e-resources, and the terminology used on the webpage compared to other university library websites (Ganski 2008). It would be interesting to expand this survey from American seminaries to seminaries around the globe to bring a level of uniformity to seminary library websites, which would greatly aid navigation and students' ability to access resources from multiple seminary libraries. In an age where inter-library loans are increasing in importance, the accessibility of library websites is an important topic that requires further analysis.

Closely tied to website accessibility is the use of Internet resources in the seminary libraries. As seminary libraries continue to expand and develop over the next decade, the Internet is likely to play a major role in expanding the worldview of the seminary. Studies should be conducted to determine what e-resources are represented by direct links from the library websites and the tools given to students to help navigate the immense amount of material available online (Ganski 2008). Another opportunity that the Internet provides is the development of distance learning programs. Jeff Groeling and Kenneth A. Boyd have examined the impact of distance education on libraries through the study of Asbury Theological Seminary's Extended Learning online program. They discovered that one challenge facing these programs will be providing Internet resources that are equivalent to the resources available to students at the library on campus (Groeling and Boyd 2009).

It is important to be apprised of the materials available to students who receive a seminary education, whether those materials come through interlibrary loans, partnerships with non-seminary libraries, or the Internet. Future studies may examine loan and book request records to see what books students are checking out or explore library holdings on significant spiritual and secular writings. These methods of analyzing library holdings will help scholars and librarians alike to understand the different perspectives that students are exposed to through the library. The worldview that students glimpse through the seminary library will impact the worldview of the communities they will serve, and ultimately, their perception of Christianity and the world gained from seminary will contribute to shaping global discussions for the next generation.

References

Ekmekcioglu, F. C., Nicholson, D., Research Support Libraries Programme, Scottish Confederation of University and Research Libraries, Scottish Collections Network Extension (Project), University of Strathclyde, & Centre for Digital Library Research. (2002). An evaluation of the current approach to collaborative collection management in SCURL libraries and alternatives to conspectus: Report and recommendations on collection strength measurement methodologies for use in SCURL libraries: Final report of the RSLP SCONE project, Annexe A.2.

Ganski, K. L. (2008). Accessibility of e-resources from theological library websites. *Theological Librarianship, 1*(1), 39–45. Online ATLA Publication.

Groeling, J., & Boyd, K. (2009). The impact of distance education on libraries. *Theological Librarianship, 2*(1), 35–44. Online ATLA Publication.

Jenkins, P. (2007). *The next Christendom: The coming of global Christianity*. New York: Oxford University Press.

Jordahl, R. (1990). The interdisciplinary nature of theological librarianship in the United States. *Libraries and Culture, 25*(2), 153–170. Austin: University of Texas Press.

Slavens, T. P. (1976). William Walker Rockwell and the development of the Union Theological Seminary library. *The Journal of Library History, 11*(1), 26–43. Austin: University of Texas Press.

Weaver, J. B. (2008). Theological libraries and "the next Christendom:" Connecting North American theological education to uses of the book in the Global South. *Theological Librarianship, 1*(2), 38–48. Online ATLA Publication.

Chapter 56
Intersections of Religion and Language Revitalization

Jenny L. Davis

56.1 Overview of Endangered Languages and Language Revitalization

Drawing on discourses from ecology and environmental activism, scholars and community members have voiced growing concern regarding the decline of "bio-linguistic diversity" through the rapid decline in the number of languages spoken throughout the world. In *When Languages Die* (2007), K. David Harrison estimates that approximately 6,900 languages are currently spoken on the planet, and that more than half of these languages are likely to become extinct over the next century. (For more on this topic, see Krauss 1992, 2007; Nettle and Romaine 2000; Grenoble and Whaley 1998; and Crystal 2000). These at-risk languages are classified as "endangered." This rapid drop in the number of languages spoken worldwide is typically characterized by a slowing or ceasing of intergenerational language transmission and by a decrease in the number of social domains in which these languages are used. Language endangerment can be attributed to a number of factors:

> Language endangerment may be the result of external forces such as military, economic, religious, cultural or educational subjugation, or it may be caused by internal forces, such as a community's negative attitude towards its own language. Internal pressures often have their source in external ones, and both halt the intergenerational transmission of linguistic and cultural traditions. (UNESCO Report 2003:2)

In this context, everything from communal shaming practices to government policies criminalizing the use of particular languages in favor of so-called "dominant" or "global" languages (English, Spanish, French, Portuguese, Dutch, etc.) have contributed to the dramatic decrease of minority language use over centuries in communities across the globe.

J.L. Davis (✉)
Deptartment of American Indian Studies, University of Illinois,
Urbana-Champaign, Urbana, Illinois 61801, USA
e-mail: Loksi@Illinois.edu

S.D. Brunn (ed.), *The Changing World Religion Map*,
DOI 10.1007/978-94-017-9376-6_56

In response to grassroots activism and increased awareness of minoritized cultures, a number of international entities and countries have formally recognized language endangerment as an issue and have passed legislation to facilitate endangered language documentation and revitalization. In 1990, for example, the United States Congress passed the *Native American Languages Act* (NALA), acknowledging that "the status of the cultures and languages of Native Americans is unique and the United States has the responsibility to act together with Native Americans to ensure [their] survival." This act stated that it was now the policy of the United States to "preserve, protect, and promote" Native Americans' rights to use their indigenous languages anywhere, including "as a medium of instruction" in schools. Thirteen years later, in 2003, the United Nations Educational, Scientific and Cultural Organization (UNESCO) put together an Ad Hoc Expert Group on Endangered Languages that asserted the importance of linguistic diversity to society at large:

> Language diversity is essential to the human heritage. Each and every language embodies the unique cultural wisdom of a people. The loss of any language is thus a loss for all humanity. (2003:1)

Thus, the group contends that language loss is not only an issue that should concern members of communities whose languages are threatened, but also is critical for all humans. That same year, the Republic of Ireland passed *The Official Languages Act*. Even though Irish had been listed as the first official language of Ireland in the 1937 Irish Constitution, this new act ensured "better availability and a higher standard of public services through Irish," which would be "principally achieved by placing a statutory obligation on Departments of State and public bodies to make specific provision for delivery of such services in a coherent and agreed fashion through a statutory planning framework" (O'Laoire 2005: 252). The Official Languages Act then demonstrated a national position that it is not enough to recognize Gaelic language symbolically; rather, direct action is required to ensure its maintenance and revitalization.

In addition to the implementation of formal language policies such as the ones described above, many communities, language activists, and linguists began various efforts to slow down—and ultimately reverse—the language endangerment process. These efforts can be divided into three different (but often related) approaches: language documentation, language maintenance, and language revitalization. *Language documentation* is the process by which a language's grammar and sound systems, conversational norms and socio-cultural genres (for example, stories, songs, religious texts) are recorded for later reference. Documentation may produce dictionaries, online databases, as well as pedagogical materials that may be utilized in language maintenance and revitalization efforts. *Language maintenance* refers to the support given to languages that still have a viable remaining speaker population in order to maintain the current domains of language use and transmission. *Language revitalization* pertains to actions through which language communities increase the number of active speakers of an endangered language via a combination political, communal, and educational means. This chapter focuses on the last of these approaches—language revitalization.

Language revitalization efforts may be taken up on the national level—by governments—or on the local level—by individuals and small groups. Methods of working toward revitalization vary dramatically from community to community; the approach taken may depend upon a number of factors, including population size, economic resources, and geographic distribution (for specific examples, see Hinton and Hale 2001). One factor, the numbers and types of existing speakers, is particularly critical; especially those classified as *native speakers*—individuals who have spoken a given language since birth or early childhood. Some programs, such as immersion and bilingual school programs, focus on creating young speakers through the formal education system; others, such as adult language classes and master/apprentice programs, aim at growing the number of adult speakers. In addition to education programs, strategies like providing public signage that includes local endangered languages has the duel effect of raising linguistic visibility and providing another mode of language use to those who have, or are acquiring, those languages. Many communities use a combination of several of these types of programs in a multi-pronged approach, including creating social events centered around language use such as language competitions or immersion camps. Endangered language communities have developed a number of media outlets to educate and entertain those interested in their languages; most recently, several communities have produced online dictionaries, while others have developed their own smart phone applications that provide language lessons or worked with companies, such as Rosetta Stone, to develop computer-based language learning materials.

Successful revitalization depends not just on increasing the number of proficient speakers, but also on combating language ideologies and practices that are detrimental to indigenous language use (Kroskrity 2000; Meek 2010). In such efforts, any number of social, economic, and political dynamics must be taken into account, as they shape what is possible. As with any social or political movement, language revitalization is inherently embedded within the contexts of the communities in which such movements occur. These contexts include everything from the day-to-day lived experiences of community members to overarching governmental systems and policies at all levels. Whether a community's leadership supports such efforts, either structurally or economically, can dramatically affect how such efforts can be executed, and who is able to participate in them. Thus, language revitalization efforts must be examined as they occur, in each unique context of development, implementation, and success.

Religious dynamics are one important context in which to explore revitalization efforts. In some cases, the connection between religious institutions, linguistic practices, and histories are overt. For example, SIL (formerly, the Summer Institute in Linguistics), a U.S.-based Christian non-profit founded in 1934, provides training and assistance in language documentation, pedagogy, translation, etc. in over 100 countries worldwide. SIL provides the following description of their mission:

> SIL's service with ethnolinguistic minority communities is motivated by the belief that all people are created in the image of God, and that languages and cultures are part of the richness of God's creation. Thus, SIL's service is founded on the principle that communities should be able to pursue their social, cultural, political, economic and spiritual goals

> without sacrificing their God-given ethnolinguistic identity. (Summer Institute of Linguistics, Inc. 2012)

The intersection between SIL's position at the intersection of language revitalization and religion is also visible in their relationship as the "primary partner organization" of Wycliffe Bible Translators, a Christian organization dedicated to translating the Bible into minority languages. In other instances, the intersections between language revitalization and religion are less visible, and only discernable through long-term engagement with a particular community. In this chapter, I draw upon ethnographic research in three geographically and linguistically different communities facing language endangerment in order to discuss the intersection between language revitalization efforts and religious practice: the Bretons of Brittany, France; the Chickasaws of Southeastern Oklahoma, United States; and the Nahuas of Hueyapan, Mexico. In doing so, I argue that religious ideologies, practices, and texts often play critical roles in endangered language revitalization.

56.2 Case Study 1: Breton in France

Located along the northwest coast of France, Brittany is home to the Breton language (Brezhoneg), which is part of the Insular Celtic branch of the Indo-European language family. It is the only Celtic language still spoken by native speakers in France, and, in fact, the only Celtic language remaining in continental Europe. While there were a million speakers of Breton at the turn of the twentieth century, there are today only approximately 200,000 speakers (Adkins 2013, citing *Ofis ar Brezhoneg* 2011). Madeleine Adkins points out that Breton is so endangered that it is no longer "exclusively, predominately, or even commonly used by the population in any city, town, or village in Brittany" because today Breton speakers are bilingual (that is, Breton and French) speakers. Revitalization efforts by community-based language activists began in the 1970s with the creation of a small number of Breton language immersion schools, or Diwan, which currently graduate a few thousand new young speakers each year. As a result of the age gap between Breton native speakers and younger Breton language learners, speakers of the language tend to fit into one of two groups: older, more rural speakers for whom Breton was the primary language spoken in their homes and in their communities, and younger, urban or semi-urban speakers who grew up in primarily or exclusively French-speaking environments and learned Breton in school.

In examining the difficulties identifying speakers from the former group during linguistic fieldwork in Brittany, Adkins shows that the complex negotiations of what "good Breton" is, and who "real" speakers of the language might be, are partially based in language ideologies that reify literary, or High, Breton that was traditionally spoken by Catholic priests in the region. These priests, given the additional authority granted by their position within the Church and the community, were often the most visible "official" speakers of Breton. Adkins (2013: 11) notes:

> … for traditional speakers of Breton who lived in rural villages, the Breton speaker with the most education and to whom everyone looked up was the village priest …. His sermons

filled with carefully crafted Breton phrases, influenced by his knowledge of French and Latin as well as literary Breton, was the model of High Breton for the community.

This iconicization of the highly educated priest as an ideal Breton speaker imbues the concept of speakerhood with the norms of being multilingual (French, Latin, and Breton) and highly educated. As such, many native speakers of Breton with whom Adkins worked evaluated their linguistic expertise as "not good" because they—like the vast majority of Breton speakers of their generation—had limited education in general, and no formal training in the Breton language. In fact, these speakers proved fluent during recorded conversations; however, unlike young learners in Breton language programs today, the older generation of native speakers never had their Breton language abilities affirmed in a formal setting. This is because the Breton language was banned from schools when these older native speakers were in school. Furthermore, the only example of written Breton that most native speakers have seen is the book *Buhez ar Sent* ('Lives of the Saints'), which many older Breton speakers have in their homes. This religious tome is written in a formal, ecclesiastic version of the language, quite distinct from the local varieties that native speakers use. In comparison to this more formal, written variety of Breton, native speakers often evaluate their everyday, spoken Breton, as lacking. In other words, ideologies that associate "real" or "good" Breton with the language's written, ecclesiastic form rather than its everyday, spoken register, contribute to native Breton speakers' self-evaluations of their language abilities as not good (enough).

The association of the Breton language with Roman Catholicism extends to French cultural stereotypes about Breton people. These negative stereotypes often frame Breton people as backward, both in their occupations (farming, fishing) and religious practices (devoutly Catholic). By extension, speaking the Breton language, as it is strongly associated with traditional Breton culture, is viewed by some being as synonymous with an unsophisticated, rural, and overly religious nature. For example, a contemporary online Breton language web series entitled *Ken Tuch'* pokes fun at a number of fictional personas, including a naïve, hyper-traditional Catholic woman named Katell (Adkins and Davis 2012). This online comedy sketch series is designed primarily for the younger generation of Breton speakers, and Katell's prominence as a key Breton persona in the series demonstrates the persistence of this (usually negative) stereotype of Bretons as rural and overly religious. Those seeking to revitalize the language are thus tasked with changing such negative ideologies—within the community and in broader contexts—and this is accomplished, in this case, through humor.

Successful documentation and revitalization of endangered languages requires the identification and valorization of the language in question, and of the remaining individuals who are still able to speak it. While the intertwining of the Breton language and Catholicism may reinforce stereotypes within contemporary revitalization efforts, it also facilitates such revitalization in some quarters: many of the Breton language bilingual programs for children are now offered in Catholic schools. In sum, the long history of Catholicism within the Breton community has led to inhibiting language ideologies that identify "good" Breton as that spoken by formally educated Catholic priests, negative stereotypes held by outsiders about the Breton community and language, and Catholic educational institutions actively promoting bilingual language learning for new generations of Breton language learners.

56.3 Case Study 2: Nahuatl in Mexico

Within the nation of Mexico, 64 indigenous languages are recognized as national languages under Mexico's *Ley General de Derechos Lingüísticos de los Pueblos Indígenas* ("General Law on the Linguistic Rights of Indigenous Peoples") (2003). One of those indigenous languages, Nahuatl, has been spoken in Central Mexico by the Nahua people (also referred to as Aztecs) from as early as the seventh century C.E. Nahuatl is from the Nahuan branch of the Uto-Aztecan family.

Magnus Pharoah Hansen's research within the Nahua community of Hueyapan (2010a, b) provides an intriguing example of a more recently active religious organization's decisions to validate the indigenous language of Nahuatl by making it the official language for religious practices. These actions have proven to be dramatically influential in the resurgence of daily use of the Nahuatl language—a resurgence that is currently limited to the members of one congregation, but nonetheless significant. Within the community of Hueyapan, Mexico, located in the northeastern corner of the state of Morelos in central Mexico, two languages are spoken: Nahuatl and Spanish. Spanish, the prestige language, is spoken by everyone and is the language used in public interactions and spaces; as a result, Nahuatl is spoken almost exclusively in the private sphere (Hansen 2010a: 126).

Hansen notes that—much like that in many other Nahua communities in Mexico, and indigenous communities more generally—the sociolinguistic dynamic in Hueyapan is such that the Nahuatl language is in a rapid process of decline:

> The use of Nahuatl is reserved almost exclusively for the private sphere, and the language is often used as a code of intimacy and solidarity among friends…A speaker's level of proficiency is generally correlatable with age: elderly speakers (above 60) are often Nahuatl-dominant. Middle-aged speakers (40–60) are often fully bilingual. Adult speakers (30–40) are often Spanish-dominant but able to converse in Nahuatl. Youth (below 30) often have only passive skills in Nahuatl; some could probably be described as semi- or quasi-speakers (Dorian, 1977; Flores Farfan, 1999), and the majority do not have any Nahuatl skills at all. (Hanson 2010: 126)

In contrast, within one particular subcommunity of the town—the congregation of Jehovah's Witnesses—the language has been experiencing a revival within the last 5 years, to the degree that congregational meetings are now held primarily in Nahuatl, and young Witnesses are probably the only group of youths in Hueyapan to communicate daily in Nahuatl.

Hansen noticed a remarkable shift in language use—religious services within the local Jehovah's Witness congregation were switching *away* from Spanish *toward* Nahuatl: after realizing that a great majority of the congregation spoke Nahuatl, leaders declared the indigenous language the official language of the congregation.

Although the use of Spanish is not formally forbidden, Nahuatl is the preferred language for all activities (for example, sermons, door-to-door ministry, and community discussions). This official shift in language status has been impactful both in terms of language practice and in the overall prestige of Nahuatl speakers within the community. As Hansen (2010a: 131) notes:

> Although use of Nahuatl in the first year was probably less than 10–15 %, it had a gradual and ultimately dramatic effect on the prestige mechanisms that operated within the congregation. While the use of Spanish was never prohibited or explicitly discouraged, little by little the official change and the implicit expectations of speaking Nahuatl in the congregation caused a reversal of the prestige mechanisms in the little community.

The arrival of a new elder who was fluent in Hueyapan Nahuatl increased the amount that the language spoken even more, due to his ability to translate between Spanish and Nahuatl, as well as to facilitate theological discussions within the congregation.

Hansen demonstrates that this situation arose as a result of both the general sociolinguistic background in Hueyapan and the specific value system and social dynamics that exist within the community of Jehovah's Witnesses. First, the hierarchical structure within the organization meant that once the group decided to make Nahuatl its official language, members were expected to work toward making that decision a reality. Second, the language philosophy of Jehovah's Witnesses focuses on using local languages, even when a more "global" language is available. In this instance, that linguistic ideology allowed them to better engage with and establish trust with Nahuatl speakers in the community.

The effects of this religious community's language policy are significant, if limited in scale. Because the social structure of Jehovah Witness is such that the majority of the organization's ministry and teaching is done by young, unmarried men, these individuals are most affected by the congregation's shift to Nahuatl as the official language. This shift forces them to increase their fluency in Nahuatl, both for the purpose of daily conversation and for the discussion of theological concepts. In addition, the shift creates both space and time in which Nahuatl is regularly used outside of the private sphere, thereby locally reversing the widespread trend in Central Mexico.

56.4 Case Study 3: Chickasaw in the United States

In my own research with the Chickasaw Nation in the southern U.S. state of Oklahoma, religious practices and genres are an important setting for language revitalization—specifically, the evangelical Protestantism practiced throughout much of Oklahoma provides a context in which Chickasaw is valorized and used. Religious texts—especially those sung or performed—emerge as a primary set of genres through which the language is learned, practiced, and presented to the community, by both adults and children. Current participants in the Chickasaw language revitalization process therefore include those who identify quite strongly with Protestantism. For them, speaking Chickasaw can be a context in which to practice Christianity. In addition to these Protestant speakers, individuals who identify either with a pre-Christian traditional spiritual practice or with no particular spirituality at all are active users of Chickasaw. These two groups are able to participate in Chickasaw language activities and envision the language as having a role in their futures and daily lives. Thus, the Chickasaw language context demonstrates how language revitalization efforts can result in a language being incorporated into communal life in such a way that it serves a truly functional purpose in the community.

The Chickasaw Nation is one of the smallest of the "Five Civilized Tribes," relocated as part of the Trail of Tears marches from the southeastern United States to Oklahoma from 1837 to 1838. Before removal to Oklahoma, the Chickasaw tribe occupied areas in Alabama, Tennessee, and South Carolina. After removal, the Chickasaw Nation was relocated to southeastern Oklahoma to an area of that encompasses thirteen counties. The Chickasaw language and its close sibling Choctaw make up the Western Branch of the Muskogean language family, which also includes Muskogee (Creek), Alabama, and Seminole, among others. Currently, as the result of various socio-cultural changes within the community, the Chickasaw language has under 100 fluent speakers remaining, and. the youngest native speaker of Chickasaw is over 50 years old.

Although the first interactions between the Chickasaw and Europeans were with the Spanish, religious conversion to Christianity occurred later, through Protestant efforts pre- and post-removal to Oklahoma, first by Presbyterians in the 1790s, and then by Methodist and Baptist missionaries arriving in the first decade of the nineteenth century (Atkinson 2004). Protestant missionaries, by necessity, had the dual roles of converter and translator—for, unless the missionaries translated their texts into the language of the Native American communities, the indigenous population would not have the means by which to engage with the religious texts themselves. The missionaries believed that without direct engagement with the texts, the Chickasaw people could not be saved. Many texts, such as the Bible, were translated into Choctaw by the Protestant missionaries who sought to convert the much larger, neighboring Choctaw tribe;[1] given that the Choctaw and Chickasaw languages are often considered to be mutually intelligible (especially the older forms of both languages),[2] the Chickasaw people had access to the Bible via the Choctaw translation. In addition, hymns were later translated into Chickasaw itself, and sermons and many religious activities were conducted in the Chickasaw language.

Religious figures who worked with the Chickasaw Nation were therefore necessarily language students and lay linguists. Several of these figures also play prominent roles in the narrative of the Chickasaw language; for example, the language's first dictionary, published in 1973, but collected several decades earlier, was written and collected by the Reverend Jesse Humes, who was known for his passion for documenting and preserving the language. As with many other tribes in North America, the

[1] See Hanks (2010) for an in-depth discussion of similar processes of linguistic and religious shift among the Yucatec Maya.

[2] See Munro and Gordon (1982) for an in depth discussion of the topic.

Chickasaws saw a sharp decline in the use of the Chickasaw language in the decades during and after the boarding school era. However, during those decades of decline, religious revivals—usually several days to a week long—as well as regular services at churches such as Oka Kapassa "Cold Water" Springs—were conducted in the Chickasaw language. In that era, religious services existed as one of the few functions—and the only public function—in which the language was guaranteed to be used. This resulted in the genres of prayer and hymns serving as important communal opportunities for speakers to use the language over the past 50 years.

Today, Chickasaw language use continues to be intertwined with Christianity. Many of the people who regularly attended Chickasaw-language religious services in decades past later became members of the Language Committee, a group of fluent Chickasaw speakers brought together once a month to provide linguistic expertise for translations and to assist with word elicitations for various projects within the Chickasaw Nation. Each Language Committee meeting, consisting of approximately 35 speakers and a few other interested individuals, begins with the singing of the selected hymn for each meeting and an opening Christian prayer in Chickasaw, reinforcing the religious orientation of the group. Religion can also be a motivating factor for individuals participating in Chickasaw language revitalization activities. This was the case for "James" (a pseudonym), a 72-year-old participant in a Chickasaw community language class. During a discussion of what each student hoped to take away from the class, James volunteered that "learning the Lord's Prayer in Chickasaw" was his primary goal for the course. Having grown up in a home where Chickasaw was spoken, James was a fairly fluent speaker in childhood, but he has since forgotten much of the language. When asked why he was interested in learning the Lord's Prayer from the class, he responded that it seemed like a way he could "really use the language" in his daily life.

Not surprisingly, Christian texts have been integrated into various language revitalization programs in the tribe; some were translated quite recently into the language, specifically for the use in language revitalization contexts. For example, the children's language club, *Chipota Chikashsha Anompoli*, "Children speaking Chickasaw," which meets once a month, competed in the 2009 University of Oklahoma indigenous language competition with the song *Jesus a Hullo*, the popular Christian children's song "Jesus loves me," translated into Chickasaw. Members of this club learn aspects of the language in the context of organized activities such as bowling or fishing. Some of the young members also attended the Family Language Immersion Camp a few months later that summer, and were asked to lead the group in singing *Jesus a Hullo* during the time specified for "Chickasaw singing." During the rest of the singing activity, parents and caregivers joined their children in singing hymns in Chickasaw.

Language revitalization materials do not consist solely of Christian texts; in congruence with efforts described in many indigenous communities (Meek 2010; Ahlers 2006), they also include word lists, traditional cultural tales such as "How Rabbit Lost His Tail" and "Spider Brings Fire," grammatical explanations, and examples of conversation. However, the consistent inclusion of sung and spoken Christian texts in these activities demonstrates that current participants in the language revitalization process include people who identify quite strongly with Protestantism—for whom speaking in Chickasaw is a way to practice their Christianity—as well as those who identify with a pre-Christian indigenous spiritual practices and non-believers. Thus, the Chickasaw language revitalization movement demonstrates that participation and interest in indigenous language revitalization is not necessarily limited those with an interest in simply reifying or reimagining pre-colonial culture and linguistic practices or 'traditional' genres, but also those "Indians in unexpected places" who are active members of Protestant churches (Deloria 2006).

56.5 Conclusion

While language revitalization can be interwoven with any number of social practices, in this chapter I have focused on one such aspect of social and personal life—religion. Religion's intersection with language revitalization is often intertwined with the both the specific history of a given language and with the broader community in which it is, or was, spoken. These intersections may be conspicuous, as in the case of a religious congregation making policy changes in favor of using endangered languages; or it may be more subtle, such as Breton speakers' evaluating anything less than the formal, High Breton of Catholic priests as "not as good enough." In some cases—such as the context of the Bretons in France or the Chickasaws in Oklahoma—the intertwining of language and specific denominations (Catholicism and Evangelical Protestantism, respectively) in language revitalization efforts represents continuity with previous cultural norms. In other cases, however, such as that of the Jehovah's Witness community of Central Mexico, new religious groups—and the linguistic philosophies that they adhere to—emerge as critical in language reclamation.

Not surprisingly, then, the products of both religious practices and language revitalization are often connected. On the one hand, texts and genres incorporated into language revitalization efforts carry not just the primary function of language learning, but also underlying themes of religious (and cultural) beliefs and practices. On the other, religious activities and writings conducted in endangered languages can carry the secondary effect of encouraging language revitalization. It is well known that language loss is the result of social and cultural shift; based on the research findings discussed above, we should examine both how reclaiming and revitalizing language is both the result, and, just as importantly, the cause of societal change. Meek (2010) explains that "language revitalization is more than just linguistic rehabilitation; it is a social transformation" (2). Current intersections in language revitalization and religion not only reflect linguistic and cultural histories—they also shape the future contexts in which endangered languages will be used, as well as who will use them.

Acknowledgement Parts of this chapter were presented at the 2010 meeting of the American Anthropological Association in New Orleans. This chapter was completed while I was a Henry Roe Cloud *Henry Roe Cloud* Dissertation Writing *Fellowship in Native American and Indigenous*

Studies at Yale University. I am grateful for the feedback of Madeleine Adkins, Rusty Barrett, and Lal Zimman; thanks also go to this volume's editor, Stan Brun, for his invitation to explore this topic.

References

Adkins, M. (2013). Will the real Breton please stand up? *International Journal of the Sociology of Language, 223*, 55–70.

Adkins, M., & Davis, J. (2012). The naïf, the sophisticate, and the party girl: Regional and gender stereotypes in Breton language web videos. *Gender and Language, 6*(2), 291–308.

Ahlers, J. (2006). Framing discourse: Creating community through native language use. *Journal of Linguistic Anthropology, 16*(1), 58–75.

Atkinson, J. (2004). *Splendid land, splendid people: The Chickasaw Indians to removal*. Tuscaloosa: University of Alabama Press.

Crystal, D. (2000). *Language death*. Cambridge: Cambridge University Press.

Deloria, P. (2006). *Indians in unexpected places*. Lawrence: University of Kansas Press.

Grenoble, L., & Whaley, L. (Eds.). (1998). *Endangered languages: Language loss and community response* (pp. 161–191). Cambridge: Cambridge University Press.

Hanks, W. (2010). *Converting words: Maya in the age of the cross*. Berkeley: University of California Press.

Hansen, M. P. (2010a). Nahuatl among Jehovah's Witnesses of Hueyapan, Morelos: A case of spontaneous revitalization. *Small Languages and Small Language Communities 65. International Journal of the Sociology of Language, 203*, 125–137.

Hansen, M. P. (2010b). *In the beginning was the word: Ideologies of language and religion in a rural Mexican community*. Paper Presentation at the American Anthropological Association, New Orleans (17–21 Nov).

Harrison, K. D. (2007). *When languages die: The extinction of the world's languages and the erosion of human knowledge*. New York: Oxford University Press.

Hinton, L., & Hale, K. (Eds.). (2001). *The green book of language revitalization in practice*. San Diego: Academic.

Krauss, M. (1992). The world's languages in crisis. *Language, 68*(4), 4–10.

Krauss, M. (2007). Classification and terminology for degrees of language endangerment. In M. Brenzinger (Ed.), *Language diversity endangered*. Berlin: Walter de Gruyter.

Kroskrity, P. (Ed.). (2000). *Regimes of language: Ideologies, politics, and identities*. Santa Fe: School of American Research Press.

Ley General de Derechos Lingüísticos de los Pueblos Indígenas. (2003). *Diario Oficial de la Federación*. Issued by the Cámara de Diputados del H. Congreso de la Unión.

Meek, B. (2010). *"We are one language:" An ethnography of language revitalization Northern Athabaskan Community*. Tucson: University of Arizona Press.

Munro, P., & Gordon, L. (1982). Syntactic relations in Western Muskogean: A typological perspective. *Language, 58*, 81–115.

Nettle, D., & Romaine, S. (2000). *Vanishing voices: The extinction of the world's languages*. Oxford: Oxford University Press.

Ó Laoire, M. (2005). The language planning situation in Ireland. *Current Issues in Language Planning, 6*, 251–314.

Summer Institute of Linguistics, Inc. www.sil.org/sil/. Retrieved February 12, 2012.

UNESCO. (2003). *Language vitality and endangerment*. Paris: Document by UNESCO Ad Hoc Expert Group on Endangered Languages.

Chapter 57
Bible Translation: Decelerating the Process of Language Shift

Dave Brunn

57.1 Introduction

In today's world there are approximately 7,100 living languages (Lewis et al. 2013). Some of those languages are edging precariously close to extinction—with only a few remaining native speakers. Many of the languages that are most vulnerable are minority languages that have never been put into writing. Perhaps no one is making a greater contribution on a worldwide scale toward preserving these vulnerable languages than Bible translators.

There are two issues at play here: (1) Language Change and (2) Language Shift. "Language Change" speaks of the inevitable modifications that occur over time in every living language. "Language Shift" is the more serious phenomenon where members of a speech community gradually abandon their native language and "shift" to another language they perceive to be more prestigious. When a Bible translation is published in a language that has previously remained unwritten, it has the potential of impacting both language change and language shift.

The direct connection between Bible translation and language stability has been substantiated through centuries of historical evidence. Every serious history of the German language, for example, includes an explanation of the significant influence Martin Luther's sixteenth century translation of the Bible had on modern High German. Also, in the case of the Insular Celtic languages, the robust development of a print culture in the Welsh language (Cymru) beyond that of its peers can clearly be

D. Brunn (✉)
Language and Linguistics Department, New Tribes Missionary Training Center, Camdenton, MO 65787, USA
e-mail: dave_brunn@ntm.org

S.D. Brunn (ed.), *The Changing World Religion Map*,
DOI 10.1007/978-94-017-9376-6_57

linked to Bible translation, as outlined in John T. Koch's Historical Encyclopedia of Celtic Culture (Koch 2006):

> Translating and publishing the scriptures in the early modern period was a momentous event in the history of many European languages. The history of the CELTIC LANGUAGES demonstrates that vernacular versions had to be accepted and disseminated by the Church if a print culture in the vernacular was to arise, a development achieved most fully in Wales (CYMRU). The majority of the populations of Ireland (ÉIRE) and Brittany (BREIZH) remained Catholic, and the Latin Bible predominated until recent times; Bible translations, where they existed, did not gain wide currency. Church and state in Scotland (ALBA) actively discouraged the use of SCOTTISH GAELIC Bibles in the HIGHLANDS. A complete MANX Bible was not published until 1775, too late to replace English as the language of Anglican church services, and the Cornish had accepted English as the language of religion by the 17th century.

57.2 Bible Translation Today

Wycliffe Bible Translators ("Wycliffe" 2013) estimates that 518 of the world's living languages have the entire Bible, and an additional 1,275 have the New Testament. There are many more languages which have at least some translated Scripture, but nearly 2,000 languages have no Scripture at all ("Wycliffe" 2013). Approximately seventy-five percent of these 2,000 languages are located in three areas of the world: (1) Central Africa and Nigeria, (2) Mainland Southeast Asia, and (3) Indonesia and the Pacific Islands ("Wycliffe" 2013).

Translating the Bible into a minority language is no small undertaking. It usually requires a commitment of several years. And in order for a translation of the Scriptures to help preserve a vulnerable minority language, there must be an accompanying literacy program. For 21 years (1980–2001) my wife and I had the privilege of living and serving in the South Pacific nation of Papua New Guinea, which is made up of approximately 600 islands. My primary responsibility was Bible translation; and my wife taught literacy.

57.3 Bible Translation in Papua New Guinea

Papua New Guinea is land of remarkable linguistic diversity. Although the population is only 6.7 million (Lewis et al. 2013), there are an estimated 836 living languages spoken there (Lewis et al. 2013), We lived on the Papua New Guinea island of New Britain, which is about the size of the U.S. State of Vermont, and yet has over 40 distinct language groups. The language group we worked with is called Lamogai (*lah-moh-guy*). Before we began learning the Lamogai language, we took time to learn Melanesian Pidgin, the *lingua franca* of Papua New Guinea, known locally as *Tok Pisin*.

During the years that we served the Lamogai people through translation, literacy and church planting, it was rewarding to know our efforts were adding stability to this vulnerable, previously-unwritten language. Initially, however, I unwittingly embraced a strategy that was counterproductive: I became overzealous in my effort to preserve the "pure" local vernacular.

57.4 Preserving the "Pure" Lamogai Language

When I first started studying the Lamogai language, I was urged by my language-learning consultant to stop using the trade language, *Tok Pisin*, as soon as possible. I can still remember his words: "Don't say anything in *Tok Pisin* that you already know how to say in Lamogai."

I took his exhortation seriously. A short time later, I determined I would no longer use *Tok Pisin* in any of my day-to-day interaction with the Lamogai people. From that time on, all my conversation with them would be done exclusively in the Lamogai language. When I explained this to my Lamogai friends, they eagerly supported my decision and they promised to help me with it. However, in their enthusiasm they took things to an extreme—replacing *all* borrowed words with "pure" vernacular words, many of which were either obsolete or contrived. In cases where no appropriate Lamogai word (either archaic or contemporary) could be found, a term would be invented. For example, the people of our village told us to refer to "money" as "tree leaves," instead of using the *Tok Pisin* word "*mani*." However, when people from other villages heard us use this term, they almost invariably interpreted it literally as "leaves."

Eventually, *I* was the one who was pushing for excessive vernacular purism. My desire to speak pure, uncorrupted Lamogai, was spurred on by the twinge of guilt I felt each time I used a borrowed *Tok Pisin* word. As a result, my speech became filled with archaisms and invented terminology. I began referring to *rice* as "ant eggs"—and *shoes* were now called "foot skins." I was certain that the next time my language-learning consultant came for a visit, he would be impressed by the fact that I was learning all the "correct" Lamogai terminology.

The people in our immediate village grew accustomed to our frequent use of archaisms and invented words. However, I found that when I conversed with visitors from neighboring Lamogai villages, they often did not know what I was saying. They were unfamiliar with many of the words and phrases in my excessively pure Lamogai idiolect. For some reason, it did not dawn on me that even the mother-tongue speakers from our own village rarely used many of these "pure" Lamogai words, except when they were speaking to me!

At times, when I was recording mother-tongue speakers giving text material on tape, I found myself getting a bit annoyed at the number of *Tok Pisin* words that they mixed in with the pure Lamogai—like *pasim* "to fasten" or *gat* "to have." I even found myself correcting them when they used too many words that were borrowed from the trade language.

A few months later, when my language-learning consultant returned for a follow-up visit, I was surprised, and a bit embarrassed to find that he was not nearly as impressed with my "pure" Lamogai as I had hoped he would be. He gave me just the kind of objective input I needed to get back on track. He helped me realize that Lamogai, as I was learning to speak it, was very different from the way it is spoken by the mother-tongue speakers. I am sure that deep down inside, I knew it all along. But I was hesitant to face up to the fact that present-day Lamogai includes a significant number of borrowed words, including *pilim* (*peel-eem*) 'to feel,' *save* (*sah-veh*) 'to know' and *tingting* 'thinking.' These words made the language seem so unexotic—and so unromantic. It was then that I began to ask myself, "What is my real objective in translating the Scriptures into Lamogai?" Ultimately my goal was to make the Bible accessible to the Lamogai people in the language they know best—the common language they use in their daily interaction with each other. Additionally, I was striving to see Lamogai become well established as a written language. But I had adopted a strategy that ultimately may have prevented me from reaching my goals.

57.5 What Is the Real Objective?

Several years ago, I met a mother-tongue translator who was translating the New Testament into his own language, using the *Tok Pisin* Bible as his base. He told me he had received training in translation principles, and was now working to produce a translation in his native tongue that would be clear, natural, and true to the original.

We talked for over an hour discussing various aspects of Bible translation. In the course of our discussion, there was one issue that gave me grave concern: He adamantly stated that he would not allow *any* borrowed words to be included in his translation. He explained how he spent hours talking with the oldest men and women in the village trying to find the best vernacular word for each context. Many of the words they came up with had long since fallen into disuse. I asked him how he expected the readers to understand his translation if it included so many unfamiliar words. "Oh, that's easy," he said, "We will create a dictionary to go along with the translation so the readers can look up the words they do not know."

He then proceeded to tell me that a few weeks earlier, he had stood up in church on a Sunday morning and read a Bible verse which included some of these old, unfamiliar words. As soon as the church service was over, some of the younger men came up and asked him what the words meant. Then, with a sly grin, he said, "I told them they would have to find that out for themselves."

It was evident to me that one of his primary aims in doing Bible translation was to revive the old, pure language forms that had been spoken by his ancestors. In his translation strategy, the preservation (or in many cases, the resurrection) of pure, vernacular terminology took precedence over the clear communication of the meaningful concepts of the Bible. His zeal led him to believe he could single-handedly turn back the clock in the process of language change.

I do not fault him in his passion for vernacular purity. I believe he was sincere in his desire to preserve his beloved native tongue. However, his effort to reverse the inevitable process of language *change* may actually have accelerated the more serious problem of language *shift*. I believe he was unwittingly driving the members of his speech community away from their own language. Since his translation of the Bible was impossible to understand, the readers would likely abandon it and shift toward the trade language Scriptures.

57.6 Is It Possible to Stem the Tide of Language-Change?

Published Scripture can play a significant role in helping to stabilize a language, and slow down the process of change. However, it is unlikely that the change process can actually be reversed to the point that the language would return to a purer, more-antiquated form. In some cases, many of the borrowed terms that would need to be eliminated may already have been in common use for decades. We found that even the oldest Lamogai people living in our village (who could not speak the trade language) used many borrowed trade-language words in their everyday speech.

It is unrealistic for a translator to expect that his or her translational choices will dictate the future course of the language. Instead, present-day, common-language vocabulary must dictate translational choices. A minimally-used translation which attempts to "turn the clock back" will do much less to counteract the process of language shift than a widely-used translation which accepts the language where it is today. Ironically, some of those who have worked the hardest to oppose the process of language shift may actually have accomplished the least. Translated Scriptures that cannot be understood by the average reader without an accompanying dictionary, will probably never be widely used. Rather than struggle against a torrent of unfamiliar, obsolete vocabulary, the average bilingual reader will likely discard the vernacular translation and turn to the more-easily understood trade-language Scriptures. This will cause the shift from the indigenous language to the trade language to accelerate rather than decelerate.

57.7 Archaisms and the English Bible

What would our English Bible sound like if it were not allowed to contain any word that had its roots in another language, including French, German, Greek, or Latin? All English translations—even the very oldest ones—include many words that came from other languages. The origin of these words does not make them inferior. The primary criterion for accepting or rejecting a particular word is how clearly it communicates the intended meaning of the original text to the present-day target audience.

The majority of English-speaking Christians who read the Bible regularly would not likely choose a version that is characterized by obsolete words or invented phrases. That is why new English translations are being produced all the time. I find it interesting that the American Bible Society completed the idiomatic *Today's English Version* (also called the *Good News Bible*) in 1976, and then published a new idiomatic translation, the *Contemporary English Version* a mere 19 years later.

Why do translators sometimes favor "pure" vernacular words over words which have been borrowed from another language? In my experience, I had an ideal in my mind of what an indigenous, minority-language Bible translation should look like. Frankly, I was embarrassed by the inclusion of so many borrowed words in our translated materials. Maybe I was afraid others would think I was lazy—that I had not put forth the effort necessary to find the true, vernacular terminology. Or perhaps I was afraid others would think the language I was working with did not sound very interesting or exotic. I must admit that I was sometimes guilty of allowing, or even pushing for, the inclusion of some outdated forms in the Lamogai translation.

One time I was rechecking a portion of Scripture with a couple of mother-tongue speakers, and we came across the Lamogai word *sudung*, which means "*to send*." One of the mother-tongue speakers in his mid-twenties spoke up and said, "Do we really have to include the word *sudung* in our translation? When we hear it, it makes the translation sound old and outdated. We never use that word in our everyday speech; and some of the younger people don't even know what it means. Can't we just use the *Tok Pisin* word *salim* instead? That is the word everyone uses and understands—even the oldest people in the village."

I knew he was right. I knew the word *sudung* was a vestige of my former push for extreme, vernacular purism. At that point I had to ask myself, "Do I really want to give the Lamogai people a Bible translation that sounds old and outdated? (Is that the kind of translation I would want for myself?) Or even worse, do I want to give them a translation that includes words which are not understood by the majority of Lamogai readers?"

Sometimes, linguists and translators have acted as though there is something *sacred* about pure vernacular words and phrases—even those that are no longer used. The indigenous languages of the world are more than just quaint linguistic systems to be preserved and studied in their purest original forms. These languages are the primary means of communication for the people who speak them.

57.8 Which "Borrowed" Words Are Acceptable?

If archaic words are to be avoided, does that mean I have the liberty in translation to use whatever national or trade-language words I choose to use? Absolutely not! When my language-learning consultant helped me leave behind my extreme vernacular purism, he told me I should feel free to use any borrowed, trade-language words that the mother-tongue speakers use in their everyday speech, to the extent that they use them. The mother-tongue speakers are the ones who have the right to determine which words can be borrowed, how often, and in what contexts.

Virtually every language in the world uses words that are borrowed from other languages. Many of the borrowed words in English are used so frequently that we do not think of them as borrowed words at all. We found this to be true in the Lamogai language as well. There are a number of borrowed, trade-language words which, through constant use, have become completely assimilated into Lamogai—even accepting full Lamogai inflection and derivation. In many cases, the *borrowed* word is now the *word of choice* for most contexts, since it is more commonly used, and more universally understood than its "pure" vernacular counterpart. We came to realize that some of these borrowed words had become an integral part of the Lamogai language before many of the present-day Lamogai speakers were even born.

If we had insisted on including archaic vernacular forms instead of using common borrowed words, we would have prevented the mother-tongue speakers from reading the Scriptures in the language they feel most comfortable with. We would have given them a translation that they would find cumbersome and difficult to understand. Also, our translation and publication efforts would not have made a significant contribution toward preserving the Lamogai language for future generations.

It could be said that old, obsolete words are actually *borrowed* from the language of a previous generation—like "thee" and "thou" in English. Many of these archaisms are indeed *foreign* to the average speaker of today. Do these old, unfamiliar words really deserve to displace commonly used words which happen to have their origin in another language?

A translator's ultimate goal is to transfer meaning from the source language into a form that is crystal-clear in the receptor language. Any word or phrase that does not communicate clearly to the majority of receptor-language speakers needs to be disqualified—like the archaic Lamogai word *sudung* ("to send"), that I had initially insisted on using in translation. On this basis, the foreign etymology of any given word does not necessarily demand its rejection, but the archaic nature of a word almost always does.

57.9 "Spiritual" Languages Versus "Common" Language

The original language of the New Testament is called *koine* or "common" Greek. It was truly the language of the people. The original message of the Bible was not encapsulated in a lofty, formalized language. The New Testament was written in a language that was clearly understood by the common person of the day. The average first-century Greek speaker could pick up the writings of the New Testament and say, "This book speaks the very same language I speak!" This no doubt helped reinforce the awareness that the message of Scripture is relevant to the everyday affairs of life. In the same way, a balanced approach to translation which produces a *relevant* Bible translation in a minority language will be much more likely to contribute to language preservation than one that is perceived to be outmoded.

57.10 The Path Toward Language Preservation

Bible translators all around the world are working tirelessly to unlock the linguistic mysteries of the world's remaining unwritten languages. But if those languages are to survive long-term in a written form, they need to have published reading materials which are distributed widely and used regularly. Any widely-read, published materials should do the job. But in many places around the world, Bible translators are the ones who are investing the time and effort necessary to produce written materials for these newly literate people groups.

New life is breathed into a faltering minority language when it is given its own translation of the Scriptures. And if that translation is clear, up-to-date and easy to understand, the speech community will have a new, compelling impetus for retaining their native tongue rather than quickly shifting to another language that they perceive to be more prestigious.

References

Koch, J. T. (Ed.). (2006). *Celtic culture: A historical encyclopedia, 206*. Santa Barbara: ABC-CLIO, Inc.

Lewis, P. M., Simons, G. F., & Fennig, C. D. (Eds.). (2013). *Ethnologue: Languages of the World, Seventeenth edition*. Dallas: SIL International. Online version: http://www.ethnologue.com

"Wycliffe". (2013). Wycliffe Bible Translators. www.wycliffe.org/about/statistics.aspx. Accessed 20 Mar 2013.

Recommended Readings

Barnwell, K. (1974, 1980). *Introduction to semantics and translation*. Horsleys Green: Summer Institute of Linguistics.

Bastardas-Boada, A. (2004). Linguistic sustainability for a multilingual humanity text based on the plenary speech for the X Linguapax Congress on 'Linguistic diversity, sustainability and peace'. Barcelona: Forum.

Beekman, J., & Callow, J. (1974). *Translating the word of God*. Grand Rapids: Zondervan.

Brunn, D. (1998). The pure vernacular: Are we producing a translation that is understandable today? *The Bible Translator (United Bible Societies), 1998*(49/4), 425–430.

Brunn, D. (2013). *One Bible, many versions: Are all translations created equal?* Downers Grove: InterVarsity Press (Academic).

Fee, G. D., & Strauss, M. L. (2007). *How to choose a translation for all it's worth*. Grand Rapids: Zondervan.

Katubi, O. (2006, January). *Lampungic languages: looking for new evidence of the possibility of language shift in Lampung and the question of its reversal*. Paper presented at the Tenth International Conference on Austronesian Linguistics, Palawan, Philippines.

Larson, M. L. (1998). *Meaning-based translation: A guide to cross-language equivalence*. Lanham: University Press of America.

Nida, E. A. (1947). *Bible translating*. London: United Bible Societies; Rapids: Zondervan.

Nida, E. A., & Taber, C. R. (1982). *The theory and practice of translation*. Leiden: Brill.

Scorgie, G. G., Strauss, M. L., & Voth, S. M. (Eds.). (2003). *The challenge of Bible translation*. Grand Rapids: Zondervan.

Tehan, T. M., & Nahhas, R. W. (2008). Mpi present and future: Reversing language shift. *The Mon-Khmer Studies Journal, 38*, 87–104.

Chapter 58
Archaeology, the Bible and Modern Faith

John T. Fitzgerald

58.1 Introduction

The English word "archaeology" derives from the ancient Greek word *archaiologia*, a compound word indicating discourse (*-logia*) about ancient (*archaio-*) times and topics. It occurs as early as Plato, where the sophist Hippias tells Socrates that the Spartans "are very fond of hearing about the genealogies of heroes and humans, about the foundations of cities in ancient times, and, in short, about antiquity (*archaiologia*) in general" (*Hippias Major* 285d; trans. Fowler (1977: 353), modified). The word is also used in the titles of several ancient historiographical works, including Dionysius of Halicarnassus's *Roman Antiquities* (*Rhōmaikē archaiologia*) and Flavius Josephus's *Jewish Antiquities* (*Ioudaikē archaiologia*). As these uses of the Greek term indicate, the word originally indicated ancient history or antiquities in general. In the nineteenth century it acquired a more specialized meaning, with archaeology used to indicate the scientific recovery and study of the whole of material culture from the past. This material culture was understood to include such things as domestic articles, tools, religious artifacts, and all other physical objects.

As archaeology has developed, it has become increasingly more specialized, with certain fields evolving into separate domains of study. These disciplines include epigraphy (inscriptions), literature (literary texts), numismatics (coins), and papyrology (papyrus and similar materials, such as parchment), with each field devoted to the study, editing, and interpretation of the writing as well as of the physical objects on which they are found. Because of the development of these specialized fields, archaeology today is often depicted as concerned with a more limited range of material culture, having a focus on such things as visual works of art (mosaics, murals, painted vases, sculpture, etc.) and monumental

J.T. Fitzgerald (✉)
Department of Theology, University of Notre Dame, Notre Dame, IN 46556, USA
e-mail: john.t.fitzgerald.105@nd.edu

S.D. Brunn (ed.), *The Changing World Religion Map*,
DOI 10.1007/978-94-017-9376-6_58

architecture (public buildings, temples, etc.), but also encompassing the study of textiles, bones, pollen, and various micro-organisms. Physical objects containing writing, though discovered by archaeologists, are often studied most intensively by specialists in other disciplines.

The explosion of knowledge in modern times has made such specialization inevitable, but the goal of all fieldwork by the "dirt archaeologists" who conduct excavations is the recovery of the entire range of material culture preserved at the sites. What is discovered at a particular site may be studied by numerous disciplines, and these discoveries will have implications for our understanding of ancient cultures, including their history and religions. Some of these discoveries will confirm and strengthen our prior understanding of our cultural and religious heritage, whereas other finds will challenge those previous understandings.

This concentration on material culture, including texts, relevant to our understanding of ancient Israel, early Judaism, and formative Christianity does not entail any geographical restrictions. One cannot, for instance, confine "biblical archaeology" to discoveries made in the Holy Land. The biblical story involves the whole of the ancient Mediterranean world, with the mythical Garden of Eden, for instance, located in Mesopotamia, and the Roman Empire of the first centuries C.E. encompassing more than 30 modern countries. Furthermore, biblical scholars make use of all kinds of archaeological discoveries to enhance the contemporary understanding of the Bible and its worlds.

58.2 The Importance of the Dead Sea Scrolls

There are many archaeological discoveries relevant to the study of the Bible, early Judaism, and formative Christianity in the pre-Constantinian period, such as the inscription celebrating the building that Pontius Pilate dedicated to the Roman emperor Tiberius in Caesarea Maritima when he was stationed there as the prefect of Judea and Samaria; the Caiaphas ossuary, which almost certainly once contained the bones of Caiaphas, the Jewish high priest who is depicted as presiding at the trial of Jesus before the Sanhedrin; the Gallio inscription, which provides a fairly secure date for the time when the apostle Paul appeared before the proconsul Gallio in Corinth (Acts 18:12–17); Masada, the mountaintop fortress where the First Roman-Jewish War was brought to a definitive end after the fall of Jerusalem and the destruction of the Second Temple in 70 C.E.; the letters of Bar Kokhba, leader of the Second Roman-Jewish War (132–135 C.E.); the Avercius epitaph, the earliest Christian inscription that can be approximately dated (to the period 196–216 C.E.); and the Nag Hammadi library, which gives us numerous Gnostic documents that enhance our perception of the various configurations assumed by Christianity in its early centuries.

Although a survey of the relevant archaeological discoveries might be informative, it is preferable, for the purposes of this article, to concentrate on the Dead Sea Scrolls (DSS), which are the best known and arguably the most important archaeological discovery of the twentieth century, at least as far as the study of Second Temple Judaism and early Christianity is concerned. For that reason, emphasis in this article will be given to these scrolls, which were discovered from 1947 to 1956 in 11 caves located in the area of Khirbet Qumran, which is on the northwest side of the Dead Sea. In discussing these scrolls, I shall draw on my own research, plus that of my colleague, James C. VanderKam, whose most accessible discussion is his *The Dead Sea Scrolls Today* (2010).

58.2.1 Discovery of the Dead Sea Scrolls

According to the most reliable account, the process of discovering the scrolls began when three Bedouin shepherds were tending their flocks in the area of Qumran. They found the first three of the scrolls in a jar that stood about 2 ft high, with two of the scrolls wrapped in linen. The jar was in a cave, subsequently called Cave 1 to indicate that it was the first cave where scrolls were found. Of the three scrolls, one was a copy of the book of Isaiah, another was a commentary (called a *pesher*) on the book of Habbakuk, and the third was a book giving rules for a religious community (initially called the *Manual of Discipline* but now usually referred to as the *Community Rule*). Four additional scrolls were subsequently found in the same cave: the *Genesis Apocryphon* (containing fictitious expansions of narratives found in the biblical book of Genesis), a partial copy of Isaiah, a collection of hymns of thanksgiving (*Hodayot*) known as the *Thanksgiving Hymns*, and the *War Scroll* (depicting a final eschatological battle between "the sons of light" and "the sons of darkness").

When knowledge of these seven scrolls became public and their immense archaeological value was realized, efforts to find additional scrolls were undertaken. Ten more caves containing scrolls and/or fragments of scrolls were located (called Caves 2–11), with the two-chambered Cave 4 containing the largest number of fragments of various manuscripts. Altogether, approximately 800 different manuscripts comprise the DSS, which have immense importance for four different areas of research.

58.2.2 Importance of the Scrolls for the Text, Textual History, and Canon of the Hebrew Bible

First, as the two copies of Isaiah found in Cave 1 suggest, many of the DSS are copies or partial copies of biblical books. The discovery of these biblical DSS has had a revolutionary impact on studies of the text of the Hebrew Bible (Old Testament). Prior to the discovery of the DSS, the oldest copies of the Hebrew Bible dated from the medieval period. The Hebrew text found in these medieval copies is known as the Masoretic Text (MT) because it was produced by a group of Jewish scholars called Masoretes who were active from the seventh to the eleventh centuries C.E. They not only copied the ancient consonantal Hebrew text of the Bible but also added diacritical signs and notes intended to standardize the pronunciation and cantillation of the text, and thereby to fix its meaning. The three oldest surviving copies of the MT are the Cairo Codex of the Prophets (dating from 895 C.E.); the Aleppo Codex (from ca. 925 C.E.), which is the most authoritative text in the Masoretic tradition; it was complete until 1947 but now, unfortunately, is missing many of its leaves (including almost all of the Pentateuch); and the Leningrad Codex (from 1008 C.E.), which is now the oldest complete manuscript of the entire Hebrew Bible.

The discovery of the DSS, which date from approximately 250 B.C.E. to 68 C.E., dramatically changed this situation because scholars now had access to pre-Masoretic biblical manuscripts that were in some instances more than a thousand years earlier that the oldest Masoretic texts. Furthermore, approximately one-fourth of the DSS manuscripts are copies of biblical books. Of the books that today comprise the Hebrew Bible, there are portions of all of the books except Esther and possibly Nehemiah (though the discovery of a DSS fragment of Nehemiah was announced in 2012). There are also copies of four books that are found in the Septuagint, the most famous Greek translation of the Hebrew Bible: Tobit, Sirach (Ecclesiasticus), the Letter of Jeremiah (=Baruch 6), and Psalm 151. In addition, there are Aramaic translations (called "targums") of portions of Leviticus and Job, as well as small parchments containing passages from Exodus and Deuteronomy; these were either placed in small boxes and tied to the head or left arm as *tefillin* (phylacteries) or attached to the doorpost of houses as *mezuzot* (Deut 6:9). As this evidence indicates, the people who placed these biblical scrolls in the caves were highly interested in Scripture and devoted themselves to its interpretation.

When the texts of the DSS are compared with the MT, results vary. Sometimes the Qumran texts (the DSS) and the MT are identical or nearly identical. This is the case, for instance, when one compares the MT of Isaiah with the great Isaiah scroll found in Cave 1. Indeed, many of the biblical DSS belong to the same textual tradition that later became enshrined in the MT and are thus referred to as "proto-Masoretic" texts. At other times, however, there are significant differences, which show that the MT was not the only textual tradition in the last three centuries B.C.E. and the first century C.E. When the Qumran texts differ from the MT, the former sometimes agree with the readings found in the Septuagint (abbreviated LXX). These instances prove that the translators of the LXX were translating a different Hebrew text from that found in the MT. For example, Exod 1:5 in the MT indicates that 70 of Jacob's descendants journeyed to Egypt, whereas the LXX and the DSS (4QExod^{a}) say that 75 people made the journey, which is also the number found in Acts 7:14. Similarly, the MT of 1 Sam 17:4 gives the height of Goliath, the Philistine warrior slain by David, as "six cubits and a span," indicating that he was a truly gigantic man who was 9 ft, 9 in., tall. The LXX and the Qumran text (4QSam^{a}), by contrast, give his height as only "four cubits and a span," that is, a "mere" 6 ft, 9 in. In this case, the smaller height given for Goliath is almost certainly the more original one. The MT's addition of three feet to Goliath's stature was designed to make him truly gigantic and thus to enhance the magnitude of David's victory over the Philistine.

At other times, a Qumran text gives a new reading that was previously unknown, yet has a strong claim to being part of the original biblical text. The longest and most important of these new readings is found in the Qumran scroll of the books of Samuel (4QSam[a]) in connection with the story of the siege of Jabesh-Gilead by Nahash, the king of the Ammonites, who wanted to gouge out the right eye of the men of the city as the price of a peace treaty with them. In the MT, the siege provides the first occasion for Saul, the newly proclaimed king of Israel, to prove his worth by rescuing the men of the city, but no reason for the siege is given. Why Jabesh-Gilead, and why such a horrible condition for a peace treaty? The answer is provided by the Qumran scroll:

> Now Nahash, king of the Ammonites, had been grievously oppressing the Gadites and the Reubenites. He would gouge out the right eye of each of them and would not grant Israel a deliverer. No one was left of the Israelites across the Jordan whose right eye Nahash, king of the Ammonites, had not gouged out. But there were seven thousand men who had escaped from the Ammonites and had entered Jabesh-Gilead.

Evidence supporting the presence of this explanation in the autograph (original scroll) of Samuel is provided by the Jewish historian Josephus (*Jewish Antiquities* 6.68–71). Its absence from the Masoretic manuscript tradition probably stems from an accidental scribal omission of the explanation. The first Bible translation to restore the omitted material and include the explanation was the New Revised Standard Version.

As these illustrations suggest, the biblical manuscripts among the DSS have been of immense value to those who are concerned with the text of the Hebrew Bible. Three of its major contributions are as follows. First, the Qumran scrolls have proved that no one manuscript or manuscript tradition preserves the original biblical text in all instances. The DSS have confirmed the readings of the MT in most instances and provided different and often superior readings in other instances. These different readings are sometimes unique to Qumran and at other times identical to the readings of other ancient witnesses, such as the Septuagint or the Samaritan Pentateuch. Consequently, the DSS have been vital to the goal of establishing the original text of the Hebrew Bible.

Second, the DSS have proved enormously helpful in establishing that the canon of the Hebrew Bible—that is, the books included in the Bible—was not yet closed by the first century C.E. and that there were even different and competing versions of books that eventually did become canonical. For example, the people at Qumran regarded the books of *Jubilees* and parts of *1 Enoch* as authoritative (compare Jude 14–15), though these books were eventually excluded from the canon. Competing versions of biblical books were already known prior to the discovery of the DSS. For instance, the version of Esther found in the Hebrew Bible (and also in Protestant Bibles) is the shorter and original version of the book. It is also a fairly secular story in that God is never explicitly mentioned. A longer and more explicitly theological version was produced, and this is the version found in the Septuagint (and in Roman Catholic Bibles). The Qumran community appears to have rejected Esther and not to have regarded it as authoritative, but they did esteem the book of Psalms and even had two different versions of it. One version is proto-Masoretic, whereas the other has a completely different sequence of psalms from that found in the MT as well as nine texts not found in the Masoretic Psalter. These include psalms previously known from other ancient versions of the Psalter (Psalms 151, 154–155) and four unique compositions previously unknown (Plea for Deliverance, Apostrophe to Zion, Hymn to the Creator, and a prose section called David's Compositions). As this evidence suggests, neither the text nor the canon of the Hebrew Bible was fixed at the time of Jesus and the early church but was still in flux, especially in regard to the section of the Hebrew Bible known as the Writings (*Kethuvim*).

Third, the Qumran biblical scrolls are helpful in charting the history of the Bible's textual development. Naturally, the textual evidence provided by the DSS is interpreted in different ways by textual critics. Many scholars, for instance, see in them confirmation of the theory that many of the textual differences arose because the biblical text was copied and edited in different geographical areas. This is the "local text" theory, which holds that certain texts of the Pentateuch have a "family" resemblance because they originated in three distinct geographical regions—Palestine (home of the Samaritan Pentateuch), Egypt (where the Hebrew Bible was first translated into Greek), and Babylon (home of the tradition represented by the MT). All three types of texts are found at Qumran, which, for advocates of this theory, indicates that the Egyptian and Babylonian text types were already present in Palestine prior to the rise of Christianity. Other scholars reject the local text theory and see in Qumran simply evidence for textual plurality prior to the later standardization by the Masoretes.

58.2.3 Importance of the Scrolls for Early Jewish Biblical Interpretation

In addition to their importance for the text, textual history, and canon of the Hebrew Bible, the DSS are indispensable for understanding how the biblical text was being interpreted prior to the rise of both early Christianity and Rabbinic Judaism. This is the chief contribution of the targums and biblical commentaries found at Qumran. To begin with, the Qumran targums to Leviticus and Job largely ended a scholarly debate about when the practice of translating the Hebrew texts into Aramaic (the language spoken by Jesus and most Jewish people in Palestine during the later Second Temple period) began, and when these translations were first written down. Prior to the discovery of the DSS, the oldest targums dated from periods after the advent of Christianity, and some scholars regarded the development of targums as a late phenomenon. The Qumran targums, dating from the second century B.C.E. and the first century C.E., proved that this was a pre-Christian practice. It originated in Jewish synagogues, where the Hebrew text was first read and then an oral translation into Aramaic was provided so that the Aramaic-speaking audience could understand the text. Subsequently, written versions of these translations were produced, and the DSS provide early evidence of this practice.

The commentaries on the Bible found at Qumran are of two basic types. The first type is the "running commentary," with the commentator typically quoting the biblical text, then giving his interpretation of the text. The three best examples of this type are the commentaries on Habakkuk (1QpHab), Nahum (4Q169), and Psalm 37 (4Q171, 173). The second type is the "thematic commentary," which draws thematically related texts from multiple biblical books or from different sections of one book to discuss a biblical theme or individual. A particularly interesting example is the Melchizedek Text (11QMelch). Melchizedek is mentioned in the Hebrew Bible in two places: Gen 14:17–24 and Psalm 110:4. In the former passage, he is described as the "king of Salem" (Jerusalem) and the "priest of God Most High"; in that capacity he blesses Abram (Abraham), who gives him a tithe. In the latter passage, the Israelite king of Jerusalem is given priestly prerogatives by quoting an oracle that says, "You are a priest forever according to the order of Melchizedek." In short, the psalm shows how archaic Jebusite traditions about Melchizedek were appropriated by the Israelites after David conquered the Jebusites and assumed control of Jerusalem. The Melchizedek tradition thus became a part of Israelite royal ideology, and the New Testament (NT) author of Hebrews applied it to Jesus (Heb 5:6, 10; 6:20–7:10, 15–17), thereby giving him priestly powers. The Qumran scroll, by contrast, turns Melchizedek into an angelic figure who will serve as God's instrument for judgment and the destruction of Satan at the end of time.

The authors of the two types of biblical commentary and other Qumran texts share a common two-fold apocalyptic assumption: They believe that biblical texts, especially prophetic texts, are speaking of the "latter days" of human history, not of the times when the prophets lived. Furthermore, the commentators see themselves as living in the times when these prophecies of the future are being fulfilled. For example, the book of Nahum concerns the fall of the Assyrian capital of Nineveh in 612 B.C.E. The prophet Nahum cackles with glee as he describes the sacking of the city, which was ripe for plunder because of the enormous wealth the Assyrians had assembled as they spread their hegemonic empire. The symbol for Assyrian power was the lion, which often appears on Assyrian reliefs. In a closing taunt, Nahum turns this lion symbolism against the Assyrians by referring to the lion's violence and the abundant food assembled from its victims: "The lion has torn enough for its whelps and strangled prey for its lioness; he has filled his caves with prey and his dens with torn flesh" (Nah 2:12). The author of the Nahum commentary totally ignores the historical meaning of the biblical text and reinterprets it so that it prophetically describes an event in his own time: "Interpreted, this concerns the furious young lion [who executes revenge] on those who seek smooth things and hangs men alive, [a thing never done] formerly in Israel. Because of a man hanged alive on [the] tree, He proclaims, 'Behold I am against [you, says the Lord of Hosts']" (Vermes 1962: 232). The Qumran interpreter regards the lion not as Assyria and its kings, but as a Maccabean Jewish king of Judea by the name of Alexander Jannaeus (103–76 BCE). He was detested by many of his Jewish constituents, who revolted against him. Jannaeus crushed the revolt and took vengeance on the Pharisees ("those who seek smooth things") for their role in the uprising by crucifying ("hanged") 800 of them at one time. Mass crucifixion was unprecedented in

Palestine when he did this ("[a thing never done] formerly in Israel"), and Jannaeus' use of this excruciating method of execution was designed to have a chilling effect on the dissidents among his people. Subsequently, crucifixion as a means of execution was especially associated with suspected revolutionaries and others whose words and deeds might function to destabilize the government.

58.2.4 Importance of the Scrolls for Understanding Early Judaism

A third area of research illuminated by the DSS is Judaism during the Maccabean (167–63 B.C.E.) and early Roman periods (63 B.C.E. – 70 C.E.). Indeed, thanks to the DSS, scholars have a better understanding of Judaism during these two periods than we have for any other religion in antiquity. This is the chief collective contribution of the non-biblical texts from Qumran. These non-biblical texts fall into three basic categories. First, the Qumran library contains copies of previously known works, which were also prized by people who had no association with Qumran. Examples include *The Damascus Document*, *Jubilees*, and *1 Enoch*. Second, the DSS contain copies of many previously unknown works that appear to be general in nature and likely were penned by authors who did not live at Qumran. The *Genesis Apocryphon* from Cave 1 is an example of a work that may have been produced elsewhere; it tells how Noah's father, marveling at the unparalleled magnificence of his son, suspected that his wife had been impregnated by a fallen angel (which is congruent with a widespread ancient interpretation of Gen 6:1–4). Third, other DSS give the theology of the community that lived at Qumran and thus are sectarian in nature, espousing a particular theological viewpoint and interpreting Mosaic laws in distinctive ways. The *Community Rule* is a well-known example of this category, as is a legal text known as 4QMMT or *Some of the Works of the Torah*, which its editors have suggested is a letter from the group at Qumran to their religious opponents in Jerusalem, including the high priest, setting out the ways in which their interpretation of biblical law was different.

This new material has given scholars a much more nuanced understanding of Jewish theology, eschatology, worship, wisdom traditions, debates about the Jewish calendar, and legal disputes than was possible prior to the discovery and publication of the DSS. This has had a transformative effect on how this period of Jewish history is interpreted. Older scholarly and popular depictions of Judaism generally did not do justice to the great variety of religious expressions during these periods, and there was a tendency to see Jewish life and thought hermetically sealed off from the larger Greco-Roman world of the time, and to posit a basic continuity between pre-70 C.E. Judaism and the Rabbinic Judaism of later times. Thanks to the DSS, other discoveries, and ongoing research, contemporary depictions of the Maccabean and early Roman periods emphasize diversity of life and thought both religiously and politically. In addition, there is a much greater awareness of the importance of

Hellenization during this period. "Hellenization" refers to the interactive process by which Greek customs, ideas, institutions, practices, and terms spread into non-Greek regions and, to varying degrees, were not only appropriated by some indigenous individuals and groups but also resisted and rejected by others. Hellenization was a phenomenon that divided the Jewish community, but its impact was felt even in isolated places like Qumran.

58.2.5 Importance of the Scrolls for the Study of Earliest Christianity

The fourth and final area to which the DSS have contributed is the study of the New Testament and earliest Christianity. This is not because any NT document has been found at Qumran or that figures mentioned in the NT (such as John the Baptist, Jesus, and James the Lord's brother) spent time at Qumran or were former members of the Qumran community, though sensationalists have occasionally advanced such theories. The true contribution is rather two-fold. First, because we now have a much better understanding of Judaism at the time of Jesus and his followers, we have a more accurate understanding of the Jewish context in which Jesus' ministry was conducted and in which formative Christianity developed. As a result, the Jewish heritage of Jesus and the early Jesus-movement is more firmly established than ever before. Christianity began as a Jewish sect that was distinguished from other Jewish sects primarily by its conviction that Jesus of Nazareth was the Messiah. Like other Jewish sects, the Jesus sect had points of agreements and disagreement, not only among themselves but with other sects. Because of the DSS, scholars can more accurately identify these points of continuity and contrast, of agreement and disagreement, between the views of Jesus and his disciples and those of their Jewish compatriots.

Second, the DSS enable us to make certain comparisons of the earliest Christians with the sectarians at Qumran, who are most often identified as Essenes. Both groups were apocalyptic sects that used some of the same terminology to refer to themselves ("sons of light") and outsiders ("sons of darkness") and saw ancient prophetic oracles as being fulfilled within their own lifetime and in their own religious community. In addition to using some of the same terminology (including Pauline phrases such as "the righteousness of God" and "works of the law"), they shared a particular fondness for three biblical books (Psalms, Deuteronomy, and Isaiah), and even for some of the same passages in certain books, such as Isa 40:3, which the Qumran sect saw itself as fulfilling, and which the NT associates with John the Baptist. Certain similar practices are attested for both Qumran and early Christianity, including the sharing of property, common meals, and a concern with Messianism and the Messianic age. Of course, there were significant differences. The author of the *Community Rule* expected the future coming of "the Prophet and the Messiahs of Aaron and Israel" (9.9–11), that is, three different individuals who would exercise distinct functions. Early Christianity saw in Jesus of Nazareth the fulfillment of all of these messianic hopes, identifying him as the Davidic Messiah, the prophet like Moses, and a priest (though one like Melchizedek rather than Aaron).

58.3 Four Trenchant Problems Affecting Archaeology and Modern Faith

As the preceding discussion of the DSS has demonstrated, archaeological discoveries have the potential to revolutionize the study of various subjects intimately connected with the Bible. At the same time, the work of professional archaeologists, who follow strict scientific guidelines, secure necessary permissions to carry out excavations, abide by national and international law regarding the objects they uncover, adhere to strict ethical principles designed to preserve the world's cultural heritage, and commit themselves to publishing and making accessible their findings, is adversely affected by at least four problems. The first of these is the extraction and/or removal of an artifact from a site in a manner that does not follow standard archaeological procedures, such as establishing the stratigraphy of a site. This happens most frequently when a site is discovered by looters who, by removing objects of perceived value, destroy the context in which these objects were preserved. A second problem is the unauthorized removal of an object from the country where it was discovered, which typically involves some violation of national and/or international law. Smuggled artifacts are almost always illegal ones.

A third problem is unprovenanced artifacts, that is, antiquities whose place and date of discovery are unknown, and thus are undocumented. The year 1970 typically plays a large role in discussions of unprovenanced antiquities, for this was the year when UNESCO (United Nations Educational, Scientific and Cultural Organization) adopted its Convention on the Means of Prohibiting and Preventing the Illicit Import, Export and Transfer of Ownership of Cultural Policy. An example of a pre-1970 unprovenanced antiquity is the Nash Papyrus, which contains the Ten Commandments. Likely dating from the second century B.C.E., it was the oldest Hebrew manuscript fragment of any part of the Hebrew Bible until the discovery of the DSS. It was acquired in 1898 by W. L. Nash of the Society of Biblical Archaeology and donated to the library of Cambridge University, long before 1970. The UNESCO Convention has little effect on such pre-1970 antiquities other than noting their inherent limitations as unprovenanced artifacts. By adopting its 1970 convention, UNESCO was seeking primarily to prevent the host of future problems that inevitably arise when objects are not found *in situ* (undisturbed in their original site) and removed scientifically.

A fourth problem is that of forgery, which occurs when unscrupulous individuals and dealers who operate on the black market seek to reap financial gains by selling forged artifacts, typically at astronomical prices. When forged artifacts are treated as genuine works from antiquity, the historical record is falsified and our understanding of the past is contaminated. Unfortunately, forgers have become so skillful that in certain cases it is unclear whether the artifacts are genuine. One prominent example in this regard is the so-called "ivory pomegranate" that the Israel Museum in Jerusalem acquired from a collector in 1988 for $550,000. At the time of purchase, it was believed that this ornamental artifact, which bears the inscription "Sacred to the Priest of the House of God," once adorned the scepter of the high priest in Jerusalem and thus constituted archaeological evidence for the existence of Solomon's temple. Subsequent study of the artifact, however, has made it highly likely that the inscription is a forgery, which means that there is no basis for linking the pomegranate with Solomon's temple.

All four of these problems have enormous implication for modern faith as it seeks to take account of archaeological data, and these problems are compounded when discoveries are announced and spread by a media that is prone to sensationalism and much more likely to publicize the speculations and theories of individuals rather than the doubts and more cautious assessments of the vast majority of specialists. When individuals do not follow established protocols for establishing the authenticity of artifacts, the announcement of spectacular finds places both scholars and the populace as a whole in a quandary. Lay people do not know what to believe, and specialists are unable to comment on these alleged discoveries. Three recent "discoveries" will show the difference. The Gospel of Judas is an apocryphal gospel that appears to portray Judas Iscariot, the "betrayer" of Jesus, in a positive light. Before its discovery was publically announced, this work was subjected to intense scientific scrutiny, which proved its authenticity beyond any reasonable doubt. Debate could thus certain on its interpretation. By contrast, the James ossuary (with the inscription "James, son of Joseph, brother of Jesus") and the so-called "Gospel of Jesus' Wife" (because the papyrus fragment contains the line, "Jesus said to them, 'My wife' ...") were announced without a sufficient peer-review process that could have established the authenticity of the artifacts or proved that the ossuary inscription and the writing on the papyrus were forgeries. This much is certain: Both the James ossuary and the Gospel of Jesus' Wife, if genuine, are unprovenanced artifacts whose owners did not want to have their identities revealed. Although the ossuary and the papyrus themselves are indeed ancient, the writing that appears on these artifacts is now widely suspected to be forged. Yet even if these texts are finally deemed genuine, they will forever carry the qualifying asterisk of being unprovenanced.

58.4 Closing Comments

As this discussion indicates, peoples of all faith traditions as well as no faith traditions should insist that individual scholars follow professional guidelines when announcing and discussing archaeological artifacts, and they should also enquire whether the provenance of particular artifacts is known and whether the announcement of new discoveries has been preceded by a scholarly peer-review process that authenticates the artifact. Only then can public confidence in archaeological announcements be restored.

References

Fowler, H. N. (Trans.). (1977). *Plato: Cratylus, Parmenides, Greater Hippias, Lesser Hippias.* Loeb Classical Library 167. Cambridge: Harvard University Press.

VanderKam, J. C. (2010). *The Dead Sea Scrolls today*. 2nd ed. Grand Rapids: Eerdmans.

Vermes, G. (Ed. & Trans.). (1962). *The Dead Sea Scrolls in English*. Baltimore: Penguin.

Part VI
Business, Finance, Economic Development and Law

Chapter 59
Belief Without Faith: The Effect of the Business of Religion in Kano State, Nigeria

Ibrahim Badamasi Lambu

59.1 Introduction

In all cultures human beings make a practice of interacting with what are taken to be spiritual powers that may be in the form of gods, spirits, ancestors or any kind of sacred reality with which humans believe themselves to be connected. Sometimes a spiritual power is understood broadly as an all-embracing reality and sometimes it is approached through its manifestation in special symbols. It may be regarded as external to the self, internal, or both. People interact with such a presence in a sacred manner, that is, with reverence and care. *Religion* is the term most commonly used to designate this complex and diverse realm of human experience. It is an integrated pattern of human knowledge, belief, and behavior that depends upon the capacity for symbolic thought and social learning (Adamu 1999 and Paden 1973) and as discussed in Lambu (2011). It can also mean the sets of shared attitudes, values, goals, and practices that characterize an institution, organization or group. In a nutshell, religion and culture are two inseparable entities that define human existence in space. It is the cultural beliefs (religion) that guide people to obey the society's norms, values and code of operation.

In recent years, or at least in the past 30 years, the "resurfacing of religion" has been a major issue in the globalized society. There were speculations in the past that secularization was an inevitable consequence of modernization, but nowadays that assertion has turned out to be untrue and fallacious. Even in technologically developed nations like the U.S., United Kingdom, France and others, which remain among the most secular societies in the world, religion has re-entered the public domain. And it has become ever more apparent that a country like the U.S. is only secular in coastal pockets. For some many societies, the return of religion has led to

I.B. Lambu (✉)
Department of Geography, Faculty of Earth and Environmental Sciences,
Bayero University Kano, P.M.B. 3011, Kano, Nigeria
e-mail: iblambu@yahoo.com

S.D. Brunn (ed.), *The Changing World Religion Map*,
DOI 10.1007/978-94-017-9376-6_59

a re-evaluation of their communities, a very high percentage of which are religious or spiritual in nature. A historical understanding of beliefs is very important even if they have been transmuted into "sects" or "religious cultures." The fact always is to acknowledge that for the great majority of human history religion has been the main medium through which people found meaning in life.

59.2 Religion: A Faceless Coin

It has proven impossible for experts to agree on a single definition of the term "religion." Over the last century and a half, the most intelligent minds have also failed to draw conceptual boundaries between "religion" on the one hand and society, culture, history, politics and economics on the other hand. Furthermore, the boundary between any two religious traditions is also fuzzy at best; historically, no major religion has developed in complete isolation from the rest of the world. Thus all religious traditions are products of syncretism, interaction as well as genuine innovations. A series of interactions between beliefs in space leads to spatial variations in religious behavior. If the concept "religion" is a slippery and unstable so as to provide a single, objectively verifiable definition, then the more complex notions of "religion of peace" and "religion of sacrifice" pose even greater challenges. Neither term is a precise concept that can be employed in an unambiguous or unbiased manner as both have originated in highly contentious debates over power, authority, and identity, and continue to be contested in a variety of ways. It is the result of such ambiguities that some people use religion to suit and gain some personal or group advantages.

A historically informed perspective does not allow us to treat any religion as if it were a static and monolithic object. No religion speaks with a single voice, and every religious tradition is characterized by a diversity of beliefs, attitudes, and expressions and that diversity tends to increase with the passage of time. To describe any religion as being solely this or exclusively that, one must reduce its inner complexity to an artificial simplicity, as well as its ever-changing character to a fixed caricature or stereotype. This reduction is itself an act of error that may take many dimensions. The simplest may be misleading those who assumed genuineness and originality of such alterations. The resulting image is almost entirely a product of the reductionist enterprise, bearing little resemblance to the dynamic and complex lived reality of the religious tradition.

Religions differ in how they divide the geographical space of earth (*and beyond!*). That is why there is a close relationship between religion and the environment. Culture that encompasses beliefs defines a relationship between nature, humans and spaces that allows for interaction, assimilation and diffusion. Simple ethnic religions often ritualize their living space through myths about ancestors, shrines and tombs. Whether or not a religious system produces a mythical geography that corresponds to its physical reality, it will have some sacred connection to elements of the original territory. For example, a common biblical term for a place of eternal fiery punishment is *gehenna*. This is a variant rendering of *Hinnom*, a valley south of Jerusalem, where child sacrifices by fire were conducted by the heathen priests. The Mount Ara'fat in the Kingdom of Saudi Arabia was the point of rescue of Ishmael since the time of Prophet Abraham. Now it remains an important point in the Muslims pilgrim rite.

As religion becomes more complex, societies and spaces are divided and often separated into sects and denominations with intense influence from the outside world. The divided space may then be divided into varying degrees of holiness or separation. In an ethnic religion, a national land itself may become holy as found in Zionism and other nationalist movements. However, in Nigeria, or even precisely Kano, the religions of Islam and Christianity are polarized. With the coming of Western missionaries that brought Christianity in the early 1900s and met the pre-existing religion of Islam, this changed the entire landscape in which the area began to experience a serious religious conflict.

59.3 Distribution of Places of Worship and Sectarian Affiliation

Sacred places may be found in lakes, rivers, rocks, mountains, and groves, possibly in association with a particular person or event. Unusual physical features that are the birthplace of a religion or religious figure, may also contribute sacred places. Shrines may be constructed and becoming a focal point within the religious system. Eventually sacred places may become religious centers as religious systems evolve. The sacredness of a religious site might be transferred to another, conquering religion. Churches were often built over pagan sites and mosques were often built over destroyed Hindu temples.

Some religious centers may rise to become preeminent because of their intense sanctity as religious capitals. Before the destruction of Jerusalem by the Romans, Jerusalem was the religious capital of Jerusalem, the center of the temple cult. Today, Jerusalem is sacred to Jews because of the importance it has in the Jewish faith, history, ritual, and identity. Every Passover is ended with the proclamation, "Next year in Jerusalem!"

Holy places, shrines and religious centers may become places of pilgrimage. Even today, millions of people trek each year to the Ganges River to bathe in its sacred water. The *hajj*, the pilgrimage to Mecca, is the sacred obligation of every Muslim once in life. The yearly pilgrimage puts great demands on the infrastructure of the region. Pilgrimages may also be responsible for the spread of ideas, increasing trade, transmitting disease, and altering existing traffic patterns. More complex religious systems may even divide space into hierarchical territories. Other religious bodies may be more or less autonomous.

59.4 Religious Factions and the Polarization of Beliefs

Religion in Nigeria, and precisely in the major urbanized centers like Kano and Lagos, experiences a proliferation of sects and denominations. In the Islamic religion, two major divisions were initially recognized, that is, the Sufism of Kadiriyyah (Qadiriyyah) and Tijjaniyyah movements. Later with the transformation of Ulama status, that is, the traditional Ulama (which is read in the informal traditional Tsangaya schools) and Modern Ulama (Muslims in Arab countries read this as part of their formal school education) resulted in the evolution of another movement known as IZALA. Izala is an abbreviation of Izalatul bidia waikamatussuna which means "The movement for removal of innovations in religion and restoration of right or correct deeds." The next identifiable movement manifesting itself in Nigeria, which had existed for a long time, was a disguised form as "Muslim brothers." Later it was included in the Shiite movement. In a nutshell, the polarization of Izala between the two ancient movements, Kadiriyyah and Tijjaniyyah, led to alliance between the latter quite unlike the present. In the past the Kadiriyyah and Tijjaniyah were not in on good terms with one another, but as IZALA evolved, the later formed an alliance or friendship. This alliance gave birth to the concept of Darika (Tariqa) movement (For the sake of this paper, it is important to note that each sect of Tijjaniyah, Kadiriyyah and Izala has many sub-divisions). That is, the Darika can be separated to several constituents. Another important religious movement has existed since the period of colonial masters is the Mahdiyyah movement. Its presence has not had a significant effect in the religious geography of Nigeria and Kano in particular. Kano is a place of high religiosity in Nigeria where the entire landscape is religion (Fig. 59.1). It should be noted that each sect has subdivisions and studying these separate sects would be another paper of its own. Lastly, another identifiable movement in the study area, though dormant, is Ahmadiyyah. It originated in Pakistan where Ahmad Gulam, its founder, claimed to have had a revelation and that his movement was considered as truly Islamic. From this background, we observe that the Islamic faith had several major sects or divisions.

Christianity can trace its initial history in Kano from 1905 with the advent of colonial invaders (Paden 1973). The Christian Missionary Society (CMS) contingents arrived in Kano and were expelled in 1900, but later moved and settled at Zaria (is a town near Kano state) with established mission in 1905 (Crampton 1979; Wakili 2009). Resistance to Christianity in the towns and cities made the mission focus on rural areas especially, among the Maguzawa (considered pagans) and at the outskirts of the main city of Sabon Gari (Wakili 2009). Uses of Western medical services (missionary schools and clinics) were adopted by the Protestant mission churches like the Sudan Interior Mission (SIM), the Sudan United Mission (SUM) and Church Missionary Society (CMS). The use of the local language, Hausa, was required to ensure success in the preaching and other missionary undertakings like dispensary services (Wakili 2009).

Indigenization of Sudan Interior Missionaries (SIM) was successful by renaming it the Evangelical Churches of Western Africa (ECWA). It was dominated by Hausa

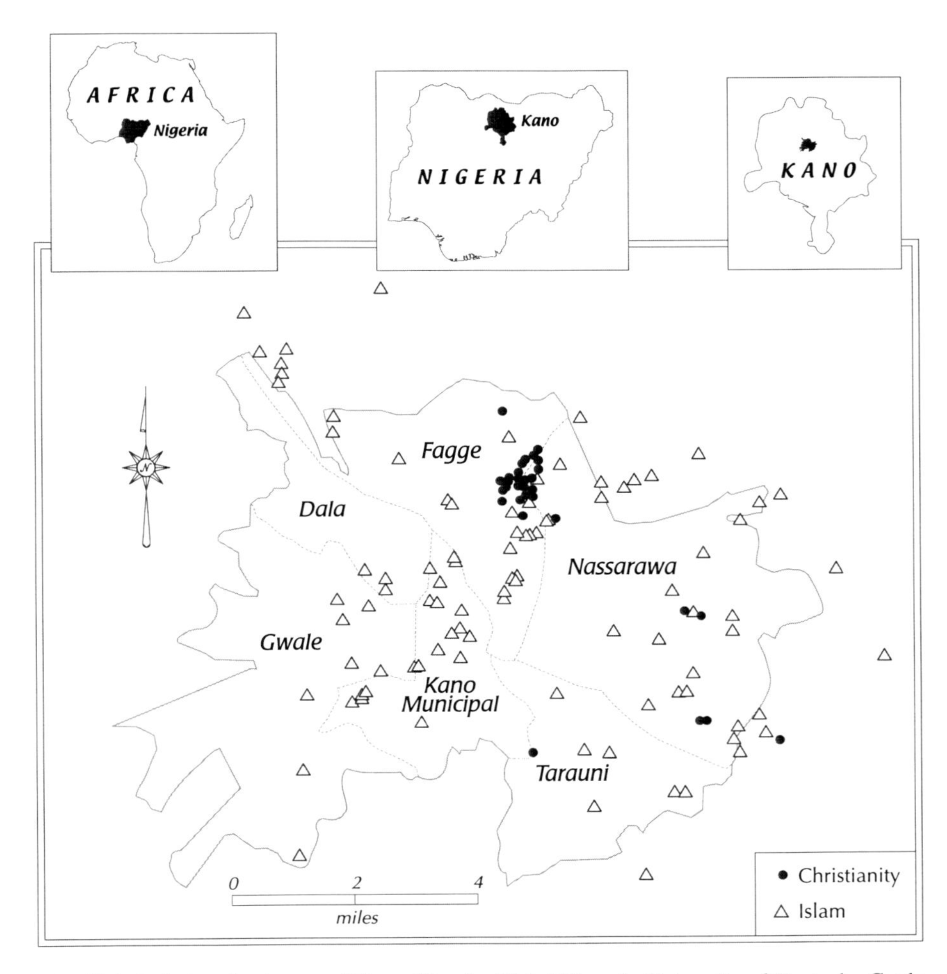

Fig. 59.1 Religious landscape of Kano (Map by Dick Gilbreath, University of Kentucky Gyula Pauer Center for Cartography and GIS; commissioned by the editor)

Christians and other northern minorities. Kano as the economic capital of the North receives the highest number of immigrants from rural areas. Thus the total number of Christians and churches is much higher in Kano than for any other city in the entire North. In 1966, there were over 56 churches in the city of Sabon Gari, but now the figure is probably twice that number. Roman Catholics are dominated by Igbo migrants while the orthodox Protestants are populated mostly by Yoruba people. The northern Christians identified with ECWA. The Pentecostals are more of a trans-ethnic mix than other denominations. They consist of people from different ethnicities. Intense interdenominational differences and polarization led to Christians eventually uniting their believers, irrespective of differences in defense of interests and aspirations, to form the Christian Association of Nigeria (CAN) in

1976. CAN is more than a strong religious association as its members have strong political interests as well (Ibrahim 1989).

There are several distinguishing features of the contemporary religions landscape of Kano.

1. Despite the fact that Christianity started in Nigeria before colonialism through missionaries, and mainly in the southern part of Nigeria, its evolution in Kano coincided with colonial conquest of the area. This fact is evident in the air-tight nature of the culture where only conquest ensures apparent submission. In many instances there was fierce fighting between the Habe and Fulani as well with the white men over the protection of their own cultural values. Among the old churches in Kano is the First Baptist Church commissioned in 1926. Initially, there were two major divisions in the Christianity in Kano, the Catholics and Protestants. Subsequently, changes that started in some other places like Lagos and Kaduna eventually came to Kano and contributed to the evolution of some divisions as confirmed by all Christian respondents in this study. The Christian sects/movements in Kano metropolis are comprised of six major sects or blocks which are part of the Christian Society of Nigeria (CSN); it also has Catholic churches (see Fig. 59.1).
2. The Christian Council of Nigeria (CCN) includes Anglicans and Methodists.
3. The Pentecostal Fellowship of Nigeria (PFN) is a group that includes New Generation Churches, COCIN and Deeper Life sects.
4. The Christian Pentecostal Fellowship of Nigeria (CPFN) is another group of Protestant churches.
5. The Organization of All African Institute of Churches (OAIC) consists of Celestial, Serephen and Cheroben churches.
6. ECWA/TECAN is a block that recognizes Evangelist Churches like Fisher's evangelical Church and the Shalom Evangelical Church.

Religious places of worship are cited erratically over and above the population demand where the motives behind the proliferations are ulterior. Religious leaders use the religious places to make money and get undue favor that is the "business of religion." Why there are only four Friday mosques in Kano municipal with population of 365,525 (NPC 2006)? Why do Fagge and Nassarawa with smaller Local Government Areas (LGAs) in the state have 15 Friday mosques each despite their smaller populations? Why are more than 90 % of the churches in the Fagge Local Government Area and the remaining churches are located in Nassarawa, despite the fact that the first colonialists, who brought the mission of Christianity, settled in the area?

59.5 Religion and Development

The concept of development is a many sided coin. It can be looked at from individual perspectives such as increases in skills and capacity, creativity and initiatives as well as freedom and self-discipline in materials and the general well-being of humankind. At a societal level, development entail issues like equity, fairness and distributive justice in dealing with public affairs. Regardless of how one perceives the concept, the most important goals in development entails building of a better society that is characterized by greater human and material prosperity (Lambu 2011). It is of interest to note that the concept of development is becoming fuller, broader and integrates sustainability issues. Proper religious education should strengthen the pillars of development. Having seen both how religion and science are interrelated and also the impact of religion in the political arena of the world, we observe that religious education is one of the most powerful ingredients contributing to development. Both Islam and Christianity cherish peace, love, kindness and sacrifice and at the same time condemned cheating, vandalism and all sorts of vices through the Holy Books of these religions. Patience, tolerance and fairness occupy conspicuous places in the cited Scriptures and all believers are mandated to observe and practice them in all undertakings throughout their entire life and to obtain eternity. Islam teaches the search for knowledge and even describes some facts that only in recent time how science and scientists acknowledged, such as earth rotation and revolution and meiosis and mitosis in the fertilization process. Christianity enshrines the creed of love and sacrifice to the extent that violence is not tolerated. Display of miracles and deliverance are religious values that encourage positive thinking for societal progress, which is concurrent with the scientific way of making life productive and more meaningful.

Christians teach that God is almighty in dominion over all that is in heaven and on earth, righteous in judgment over good and evil, beyond time and space and change; but above all they teach that "God is love." Christians and Muslims educate their followers that the creation of the world out of nothing and the creation of the human race were expressions of that love and so was the coming of the Day of Judgment. The classic statement of this trust in the love of God came in the words of God. For example, "Look at the birds of the air: they neither sow nor reap nor gather into barns, and yet your heavenly Father feeds them. Are you not of more value than they?" (Matthew 6:26). Business and economic undertakings are clearly spelled in the holy Books of Muslims and Christians, that is the Qur'an and Bible. Issues of alms (Zakat), charity (Sadaqat), tenth (10 %) or Ushra and assistance are encouraged in several verses of these books. What remain for the believers to acquire the education and integrate it in the daily affairs of their life.

59.6 Globalization, Modernization and Religion

Society today is facing numerous challenges that attack its stability, security and even its survival or continuous existence. Modernization and globalization alter the existing social forms from material to non-material components such as family set up, art and values, that is, beliefs. The long inherited Gandu system with its communal family or extended family has gradually faded away leaving behind the solitary lifestyles of the modern world that lacked any meaning in social sustainability

(Lambu 2011. Gandu is an extended Nigerian family arrangement where the whole kinship eat from a single pot. In the past, the most elder member of the family assumed the leadership of the household as dictated by religion and social heritage. The elder managed the social, economic and religious affairs of the members. Nowadays, families are no longer operating as a system for the success of the whole. Rather the parts are broken and stand alone. This breakup in the family system exposes youth to the economic vagaries which in the Gandu system were taken by the elderly (Lambu 2011). Consequently, all social ties started degenerating where the role of traditional authority faded to an image of its shadow. Whenever there is any conflict, the immediate people (elders) and local authority cannot prevent or protect the worse from happening. In religious education, the adoption of sons and daughters and assisting those in orphanage is encouraged. But in the contemporary situation, people have less concern for their own children, the less privileged are often left at the mercy of nature. The unfortunate scenario of children neglects nurses or breeds hoodlums that attack the existing peace and harmony.

Interviews (in-depth interviews) were held at purposively sampled places of worship in Kano to gather the responses of religious stakeholders. Ten percent were sampled which is considered adequate (Lambu 2011). The religious leaders interviewed expressed concern about the effect of Western culture on Nigeria's our religious teaching and education. They noted that "The moral decadence and rowdiness in our society is not related to a failure of religious education, but a planned move by Western nations through their media to destroy our spiritual social system established from long inherited and noblest culture that ensures peace, happiness and progress. There is need for faith in everyone's life," said a reverend (whose identity is not disclosed). He continues: "But when faith does not provide methods for challenging the power of an unjust social and economic system, or impedes them, it loses its way and becomes a tool of mass repression rather than spiritually liberating. We must recognize that Christ was philosophically and pragmatically closer to Karl Marx than to Adam Smith. In other word we prefer the justice for all, a born-again society. Other religious leaders that contributed to the study asserted that poverty contributes in great measure to a destruction of religious teaching. Hungry and heartbroken youths with paltry or token amounts of money attack, maim or commit any number of atrocities.

59.7 Religious Education and Its Challenges

Many religious leaders interviewed expressed concern about the effect of Western culture on our religious teaching and education. "Immoralities through films, books and internet programs have destroyed our religious way of life," responded one religious leader. Observations across religious schools revealed that boys aged 12 and above boycotts schools and resort to football, TV centers, snooker and other imported games. About 70 % of religious stakeholders described football as a great destroyer of our youths. On the atrocities and other vices, 65 % attributed the acts to Western films where youths learn murder, theft, alcoholism and sexual misdeeds. Improper dresses, immoral gatherings and aimless loitering consume substantial part of the life of our teaming youths.

59.8 Religious Polity and Monopoly

The existence of Islam and Christianity in present day Kano creates steep competition and rivalry among followers of the two religions. Religious places are established erratically far above the justification of simply numbers (population). All sorts of forms, tactics and ploys are used in the name of religion in order to attract and maintain followers. False deliverance or promises and miracles are occupying the most precious part of the religious activities in the mosques and churches. The use of instigating and inciting violence and hatred against believers or followers of other religious followers or sectarian affiliates causes mayhem.

Yet these two major world religions, according to some studies, share more areas of commonality than differences in their code of teaching. Love, sacrifice, kindness, patience, peace and tolerance are the fundamental values of Islam and Christianity. Murder, alcoholism, vandals and lootings are all discouraged by both beliefs. Over 90 % of Nigeria's population are the followers of Islam and Christianity; pagans and animists account for less than 10 %. If the followers of the major religions adhere strictly to the teaching of their religions, most of the social problems could have been averted.

Recent religious conflicts and crises in 2011 include post-election violence in northern Nigeria, the perpetuated Jos Plateau ethno-religious episodes from 1999 to 2012 and the bomb-blast syndrome of 2011–2012 all lacked a religious base or justification. Numerous questions emerged from these events, including: is God revising the Qur'an and Bible? Have the teachings of Islam and Christianity been repealed? What is the significance of the Friday and Sunday sermons? The faith in God has faded so fast that the belief is completely lost and people use religion to amazed wealth, gain political favour and social influences. In this way, religion has become a business instead of spiritual relationship that convey a divine message for humanity. Places of worship are regarded as money-making shops where rituality exists without spirituality. The religious values and preferences in the two religions are rooted from two source, the Qur'an and the Bible, with modifications due to modernization which change the entire spiritual landscapes. High taste, greed, pride and "showing off" accentuate the desire for some to amass wealth at all costs. Religious scholars use religion as a means of money-making through exploitation and marginalization of followers in form of illegal and illegitimate collections in the name of religion. Religion is now being used in political campaigns by aspiring candidates of different sides to generate support for their positions.

59.9 Conclusion

It is time to bring to the attention of the religious leaders of the need to stick to the pure teaching of their religions and educate the followers regarding the precise tenants of Islam and Christianity so as to transform the entire Nigerian nation into an egalitarian society free of anarchy, rancor and immorality.

On the side of the political leaders, there is need for integrating religious values into the affairs of governance by incorporating religiously pious and erudite people to guide and patronize public affairs. "Divide and rule" maneuvers must be stopped and ways sought to integrate society must be pursued with vigour.

Religious followers of all faiths must abide by the empirical guides of their faiths and respect the ethics of peace, love and sacrifice by mutual understanding, caring and dedication. Humans are created equally by religious thoughts where murder, theft and violence are forbidden and discouraged in an attempt to uphold the dignity of humankind by safeguarding life, property and dignity The entirety of humanity is united under the umbrella of being servants to the Supreme Being thereby integrating the society to pursue a common goal that is the eternal life.

The root of our societal problem stem from socio-economic and socio-political whims, including need of power, amazing wealth and affluent life, at all cost. All societal problems ranging from violence, corruption, immorality are not in any way thought or harbored by the religion of Islam or Christianity. In fact, there are more areas of similarities than difference between these religions. The results from my interviews confirm that the violence that people call "religious crises" is not only fallacious but deceptive, as it has no root from religious beliefs.

The religions educate their followers to be hardworking, dedicative, honest, and lovely. All these are ingredients that promote social integration, posterity and development. To understand the real genesis of the social problem in Nigeria, one has to direct the blame on the effects of globalization of the modern world that seeks to fight the teaching of religion in the name of technological advancement or secularization. Evidence on the ground has confirms the failure of secularism as result of current resurgence and upsurge of religion in the political arena across the world these developments make religion a global issue that should not be allowed to be used recklessly. Continued neglect about religious phenomena may spell doom to entire humanity due to the fluidity of the society as result of communication and transportation.

There is hope that as religion penetrates our lives, as is now evidenced in the political and social spheres of Kano and Nigeria at large, society will be changing for the better or the worse economically, culturally and politically depending on the way and manner religious matters are handled. Unless mutual respect, fair play and honesty are going to be the fulcrum of social life through religious indoctrination, the whole nation should forget about peace, harmony and even development. In this way the Western propagandas that are imported into our lives can be eradicated and quenched.

In the end, there is need to stress that researches the religious behaviors, trends, pattern and distribution are not only desirable, but a necessity in order to understand the way and manner in which the society is changing especially with the issue of globalization where many satanic thoughts are misconceived as Godly or where divine thoughts that are not only blackmailing religion, but detrimental to its development. Knowing where what happens and why are of interest to geography and to all humanity so that Planet Earth will continue to give humans maximum shelter before their journey towards the Day of resurrection.

References

Adamu, M. U. (1999). *Confluences and influences: The emergence of Kano as a city state*. Kano: Munawwara Book Foundation.

Adnan, A. (2000). *Groundwater condition in parts of Kano metropolis*. M.Sc. unpublished thesis. Department of Geography, Bayero University, Kano.

Crampton, E. P. T. (1979). *Christianity in northern Nigeria*. London: Geoffrey Chapman/The Chaucer Press Ltd.

Falola, J. A. (2003). *Inequality its meaning and its reasons*. Paper presented, Department of Geography, Bayero University, Kano.

Ibrahim, J. (1989). The politics of religion in Nigeria: The parameters of the 1987 crisis in Kaduna State. *Review of African Political Economy, 45*(6), 65–82.

Lambu, I. B. (2011). *Culture and nature: The co-actors in climate change*. Sokoto: Usmanu Danfodio University 52nd Association of Nigerian Geographers (ANG).

National Population Commission. (2006). *National head count, Nigerian population census*. Abuja.

Paden, J. N. (1973). *Religion and political culture in Kano*. Berkeley/Los Angeles: University of California Press.

Wakili, H. (2009). *Religious pluralism and conflict*. Kano: Centre for Research and Development (CRD) Research Report No 2.

Chapter 60
Economic Development and Cultural Change in Islamic Context: The Malaysian Experience

Samuel Zalanga

60.1 Introduction

The main goal of this chapter is to examine the role of leadership in public policy reform and the implementation and state bureaucratic/institutional capacity reforms that took place in Malaysia under the leadership of Prime Minister Mahathir Mohamad and the ruling elite coalition that supported him. The chapter is a step towards explaining variations in the process through which ruling elites shape the success or failure of development policy and bureaucratic/institutional capacity reform in developing countries. I primarily treat variation in the nature, quality, and type of leadership and ruling elites' coalition as one of the main explanatory factors accounting for variation in the degree of progress a country is making in the pursuit of economic development and cultural change. Yet it must be noted that the ruling elites in their process of making decisions are constrained or empowered by social processes at the individual, local, institutional, national, and international levels. I am broadly interested in articulating how the intersection of all these social processes provides a richer and deeper understanding of the process that leads to success or failure of development policy formulation and implementation.

Malaysia under Mahathir Mohamad is an excellent case study because of the unique challenges it had faced as a country. Malaysia was a traditional Islamic society in the pre-colonial period. The predominant ethnic group in the pre-colonial region and to this day is the Malay. Later Indians and Chinese migrated to the Malay Peninsula. The British later colonized Malaysia. Britain restructured the economy of the region to primarily serve her national interest. By and large, this meant making

S. Zalanga (✉)
Department of Anthropology, Sociology and Reconciliation Studies,
Bethel University, St. Paul, MN 55112, USA
e-mail: szalanga@bethel.edu

S.D. Brunn (ed.), *The Changing World Religion Map*,
DOI 10.1007/978-94-017-9376-6_60

Malaysia a producer of raw materials for British industries and a market for British manufactured goods. Consequently, Malaysia became a producer of primary products such as rubber and tin. The country achieved independence from Britain in 1957 (that is, Merdeka) as a multi-ethnic society, which is reflected in the name of the country in the sense of combining "Malay" and "Asia." Malaysia faced significant socio- political and economic development challenges in the postcolonial period. The Malays represent the largest ethnic group in the country and politically the most influential, but economically, they were very weak and less successful compared to the Chinese community at the time of independence from Britain. Thus economic power at independence was largely in the hands of the Chinese business community, while political power was solidly in the hands of the Malays (Jesudason 1990).

In May 1969, there was race riot in the country when the Chinese community appeared to have made some political gains in national elections in addition to their established economic dominance. This political inroad by the Chinese scared the Malays that they may potentially lose political power and with no strong economic base in the country they would be marginalized in their own "native land" by a group they perceived as originally immigrants. For the sake of political stability, the Malay political elites decided to collaborate with the leaders of the two other large racial-ethnic groups in the country (that is, Chinese and Indians) to form a grand political coalition that united the three dominant "racial-ethnic groups" into one political party known as Barisan Nasional (that is, National Front). The country also embarked on pursuing an ambitious program of economic restructuring known as the "New Economic Policy" (NEP), which is a kind of systematic affirmative action program aimed at redistributing wealth and opportunities in the country such that no one racial-ethnic group would become extremely wealthy while the others remain in abject poverty (Jesudason 1990). Often such restructuring programs generate great backlash (Antoine 1991: 81–93). The Chinese complained that they were punished for their economic and business ingenuity and success. The Malays believed that it was necessary for them to succeed in their "native land" and that shared prosperity is in the interest of all racial-ethnic groups in the country because it reduces the potential of political conflict and violence, which would otherwise create a perpetual state of insecurity for all. Some technocrats complained that the implementation of the New Economic Policy was ineffective because it led to inefficiency in the allocation and use of scarce resources and cronyism in public policy implementation. It is within the context of this political and economic challenge and turmoil in Malaysia that Mahathir Mohamad emerged as Prime Minister in 1981 and ruled the country up to 2003 when he retired. He led the country during the heydays of neo-liberal economic reforms which was spreading all over the world under the auspices of the United States, Britain and multi-lateral institutions controlled by the Western world, such as the World Bank and the International Monetary Firm (Harvey 2005).

The first part of the chapter examines the values, vision, and commitment of Prime Minister Mahathir as a leader. This information was gleaned out primarily from his own publications, which he used as medium to disseminate his views and galvanize support among the Malaysian population in general and the Malay people in particular. I demonstrate that the ideas for reforming and transforming

Malaysian society were deeply rooted in Dr. Mahathir Mohamad's mind long before he rose to the position of Prime Minister. In the second part of the chapter, I highlight the Islamic context of Malaysia's struggle for economic development and cultural change in a globalized society dominated by Western nations who set the rules of the game. The third section of the chapter draws some lessons and conclusions from the analysis of material presented in the first and second sections of the chapter.

60.2 Leadership and Development: The Emergence of Mahathir Mohamad as Prime Minister of Malaysia

Tun Razak, the second Prime Minister of Malaysia, died in mid-January 1976 at the age of 53 (Mauzy 1983: 101). Tun Hussein Onn succeeded him. Tun Hussein Onn ruled as Prime Minister from mid-January 1976 to July 1981 when he retired from office. Dr. Mahathir Mohammed, who became the president of United Malays National Organization (UMNO), the Malay party that is the dominant partner of the ruling elite coalition in Malaysia (Means 1991: 82–109) succeeded him as Prime Minister. Dr. Mohamad served as the Malaysian Prime Minister from 1981 to 2003. He had decisively influenced and shaped Malaysian politics and economic development, such that one cannot have a real understanding of the development of modern Malaysia without closely examining his ideas and vision

A brief examination of his ideas/vision and how he has changed the trajectory of the country's economic development and state bureaucratic capacity serve the goal of this chapter in one important respect. It helps me demonstrate that ruling elites play a decisive role in determining the course of development policy formulation and implementation, since their control of public institutions grants them institutional power. This observation may sound mundane, but it is important in the world of neoliberal hegemony where public choice theorists have put forward a concerted attack on the role of government in promoting economic development and cultural change (Grindle and Thomas 1991; Buchanan 1954). In concrete historical reality, once the leaders of a ruling elite coalition of a country have power or they can mobilize their followers or the citizens of their country based on their personal ideas, visions, values, and conceptualizations of the strategies needed to bring about rapid development in society, the bureaucracy, and the public at large. On the other hand, if the elites who control public institutions have values and ideas that are inimical to national development or have incorrectly conceptualized the strategies for developing their nation, they would then constitute a major obstacle to national development. It is with this general theoretical insight in mind that I provide an analysis of Prime Minister Mohamad's role in leading Malaysia. I intend to briefly examine themes relevant to the main concern of this chapter from two books written by Dr. Mohamad to express his ideas, values, vision and strategies for transforming Malaysian society. Both books were written before he became Prime Minister. Furthermore, much of what he did after becoming Prime Minister was to translate those ideas and values into state institutions that came directly under his official control. He was supported in this by the ruling elite coalition (that is, Barisan Nasional). I begin by analyzing the relevant themes from two of his books, *The Malay Dilemma* followed by *Challenge.*

60.2.1 Mahathir's Interpretation and Diagnosis of Malaysia's Development Problems and the Way Forward: Relevant Themes from the Malay Dilemma and the Challenge

In my assessment, the subject matter of *The Malay Dilemma* (Mahathir bin Mohamad 1970) can be divided into three broad themes. First, it attempts to lay bare the problem of the Malay people in Malaysia, that is, their economic and socio-cultural backwardness. Indeed, the book provides detailed descriptive information about the problems. What the book sees as a very critical part of the problems was that the Malays were economically and socio-culturally backward in their own homeland. The second theme in *The Malay Dilemma* was explaining the causes of Malay economic and socio-cultural backwardness. Mahathir ventured into history in order to account for the economic and socio-cultural backwardness of the Malays. The third theme of *The Malay Dilemma* was putting forward proposals and strategies for remedying and alleviating the economic and socio-cultural backwardness of the Malays.

In *The Malay Dilemma,* the causes of Malays' backwardness were conceptualized on two broad levels: the individual and the community. At the individual level, Mahathir asserted that the Malays were fatalistic in their way of thinking. They lacked the psychological and social characteristics necessary for capitalist accumulation (that is, the spirit of capitalism), and their value system was an impediment to their progress (Mahathir bin Mohamad 1970: 158–159). He asserted that:

> By and large, the Malay value system and code of ethics are impediments to their progress. If they admit this, and if the need for change is realized, then there is hope; for as in psychiatry, success in isolating the root cause is in itself a part of treatment. From then on, planning a cure would be relatively simple. (Mahathir bin Mohamad 1970: 173)[1]

At the community level, the Malay people lacked effective leadership. Because the Malay rajas (the traditional political elite) did not take the issue of Malay development seriously, their followers resisted change and retreated into the countryside as the aggressive Chinese dominated the urban centers and the economy. In a subtle critique of the Malay traditional leadership, Mahathir asserted that:

> The reason why animist Malays became Hindus is because their rajas (i.e., traditional political elite) became Hindus. Later when the rajas became Muslim the 'rak'ayat' became Muslim. The political rajas of today can therefore institute change if they themselves are

[1] Mahathir Mohamad is a medical doctor by profession. He attended the University of Singapore under British colonial rule. In my assessment, his professional training probably explains his use of medical language and mode of reasoning to analyze social issues.

> willing to change. Such a change would spread rapidly. If the indicators are that there would be a change in the value system and ethical code, then the leaders can lead the way with the certainty that they will be followed by the masses. In a feudal society, if the leaders fail, then there is little hope for the masses. (Mahathir bin Mohamad 1970: 173)

It is evident from the above quotation that Mahathir Mohamad believed leadership to be central to the development and transformation of a developing society.

At the community level, Malays manifest certain social characteristics that hinder them from operating in a "competitive self-seeking" capitalist economy. These social characteristics are an integral part of the Malay character, which are ordinarily attributes of people seen as well-cultivated, and of "good breeding" and "good manners" (Mahathir bin Mohamad 1970: 116). Mahathir asserted vehemently that the fatalistic orientation of the Malays, their failure to value time, and their lack of a long-term perspective in making investments and then reaping the benefits, and their lack of appreciation of the value of money and property, and what money can accomplish in a global capitalist system are at the center of any explanation of their economic and cultural backwardness (Mahathir bin Mohamad 1970: 167–169). Indeed, in his scrutiny of the causes of Malay backwardness, Mahathir even asserted that marriage patterns and child rearing practices of the Malays were important causes of their backwardness (Mahathir bin Mohamad 1970: 29).

Another major explanation of Malay backwardness was the presence of the Chinese in Malaysia. For Mahathir, a critical explanation of Malay economic and cultural backwardness was the Chinese aggressive pursuit and monopolization of the Malaysian economy in a manner that blocked possibilities for Malays. Mahathir conceded that the Chinese were inherently good business leaders (1970: 84) and that they conducted their business with strict economic rationality that excluded emotional concerns (1970: 39). The insatiable desire of the Chinese is shown by the fact that as the government granted them concessions, they requested more. Yet he acknowledged that they were clever enough not to provoke the Malays, as they knew their population was comparatively smaller (1970: 39). Throughout *The Malay Dilemma* Mahathir analyzed business strategies that exposed the Chinese manner of capitalist accumulation as one of the major hindrances to Malay economic development. It seems Mahathir's problem was the exclusionary nature of the Chinese accumulation strategy, which made it a zero sum game.

One last important explanation for the persistence of the post-independence Malay economic and cultural backwardness was the official orientation of the government of Malaysia under Prime Minister Tunku Abdul Rahman, that is, the Alliance Government. It is worthwhile to quote Mahathir at length on this issue:

> What went wrong? Obviously, a lot went wrong. In the first place, the Government started off on the wrong premise. It believed that there had been racial harmony in the past and that the Sino-Malay cooperation to achieve independence was an example of racial harmony. It believed that the Chinese were only interested in business and acquisition of wealth, and that the Malays wished only to become government servants. These ridiculous assumptions led to policies that undermined whatever superficial understanding there was between Malays and non-Malays. On top of this the government, glorifying in its massive strength, became contemptuous of criticisms directed at it either by the opposition or its own supporters. The gulf between the government and the people widened so that the government was no longer able to feel the pulse of the people or interpret it correctly. It was therefore unable to

> appreciate the radical change in the thinking of the people from the time of independence and as the 1969 elections approached. And finally when it won by such a reduced majority the Government went into a state of shock which marred its judgment. And so murder and arson and anarchy exploded on 13 May 1969. That was what went wrong. (Mahathir bin Mohamad 1970: 15)

In the above quotation, Mahathir seems to articulate several issues about development policy in Malaysia. First, that it was wrong to base development strategy on the ethnic specialization of roles and functions that was created during the colonial period. Second, a government stands the risk of creating political catastrophe if its development policy is premised on a distorted perception and conceptualization of the development problem. This draws attention to the need for an effective mechanism that ensures accurate assessment and conceptualization of development problems. Third, Mahathir felt there was danger in having a government that is unresponsive to the yearnings of the people. The government, he seemed to suggest, needs to have an effective means of understanding the changing aspirations of its people so that it can act accordingly. Having seen Dr. Mohamad's evaluation of the causes of Malay economic and socio-cultural backwardness in *The Malay Dilemma*, we can now briefly examine the solution he provided.

Throughout the first dozen of years after independence, the postcolonial state helped assist the Malays to improve themselves. However, according to Mahathir, what was needed was not simply helping or assisting the Malays, but a total restructuring of the economic and political sphere of Malaysian society, so that the Malays could achieve equal status with immigrant groups. He also asserted that: "harsh punitive measures should be meted out to those who impeded the elevation of the Malays to an equality with the other races" (Mahathir bin Mohamad 1970: 60). This solution suggests an element of desperation on the part of Mahathir concerning the development of the Malay people. Another recommendation suggested not only the urgency of the problem, but also the need to thoroughly and accurately diagnose the problem. He asserted that what was needed was an equitable solution which did not discriminate against anyone, and yet gave the Malay his/her place under the sun. "The Malay problem must be enunciated, analyzed and evaluated so as to enable us to find a solution. The problem must be faced, and it must be faced now before it is late" (Mahathir bin Mohamad 1970: 121). At another level, he supported the constructive protection of Malays. Although not a decisive solution, Mahathir supported the idea of "appointing Malays to the position of directorship in large companies owned by non-Malays so that the Malays can develop skills in business and raise their levels of business accomplishment in the future to higher levels" (Mahathir bin Mohamad 1970: 31). Mahathir ruled out the option of allowing all Malaysians to compete on a meritocratic basis, because equality of opportunity ignores inequality of condition and past history of deprivation. Under such conditions, since the social space for competition is not a level playing field, inequality would continue to proliferate and intensify between backward Malays and successful non-Malays (Mahathir bin Mohamad 1970: 94).

Mahathir vehemently defended the positions reserved for Malays in the civil service, at the ratio of 4:1.[2] According to him, without such a provision, Chinese would constitute the entire Malaysian civil service. He defended quotas for Malay students' scholarship along the same lines. In defending the whole gamut of affirmative action programs for Malays, he asserted that the "laws do not make people equal" they "can only make equality possible" (Mahathir bin Mohamad 1970: 9). In other words, Mahathir did not expect the Malays to complacently withdraw from struggling to improve themselves because of affirmative action programs initiated by the government. He believed that the laws protecting Malays were simply aimed at creating an enabling environment, not guaranteeing outcomes. According to him, the fundamental solution to Malay backwardness lay in the Malays waking from their cultural and economic slumber and changing their ways of life in order to effectively participate in the fast, competitive, and changing Malaysian society and the global economy.

The second book containing Dr. Mahathir's essential ideas, values, and a conceptualization of Malaysia's development was *The Challenge*. While the "The Malay Dilemma" was written when Malays were simply trying to survive the monopolizing tendencies of the Chinese, *The Challenge* (1986, English version) was written after the Malays had established political hegemony. Beyond that, through state intervention (i.e., sponsored social mobility), the Malay community was becoming an economically dominant force in Malaysian society.

Between 1970 and the early 1980s when Dr. Mahathir wrote *The Challenge*, several changes had taken place in Malaysia society. First, under the governments of Prime Minister Tun Razak and Tun Hussein Onn, there had been continuous implementation of the New Economic Policy (NEP). The state, contrary to its pre-1970 situation, had expanded its role and intervened vigorously in the economy on behalf of the Malays. Private Malay capital had during this period successfully ventured into strategic sectors of the Malaysian economy, hitherto an exclusive reserve of Chinese and foreign capital. Consequently, Chinese economic hegemony, which was a major concern in *The Malay Dilemma*, became insignificant. Furthermore, because of the NEP, many Malays became urbanized and succeeded in acquiring new skills, as many of them were employed in the free trade zones (Jesudason 1990: Chap. 4). The Malaysian government had also during this period sponsored many Malay students' acquisition of higher education, both at home and abroad. Given the backwardness of the Malays circa 1969, these achievements were remarkable and a cause for celebration by the mid-1980s (Means 1991: 26).

However, in The Challenge, Dr. Mahathir was still not satisfied by the level of Malay accomplishments. By the mid-1980s, Mahathir saw many types of problems and "confusions threatening the Malays." In the introduction to the book, Mahathir asserted:

> The Malays have emerged from a long period of backwardness only to be pulled in different directions by conflicting forces, some of which seek to undo whatever progress that has been made and plunge the entire community back into the dark ages. (Mahathir bin Mohammed, 1986: Introduction)

[2] For more on this issue, see Means (1991: 26).

He saw new conflicting social forces which had the potential of undermining Malay values, ethics, and attitudes in the area of religious beliefs, Western education, politics and the economy. In *The Challenge*, Mahathir defined what he thought was a genuine future for the Malays and Malaysia. For the Malays, he wanted to assert that opportunities were open to them, but that they needed to make the changes necessary in themselves in order to take advantage of those opportunities. Although *The Challenge* (1986) contains many chapters, two broad themes ran through the entire book. The first theme was Mahathir's conceptualization of Islam and Islamic resurgence, and his interpretation of social issues through that prism. The second theme was the decline of the West, and the consequences of that for the social, economic and political development of postcolonial societies. I intend to summarize the essential issues of each theme.

60.2.2 Conceptualization of Islam and Islamic Revival

Mahathir asserted that Islam as practiced in Malaysia was misused and misinterpreted. This was done in a manner that was having a destructive effect on Malaysian students, intellectuals, government officials and the Malays in the countryside. One way this was happening was that some Muslim groups were evangelizing the notion that in order to be spiritual, one had to avoid the acquisition of material objects, because that would make them morally decadent.[3] Mahathir responded that if Muslims avoided material acquisition, they would have to beg other people for material things, and Islam would ultimately be discredited if it became a religion of beggars. Mahathir's critique of this group of Muslims is summarized in this quotation:

> If faith in Islam and spirituality is to be preserved, reality must strengthen that faith. When someone is suffering and forced to forget his self-respect, it is very difficult for him to believe that he actually lives a happy life and is more fortunate than his neighbor who has no religious faith but is very prosperous. For him, reality conflicts with faith. His mind cannot accept a claim that is manifestly untrue.
>
> If he is forced to beg for help from or be dependent on the charity of his atheist neighbor, his faith will weaken. If the others in his circle are suffering as well and cannot help him, his faith will weaken even more. Finally, if in this precarious state his faith is undermined by materialists, who give him aid, that faith will collapse. (Mahathir bin Mohamad 1986: 72)

Mahathir condemned Muslims who evangelized the idea that Muslims should avoid all Western education and secular learning. He criticized that position on the grounds that refusal to change and modernize spelled doom for the Islamic religion because such Muslim countries would be permanently dependent on non-Muslim societies for sophisticated technology, and scientific knowledge in general. The solution for him was not to avoid modern Western education entirely, but to reconcile and integrate Western education with the tenets of Islam. Mahathir strongly disagrees with those who teach that Muslims should be intolerant of believers of other faiths.

---

[3] For more on Malaysia's Islamic Movements, see Jomo and Cheek (1992).

He maintained that Islam does not accept that attitude or behavior (Mahathir bin Mohamad 1994: 15–26).

It is evident that in *The Challenge*, Mahathir's Malay nationalism, which was very clear in *The Malay Dilemma*, metamorphosed into a broader nationalism, a nationalism of the Islamic community in general. By the early 1980s, Mahathir was reconfiguring his Malay nationalism from its obsession with Chinese economic dominance and success. The reconfigured Malay nationalism was not now focused on appealing for ethnic solidarity, but to a wider community (that is, the Muslim community). The main threat to the Malays in 1986, as perceived by Mahathir, was not the Chinese or other immigrants. Although he conceded that others could have caused the Malays problems, he believed that ultimately the Malays themselves would have to change and shape their destiny. He therefore counseled them on the need to realize the task ahead of them. He asserted that "God will not change the fate of a nation unless that nation itself strives for improvement" (Mahathir bin Mohamad 1986: 3).

An important point Mahathir made about the developmental destiny of Malays was that their ability to improve their condition was inextricably tied to their values. He maintained that for the Malays to deal with their problems head-on and emerge victorious, they needed to have good Malay values. The good values he referred to are exemplified by this quote, when he asserted that Malays must:

> observe rules and customs like proper attire, decent behavior, and reverence for religion, marriage, the family, work, mutual respect, and honesty. (Mahathir bin Mohamad 1986: 101)

The Malay must also attribute the highest respect for "industry, efficiency, honesty, discipline and other good values" (Mahathir bin Mohamad 1986: 31). Doing this was sine-qua-non for the achievement of progress. However, Mahathir strongly believed that the major threat to the Malays positively changing their behavior was the negative influence of a declining Western society. Before proceeding to outline Mahathir's argument concerning the threat from the West, I must assert that there appears to me to be a tension in Mahathir's conceptualization of development between tradition and modernity. He was sometimes ambivalent about certain elements of the two situations and processes.

On a critical note, Mahathir's position on these issues is a characteristic of human behavior that is expressed in varying degrees among individuals, across societies and historical periods. In an attempt to explain this dimension of human behavior, Smelser asserted the following:

> The issues of affect and valence open the door to another psychological foundation of behavior – human ambivalence, i.e., the simultaneous existence of attraction and repulsion, of love and hate. Ambivalence is on people, objects, and symbols. The nature of ambivalence is to hold opposing affective orientations toward the same person, object or symbol ... a truer account would consider the continuing ambivalence toward both old and new, just as colonies that have 'exited' an empire experience the former colonial power. (Smelser 1998: 4, 5, and 12)

Smelser continued to assert that even the legacies of historical events, such as the French Revolution, elicited ambivalent and contradictory attitudes among people. He maintains that in the process of resolving ambivalent and contradictory attitudes, situations, and conditions, institutions are created as coping mechanisms. The operation of these institutions serves as a mode of adaptation, which helps in accounting for the form that social change takes. It does appear to me that a higher level of ambivalence and the desire to resolve it is manifested in the worldview of Mahathir Mohamad and his ruling elite coalition, which creates a constant desire to think of institutional solutions to development problems. By implication, societies in which the leaders experience low levels of ambivalence will on the average feel satisfied with the status quo and pursue development at a very slow pace.

60.2.3 The Decline of the West and Its Consequences on the Social and Economic Development of Postcolonial Societies

After examining postwar Western European societies, Mahathir, rightly or wrongly, concluded that contemporary Western society had many undesirable social values, some of which he thought were a perversion of the social order (Mahathir bin Mohamad 1986: 103). Mahathir, however, expressed his profound appreciation of past Western European values that had enabled the Western world to transform their society from agrarian to a modern industrial one in a manner that made it the dominant civilization in the world. However, in the postwar period, the values that had originally elevated Western Europe were discarded without new viable ones replacing them. The result of this was social flux and confusion. Some Western values that Mahathir appreciated were "orderliness, discipline, and firm social organization" (Mahathir bin Mohamad 1986: 47). These values were rendered untenable because "priority, devotion and adulation are given to 'basic rights'" (Mahathir bin Mohamad 1986: 101). The pursuit of these new values brought about many unpleasant behaviors, such as frequent workers' strikes, self-seeking workers, and socialism, which in Mahathir's way of thinking had led to anarchy. Other evils brought about by the new values were students' demonstration against the laws of the country, demonstrations against the Vietnam War, too much concern and concession for minority social rights (for example, nudity, cohabitation, male prostitution and homosexual marriage (Mahathir bin Mohamad 1986: 102).[4]

[4] It is evident from Mahathir's ideas in this paragraph that he has very elitist and socially conservative values. The reader might think that these culturally conservative views are not particularly relevant to the issue of economic growth, but in my assessment, they help us understand Dr. Mahathir's views on what kind of social environment produces economic growth. Of course, we may disagree with him, but my point here is not to agree or disagree, but to demonstrate how his worldview as a leader shaped the process of development in Malaysia. First, the ideas expressed in this paragraph illuminate Mahathir's suspicion and ambivalence about the appropriateness of the Western European development experience for developing countries like Malaysia. Second, they portray him as someone seriously concerned about order and hierarchy in society. In governance, these ideas translate into an emphasis on technocratic, authoritarian and command-type development planning

Thus even though the West had spectacular values that enabled her to achieve a glorious past, for Mahathir, the postwar value transformation in the West no longer made it a role model for postcolonial societies (Mahathir bin Mohamad 1986: 47). Because of the prevalence of a sophisticated modern communication system, the "decadence" of Western values constitutes a new process and mechanism for the re-colonization of postcolonial societies. The effective modern communication system enhances the faster diffusion of decaying Western socio-cultural values. Mahathir did not see the diffusion of Western socio-cultural values as completely bad, as some of the values have helped transform Malay society positively. However, because of a continuous and unquestioning acceptance of Western values by Malays, Mahathir argues that by the 1980s, Malay values were in a state of flux, and the danger of that situation was that any moral entrepreneur could establish new values among the Malays.

Unfortunately, according to Mahathir, the Malays, in emulating Western values, copy the form rather than the substance of Western civilization (Mahathir bin Mohamad 1986: 55). Thus in this process only the destructive Western values are learned and adopted, rather than the effective ones which laid the foundation of Western civilization. Yet even in its decline, the West constitutes a serious danger because it tries to continue to perpetuate its economic hegemony. The aim of this is to avert a situation in which the East emerges as more dominant than the West.

Whatever one thinks about Mahathir Mohamad's assessment, one thing is clear. He was a leader who always thinks globally while acting locally. Many of the issues he raised above are indeed debated in the social science literature on North-South relations, as they affect the development of countries in the South. In the next section, the Islamic context of Malaysia's struggle for economic development and cultural change is examined.

60.3 The Islamic Context of Malaysia's Struggle for Economic Development and Cultural Change

60.3.1 Early Indicators of Influence of Islam on Malaysian Public Affairs

There are indications that Prime Minister Mohamad's penetrating critique of the state, society, and culture of Malaysia in general but Malay people in particular has had a long historical genealogy in Malay intellectual history. Anthony Milner (1995) has provided an excellent documentation of Malay intellectual history in his book entitled: *The Invention of Politics in Colonial Malaya*. Using editorial comments

and implementation. Such a style of leadership helped to bring rapid development to Malaysia, but the success of the strategy would boomerang, as in the late 1990s, the Malaysian people begin to agitate for the liberalization of their society and social institutions.

of Malay newspapers and Islamic journal articles published as far back as 1908, he articulated the passion of Malay intelligentsia in terms of how they saw their culture and society in relation to other people in the world. There are several themes that run through the pages of their newspapers and journals (Milner 1995). Some of the themes are:

1. A trenchant critique of the traditional Malay political elite for their lack of foresight, and exemplary leadership. The critique is similar to the critique of the ancient regime in France (Milner 1995: 10–30).
2. That when a society is beholden to corrupt leaders that are ignorant and lacking foresight, they and their people would lose their sovereignty or autonomy to other more progressive societies.
3. That when a people find themselves under the leadership of such corrupt and visionless leaders, they have a social responsibility to themselves and history to organize and get rid of the leaders.
4. They documented the social consequences of colonial subjugation on the life chances, culture and the institutions of Malay people after they lost their independence and sovereignty because of incompetent ruling elites who are primarily more concerned with personal aggrandizement and sumptuary laws that distinguished them as ruling elites from ordinary people.
5. They articulated and elevated the importance of history as an epistemological tool for learning what happened in the past and on that basis plan for the future. In this respect, they maintained that looking at history, when a society is infused with parochially-minded people who are concerned just about their selfish interest, such a society would not receive the benefits that God promised his followers.
6. The Malay intelligentsia was never hesitant to acknowledge ideas, institutions or virtues in other cultures that they thought were more effective or superior in terms of practical adequacy and solving problems. They insisted that such innovations should be adopted and adapted and the creativity of the originators be seen as a source of inspiration.
7. Finally, the Malay intelligentsia made a strong case for the pursuit of knowledge, broadly defined, and conceptualizing it as a tool for social liberation and empowerment. Unlike the situation in many Muslim countries, the Malay intelligentsia called for the pursuit of knowledge in all areas, namely, psychology, biology, economics and commerce.

Milner further documents that the traditional Malay ruling elite was faced with critique and demand for reform not only from the liberal wing of the Malay intelligentsia (Milner 1995: 31–58), but also from the Islamic wing of the Malay intelligentsia, which also called for reform but framed its critique in religious discourse (Milner 1995: 137–166). The overall effect of these campaigns demanding for reform is that there is a long historical tradition expressing a strong demand for reform and change in Malaysia even before the country became independent.

The main explanation for this unusual and broad-based demand for reform in Malaysia has to do with the geography and location of peninsula Malaysia.

The peninsula was a cosmopolitan place in the late nineteenth and early twentieth centuries because of its geographical and strategic location that favored trade and the mixing of people, ideas, and cultures between the Far East and the West. The Malay people were overwhelmed by the Dutch, the British, the Japanese, Chinese and Indian immigrants. By and large, they found themselves being left behind if not defeated in the quasi-Darwinian struggle for survival in the peninsular during the colonial period. Consequently, the cultural elites of the Malay people arrived at the conclusion, both from the liberal or Islamic tradition that reform was necessary and inevitable if the Malay culture and religion were going to survive under multiple forms of cultural invasion by people that seemed to be more effective than the Malay in the struggle to improve their life chances. What this means is that unlike many other societies where there is serious debate on whether there is need for reform or not, in Malaysia, the debate as early as the twentieth century was on what was the best strategy for reform and the pace of the reform, since the reform was seen as a necessary condition for the survival of Malay culture, religion and society. More importantly, the call for reform was not just framed in the form of trenchant critique of the traditional political elite, but there was also serious exchange of ideas between the liberal and Islamic factions of the Malay intelligentsia on the best way forward, which surely benefited the institutions and political climate of Malaysia in terms of thinking strategically about the best way to reform and move the society forward (Milner 1995: 167–192).

60.3.2 Prime Minister Mohamad's Vision of Islam in the Modern World

In general, Prime Minister Mahathir Mohamad had a view of religion that was dynamic, adaptive and that could be reinterpreted to cope with contemporary challenges posed by modernity. While he took the past and the heritage of Islam seriously, he did not believe that a religion that remained stagnant will continue to provide adequate and meaningful guidance to the social and existential challenges of people living in contemporary society characterized by hyper-modern processes of change. In this respect, the Prime Minister engaged in the hermeneutic of suspicion with regard to how some Islamic scholars in the past had interpreted Islamic Scriptures and tradition, resulting in Islamic civilization losing its hitherto leading role in the area of scientific and intellectual accomplishments. This position of the Prime Minister was succinctly articulated in his speech at the Oxford University Center for Islamic Studies, in the United Kingdom. Prime Minister Mohamad (1966) provided an incisive critique of traditional Islam as follows:

> If Islam appears rigid and doctrinaire, it is because the learned interpreters make it so. They tended to be harsh and intolerant when interpreting during the hey-day of the Muslim Empires. And they and their followers brook no opposition to their writs once they were made. And so, long after the Muslims have lost their predominant position, long after the

> world environment has changed, the Muslims were exhorted to adhere to interpretations, which are no longer adequate or relevant or practicable.
>
> What Muslims must do is to go back to the Quran and the genuine 'hadiths,' study and interpret them in the context of the present world. It is Allah's will that the world has changed. It is not for man to reverse what has been willed by Allah. The faithful must look for guidance from the teachings of the Quran and 'Hadith' in the present context. Islam is not meant only for 7th Century Arabs. Islam is for all times and for every part of the world. If we Muslims understand this, then there will be less misunderstandings among us. If the non-Muslims appreciate the problems that the Muslims have in trying to adjust to modern changes, then they will not misunderstand Islam and the Muslims as much as they do now. And the world will be a better place if all these misunderstandings are removed. (Mohamad 1996: 21–22)

It is obvious that in the process of spearheading economic development and cultural change in a Muslim society and in the modern world, Prime Minister Mohamad had come to realize that religion as the source of many peoples' values and worldview is a critical component of successful social change. As many scholars have argued, one of the main contributions of Weber's *The Protestant Ethic and the Spirit of Capitalism* is illuminating the fact that capitalist accumulation cannot take off effectively without some appropriate change in values and ethics that see such venture as legitimate and worthwhile. In effect, no matter what a government does to promote economic development and cultural change, if there is no appropriate and commensurate change in people's values and worldview, the project will woefully fail (Long 1977).

60.3.3 The Political Economy of Religion, Economic Development and Cultural Change in Malaysia's Modernization

Malaysia is a country that has gone through rapid process of modernization, economic development and cultural change. While the country has tried to adapt and enculturate the forces of modernization to fit her cultural heritage, this has not been easy. The most dominant forces guiding the process of modernization in the world are Western in their incarnation. Even though on the surface, the institutions and forces of globalization appear to be neutral and innocuous, the reality is that culturally, and substantively, they project Western cultural values and entailed a social transformation or adaption of one' culture so as to enable the country to favorably compete in the global marketplace dominated by the Western world. Given that the Western tradition is Judeo-Christian and individualistic even in its secular incarnation, often the spread of globalization is considered more or less the same as the spread of Americanization among many countries in the Global South, especially Islamic countries.

Globalization anticipates and presupposes a society with individual rights and freedoms. Yet when Muslims feel overwhelmed by the process of modernization and social change spearheaded by Western nations and global corporations, they naturally react defensively by becoming either fundamentalist, traditional, de-secularized or reformist (depending on context) in their thinking and political orientation. Such defensive reaction becomes a critical and politically contentious issue that can destabilize peace and stability of a nation, which is crucial for a healthy investment climate. And healthy investment climate is a virtue in the neoliberal global marketplace.

With regard to defensive reaction, for instance, because of the strong influence of Islamic Non-Governmental Organizations (NGOs), an Islamic political party (that is, PAS), and traditionally-oriented practicing Muslims, there are constitutional provisions in Malaysian law, which declares apostasy by a Muslim believer a criminal offense. The Malaysian constitution assumes that once someone is a Muslim, the person must always be a Muslim. Muslims who change their religion have to apply for constitutional approval, which is often denied. The apostates are taken to an Islamic rehabilitation center for re-education. Islam in this case does not approve of certain individual rights of a person to choose. Similarly, there are constitutional laws that make proselytizing illegal, while making legal, the moral policing of peoples' ordinary behavior such as alcohol drinking and interaction with the opposite sex (Tamney 2012: 343–376).

To complicate things further, there are many NGOs that have emerged in Malaysia that are campaigning for individual rights and freedoms, for example, Sisters in Islam. Indeed, even among Muslim scholars, there is serious debate and contention on the proper and appropriate interpretation of scriptures, given that some Islamic scholars who are considered liberal in orientation maintained that Islamic law does not apply to non-Muslims and that there is no compulsion in religion according the Holy Quran. This situation indicates that globalization is not just manifested in the form of economic transactions, but there are also transnational cultural exchanges in the form of ideas, religious discourses and new intellectual horizons (Corten and Marshall-Fratani 2001).

One important lesson from the preceding analysis is the relevance of William Ogburn's cultural lag theory to explaining the uneven pace of economic and material cultural change on the one hand, and the social-cultural changes in the sphere of non-material culture on the other hand (Volti 2009). Malaysians have by and large embraced economic and material innovations and changes in their economy and society, but many have been very resistant to embracing social-cultural changes that are considered to be correlates of a modern liberal democratic society. If a person is not free to choose and change his or her religion, this would be considered a major cultural lag in a society that is committed to honoring the human rights of all people.

One may rush quickly to conclude that the main purpose for resisting social-cultural changes that accompany liberal democracy and globalization is the high degree of religious piety on the part of Malaysian Muslims and the neocolonial implications of embracing neoliberal globalization hook, line, and sinker. There is some truth to that, but this is definitely not an adequate explanation given the political economy of Malaysian politics. The Malay people constitute over 60 % of the Malaysian population and constitutionally, to maintain their political dominance in the society, they need not only maintain a high fertility rate but be sure that Malays who are constitutionally and religiously defined as Muslim, remained Muslim until they die (Tamney 2012: 348). Traditionally, the most successful capitalists are Chinese and have a preponderant influence in the economy while the Malay predominate the political institutions of the country.

In a liberal democracy where people are assumed to be free to pursue their interests in a dynamic free market economy that sorts people out not based on their degree of religious piety but material and social interests, it is easy to see how this can be a recipe for dissolving Malay identity and its presumed monolithic cultural cohesion. Left on their own terms, the dynamic realities of neoliberal globalization and market economy will dissolve Malay identity into small pieces and initiate the unraveling of Malay political dominance since there will no more be a single Malay identity around which to mobilize all Malays. In this respect, fighting for conservative or fundamentalist Islamic religious values are a proxy for the struggle to maintain political control, dominance and privilege by the Malay people and elite in particular (Tamney 2012: 365–367). Thus without devaluing the religious commitment of many Malay people, on a critical note, one can assert that religion is here being used as a tool for a broader goal than merely a struggle for salvation or life hereafter.

Interestingly, this multi-purpose use of religion is not unique to Malaysia. All the above strongly suggests that religion cannot be assigned one role across historical time and space. The preceding statement is indeed buttressed by more contemporary research on religion in society. For instance, Wuthnow (1980) identified six broad ways in which contemporary religious movements could be conceptualized vis-a-vis social change. The six types are: revitalization, reformation, military, counter-reform, accommodation, and sectarianism. Similarly, after surveying a large number of comparative historical studies on how religion is used for political legitimation, Lincoln (1985) identified four ideal-type categories of religious legitimation. These are: religion of status quo, religion of resistance, religion of revolution, and religion of counter-revolution. The existence of each of these will have a different implication for economic development and cultural change.

Along the same line of reasoning, one can deduce several ways in which religion shapes public policy in the contemporary world based on insights from the work of Billings and Scott (1994). These are:

1. It influences the legitimacy of certain public policies by sanctioning them.
2. It shapes political constituencies, coalitions, and the levels of political participation of people in elections.
3. Religious groups and organizations may help people to see the connection between conservative religious beliefs and public policy.
4. Religious political mobilization could produce mass cognitive restructuring and the reorientation of people's mindsets, predisposing them to support or oppose certain public policies.
5. In developing societies, in particular, religion assumes a very strong capacity when it becomes the cultural medium of political protest against internal colo-

nialism by marginalized cultural groups or against external colonialism by subordinated national groups.

These are by no means the only ways religion shapes public policy, but they serve my purpose here to draw attention to the strong connection between religion and the process of economic development and cultural change in Malaysia, and by implication, in other parts of the world.

60.4 Summary and Conclusion

In summarizing some key themes from the chapter, I want to highlight some important lessons about leadership, economic development and social change in the non-Western world, with particular reference to a society that is predominantly Islamic in the Far East.

Prime Minister Mohamad makes the quality of leadership central to any project of social transformation in a developing society. He recognizes that in a society that is traditionally hierarchical and overly conscious of social status, if the initial push for desirable social change and social transformation does not come from the top, in this case "the rajas" and the elites, the society will continue to stagnate. Mohamad's counsel is indeed valid beyond Malaysia and the Far East. Whether it is in Africa, the Middle East or Latin America, one can only diminish the central role of the quality of leadership at the detriment of the real potential for desirable social transformation. The quality of leadership is not the only variable that matters when it comes to the challenge of economic development and cultural change of a society, but it is key and indeed central. Without high quality leadership, even when desirable opportunities are available for economic development and cultural change, they are not taken advantage of, which means wasting opportunities (Samatar 1999).

The second lesson from Prime Minister Mohamad's social thought and leadership experience is the critical question of how the postcolonial state constitutes itself and its role in relation to civil society. The postcolonial state that assumed power immediately after independence in Malaysia naively tried to continue the economic development policies and racial-ethnic division of labor that was instituted by the British to pursue their colonial objectives. This was not sustainable and after the May 1969 racial-ethnic riot, the New Economic Policy (NEP) was implemented in order to rectify the colonial legacy of racial-ethnic division of labor. But restructuring the economy and society which was what NEP was, is not an easy task in any society as it creates many social divisions and grievances in society. Managing the tension and conflict that characterize such efforts requires commitment and statesmanship; otherwise, failure can easily result in the problems constituting themselves into centrifugal forces that can derail meaningful progress in the process of desirable social transformation of a country. Leadership and statesmanship is necessary to balance social forces and keep the ship of state in disciplined focus as it pursues the project of economic development and cultural change.

Third, Prime Minister Mohamad demonstrates the challenging problem of the pursuit of equality in nation building in the postcolonial period. The Mohamad that vehemently critiqued Chinese monopoly and domination of the commanding heights of Malaysian economy also was the same person who took the risk of losing an election because of his insistence on the policy, which vehemently maintains that Malay business owners cannot expect to continue receiving state subsidy while they remain inefficient indefinitely in the name of affirmative action in support of "Bumiputra," that is, sons and daughters of the soil (Malaysia 1984: 11; Jomo 1990: 211–218). While neoliberalism has its own theoretical, empirical, moral, and ethical problems with regard to its philosophy and strategy of development (Harvey 2005; Sachs 2005: 70–87), often people rush in glossing over the fact that in many postcolonial societies, under import substitution economic development policies, many corporations and businesses were funded or subsidized based totally on political and social concerns and criteria (Todaro and Smith 2009: 618–635). In doing so, the postcolonial states ignored concern for sustainability and the social cost of such policies to the ordinary population. Unfortunately, the social concerns were elitist, to the extent that they were not inclusive of the welfare and life chances of the masses who are the great majority. From Prime Minister Mohamad, we can learn the important lesson that no nation can totally ignore the question of relative equality in the distribution of scarce resources and opportunities among its citizens. Neither can it naively in the pursuit of equality ignore questions of efficiency and sustainability in the use of scarce resources and opportunities. This is a struggle that all modern societies have to confront, be they in Asia, the Western world, Africa, the Middle East, or Latin America (Marshall 2008; Seers 1969).

Fourth, although Prime Minister Mohamad is not a Durkheimian scholar, he highlights an important lesson for all leaders spearheading a project of rapid development and social change in a manner that is consistent with Durkheim's theory of anomie (Coser 1977: 132–136). Emile Durkheim, who is a French sociologist, is of the view that the norms and values of a society are central stabilizing forces that create social solidarity and a shared moral community called society. The norms and values regulate human behavior and social interaction, and they are internalized by human beings. Their existence is not just external as represented by positive and negative sanctions. The problem as Prime Minister Mohamad has identified for his country was the rapid process of modernization, development and social change that Malaysia went through under his watch brought about great moral and ethical confusion to his people. The traditional Malay norms, values and ethics were destabilized, while effective new ones were not immediately put in place to replace them. The social institutions traditionally charged with the process of socializing people and maintaining social control were weakened because they were themselves going through social transformation. The society was in a state of flux and this created a state of quasi-anomie. This situation was further worsened by the fact that as the society and culture were trying to reconfigure themselves and their bearing in a new social equilibrium, Western influence through the mass media was penetrating the hearts and minds of people in the society intensively at a time when people had no strong moral and ethical base and compass through which they could judge or evaluate what they should accept or reject. Mohamad's observation on this issue is a very poignant and thoughtful caution to all societies embarking on a rapid project of economic development and cultural change. Such societies must anticipate the possibility of socially anomic situations and plan systematically on how to mitigate their impact in advance.

Fifth, a careful reading of Prime Minister Mohamad's social thought and the public policies he implemented would lead one to the conclusion that he must be commended beyond many postcolonial leaders for openly grappling with the challenge of religion and modernity. Mohamad was very critical of religious scholars and believers (in this case Islam) who apply religious beliefs and rituals to the total detriment and negligence of social and empirical reality that the religion is supposed to directly have an impact on. Religion for him is not just about the life hereafter but it also has relevance for social reality here and now. On this concern, Prime Minister Mohamad broadly speaking, shares similar intellectual and theological commitment with the Christian liberation theologians of Latin America (De La Torre 2008: 91–112). Mohamad said that his Malay people were committed Muslims and Islam is not just supposed to prepare them for life after death but also for the here and now. They considered themselves to be committed to a faith that privileges them over "pagans" and "atheists." Unfortunately, they live in abject poverty and suffer numerous negative life chances compared to many of their neighbors who were not Muslims. This makes the Muslims with their presumably superior and enlightened religion to beg from atheists and pagans who are more prosperous and living happier lives. The empirical question for him was whether under such conditions, religion still matters if it is unable to deliver human aspirations that are consistent with the ethical and moral claims of the faith? This observation entails all religious people need to self-interrogate their beliefs, but not to necessarily abandon them as some would like to think, given that Prime Minister Mohamad is a committed believer himself.

If people continue on the same path, then after sometime, the religion will become irrelevant to the struggles faced by the people at any historical moment, to the extent that people can achieve what the religion promises on earth through other means. He was calling on his people to juxtapose the moral and ethical claims of their religion on the one hand, with the social and empirical reality of the here and now and not just life after death, on the other hand. This call by Prime Minister Mahathir is relevant not just for the Malay Muslims of Malaysia but Muslims in other parts of the world, and indeed Christians, Hindus, Buddhist, and believers of Judaism too. In Sub-Saharan Africa, for instance, Pentecostalism is spreading very fast but poverty and human misery is equally widespread (Maxwell 2006; Mihevc 1995). It almost seems like prayers and worship have no effect on social reality except as coping mechanism. The rich in Africa continue to exploit the poor, and the powerful continue to oppress the weak with no injury to the social conscience of the rich. Corruption and embezzlement of public funds is destroying the basis for progress, hope and productivity. Yet, this is happening in countries that are overwhelmingly Christian or Islamic and the social ethics of the two religions are supposed to directly address such moral frailties. Mohamad asks: what is the connection between religion and social reality? This is a question that people and countries beyond Malaysia have to answer.

Sixth, Prime Minister Mohamad is a proud "Third World" citizen in the sense that he is a man confident in the ability of all human beings to rise to the level of self-actualization if they can muster the courage and the discipline to do what is necessary. Yet, he is also a man with great self-esteem, who does not hesitate to critique his own people or culture publicly when there is valid reason to do so. Unlike many Third World leaders, he recognizes the moral and ethical problems of capitalism, but he was clearly committed to understanding what it will take his country and people to succeed along the capitalist path of development. He never thought that socialist development strategy was an option for his country, but he was equally not a naïve embracer of everything Western, or a capitalism that was dominated by the Western world and that has run amok.[5] His courage and insight was in getting his vision clear and creating effective institutions that would enable Malaysia have a good standing in the global economy. Some of the incisive critique Mohamad made of the reasons for Malay economic and social backwardness in the modern world would be considered racist if uttered by a Westerner. He recognizes that once a society has decided to pursue the capitalist path of development, it is the responsibility of its leadership and elites to create the institutions, human capital, appropriate business practices, and the social and technical skills necessary for success.

Seventh, Prime Minister Mohamad has been legitimately criticized as an authoritarian leader. But compared to many leaders in Africa who were dictators and authoritarian, his 'benevolent' authoritarianism has transformed Malaysia far better than any African country excluding South Africa, which is a unique situation. Many African countries would wish to have a benevolent, purposeful, and transformative authoritarian leadership if that would truly enable them to cross the industrial divide and improve their human development index as Malaysia did. Moreover, some would argue that at their early stages of development, all the successful and advanced capitalist societies of today including the U.S. had what was more than authoritarian leadership during the era of primitive accumulation. Slavery in the U.S. denied the humanity of Black people by treating them as property and machine for production, that is, chattel. The Gilded Age at the end of the nineteenth century in the United States was a terrible era in terms of how robber barons exploited their workers (Foner 1998: 115–138). All these should not excuse any leader to be authoritarian or oppressive as the fact that these situations happened is not tantamount to saying they must necessarily happen or are the only strategies that countries can implement in order to develop. Human history is complex but we must be careful so as not to convert a historical fact into a historical necessity. It is now over 50 years since many African countries got independence from colonial rule, and yet, many of such countries have been notorious for having impressive record of dictators and authoritarian leaders while their human development indicators have worsened in most cases and

[5] For more details, watch the documentary entitled: "Commanding Heights: The Battle for the World Economy" by Daniel Yergin, where in episode three of the program, Prime Minister Mahathir Mohamad articulately expresses this line of thinking (www.pbs.org/wgbh/commanding-heights/hi/story/index.html).

their institutional capacity is still very weak. This postcolonial state failure is the major explanation for the lack of peace and development in Sub-Saharan Africa. For this reason, Prime Minister Mohamad is a lesson in leadership for many countries in Africa, the Middle East, and Latin America, notwithstanding his shortcomings.

As a leader, Prime Minister Mohamad demonstrates one important insight that many postcolonial leaders ignore. This insight is about the need to understand the history of Western civilization in an in-depth and sophisticated manner, because as Paulo Freire argues (Freire 1997), oppressed people cannot liberate themselves without understanding the mindset of their oppressors with a view to liberating the oppressors as part of the process of the oppressed liberating themselves. Contrary to what many assume, Prime Minister Mohamad was not anti-Western in the form of having prejudice. He indeed appreciates past Western values and achievements that have led Western civilization to ascendance in the world. His concern was with how the Western system of domination was set up in such a way that it became a club for a selected few to control and exploit the many. The logic of the global capitalist economic system is structured in such a way that it favors the West and makes it extravagantly difficult for non-Western countries to make inroads. He considers this unfair and unjust. He is critical of double standards in the way Western nations on the basis of certain presumably universal human values treat their citizens in a manner that is humanly dignified, while treating the citizens of many non-Western nations as second class human beings. For any serious leader that wants to deal with the West from the perspective of the non-Western world, he or she needs to have a sophisticated grasp and understanding of the good, the bad and the ugly in the history of the West as Prime Minister Mohamad tried to do. This helps to avoid unconstructive dialog or unwarranted wholesale critique or condemnation of Western civilization.

Thus, Mohamad combines a trenchant critique of traditional Malay culture that keeps the people socially and economically backward in the modern world, while also cautiously embracing the good from the West, and critiquing what is oppressive about Western civilization as well. It is a balanced and complex role of statesmanship to manage the ship of state of a developing country in a modern world dominated by the West. Prime Minister Mohamad exemplifies some elements of effective leadership qualities in a modern Islamic context that are worth reflecting on, adopting and adapting in other developing countries beyond Southeast Asia. He has demonstrated the capacity and potential of an Islamic nation and society rooted in Islamic culture to transform itself, contrary to Samuel Huntington's argument (1984: 193–218). Huntington is of the view that democracy is not suited to Eastern cultures and societies because of their religion and culture. Yet, looking at the history of the Southern region of the United States, Christianity was used to create and justify an authoritarian White supremacist government that systematically used so-called democratic institutions to enshrine the legal inferiority of Blacks (Katznelson 2005). If the South which is Christian and upheld slavery is still democratic, one has to legitimately ask, what is the meaning of democracy under this circumstance? They key lesson here is that depending on who is reading a holy book and the hermeneutical technique they used for interpreting it, any religion can be used to justify violence and oppression (Jenkins 2009).

Another important lesson from this study is the insight that suggests the tradition of reform in Islam varies from one culture to another, one historical context to another, and is also significantly influenced by the geographical location of the believers and the mode and nature of social-cultural exchange of ideas, visions and opportunities that the geographical location provides. Compared to other regions of the Islamic world, Malaysia has a long tradition of Islamic reform movement owing to the struggle for survival by Islamic civilization in the midst of other civilizations that were threatening the relevance of the faith to the culture and progress of Malay people in the Malay Peninsula. If this observation is correct, then in general, one can make the broader claim that religions that have no need to compete for survival or flourishing optimally are most likely to be less dynamic and reform-oriented. Such religions are most likely to be in a long state of inertia. Of course the fact that a religious group embarks on a reform project does not in and of itself guarantees that the reform envisioned by the group would necessarily be the best or most conducive for the society in reference and the world at large.

A major lesson from this study also is that if the world comes to terms with the idea of multiple forms of modernity, it will make for greater peace and less religious acrimony and potential violence in the world. As it is currently constituted, it appears that globalization while having the potential to be globally inclusive, yet it is in concrete reality dominated by Western nations, and particularly the United States. The numerous criteria used for measuring success or progress under neoliberal globalization, which is at the core of contemporary debates on modernity are determined by Western nations in their own image, represented by corporate elite interests. This of course leads to a conception of modernity that is skewed in one direction. Yet the Malaysian experience strongly suggests that while a globalized modern world will share much in common, we have to accept the fact that modernity will never be incarnated in the same way all over the world. Western nations have to be realistic by making room for multiple forms of modernity in so far as such non-Western forms of modernity are committed to peaceful coexistence, the pursuit of collective prosperity, the human rights and dignity of all people, and social justice. Insisting on a uni-linear conception of modernity as some neoconservatives think, is a recipe for global disaster. One must note, however, that it is not only non-Western forms of modernity that have to be held to standard described, but even Western nations too have to be held accountable because they have not always been a good model for what modernity should be for all humanity.

References

Antoine, J. (1991). The resurgence of racial conflict in post-industrial America. *International Journal of Politics, Culture, and Society, 5*(1), 81–93.

Billings, D. B., & Scott, S. L. (1994). Religion and political legitimacy. *Annual Review of Sociology, 20*, 173–201.

Buchanan, J. M. (1954). Social choice, democracy and free markets. *Journal of Political Economy, 62*, 114–123.

Corten, A., & Marshall-Fratani, R. (Eds.). (2001). *Between Babel and Pentecost: Transnational Pentecostalism in Africa and Latin America*. Bloomington/Indianapolis: Indiana University Press.
Coser, L. (1977). *Masters of sociological thought* (2nd ed.). New York: Harcourt Brace & Company.
De La Torre, M. A. (2008). *The hope of liberation in world religions*. Waco: Baylor University Press.
Foner, E. (1998). *The story of American freedom*. New York: W. W. Norton & Company.
Freire, P. (1997). *Pedagogy of the oppressed*. New York: Continuum.
Grindle, M. S., & Thomas, J. W. (1991). *Public choices and public policy change: The political economy of reform in developing countries*. Baltimore: Johns Hopkins University Press.
Harvey, D. (2005). *A brief history of neoliberalism*. New York: Oxford University Press.
Huntington, S. (1984). Will more countries become democratic? *Political Science Quarterly, 99*, 193–218.
Jenkins, P. (2009, March 8). Dark passages: Does the harsh language in the Koran explain Islamic violence? Don't answer till you've taken a look inside the Bible. *Boston Globe*. Accessed 12 Nov 2012, www.boston.com/bostonglobe/ideas/articles/2009/03/08/dark_passages/?page=full
Jesudason, J. V. (1990). *Ethnicity and the economy: The state, Chinese business, and multinationals in Malaysia*. Singapore: Oxford University Press.
Jomo, K. S. (1990). *Growth and structural change in the Malaysian economy*. London: Macmillan Press.
Jomo, K. S., & Cheek, A. S. (1992). Malaysia's Islamic movements. In *Fragmented vision: Culture and politics contemporary Malaysia*. Sydney: Allen and Unwin.
Katznelson, I. (2005). *When affirmative action was white: An untold history of racial inequality in twentieth-century America*. New York: W.W. Norton & Company.
Lincoln, B. (Ed.). (1985). *Religion, rebellion, revolution*. London: Macmillan.
Long, N. (1977). *An introduction to the sociology of rural development*. Boulder: Westview Press.
Malaysia. (1984). *Mid-term review of fourth Malaysia plan, 1981–1985*. Kuala Lumpur: Government Printers.
Marshall, K. (2008). *The World Bank: From reconstruction to development to equity*. New York: Routledge.
Mauzy, D. K. (1983). *Barison Nasional: Coalition government in Malaysia*. Kuala Lumpur: Marican and Sons (Malaysia) SDN. BHD.
Maxwell, D. (2006). *African gifts of the spirit: Pentecostalism & the rise of a Zimbabwean transnational religious movement*. Athens: Ohio University Press.
Means, G. P. (1991). *Malaysian politics: The second generation*. Singapore: Oxford University Press.
Mihevc, J. (1995). *The market tells them so: The World Bank and economic fundamentalism in Africa*. London: ZED Books.
Milner, A. (1995). *The invention of politics in colonial Malaya*. Hong Kong: Cambridge University Press.
Mohamad, M. B. (1970). *The Malay dilemma*. Kuala Lumpur: Times Books International.
Mohamad, M. B. (1986). *The challenge*. Kuala Lumpur: Pelanduk Publications.
Mohamad, M. B. (1994). Improving tolerance through better understanding. In S. O. Alhabshi & N. M. H. Hassan (Eds.), *Islam and Tolerance* (pp. 15–26). Kuala Lumpur: Institute of Islamic Understanding Malaysia (IKIM).
Mohamad, M. B. (1996). *Being a speech delivered at the Oxford Center for Islamic Studies*. Oxford.
Sachs, W. (2005). *The development dictionary: A guide to knowledge as power*. London: ZED Books.
Samatar, A. I. (1999). *An African miracle: State and class leadership and colonial legacy in Botswana development*. Portsmouth: Heinemann Publishing.
Seers, D. (1969). The meaning of development. In Eleventh world conference of the Society for International Development, New Delhi, India.
Smelser, N. J. (1998). The rational and the ambivalent in the social sciences. *American Sociological Review, 63*, 1–16.

Tamney, J. B. (2012). Religion – State relations in Malaysia. In T. Keskin (Ed.), *The sociology of Islam: Secularism, economy and politics* (pp. 343–376). Reading: Ithaca Press.
Todaro, M. P., & Smith, S. C. (2009). *Economic development* (10th ed.). New York: Pearson.
Volti, R. (2009). *Society and technological change* (6th ed.). New York: Worth Publishers.
Wuthnow, R. (1980). World order and religious movements. In A. Bergesen (Ed.), *Studies of the modern world system* (pp. 57–75). New York: Academic.

Chapter 61
Unveiling Islamic Finance: Economics, Practice and Outcomes

David Bassens

61.1 Introduction

Albeit to varying degrees, economies across the globe are increasingly undergoing a process of financialization, which can be understood as a set of trends that support the growing dominance of financial logic over multiple previously un-financial domains of life (Martin 2002). Working through an amalgamation of practices and discourses, financialization makes that "the economic" actually permeates the wider society as to the point that social and economic relations can be considered mutually constitutive. This co-constitution of economy and society has long been established. Most notably, Karl Polanyi's *The Great Transformation* (Polanyi 1944) identified a an antagonistic "double movement" whereby, on the one hand, the economic (under the form of liberal market logic) moves to 'disembed' social relations, while, on the other hand, a re-embedding countermovement moves to make economic goals subordinate to society's needs. Of course, as the more recent geography literature on financialization acknowledges (Engelen 2008; Pike and Pollard 2010; French et al. 2011), economic or 'market' relations themselves have changed profoundly since the 1980s, especially since financial markets in the US and the UK have become deregulated (for example, London's Big Bang in 1987, Leyshon and Thrift 1997). This trend subsequently led to multiple transformations of the wider economy. First, financialization embodies the emergence of *finance-driven regimes of accumulation*, such as those centred on major nodes in financial markets: Wall Street, The City, but also a growing number of peers in emerging economies such as Dubai, Shanghai, and Mumbai (Sassen 2001). Second, financialization denotes the related growth of *financial engineering*, which has introduced the use of derivatives, securitization, and numerous other "innovative" products which have appeared to be

D. Bassens (✉)
Department of Geography, Free University Brussels, Pleinlaan 2,
B-1050 Brussels, Belgium
e-mail: david.bassens@vub.ac.be

S.D. Brunn (ed.), *The Changing World Religion Map*,
DOI 10.1007/978-94-017-9376-6_61

crucial chains in the dissemination of the global financial crisis (Engelen et al. 2011). Third, financialization carries the growing *influence of capital markets* over firms and households, which in the case of firms translates in the growing influence of shareholders over corporate agenda's (Clark and Wójcik 2007). Fourth and finally, financialization hinges on the *commodification* of previously non-economic aspects of life including personal inclinations such as environmentalist attitudes, the wish to invest ethically, or the conviction to perform economic and financial transactions in line with one's personal faith or religion (Maurer 2005).

One of the most intriguing and certainly one of the best developed and complex examples of such a commodification in particular and of financialization more generally, is the case of Islamic finance. This faith-based form of finance proclaims to procure banking, finance, or insurance in compliance with interpretations of the Shari'a, thus making Shari'a-compliance or the wish to invest/borrow as a good Muslim a commodifiable and marketable substance. Successfully so, since Islamic finance is growing at a rapid pace in its "heartlands," viz., the Gulf Region and Malaysia to reach a market size of about US $1 trillion (which is about 1.5 % of global banking assets (see The Banker 2012)), in addition to undergoing far-reaching processes of globalization that connect these markets to financial centers in Europe and the US. In this chapter I aim to briefly position contemporary Islamic finance as an originally faith-based practice within the dynamically intersecting map of global finance and religion, thereby respectively focusing on its origins in Islamic economics, then on its most common practices and products, and, finally, on the social and economic outcomes it may produce.

61.2 Islamic Economics and the Emergence of Contemporary Islamic Finance

In popular discourse (at least in some European countries) Islamic finance is often simplistically characterized as *halal* banking, because of its ban on the trading and consumption of pork meat, alcohol, pornography, weaponry, and others, which are considered *haram* by Shari'a scholars. Islamic finance, however, has a much broader ideological foundation than these anecdotal "exoticisms" suggest. Underneath the rise of Islamic economics lays a broad reassertion of Islam in Middle Eastern societies since the mid-twentieth-century, which was defined along post-colonial lines. As Kuran (1997: 302, original emphasis) notes: "the *economics* of Islamic economics was merely incidental to its *Islamic* character." In this view, the onset of Islamic economics was not rooted in the claimed intentions to change economic imbalances, injustices or inequalities. Instead, Kuran frames it within the specific historical situation of Indian Muslims in the 1940s, who wanted to defend Islamic civilization against foreign cultural influences. For instance, the Pakistani ideologist Sayyid Abul A'la Maududi (1903–1979), popularized the term "Islamic economics" because he saw economics as a vehicle for (re)establishing Islamic authority in a domain where Muslims were falling increasingly under the influence of Western ideas. The pledge by Islamic ideologists for a return to Islam can hence be read as an antidote to the rising influence of the West, which was increasingly experienced as disturbing (hence the introduction of the term "westoxication." Kuran 1995, 1997). Islamic economics therefore proclaimed itself to be a "Third Way" to conventional capitalism and communism, the leading paradigms during the Cold War. It was thought to avoid the inegalitarian excesses of modern capitalism, while at the same time unleashing the energies of entrepreneurs and merchants (Hefner 2006: 17). In essence, Islamic economics was aimed at freedom from colonial rule, exploitation, and oppression through a return to Islam, which stood for the elimination of poverty and the reduction of unequal distribution of wealth (Siddiqi 2007: 99–100).

In the above context Islamic economists started to refer to the Shari'a, the Right Path for Muslims, which is constituted by the Quran and the *hadiths*, the sayings of the Prophet Muhammed, as guidance for economic actions. Based on these sources they prohibited *riba*, *gharar*, and *maysir*, while proposing profit-and-loss sharing instead of interest-taking as the preferred alternative to mainstream capitalism (Iqbal and Molyneux 2005). The first feature, the prohibition of *riba* ("increase" or interest) is probably the most widely known. In Islamic thought, time does not equal money. Since money is not considered a commodity, but rather a bearer of risk (Pollard and Samers 2007: 315), there is a strong belief that money should not make money by itself. In an interest-based system, an entrepreneur borrowing money from a "conventional" bank has to pay up every term, regardless of the profits or losses he/she has made. From an Islamic point of view, the fact that one party could earn a profit while his business project partner suffers a loss is considered highly inequitable. Profit-and-loss sharing between commercial parties is therefore considered *the* alternative to the interest-based system, thus guaranteeing social harmony (Ala Hamoudi 2007). For instance, in the case of a loan, a money lender does not earn money simply by lending it to the borrower. Instead, a money lender can get an extra return in the form of a pre-agreed part of the profits, or take the loss by ratio of the investment. By prohibiting the raising of interest, the Islamic economic rationale aims to prevent the occurrence of disproportionate losses and gains, thereby aiming to enhance social cohesion.

Second, the prohibition of *gharar* is referring to acts and conditions in exchange contracts, of which the full implications are not clearly known to the parties and could lead to exposing one of the parties to unnecessary risks (Iqbal and Molyneux 2005: 14). For instance, one speaks of *gharar* if (i) the parties lack knowledge; (ii) the object does not exist; or (iii) the object evades the parties' control (Vogel and Hayes 1998: 89–90). The word *gharar* means deception or delusion, but also connotes peril, risk or hazard. In financial terms, it is interpreted as "uncertainty, risk or speculation" (Warde 2000: 59). In essence, the prohibition on *gharar* rejects any gain that may result from chance or undetermined causes. Therefore, there is a ban on the sale of an item that does not exist yet and also on contracts with uncertainty about their cost and duration (Sanhuri 1967: 31, 49). *Gharar* thus refers to the exploitation that can arise from contractual ambiguities and its prohibition, therefore, aims to eliminate or at least minimize this legal risk. The rejection of excessive risk-taking not only influences the structure of financial contracts, but also highly restricts their further trading, be it via futures, options, swaps and other derivatives (see Naughton and Naughton 2000). Related to the prohibition of uncertainty is a fundamental preference towards investing in the "real economy," which also explains why Islamic investors are attracted to "tangible" assets in real estate, commodities, and infrastructure markets.

The third prohibition, the one of *maysir*, is very much related to the abovementioned concept of *gharar*. It bans all kinds of gambling and games of chance, but equally rejects all forms of speculation (which can be thought of as gambling). Gambling is considered unnecessary for a person and for society as a whole, since no economic value is added to the wealth of society (Iqbal and Molyneux 2005: 15). Since *maysir* is often related with high levels of uncertainty, the difference with *gharar* may sometimes seem relatively vague. However, whereas a degree of *gharar* is inherent to economic activities and is thus acceptable, *maysir* is completely banned by Shari'a scholars.

While Islamic economics was established by the mid of the twentieth century, it was only since the 1970s, when massive oil revenues flooded Middle Eastern economies, that Gulf-based Islamic banks started to put Islamic economics into practice (Tripp 2006). Originally initiated by financial institutions in the 1960, the sector experienced a first *aggiornamento* as a result of massive oil money revenues in the 1970s, which allowed Islamic bankers to set up Shari'a-compliant alternatives at the national scale (Warde 2000: 73–89). Much of this initial growth took place in the Middle East and especially in the oil-rich Gulf region. By the 1980s, following the growing influence of political Islam, Islamic finance was further boosted by the Islamicization of the Iran, Pakistan and Sudan, which transformed their economies accordingly to (wholly or partially) exclude conventional financial practices. However, by the 1980s Islamic was experiencing serious "growing pains," not in the least because of historically low oil prices during this decade. However, the 1980s also saw the rise of Malaysia as an Islamic state, which led to the initiation of a dual banking system which combines conventional and Shari'a-compliant banking. Since the 1980s the context for Islamic finance has changed profoundly in light of broader trends in the financialization of the global economy. Since the new millennium, a second *aggiornamento* of Islamic finance is unfolding, this time marked by the globalization of Islamic financial practices and growing interaction with conventional finance. It is to the practices in such globalized Islamic finance sector we turn in the next section.

61.3 Islamic Financial Practice: Retail and Wholesale Markets

Grounded in a different economic rationale and claiming a different economic practice, Islamic finance discursively presents itself as an "alternative" to "mainstream capitalism" as it is practiced by the global financial sector operating from leading international financial centers. This discourse of "difference" towards "conventional" modes of accumulation and intermediation revolves around the three above-mentioned prohibitions, which are key characteristics of global finance marked by the trading of often volatile derivative products such as options, interest rate swaps, and an array of securitization products. As such, the discourse of Islamic finance stands diametrically vis-à-vis the global financial system, in which interest rates are the main "objective" risk-measurement and pricing mechanism. In practice, however, the Islamic economic rationale gave rise to the introduction of an Islamic banking model and array of Shari'a-compliant financial "alternatives" for nearly all common conventional products (Bassens et al. 2011a). Here the role of Shari'a scholars in translating Islamic economic fundamentals to concrete financial products is paramount (Bassens et al. 2011b; Pollard and Samers 2013). Their influence is operationalized though centralized bodies such as the Bahrain-based Accounting and Auditing Organization for Islamic Financial Institutions (AAOIFI), but also through so-called Shari'a boards organized at the firm level, which screen the practices of the financial institution. These cleric, but also increasingly layman, scholars are fundamental to the sector since their authority to proclaim legal opinions (*fatawa*) is what makes a product Shari'a-compliant in practice. Their opinions emerge from a close knowledge of the Quran and the *hadiths*, in addition to processes of analogy, consensus, and personal interpretation. However, the need for Shari'a screening often stimulates Shari'a arbitrage, whereby formally Shari'a-compliant products are rubber-stamped as Islamic by influential scholars (El-Gamal 2006). As will be illustrated below, this has led to a dominance of "mark-up" products, which basically mimic interest-based products, over "true" profit-and-loss sharing solutions. Importantly, this illustrates the gap between Islamic financial practices and their underlying Islamic economic rationale. This discrepancy can be best understood by looking at the most common products in retail and wholesale finance markets.

Similar to conventional markets, one of the most important segments in Islamic retail markets is the provision of home mortgages to households and individuals. In a conventional home mortgage scheme, a customer borrows a substantial amount of money from the bank as to secure the purchase of a home. Upon the sale of the house the buyer becomes owner of the home, while the bank will collateralize the loan through a mortgage on the purchased property. Over time the borrower will repay the loan though installments, thereby paying back the principal plus interest. From an "Islamic" perspective, next to the use of interest, the main problem with such a conventional scheme revolves around the question of whether the bank (the lender) actually runs a risk that justifies the profit made on its execution. As explained above, Shari'a scholars would argue that this is not the case, since the bank is not the owner of the property (thus escaping the risks of property devaluation), while the risk of default on repayment is in effect covered by the collateral. In a normal non-crisis situation (so, contrary to the U.S. mortgage crisis in 2008 when real estate prices plummeted) the bank runs virtually no investment risk and is still remunerated through the interest-based payback

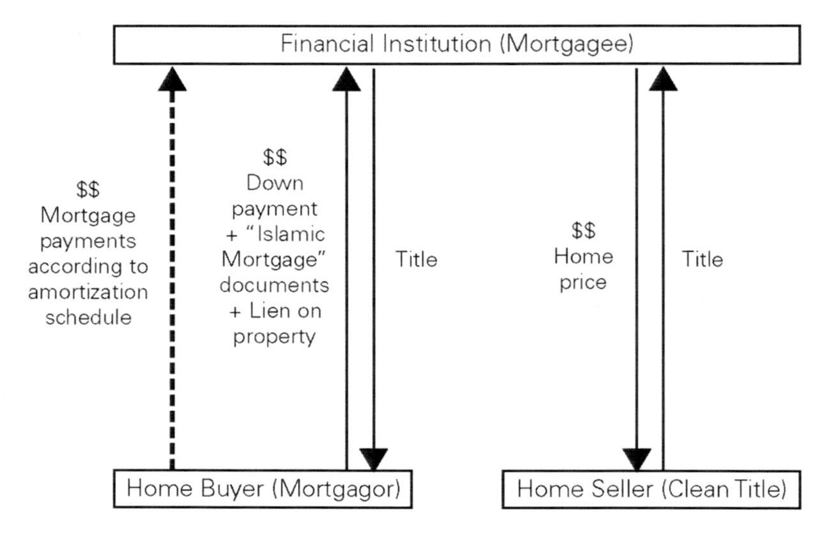

Fig. 61.1 Scheme of a typical Islamic home mortgage (Source: Diagram by Mahmoud Amin El-Gamal, 2005; adapted and used with permission)

system. The most common formal solution to this problem is to provide home mortgages via an Islamic mortgage (Fig. 61.1).

This structure differs from conventional mortgages in an important way for the bank now becomes the owner of the purchased property until the last mortgage payment is made. In many types of contracts the contract defines a diminishing proportion of ownership by the bank as installments are being made. At the final payment, the borrower becomes the full owner of the property. This means that the bank actually runs the risk associated with owning a property (and anything that might happen to it) during the time of the contract. Because of the risks involved, the bank will ask for repayments with a mark-up. For instance, on a "loan" of US $100,000, a bank could charge US $10,000 as a fee, which are then spread over and paid back through the installments.

The question then arises whether such an Islamic mortgage scheme is actually Shari'a-compliant. In this case multiple issues have been raised by Shari'a scholars as to whether the practice of claiming a mark-up is truly "Islamic." This is particularly so for practice of claiming a mark-up fee, which is most often structured to replicate existing interest rates (for example, London Interbank Offered Rate, LIBOR). The whole structure hence seems to rest on *formally* adapting an existing product to the needs of Muslim borrowers, while the substance of these products is no different than conventional ones.

Similar observations can be made as to common practices in wholesale markets, which mostly deal with the structuring of finance products by investment banks to allow the client to access capital markets in a Shari'a-compliant manner. In this context, *sukuk* or Islamic investment notes have become an increasingly popular instrument (Bassens et al. 2013). The current amount of outstanding "Islamic debt" is estimated

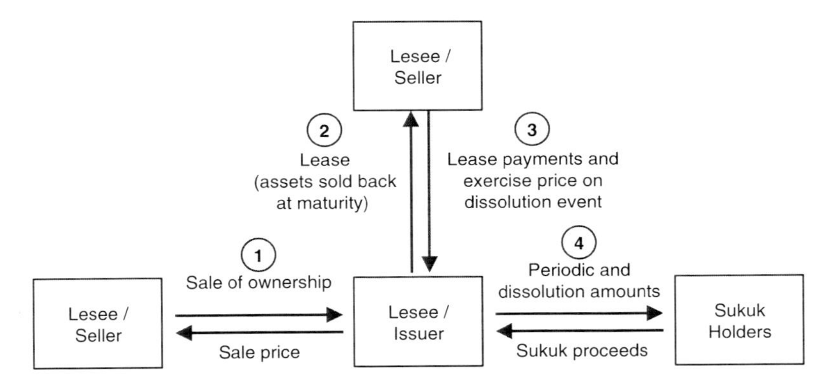

Fig. 61.2 Basic sale-and-lease-back sukuk structure (Source: Diagram by Mahmoud Amin El-Gamal, 2005; adapted and used with permission)

around US $136 billion (IIFM 2010). Mainly since the turn of the millennium, sukuk markets are growing in Malaysia and in the Gulf region. In Malaysia, *sukuk* are mainly issued to attract domestic capital, while in the Gulf region *sukuk* are vehicles to attract global capital (from Europe and the U.S.) next to regional oil-money and switch it to infrastructure and real estate developments. Lately, however, in the hope to attract surplus capital from the Gulf, European and U.S. firms (for example, General Electric, Toyota, Tesco, and others) and governments (e.g., the province of Saxony-Anhalt) have issued sukuk. Again, in theory, as with the Islamic mortgages, *sukuk* aim to avoid the Shari'a problems that surround the use of interest-generating debt in the structure and replace it with a profit-and-loss-sharing mechanism, where the investors are rewarded for the investment risk taken. In practice this is often formally structured through a so-called *sukuk al-ijarah*, which is basically a sale-and-lease-back contract (Fig. 61.2).

In this structure, tangible assets such as materials, infrastructure, or property owned by the borrower (the firm in need of finance) are sold to a purely legal entity called a Special Purpose Vehicle, which is often based in a tax haven such as the Cayman Islands, and subsequently leased-backed by the capital-needing firm. The idea behind this sale is that by doing this the performance of the *sukuk* will depend on the real performance of tangible assets or the project, instead of depending on the credit rating of the borrower, which in most cases makes its cheaper to borrow money. The initial purchase of the assets is financed through the *sukuk* proceeds made by the sukuk investors. After the sale, the lease payments are used to produce a cash flow to the investors until maturity, at which point the original sum amount is repaid. Again, several issues have been raised by Shari'a scholars, the most important ones being the following two. First, as in the mortgage scheme, lease payments are commonly priced with a reference to LIBOR or another prominent interest-rate, which violates the *riba* rule. Second, the risk taken by investors is strongly diminished by a clause called the *purchase undertaking*, which is the obligation to buy-back the assets at maturity. All in all, both aspects recognize that *sukuk* are in substance very similar to regular debt-based bonds.

61.4 Alternative Outcomes?

The close resemblance with conventional products and practices, of which the previous section has provided two egregious examples, have many left to consider Islamic finance as a recent innovation by conventional financial engineers and their legal advisors as ways to access previously untapped sources of capital from emerging markets. The paradigm of Islamic economics, which instigated the practice of Islamic finance, stands diametrically versus the global financial practice of debt trading in volatile markets. According to Islamic economics sustainable growth can only be the result of real economic growth, instead of purely financial capital accumulation. This particular line of thought shows broad similarities with neo-Keynesian ideas regarding the role and function of the financial sector in national, regional, or global economies – ideas that have been reasserted since the outbreak of the 2008 crisis. In developed market contexts as well, it is increasingly argued that the financial sector should focuses on the economically supportive role of capital intermediation, instead of being a major source of GDP in and by itself. According to neo-Keynesian lines of thinking, the financial sector could use more regulation instead of the competitive deregulation that has supported the financialization of the global economy since the 1980s, especially since current practices lead to cyclical bubbles and bursts. In line with neo-Keynesian reassertions, re-regulation is on its way under the auspices of Basel III, but it remains questionable whether its focus on increasing required capital ratios will truly affect leveraged off-balance trading of securities, derivatives, and other potentially "toxic" products. In the meantime, the experimentation with alternative paradigms may provide proof to take the necessary steps in rethinking and reorganizing global finance.

Even if the current practices of Islamic finance do not trigger sea change in the field of global finance, Islamic finance could support the slumbering public demand for the "responsabilization" of the financial sector. In retail markets, it could serve as the starting grid for a more inclusive mode of finance, catering to customers and investors that were previously un(der)serviced. On the one hand, in developing Muslim countries such as Middle Eastern and South(East) Asian economies, Shari'a-compliant alternatives to home funding schemes, car leasing schemes, insurance, and even credit cards support the growth of middle classes. On the other hand, Shari'a-compliant mortgage products also open-up finance possibilities for Muslim communities in non-Muslim countries, especially migrants with low socioeconomic status (e.g., South Asians in the U.K.) who have been "redlined" by conventional commercial banks and who could benefit greatly from Shari'a-compliant products. However, the development of these markets requires the removal of entrance barriers, be it in terms of customer awareness, legal framework, or political climate. First, even though Islamic finance may financially empower Muslim minorities, servicing to some Muslim groups (for example, North Africans and Turks, who constitute the majority of Muslims on the European continent) is hampered by the low awareness of Islamic finance products. Quite often these Muslim migrants are unaware of Shari'a prohibitions or they choose to apply the concept of *darura* ("necessity") to use the conventional banking system in non-Muslim countries. Second, legal frameworks in non-Muslim countries often make Islamic products uncompetitive compared to conventional products, for instance in the field of home financing. Islamic mortgage loans technically involve the double sale of property, that is, a first time from owner to the bank, and a second time, that is, the eventual repayment of the loan from bank to the customer. In the U.K., the government has removed the so-called double stamp tax in order to provide a level playing field. Since then a number of full-fledged banks and Islamic windows have started offering Islamic mortgages. Last, but not least, the eventual success of Islamic finance in non-Muslim countries will remain strongly influenced by the more general public and political sentiment vis-à-vis "Islamic" symbols and customs. Post-9/11 the U.S. government has been particularly suspicious of money with an Islamic name-tag (see de Goede 2003 on *hawala* remittance networks), but also in Europe recent bans on the veil of *burka*, for instance in Belgium and France, illustrate the politicization of anti-Islamic sentiment, which could work adversely on Islamic financial developments, both in retail and wholesale markets.

Next to inclusion in retail markets, Islamic finance could prove to be vehicles of gradual change in wholesale markets as well, since powerful Islamic investors can negotiate specific investment requirements. Beyond the fact that Shari'a-compliant money is not invested in businesses related to alcohol, weaponry, gambling, pork meat, or pornography, these investors also ask that underlying projects or firms should have low debt-ratios and good performance figures. Further, investors refrain from investment in volatile derivatives markets, but prefer buying into "tangible" stock and real estate markets, which produce "real" economic growth. In Muslim countries – particularly in emerging economies such as Malaysia and the Gulf states, Shari'a-compliant money is preferably invested in urban infrastructure and real estate development. Although imminent defaults in Dubai in late 2009 have illustrated the urgent need for more transparency, *sukuk* are a popular way to finance these projects. At the same time Shari'a-compliant capital flows that originate in the Gulf region are increasingly channeled into non-Muslim markets. This happens via listings on stock markets in "mainstream" financial centers (e.g., Dow Jones Islamic Market and the London Stock Exchange), but also directly through *sukuk* issued by firms and governments in non-Muslim majority countries, which aim to attract capital from oil-rich Gulf economies. Although much of the agency behind these investment flows derives from diversifying investment strategies of sovereign, corporate and private investors from Gulf economies themselves, governments in "receiving" non-Muslim countries play a crucial role in issuing landmark sukuk that can serve as of bench-mark for corporate issuances. Given the significant oil-related financial leverage power which Gulf investors have, power which is likely to remain in place in the coming decade, these investments can serve as channels for the gradual dissemination of alternative economic visions into the global financial practice.

61.5 Concluding Remarks: Finance Over Faith?

Looking at the case of Islamic finance suggests that we should at least be critical of its acclaimed alternative approach since its practices are very much embedded within or entangled with conventional financial capitalism (Pollard and Samers 2007; Maurer 2005). While Islamic economics could serve as an alternative, Islamic finance is clearly going down the path of "captured" demand for ethical or "green" investments, which have been marketized into a large number of funds, stock market indices, etc. Illustrative of this trend is the fact that with the growth of Islamic finance, the relative market share of conventional banks with Islamic windows (e.g. HSBC Amanah, Barclays, Deutsche Bank, Standard Chartered, etc.) has easily kept up pace. Fully-fledged Islamic banks have the benefit of harboring context-specific legal, cultural, and faith-based knowledge. However, the ongoing financialization of Shari'a-compliant markets (for example, through the rolling out of *sukuk* securitization structures in Gulf markets) necessitates financial expertise to operate in global markets, which is strongly skewed towards conventional banks and their teams of senior investment bankers, lawyers, and others. Although not unambiguously so, Shari'a scholars have up to now acted as "gatekeepers" of Shari'a-compliant markets (especially in the Gulf), safekeeping to varying extent the Shari'a-compliant substance of Islamic finance. Lately, however, a group of "liberal" Malaysian scholars and Shari'a experts trained at U.K. or U.S. academic institutions is gaining influence in terms of Shari'a governance (Bassens et al. 2012). As a consequence, these gatekeeping mechanisms are gradually being eroded and Islamic banking models are being replaced by more mainstream capitalist visions. In this way, the contents of these products will likely become less Shari'a-compliant than Islamic economists in the 1960–1970s hoped for.

To conclude, the current state of Islamic finance appears to be emblematic for the tensions that were described by Polanyi's double movement, namely the antagonistic relationship between economic goals and societal needs. As it stands, the case of Islamic finance shows that, out of the contention to make economic actions subordinate to societal needs – as was Islamic economists' goal, a viable financial practice can emerge. Yet, and here may lie the biggest value-added of the Islamic experiment, current mimicking practices in Islamic finance emphasize that it takes a lot more than faith and belief to reshape the purpose and role of finance in these highly financial times. Instead of fundamentally changing the practices of global finance, Islamic faith-based discourses themselves have been to a large extent commercialized and marketized under the form of Islamic finance. All things considered, Islamic finance illustrates aptly what could be the biggest irony of the current crisis deadlock, namely the contemporary 'laziness' to politicize finance and outsource the re-embedding of the economic to the financial sector itself (see Engelen et al. 2011).

References

Ala Hamoudi, H. (2007). You say you want a revolution: Interpretive communities and the origins of Islamic finance. *Virginia Journal of International Law, 48*(2), 249–306.

Al-Sanhuri, A. (1967). *The sources of authority in Islamic jurisprudence: A comparative study with western jurisprudence*. Cairo: Ma'had al-Buhuth wa-al-Dirasat al-'Arabiyeh.

Bassens, D., Derudder, B., & Witlox, F. (2011a). Oiling global capital accumulation: Analyzing the principles, practices and geographical distribution of Islamic financial services. *Service Industries Journal, 31*(3), 327–341.

Bassens, D., Derudder, B., & Witlox, F. (2011b). Setting Shari'a standards: On the role, power and spatialities of interlocking Shari'a boards in Islamic financial services. *Geoforum, 42*(1), 94–103.

Bassens, D., Derudder, B., & Witlox, F. (2012). 'Gatekeepers' of Islamic financial circuits: Analyzing urban geographies of the global Shari'a elite. *Entrepreneurship and Regional Development, 24*(5–6), 337–355.

Bassens, D., Engelen, E., Derudder, B., & Witlox, F. (2013). Securitization across borders: Organizational mimicry in Islamic finance. *Journal of Economic Geography, 13*(1), 85–106.

Clark, G. L., & Wójcik, D. (2007). *The geography of finance: Corporate governance in the global marketplace*. Oxford: Oxford University Press.

de Goede, M. (2003). Hawala discourses and the war on terrorist finance. *Environment and Planning D: Society and Space, 21*(5), 513–532.

El-Gamal, M. A. (2006). *Islamic finance law, economics and practice*. New York: Cambridge University Press.

Engelen, E. (2008). The case for financialization. *Competition and Change, 12*(2), 111–119.

Engelen, E., Ertürk, I., Froud, J., Johal, S., Leaver, A., Moran, M., Nilsson, A., & Williams, K. (2011). *After the great complacency: Financial crisis and the politics of reform*. Oxford: Oxford University Press.

French, S., Leyshon, A., & Wainwright, T. (2011). Financializing space, spacing financialization. *Progress in Human Geography, 35*(6), 798–819.

Hefner, R. W. (2006). Symposium visible hands, religion and the market: Islamic economics and global capitalism. *Society, 44*(1), 16–22.

IIFM. (2010). Sukuk report: A comprehensive study of the international sukuk market. Retrieved September 15, 2010, from www.iifm.net

Iqbal, M., & Molyneux, P. (2005). *Thirty years of Islamic banking: History, performance and prospects*. New York: Palgrave Macmillan.

Kuran, T. (1995). Islamic economics and the Islamic subeconomy. *Journal of Economic Perspectives, 9*(4), 155–173.

Kuran, T. (1997). The genesis of Islamic economics: A chapter in the politics of Muslim identity. *Social Research, 64*(2), 301–338.

Leyshon, A., & Thrift, N. (1997). *Money/space: Geographies of monetary transformation*. London/New York: Routledge.

Martin, R. (2002). *Financialization of daily life*. Philadelphia: Temple University Press.

Maurer, B. (2005). *Mutual life, limited: Islamic banking, alternative currencies, lateral reason*. Princeton: Princeton University Press.

Naughton, S., & Naughton, T. (2000). Religion, ethics and stock trading: The case of an Islamic equities market. *Journal of Business Ethics, 23*(2), 145–159.

Pike, A., & Pollard, J. (2010). Economic geographies of financialization. *Economic Geography, 86*(1), 29–51.

Polanyi, K. (1944). *The great transformation*. New York: Rinehart.

Pollard, J., & Samers, M. (2007). Islamic banking and finance: Postcolonial political economy and the decentring of economic geography. *Transactions of the Institute of British Geographers, 32*(3), 313–330.

Pollard, J., & Samers, M. (2013). Governing Islamic finance: Territory, agency, and the making of cosmopolitan financial geographies. *Annals of the Association of American Geographers (Published online)*. doi:10.1080/00045608.2011.628256.

Sassen, S. (2001). *The global city: New York, London, Tokyo*. Princeton: Princeton University Press.

Siddiqi, N. (2007). Shari'a, economics, and the progress of Islamic finance: The role of Shari'a experts. In S. Nazim Ali (Ed.), *In Integrating Islamic finance into the mainstream: Regulation, standardization and transparency* (pp. 99–107). Cambridge: Harvard Law School.

The Banker. (2012). Islamic finance's growth story is only just beginning. www.thebanker.com/Markets/Islamic-Finance/Islamic-finance-s-growth-story-is-only-just-beginning. Accessed 21 Mar 2012.

Tripp, C. (2006). *Islam and the moral economy: The challenge of capitalism*. Cambridge: Cambridge University Press.

Vogel, F. E., & Hayes, S. L. (1998). *Islamic law and finance: Religion, risk and return*. The Hague: Kluwer Law International.

Warde, I. (2000). *Islamic finance in the global economy*. Edinburgh: Edinburgh University Press.

Chapter 62
A Marriage of Convenience? Islamic Banking and Finance Meet Neoliberalization

Michael Samers

62.1 Introduction

It is disappointing perhaps that one of the most reputed sociologists, Max Weber, is responsible for some of the most erroneous assessments of the relationship between Islam and capitalism. In other words, Weber argued that "Islam" and its historical practice in Islamic societies failed to develop the necessary resources for "modern, rational capitalism." Fortunately, in the 1970s, a post-Weberian literature emerged on Islam and capitalism across the social sciences, particularly in the well-rehearsed works of Maxime Rodinson (1974) and Bryan Turner (1974 , 2010). For these authors, Weber proved to be incorrect about the relationship between Islam and capitalism (except perhaps on the subject of patrimonial structure and its effects). Said's *Orientalism* (1978) along with the ocean of post-colonial theory that ensued, assured that scholars would never read Weber in the same way, with the exception perhaps of Samuel Huntington (1993) and his followers. Besides, any casual visitor to the gleaming downtowns of Dubai or Kuala Lumpur in the twenty-first century might simply jeer at Weber's thesis, even if the processes of capitalism in the Gulf States or Malaysia have decidedly different contours than those in let us say, the UK or the U.S.

Ultimately, many of Weber's ideas concerning Islam and its relationship to capitalism have been dismissed as "orientalist," "essentialist," and so forth. As the decades have progressed, a handful of studies have reflected on such themes further (for example, Barber 1995; Karim 2010; Kuran 1997; Rudnyckyj 2009; Tripp 2006; Warde 2000, 2010; Wilson 1998, 2006) including the Huntingdonian thesis (for example, Kuran 1997; Tripp 2006). Warde (2010) in particular has taken a comprehensive and broad assessment of Islamic banking and finance within the context of

M. Samers (✉)
Department of Geography, University of Kentucky, Lexington, KY 40506, USA
e-mail: michael.samers@uky.edu

S.D. Brunn (ed.), *The Changing World Religion Map*,
DOI 10.1007/978-94-017-9376-6_62

the global economy over the last decade, but very few have discussed the relationship between Islam and what is arguably a phase of capitalism that we might wish to call "neoliberalism," "neoliberalization," or in more nuanced and spatially-sensitive terms: 'variegated neoliberalization' (Brenner et al. 2010).

To put it differently, while one part of the equation (Islam) has amassed the bulk of scrutiny, the other part of the equation – capitalism, has been implicitly left as monolithic and immutable *by Islamic scholars critical of capitalism*[1] (see Tripp 2006). I argue in this chapter however, that in order to grasp changes in Islamic jurisprudence and practice, one also has to understand transformations in the character of capitalism, a theme addressed – at least implicitly – in the chapter by David Bassens in this volume. To paraphrase Bill Maurer (2005), the practice of IBF transforms Islam itself and in turn changes (to one degree or another) what becomes or does not become acceptable doctrine and practice.

This chapter explores the relationship between so-called neoliberalization and Islamic banking and finance (henceforth abbreviated as IBF) through a study of the UK and the U.S. The choice of these countries is based on not only their arguable embrace of neoliberalization (if one is to take seriously the evidence for these processes), but also their rather sharply contrasting trajectories in terms of the promotion, enablement, legitimacy, and actual development of IBF. In doing so, my purpose is less a comparison of the two countries (although I do compare them), than it is to demonstrate the significance of neoliberalization in its contentious "entanglement" (Callon 1998) with IBF (Maurer 2005; Pollard and Samers 2007) in two different territorial contexts. I do not purport to assess all the supposed elements of neoliberalization as a general set of processes in these countries, not least because this has already been demonstrated amply (if not unconscientiously) for the UK and the U.S., and it would be nigh impossible to undertake this sort of analysis in such a short chapter, if not in an entire volume. What I tackle instead is the idea of what Peck and Tickell (2002) call "roll-forward" neoliberalization *with respect to* the development of IBF.

62.2 On Neoliberalization

Let me begin with the notion of "neoliberalization" which has been wielded *ad infinitum* by geographers and others, and which has nevertheless, rightly or wrongly, persisted in the lexicon of critical observers of political economy. Most agree that

[1] Certainly, I recognize that many prominent shari'a scholars and other observers of Islamic banking and finance have commented critically on the "western" or "global" economic and financial crisis since 2008, and have offered some proclamations as to why this "crisis" occurred (over-indebtedness' of western consumers, etc.). At the same time, there is little engagement with theory. A prominent exception is Rethel's (2009) analysis of the problems of legitimacy in the broader global political economy. She argues that practitioners of IBF have to speak to two constituencies: conventional financiers, regulators and the "global financial architecture" on the one hand, and Muslims concerned with the piousness of IBF on the other. Her work aside,, there is little attempt to systematically interrogate changes in the (the political economy of) capitalism since some of the early "modern" writers such as Chapra (1985). See Tripp (2006) for a thorough discussion of these matters.

neoliberalization (gone are the days that academics speak "vulgarly" of "hegemonic neoliberalism") comprises a set of discourses, economic and social policies, programs, and practices that have been ardently promoted by a range of policy ideologues, practitioners, and economic and financial elite (among many others) in the "advanced liberal" states. This entails both 'roll-back' and "roll-forward" neoliberalization (Peck and Tickell 2002, 2007). The former – associated especially with the 1980s – entails the deregulation of labor markets, cutting back social programs, privatizing them, "marketizing" them, or eliminating them altogether. The latter – more associated with the 1990s and beyond – and with which I am most concerned here, refers to a range of processes that include "market-making" processes (Overbeek 2003), "technocratic economic management," "capital welfare," the encouragement of entrepreneurship, and a pervasive logic of competition (ironically either advocated by governments or tacitly supported); regressive taxation, and significantly, various processes of "financialization." This last element might involve an emphasis on monetarism, maintaining low interest rates, a focus on short-term profits for shareholder value, promoting "financial inclusion," and for those who subscribe to the notion of a "neoliberal governmentality" (for example, Ferguson and Gupta 2002), the combined privatization and individualization of financial security and responsibility, such as the promotion of home ownership (for a review of financialization, see Hall 2011).

Many scholars also argue that the most intensive application of these policies or processes has occurred in a set of Anglophone countries (especially the UK and the U.S.), but also in New Zealand, for example (Larner 2003), rather than in let us say the "corporatist" "welfare states" of Western Europe. Others maintain that it experienced its sharpest manifestation in the poorest countries of the southern hemisphere through "structural adjustment programs" and the like. After all, many have noted how it may be as much a product of the "boys" from Santiago, as it is from the "boys" in Chicago. In the same vein, most now concede that neoliberalization is not simply meted out without problems of legitimacy by the power centers of London, New York, or Washington to prostrate citizens who passively accept its tenets without active protest (e.g. Leitner et al. 2007; Lockie and Higgins 2007). These then, are some of the basic points of agreement with respect to processes of neoliberalization, but it becomes more contentious from there.

Indeed, as noted above, some noteworthy contributions have expressed criticisms of its supposed hegemony or even its pervasiveness in particular countries, regions, cities, and so forth (for example, Larner 2003), and lament its overly "western" conceptualization (Ong 2006). Others refuse to view it as an 'ideal-type' and focus instead on neoliberalization as practice, or 'actually existing neoliberalism' (see Bakker 2010). Others see it as peculiarly Gramscian and structuralist (and specifically Marxist), devoid of the messiness of agency, resistance, and social action that entails all sorts of social processes that cannot be easily attributed to some overarching notion of neoliberalism (Barnett 2005; Collier 2012; Larner 2003; Leitner et al. 2007). In Barnett's case, his criticisms entail a quasi-methodological plea, calling for the hard work of figuring out who really does what to whom in terms of power and its effects, rather than assuming that neoliberalism exists *a priori.*

In short, they imagine neoliberalization with a small "n," as Ong (2006) put it (see a discussion of this in Collier 2012). The debate is then established. In one corner are the structural Marxists and Gramscians who insist that neoliberalization is, or at least diffused across the globe through "fast policy transfer" with varying characteristics from place to place. Neoliberalization, therefore, presses upon all of us (or courses through all of us) in one way or another through its various discourses and "roll-backed" and "roll-forwarded" practices. In the other corner are those who see either contradictory or paradoxical evidence for neoliberalization (for example in the size of the state in terms of expenditures, and so forth), those who see contestation and resistance in shaping neoliberalization (Leitner et al. 2007), or those who see 'neoliberal moves' as omnipresent, but not evidence of an over-bearing structure called neoliberalism or even neoliberalization (Barnett 2005; Collier 2012; Ferguson 2010; Larner 2003).

Enter Brenner et al. (2010) who seem to act as referees, and steer a middle way through this theoretical and conceptual quandary. They acknowledge the criticisms of those who prefer to diminish or pluralize the concept of neoliberalism/neoliberalization, but insist that their notion of "variegated neoliberalization" reflects a set of *processual* (rather than static) "strong discourses," "rule regimes," and successive rounds of market reform (a "roll-forward neoliberalization") that has significant implications for not only the advanced liberal states, but for the rest of the world as well. Neoliberalization exists as a set of structural market-making processes, but it is geographically *variegated*, and subject to myriad policy failures, discursive and material experiments and reformulations, as well as discursive and material contestations. Bearing in mind the myriad corners of this debate (which cannot adequately be addressed here), this chapter adopts the view of Brenner et al. (2010) and empirically assess Peck and Tickell (2007)'s call for "roll-forward neoliberalization" by exploring how IBF has been embraced and supported by governments, (Islamic) bankers and financiers, and even some community organizations alike.

My argument is this: regardless of the faith or profits that motivate Islamic bankers and financiers in these two countries (some of whom are not Muslim at all), a concern with "social justice" (*adalah*) or more concretely (*zakat*), which has been associated with the tenets of Islamic economics, is only barely present in the practice of IBF.[2] Instead, through a range of actors, the development of IBF in both countries has created at least three different but related markets. One for *sukuk* (Islamic bond) issuances, money markets, and large-scale investment and trading opportunities, another for asset management, large net worth/private client banking and a third for more modest Muslims – an allegedly interest-adverse set of consumers. This third market may be forging new kinds of financial subjects alongside *Homo Economicus* (a certain *Homo Islamicus* if you will – see, for example, Nasr 2006) that may represent a "hybrid" form of neoliberal governmentality. Let me be prudent though, the financialization associated with neoliberal governmentality is not one that steamrolls itself across the landscape of individual subjectification, and *Homo Islamicus*

[2] In other countries, such as Malaysia, zakat at least functions much more prominently in the role of some sort of redistributive justice (see for example, Karim 2010).

is as much contested and incomplete as *Homo Economicus*; both work together through Muslim consumers of IBF.

In relation to discussions of neoliberalization then, this chapter leaves aside the question of "roll-back" neoliberalization, if not only for the reasons that this could not be covered adequately in this chapter, but also on the grounds that empirically, it is difficult to deny what seems like a 30-year trend of marketized, privatized, or eviscerated social services. This chapter also side-steps a discussion of the basic features of IBF (which are outlined in David Bassens' chapter in this volume), and there is now a cavernous literature that accomplishes the same. However, some of its central features and operational characteristics will surface selectively as need arises. With this in mind, I turn now to an evaluation of 'roll-forward neoliberalization' and IBF in the UK and the U.S.

62.3 IBF in the United Kingdom

The British government has proved to be much more proactive than the United States in promoting IBF. This is unsurprising, given the City of London's (London's financial district) expertise in international banking and finance, the long-time presence of banks from the Gulf States in the UK, the desire of the British government to increase the range of financial services that the City of London provides, and that the attacks of September 11 did not happen within UK territory.

In the late 1990s, IBF began to attract greater interest among at least wealthier British Muslims, "conventional" banks such as Barclays, Citibank, and HSBC, as well as the British government (Dar 2004; FSA 2007; Her Majesty's Treasury 2008; Wilson 2000, 2011). The Bank of England set up the "Working Group on Islamic Banking" based on the assumption that Muslims were financially excluded from conventional banking because they refused to rely on interest-based banking. While Dar (2010) argues that this assumption remained incorrect, the government proceeded with this in mind nonetheless. In fact, the British government would move beyond the encouragement of IBF by beginning to modify legal constraints on IBF. The FSA (Financial Sector Authority) approved an "Islamic window" – the *shari'a*-compliant *Amanah* division of HSBC, which then began to offer "Islamic mortgages" (or more appropriately Islamic home finance) in the late 1990s through an *ijara* (a lease-to-own) contract (Wilson 2000). In using this particular contractual form, *Amanah* sought to avoid using a *murabaha* (cost-plus) transaction that would have created the problem of so-called "double stamp duty" in the UK – a tax initially on the sale of the house to a bank, and then on the reselling of the house to the purchaser (FSA 2007). A number of other small providers for retail banking and Islamic home finance, such as the West Bromwich Building Society, accompanied HSBC in servicing what providers saw as a "niche market" for banking services. Yet this niche market had to be created, as most Muslims in the UK had little idea about how IBF actually worked. n fact, HSBC held a number of "road shows" in the early 2000s that involved Imams seeking to convince their respective Mosque congregants of the religious appropriateness and value of Islamic retail banking. One might say therefore that Muslim consumers of IBF were partially "invented," resulting in a sort of *Homo Islamicus* (Pollard and Samers 2007).

The undeniable support and arguably, the liberalization of IBF gained further steam in the UK when the FSA removed the requirement of "stamp duty" for Islamic mortgages through the Finance Act of 2003. Once removed, Islamic home finance became more attractive for both Muslim customers, and for the providers of Islamic home finance (Dar 2004; FSA 2007; Her Majesty's Treasury 2008). Indeed HSBC's Amanah would be soon complemented as early as 2001 by the Prince of Wales Trust, which further established a nascent Islamic financial services industry by offering no-interest "loans" to Muslims for entrepreneurial purposes (see *BBC online*, November 7, 2001; HRH, the Prince of Wales, 19 May 2004). There are various ways of interpreting this offer by the Trust; one, it could be seen – as was officially stated – as a response to the social and financial exclusion of Muslims in the UK. The unspoken message in this speech is that the frustration," "unemployment" and "alienation" of young Muslims in British cities (as the Prince of Wales put it), might lead to social disorder. All of this might in turn be interpreted as the consequence of the neoliberal evisceration of especially urban social welfare policy alongside the growth of precarious under- or unemployment. Whatever the case may be, 2004 marked a rather prodigious year for the evolution of IBF in the UK. That is, the first Islamic bank in the "west" opened up – the Islamic Bank of Britain. Additional changes to regulation helped matters to progress further: the Finance Acts of 2005 and 2006 altered the tax requirements on three widely-used Islamic contracts: *murabaha, musharaka, and mudarabah* (Her Majesty's Treasury 2008; Her Majesty's Treasury and FSA 2008).

Like the Bahrainian or Malaysian's government's emphasis on Manama or Kuala Lumpur as hubs for IBF, the British government has sought to promote London as the center of investment for Islamic finance in Europe since the mid-2000s. In 2006, then Chancellor of the Exchequer Gordon Brown announced his plans to make London "the capital of IBF," and the City of London's marketing arm (City UK) proclaimed the flexibility of the City's financial institutions and welcoming institutional regulations, and in 2010, helped to establish the lobby group: the UK Islamic Finance Secretariat within its remit. One of the chief aims of UKIFS consisted of lobbying for a UK sovereign *sukuk*. In fact, this sparked some competitive institutional re-regulation by the French government to embrace IBF, which argued that Islamic investments could redress some of the problems of the economic and financial crisis in France, as well as steer Islamic investments away from London (see, for example, *La Tribune*, 18 April 2012).

Despite these challenges from the French government and other European governments, by 2011, the number of fully *shari'a*-compliant IFIs (5) and conventional banks with Islamic windows (17) in the UK simply dwarfed the number in France (3) and Germany (1) (UKIFS 2012). Table 62.1 provides a list of the number of Islamic financial institutions in 'western' countries and offshore centers, and Table 62.2 provides some comparative data on IBF assets across the globe, including the UK.

Table 62.1 Islamic financial institutions in 'western' countries and offshore centers

Country	Number
UK	22
US	10
Australia	4
Switzerland	4
France	3
Canada	1
Cayman Islands	1
Germany	1
Ireland	1
Luxembourg	1
Russia	1

Data source: UKIFS (2012: 5), based on data from The Banker

Table 62.2 Comparative data on IBF, by country (ranked by total assets)

	Total	Banks	Takaful (insurance)	Funds	Others	Number of firms
Iran	338.0	383.5	4.2	0.3	–	27
Saudi Arabia	151.0	147.8	3.2	–	0.0	26
Malaysia	133.4	120.4	9.9	–	3.2	39
UAE	94.1	92.5	1.5	0.0	0.1	21
Kuwait	79.7	68.9	0.1	10.6	0.0	39
Bahrain	57.9	56.2	0.4	1.3	–	33
Qatar	52.3	50.0	0.5	1.8	0.1	19
Turkey	28.0	28.0	–	–	–	4
UK	19.0	19.0	–	–	–	6
Sudan	12.1	12.1	–	–	–	13
Bangladesh	11.7	11.4	0.3	0.1	–	16
Indonesia	10.5	10.0	0.5	–	–	26
Syria	8.7	8.7	–	–	–	3
Egypt	7.9	7.9	–	–	–	2
Switzerland	6.6	6.6	–	–	–	2
Jordan	5.9	5.7	0.1	0.1	–	10
Pakistan	5.7	5.6	–	–	0.1	23
Brunei	3.8	3.8	–	–	–	1
Other countries	10.3	9.8	0.4	0.1	0.0	35
Total	1,086.5	1,047.7	20.9	14.3	3.5	345

Data source: UKIFS (2012: 3), Based on Data from The Banker

Certainly, the City of London's expertise as a center for international finance is key here to its relative dominance in this vein, as mentioned above, but it is also case that the City of London's continuing moves to create a favorable regulatory environment, such as the Finance Acts of 2007 and 2008, which altered the tax regulations on *sukuk* so that they would be similar to other debt instruments, have facilitated the hegemony of London's position. In fact, according to an analysis by Bassens et al. (2010) that measures the connectivity of cities across the world in terms of their Islamic financial services, London is ranked as third behind Manama and Tehran, with New York ranked 20th. The number of IFIs in the UK would continue to grow between 2004 and 2008 from 1 fully *shari'a*-compliant bank (the IBB) to 5 between 2004 and 2008 and by 2007 were lining up like "airplanes on a runway" (Pollard and Samers 2013) for approval from the FSA (Financial Services Authority) (UKIFS 2012). The Financial Services and Markets Act 2000, Order 2010 sought to support corporate *sukuk* issuance by establishing a "level playing field" (clarifying and simplifying regulatory issues surrounding their issuance, reducing the legal costs associated with their issuance, and eliminating many of the hurdles that stood in their way) (*Arab News*, 14 February 2010). These regulatory changes were brought about through close cooperation between an amalgam of different actors, from the FSA to the Muslim Council of Britain and the Islamic Finance Experts Group. It demonstrated the willingness of the UK government to modify existing financial regulations to nourish what they saw as a growing market, although not all of these various IBF markets were actually expanding (Pollard and Samers 2014), as I discuss further below.

Indeed, while British-owned IFIs, or international IFIs with a base in the UK have been involved directly or indirectly in – broadly-speaking – a growing global *sukuk* market (4 of the 5 fully-*shari'a*-compliant IFIs in the UK as of 2011 were international investment banks with convenient bases in London) (UKIFS 2012), Islamic retail financial services have not flourished to the extent that international investment and trade-based financing have. Even for *sukuk* markets, the evidence is a little bit mixed. While the UK government ultimately did not pursue issuing its own sovereign *sukuk,* it is still being considered, and in 2010, a manufacturer in northeastern England raised $10 million through a Dubai-issued *sukuk* listed on the Cayman Islands Stock Exchange (Reuters, January 18, 2011; UKIFS 2012). In terms of retail financial services, the reason for the somewhat stagnant market is blamed on the relatively small number of eligible or willing Muslim customers (too poor, at present too young to bank, or not averse to conventional banking), and because of the nature of its products and services, which are both relatively expensive and have been poorly tailored to the kind of "committee financing" (a form of a Rotating and Savings and Credit Association) common to Bangladeshi, Indian, and Pakistani-origin Muslims (Amin 2010; Dar 2010). In short, developing a niche market such as Islamic retail financial services in a regulatory structure that is arguably more open to *shari'a*-compliant investments does not guarantee the sustainability of those markets, and may or may not impel the British government to address the social and financial needs of Muslim communities by other means. Brenner et al. (2010) insist on this element of neoliberalization; that it is subject to routine failures which require further experimentation with market-based, communitarian-based, or even state-based initiatives.

62.4 IBF in the United States

The United States has not held the same proactive stance towards IBF as the UK, *perhaps* in large measure because of the Islamophobic conflation of IBF with "terrorist financing" following the attacks of September 11, 2001 (Ilias 2010; Warde 2008). New York or Washington, unlike London, is not encouraged as a center for IBF, and the number of IFIs (about 22 in the UK around 2010) exceeded those in the U.S. (about 15) (Pollard and Samers 2013).[3] Moreover, if we take into account total *shari'a*-compliant assets, the UK is ranked 9th, while the U.S. is not even listed in the top 18 countries in terms of assets (UKIFS 2012) (see Table 62.2). The fear that Islamic bankers and financiers have of investing in the U.S. may be based more on the *perception* that the American government is less welcoming than it actually is. Nonetheless, there is evidence to suggest that this perception changed towards the end of the 2000s (Dar 2009; Reuters 2008), and it would also be an error to think that IFIs were never welcome. In fact, in 1996, as more "contemporary" versions of IBF began to gather steam outside Malaysia, the Middle East, South Asia, and so forth, the former vice president of the Federal Reserve Bank of New York (Ernest T. Patrikis) gave a speech to the Islamic Finance and Investment Conference in New York City, eagerly supporting the emergence of IBF in the United States. As he stated:

> …the fact that Islamic banking and finance is based on religious principles is irrelevant to us. Issues of religion are not supervisory matters of concern. If an Islamic banker needs to structure a transaction in a given way, we will work to try to assist in that process. The fact that this need is grounded on religion does not affect our view of the matter. We will keep our focus on bank supervisory concerns. (Patrikis 1996, n.p.)

He went on to conclude:

> The primary message that I want you take with you today is that as bank supervisors we have an open mind on how to approach any issues you may raise. Islamic bankers have been quite ingenious in developing financial transactions that suit their needs: we bank supervisors, too, can be ingenious and will want to work with any of you should you decide that you want to engage in Islamic banking in the United States. That is not to say that all issues can be resolved to your satisfaction. But our doors are open; indeed, that is the American way. (Patrikis 1996, n.p.)

Regardless, or because of the high-minded and stereotypical, self-righteous nationalism present in the last few words of this speech, such an embrace of IBF would continue in the cautious vein of a multicultural capitalism. In fact, a request by the United Bank of Kuwait in the United States prompted two 'interpretive letters' from the U.S. Office of the Comptroller and Currency (the OCC) and the New York State Banking Department in 1997 and 1999. This in turn paved the way for *murabaha* contracts for commercial and retail banking, as well as *ijara* contracts for "Islamic mortgages." Despite that *ijara* "mortgages" differ in structure from

[3] CityUK (2010) counts only 10 IFIs for the US, while our study in addition to data from the IFSL (2010) records 15 IFIs.

"conventional" mortgages, as Maurer (2006) points out, they are made to more or less mimic conventional mortgages – not least in the similar language used in mortgage applications – in order to appeal to Muslim consumers. In any case, the Office of the Comptroller of the Currency then moved to approve many of the products associated with HSBC's "Islamic window" in New York: *Amanah*. Further facilitation of IBF markets in the U.S. – this time inadvertent – came with the Financial Services and Modernization Act (FMSA) in 1999, which dismantled the 1933 requirement that banks must be either investment, commercial or insurance-related banks but not all three simultaneously. Since Islamic banks often undertake all three functions, the FMSA permitted the establishment of a wholly *shari'a*-compliant independent "bank" (under the definition of a bank in U.S. law) (Taylor 2003) although no such fully *shari'a*-compliant stand-alone bank has been forthcoming.

Nevertheless, a number of non-bank IFIs, subsidiaries, or conventional banks with "Islamic windows" have since sprouted to address the desire of Muslims for Islamic Home Finance, including Devon Bank in Chicago, Virginia-based Guidance Residential, LARIBA in Los Angeles, and the University Islamic Financial Corporation (a subsidiary of University Bank) in Ann Arbor, Michigan. This is not simply a private affair, however. In fact, the U.S. government had been indirectly engaged in Islamic home finance through Freddie Mac, which announced in March 2001 that it would "securitize" LARIBA's mortgages, and in 2003, Fannie Mae also followed suit. This had the effect of lubricating Islamic mortgage markets with liquidity, a more than significant move since Islamic home finance proved to be expensive compared to interest-bearing mortgages, and lacking in "sufficient" liquidity (Freddie Mac 2002; Maurer 2006). Furthermore, this demonstrated the entanglement of quasi-governmental institutions with Islamic home finance, rather than these or similar institutions becoming heavily involved in let us say building public housing. Most financing as of 2010 in the U.S. consisted of Islamic home finance, although Amana funds in the state of Washington or HSBC Amanah based in New York are exceptions to this, which offer *shari'a*-compliant mutual funds and other investment products (Ilias 2010).

By 2004, recognizing IBF's global ascendancy and that the government might realize some economic or financial benefits from Islamic investments into the banking and finance sector in the U.S., President Bush appointed Mahmoud El Gamal (a Professor of Economics at Rice University) as the White House's first economic advisor on IBF. The Department of Treasury held an inter-governmental Seminar (Islamic Banking and Finance 101) in the same year to further establish ways of courting IBF. William Rutledge, the Executive Vice President of the Federal Reserve Bank of New York, speaking at the 2005 Arab Bankers Association of North America (ABANA) Conference on Islamic Finance in New York City summarized some of the reserved embrace of IBF in the mid-2000s:

> At the outset, I would like to emphasize that we—and here I am referring broadly to U.S. regulators—are open to Islamic financial products. Our mindset is to try to accommodate a variety of approaches to finance, focusing to the extent possible on the underlying substance—that is, focusing on what the implications for safety and soundness and consumer protection would be of a given product. Consistent with that approach, while we are

> committed to accommodating Islamic finance within the U.S. structure, we will hold Islamic financial institutions to the same high licensing and supervision standards to which we hold conventional ones. Although we are certainly in no position to take a stance on issues of shari'a interpretation, it is important that we become more familiar with the principles and practices unique to Islamic finance in order to make our supervisory and regulatory judgments. (Rutledge 2005, n.p.)

Since 2004, Islamic home finance has grown apace with some IFIs, such as Guidance Residential reporting some 1.5 billion in home financing sales in 2010 (Ilias 2010). However, IBF in the United States has simply not witnessed the sort of expansion that the UK has experienced, at least in investment funds and *sukuk* trading, while some Islamic financiers saw more hope in a court case in July of 2011. This particular case reinforced the legality of Islamic windows when a Michigan district court dismissed the argument that the money supplied by the government's bailout of the American International Group could not be used to support its Islamic insurance businesses (Reuters, January 18, 2011).

Despite the efforts of the US government and related institutions to facilitate IBF, the U.S. Federal government has clearly not held the same encouraging stance towards IBF as in the UK, despite that financial liberalization has been an on-going process in the United States from the early 1970s to at least 2008. When one looks beyond the national government, however, sub-national governments and agencies demonstrate how IBF is being rolled out as both an entrepreneurial and social mobility vehicle alongside financially strained social services. For example, in Minneapolis, where there are a growing number of residents of Somali origin (somewhere around 32,000 people by the end of the 2000s) as well as from other countries in East Africa, community-oriented banking in the form of Islamic home or business finance has become a significant means of home equity and entrepreneurial capital. Typically, this leads to the purchase of a modest apartment or home or starting a small business in the face of difficult challenges in Minneapolis' labor markets. At this municipal level, loans that avoid *riba* (or *reba* – interest) are provided through the New Markets Mortgage Program to predominantly, but not only Somali-Americans by a partnership between the Minneapolis-based African Development Center, the Minnesota Housing Finance Agency, the Neighborhood Development Center in St. Paul, and Chicago's Devon Bank (see Samers 2012 for a fuller discussion).

62.5 Conclusions

The debate on the relationship between capitalism and Islam continues to preoccupy Islamic scholars, jurists, bankers, and observers of IBF. Max Weber would be pleased by the vigorous continuation of the debate, but probably also surprised by the entanglements of IBF and capitalism. Esteemed sociologists aside, a more modest individual with only a popular news understanding of IBF might mistake it for something entirely "exotic" or even dangerous, given the links made by those fearful of IBF's connection with international terrorist financing. No doubt that the various prohibitions on *riba*, *maysir* and *gharar* (gambling, excessive speculation), IBF's prohibitions on investments into certain kinds of activity (pork and alcohol-related businesses, certain forms of entertainment, etc.) and some of its profit and loss sharing contracts such as *mudarabah* and *musharaka* constitute something distinctive. Yet it would be erroneous to assume that it is a system of banking entirely set apart from so-called "conventional" banking (or that the latter is not in itself religiously inflected) (see Pollard and Samers 2007; Pitluck 2012). In fact, in a close and summative analysis of the proximity of Islamic and conventional contracts, Pitluck (2012) insists on their similarities, and explores how IBF has mimicked or "married" (my words) the "global financial architecture" of the twenty-first century. This may be as much a matter of practicality and "convenience" as Islamic bankers and financiers find it difficult to operate outside conventional financial institutions, practices, regulations, and standards. In fact, even international Islamic organizations such as AAOIFI (the Accounting and Auditing Organization for Islamic Financial Institutions) has found it difficult for practical reasons to avoid mirroring the standards associated with the conventional Basel accords on financial standards (for example, McMillen 2011; Pollard and Samers 2013).

In other words, IBF is intertwined with a *specific and provisional form* of capitalism in the context of the UK and the U.S., namely "roll-forward neoliberalization." As Brenner et al. (2010) insist, this involves "rule regimes," "strong discourses," "successive rounds of market reform," "technocratic economic management" and "market-making" policies. In the UK case, this has involved a pro-active stance towards IBF which may have been strategically legitimated to potentially skeptical publics by appealing to a certain pride in London's preeminence with respect to IBF in "the west." In contrast, the U.S. government *quietly* embraced IBF in the spirit of some sort of multi-cultural capitalism, perhaps because the process of a more prominent campaign for IBF entailing a louder legitimation may have faced gargantuan hurdles in the shadow of September 11. Regardless of these questions of legitimacy, the "market-making" "rule regimes" of roll-forward neoliberalization have certainly been visible in the evolution of IBF in the UK and the U.S. If one wishes, these can be characterized as structural processes, but they are nonetheless promulgated by both state actors (national and sub-national governments) and "non-state" actors (bankers, financiers, lawyers, lobbyists, community groups, religious scholars, and so forth) eager to one degree or another to facilitate the development of IBF in their respective countries.

While this understanding of roll-forward neoliberalization could be easily dismissed as nothing more than pro-business discourses and policies that were no stranger to the supposed post-war era of social liberalism in the UK or the U.S., I would maintain that the pervasiveness of these discourses and practices suggests that capitalism has evolved in ways that encourage new forms of financialization and market-making unfamiliar to an earlier period of capitalism. No doubt such debates around these contentions will continue.

My point in this chapter, which can only be modestly demonstrated, is that if it has become starkly evident that IBF is not a "thing apart," then the object with which it is entangled needs equal attention in light of the development of IBF. What is crucial I suspect is that rather than tackling this issue from the beginning point of fear, suspicion, or the essentialism of "clash of civilizations" type discussions that occasionally haunt discussions of Islam and IBF, it might be better to begin with a close analysis of the moments of inter-connections, matters of interpretation, and what Bhabba (1994) calls "negotiations." At the same time, as all the literature on "agonistic politics" has made painfully clear, conflicts and disputes may not only be desirable but also unavoidable. Indeed, *shari'a*-compliant banking may raise some questions for devout secularists, as do broader questions around "western" liberalism and religious law. However, once again, my discussion has suggested that it might be more fruitful to begin with a thorough analysis of the untidy entanglements of economic and financial practices, instead of assuming their distance and danger. For this author anyway, that seems to be a more hopeful road than one of disengagement and foreclosure.

References

Amin, M. (2010, September 1). The UK Islamic banking scene. Originally from *Islamic Finance News*. Available at www.mohammedamin.com/Islamic_finance/UK-Islamic-banking-scene.html

Arab News. (2010, February 14). UK gears up to issue first corporate sukuk. Available at www.arabnews.com/node/337074

Bakker, K. (2010). The limits of 'neoliberal natures': Debating green neoliberalism. *Progress in Human Geography, 34*(6), 715–735.

Barber, B. (1995). *Jihad vs. McWorld*. New York: Ballatine Books.

Barnett, C. (2005). Consolations of neoliberalism. *Geoforum, 36*, 7–12.

Bassens, D., Derudder, B., & Witlox, F. (2010). Searching for the Mecca of finance: Islamic financial services and the world city network. *Area, 42*(1), 35–46.

Bhabba, H. (1994). *The location of culture*. London: Routledge.

Brenner, N., Peck, J., & Theodore, N. (2010). Variegated neoliberalization: Geographies, modalities, pathways. *Global Networks, 10*, 182–222.

Callon, M. (1998). Introduction: the embeddedness of economic markets in economics. In M. Callon (Ed.), *The laws of the markets* (pp. 1–57). Oxford: Blackwell.

CityUK. (2010). *ISLAMIC FINANCE 2010*. Available at http://www.thecityuk.com/research/our-work/reports-list/islamic-finance-2013.

Chapra, M. U. (1985). *Towards a just monetary system*. Leicester: The Islamic Foundation.

Collier, S. (2012). Neoliberalism as big Leviathan, or … ? A response to Wacquant and Hilgers. *Social Anthropology, 20*(2), 186–195.

Dar, H. (2004). Demand for Islamic financial services in the UK: Chasing a mirage? Paper available at https://dspace.lboro.ac.uk/dspace-jspui/bitstream/2134/335/3/TSIJ.pdf. Accessed 16 Oct 2014.

Dar, H. (2009). *The Chancellor guide to the legal and Shari'a aspects of Islamic finance*. London: Chancellor.

Dar, H. (2010, July 5–11). Why retail Islamic banking has not taken off in the UK. *Business Asia*. Available at www.edbizconsulting.com/articles/2010_07_05_why_retail_islamic_banking_has_not_taken_off.pdf

Ferguson, J. (2010). The uses of neoliberalism. In H. Castree, P. Chaterton, N. Heynen, W. Larner, & M. W. Wright (Eds.), *The point is to change it: Geographies of hope and survival in an age of crisis* (pp. 166–184). Oxford: Wiley-Blackwell.

Ferguson, J., & Gupta, A. (2002). Spatializing states: Toward an ethnography of neoliberal governmentality. *American Ethnologist, 29*(4), 981–1002.

Freddie Mac. (2002, April). *Opening doors for Muslim families in America*. Freddie Mac.
FSA (Financial Services Authority). (2007). *Islamic finance in the UK: Regulation and challenges*. London: Financial Services Authority.
Hall, S. (2011). Geographies of money and finance II: Financialization and financial subjects. *Progress in Human Geography, 36*(3), 403–411.
Her Majesty's Treasury. (2008). *The development of Islamic finance in the UK: The government's perspective*. London: Her Majesty's Treasury.
Her Majesty's Treasury and FSA. (2008). *Consultation on the legislative framework for the regulation of alternative finance investment bonds (sukuk)*. London: Her Majesty's Treasury.
HRH The Prince of Wales. (2004, 19 May). Speech at the Islamic Financial Services Industry and Global Regulatory Environment Summit, London. Available on-line at www.princeofwales.gov.uk/speechesandarticles/a_speech_by_hrh_the_prince_of_wales_at_the_islamic_financial_85.html
Huntington, S. (1993). The clash of civilizations. *Foreign Policy, 72*(3), 22–49.
Ilias, S. (2010). *Islamic finance: Overview and policy concerns*. Congressional Research Service. Available at www.fas.org/sgp/crs/misc/RS22931.pdf
Islamic Financial Services London. (2010). Islamic finance. Available at http://www.thecityuk.com/assets/Uploads/Islamic-finance-2010.pdf. Accessed 20 July 2011.
Karim, W. J. (2010). The economic crisis, capitalism and Islam: The making of a new economic order? *Globalizations, 7*(1–2), 105–125.
Kuran, T. (1997). Genesis of Islamic economics: A chapter in the politics of Muslim identity. *Social Research, 64*(2), 301–338.
Larner, W. (2003). Neoliberalism? *Environment and Planning D: Society and Space, 21*, 509–512.
Leitner, H., Peck, J., & Sheppard, E. (2007). *Contesting neoliberalism: Urban frontiers*. New York: Guilford Publications.
La Tribune. (2012, April 18). La finance islamique fait un nouveau pas en France avec une assurance-vie. Available at http://www.latribune.fr/entreprises-finance/banques-finance/industrie-financiere/20120418trib000694127/la-finance-islamique-fait-un-nouveau-pas-en-france-avec-une-assurance-vie.html. Accessed 16 Oct 2014.
Lockie, S., & Higgins, V. (2007). Roll-out neoliberalism and hybrid practices of regulation in Australian agri-environmental governance. *Journal of Rural Studies, 23*(1), 1–11.
Maurer, B. (2005). *Mutual life, limited: Islamic banking, alternative currencies, lateral reason*. Princeton: Princeton University Press.
Maurer, B. (2006). *Pious property: Islamic mortgages in the United States*. New York: Russell Foundation.
McMillen, M. J. T. (2011). *Islamic capital markets: Market developments and conceptual evolution in the first thirteen years*. Paper available at http://papers.ssrn.com/sol3/papers.cfm?abstract_id=1781112
Nasr, S. H. (2006). *Islamic philosophy from its origin to the present: philosophy in the land of the prophecy*. Albany: SUNY Press.
Ong, A. (2006). *Neoliberalism as exception: Mutations in citizenship and sovereignty*. Durham: Duke University Press.
Overbeek, H. (Ed.). (2003). *The political economy of European Union unemployment*. London: Routledge. www.newyorkfed.org/newsevents/speeches/1996/ep960523.html
Patrikis, E. (1996). *Islamic finance in the United States: The regulatory framework*. Remarks by Ernest Patrikis, First Vice President, Federal Reserve Bank of New York, before the Islamic finance and investment conference, New York, 23 May. Federal Reserve Bank of New York.
Peck, J., & Tickell, A. (2002). Neoliberalizing space. *Antipode, 34*, 380–404.
Peck, J., & Tickell, A. (2007). Conceptualizing neoliberalism, thinking Thatcherism. In H. Leitner, J. Peck, & E. S. Sheppard (Eds.), *Contesting neoliberalism: Urban frontiers* (pp. 26–50). New York: Guildford Publications.
Pitluck, A. Z. (2012). Islamic banking and finance: Alternative or façade? In K. Knorr-Cetina & A. Preda (Eds.), *The Oxford handbook of the sociology of finance* (forthcoming). Oxford: Oxford University Press.

Pollard, J. S., & Samers, M. (2007). Islamic banking and finance: Postcolonial political economy and the decentring of economic geography. *Transactions of the Institute of British Geographers, 32*, 313–330.
Pollard, J. S., & Samers, M. (2013). Governing Islamic finance: Territory, agency, and the making of cosmopolitan financial geographies. *Annals of the Association of American Geographers, 103*, 710–726.
Rethel, L. (2009). Whose legitimacy? Islamic finance and the global financial order. *Review of International Political Economy, 18*(1), 75–98.
Reuters. (2008, February 8). U.S. Islamic finance market underserved. www.reuters.com/article/2008/02/08/us-islamic-summit-usa-idUSN0843029520080208
Reuters. (2011, January 18). US legal win to boost Islamic finance – lawyers. Available on-line at www.reuters.com/article/2011/01/18/lawsuit-islamic-idUSLDE70G17920110118
Rodinson, M. (1974). *Islam and capitalism*. London: Pantheon Books.
Rudnyckyj, D. (2009). Spiritual economies: Islam and neoliberalism in contemporary Indonesia. *Cultural Anthropology, 24*, 1104–1141.
Rutledge, W. (2005). *Regulation and supervision of Islamic banking in the United States*, Remarks given at the 2005 Arab Bankers Association of North America (ABANA) Conference on Islamic Finance: Players, Products & Innovations in New York City, April 19. Available at http://www.newyorkfed.org/newsevents/speeches/2005/rut050422.html. Accessed 16 Oct 2014.
Said, E. (1978). *Orientalism*. New York: Vintage Books.
Samers, M. (2012). Islamic housing finance. In S. J. Smith, M. Elsinga, L. Fox O'Mahony, O. S. Eng, S. Wachter, & A. B. Sanders (Eds.), *International Encyclopedia of housing and home* (Vol. 4, pp. 130–138). Oxford: Elsevier.
Taylor, J. M. (2003). Islamic banking – The feasibility of establishing an Islamic bank in the United States. *American Business Law Journal, 40*(2), 385–414.
Tripp, C. (2006). *Islam and the moral economy*. Cambridge: Cambridge University Press.
Turner, B. S. (1974). Islam, capitalism and the Weber. *British Journal of Sociology, 25*(2), 230–243.
Turner, B. S. (2010). Revisiting Weber and Islam. *British Journal of Sociology, 61*(Suppl s1), 161–166.
UKIFS (UK Islamic Finance Secretariat). (2012). *Islamic Finance 2012*. London: The City UK/ UKIFS.
Warde, I. (2000). *Islamic finance in the global economy*. Edinburgh: Edinburgh University Press.
Warde, I. (2008). Interview with Ibrahim Warde, in Akhtar et al. (2008) Understanding Islamic finance: Local innovation and global integration. *Asia Policy, 6*, 1–14.
Warde, I. (2010). *Islamic finance in the global economy* (2nd ed.). Edinburgh: Edinburgh University Press.
Wilson, R. (1998). Islam and Malaysia's economic development. *Journal of Islamic Studies, 9*(2), 259–276.
Wilson, R. (2000) Challenges and opportunities for Islamic banking and finance in the west: The United Kingdom experience. *Islamic Economic Studies, 7*(1 & 2) (1999 and 2000), 35–59.
Wilson, R. (2006). Islam and business. *Thunderbird International Business Review, 48*(1), 109–123.
Wilson, R. (2011). *The determinants of Islamic financial development and the constraints on its growth*. 4th IFSB lecture on financial policy and stability, Amman, Jordan. Available at www.ifsb.org/docs/FINAL_-_ISFB_4th_Public_Lecture_Rodney_Wilson.pdf

Chapter 63
Pious Merchants as Missionaries and the Diffusion of Religions in Indonesia

Chad F. Emmett

63.1 Introduction

The varied religious landscape of Java is a testament to Indonesia's rich religious history. On the plains of Central Java that surround tempestuous Mount Merapi, the magnificent Hindu temple of Prambanan and Buddhist temple of Borobudur stand as silent reminders of the first post-animist religions to reach the fabled Spice Islands. On the northern coast of Java, the Grand Mosque at Demak and the Protestant church in old town Semarang represent some of the first places of worship established for Muslims and Christians. In Surakarta (Solo), a much later addition to the landscape is the first chapel of The Church of Jesus Christ of Latter-day Saints (Mormon) to be built in Indonesia (Fig. 63.1).

From Hinduism to Mormonism, migrants and missionaries were certainly part of the process, but merchants also played a key role. Often times it was pious merchants, willing to share their beliefs, who first paved the way for later, more formalized forays from the likes of Brahmans, Sufis, and Jesuits. These merchants brought about religious change through example, teachings, marriage and perceived or real opportunities. Their efforts, intentional or not, were not centrally coordinated. They seemed to just happen as part of the trading process. Not all merchants were missionary minded. In some cases the bad examples of certain traders proved to be a negative influence which turned locals away for the incoming religion.

A merchant is generally defined as a person involved in trade or commerce. For most of the early merchants, that trade was primarily in spices, followed later by other commodities, such as tea, coffee, and rubber. In the case of this article, and in particular with the late coming Mormons, that trade will also include intangible commodities of expertise, education and entertainment.

C.F. Emmett (✉)
Department of Geography, Brigham Young University, Provo, UT 84602, USA
e-mail: chad_emmett@byu.edu

S.D. Brunn (ed.), *The Changing World Religion Map*,
DOI 10.1007/978-94-017-9376-6_63

Fig. 63.1 In 1986, after years of meeting in home-churches, the first LDS Church in Indonesia was built in Surakarta. This chapel houses two of the four LDS congregations (wards) of Solo (Photo by Chad F. Emmett)

Indonesia is noted as the country in the world with the largest population of Muslims. While Muslims make up approximately 85 % of the population, Catholicism, Protestantism, Hinduism, Buddhism and Confucianism are also all officially recognized religions.[1] The coming of these religions to Indonesia provides an excellent example of the lasting impact of the various types of religious diffusion. Relocation diffusion involves "the planting of a religion in a new area" and is usually undertaken with "deliberate intent" by either migrants seeking religious freedom or missionaries seeking converts. In other instances *relocation diffusion* is not deliberate and can happen "incidental to the purpose of the move." It can also be temporary (Park 1994: 138–142). Merchant missionaries in Indonesia most often fall in the non-deliberate category. Diffusion can also occur via *expansion diffusion* in which a religion spreads within the same area. This can happen via *hierarchical diffusion*

[1] When Indonesia declared its independence in 1945 one faction wanted to create a state based on Islamic law. Another faction led by Sukarno and other nationalists was more pragmatic. They realized that because of the rich religious history of the islands (thanks to the many merchant missionaries), no one religion should dominate. This means, for example, that it is legal for Christian missionaries to teach Muslims and for Muslims to change religion. Legalities aside, religious minorities in Indonesia still have difficulties, including the silencing of Ahmadiyyas and the burning and bombing of Christian churches (Harsono 2012). Many Christian congregations are finding it difficult to gain permission to build new churches. In the Jakarta suburbs of Bekasi and Tanggerang, LDS congregations, like many other Christian congregations, have not been able to secure multi-level bureaucratic permissions to build a typical chapel and so the congregations meet in renovated store/homes (*ruko*) hidden in the remote reaches of strip malls (Emmett 2009).

where important rulers and leaders convert first or via *contagious* or *contact diffusion* where the process happens at a more intimate level via everyday contacts with family, friends and neighbors. Conversion through intermarriage is a part of this process (Park 1994: 100; Sopher 1967: 89).

This chapter will focus on the role played by merchants in spreading religions to Indonesia. Some of these merchants came for just a season, while others came to stay and to even marry. Most were not concerned about spreading religion, but it happened nonetheless through their example, their discussions and their everyday interactions. In some instances merchants were quite deliberate in their efforts to proselytize as they shared their beliefs with top leaders and next door neighbors.

Historical records, as noted below, are non-existent on the spread of Hinduism and Buddhism and very limited on the spread of Islam. Examples from more recent eras, particularly among Christians, give dialogue and description to a diffusion process that in many ways has probably not changed much over the centuries.

63.2 Hinduism and Buddhism

Little is known about the arrival of Hinduism to the archipelago of Indonesia other than the fact that it is "thought to have been introduced to Indonesia at the beginning of the first millennium through trade contacts with India" (Proudfoot 1998a: 42). It is a similar story for Buddhism, which spread throughout Southeast Asia alongside Hinduism. The first know recorded mention of Buddhist merchants in maritime Southeast Asia are found in fifth century inscriptions (Proudfoot 1998b: 50). The process by which these traders introduced Hinduism and Buddhism to Indonesia is not known.

In his book *The Indianized States of Southeast Asia*, G. Coedes (1968: 22) quotes Gabriel Ferrand's "hypothetical reconstruction" of how he thinks Java was Indianized. He explains:

> The true picture must have been something like this: two or three Indian vessels sailing together eventually arrived at Java. The newcomers established relations with the chiefs of the country, earning favor with them by means of presents, treatment of illnesses, and amulets. . . . The stranger must be or pass for a rich man, a healer, and a magician. No one could use such procedures better than an Indian. He would undoubtedly pass himself off as of royal or princely extraction, and his host could not help but be favorably impressed.

These migrants then learned the language and married the daughters of the chiefs. These wives would often convert and in doing so they "became the best propaganda agents for the new ideas and faith" (Coedes 1968: 22).

Coedes goes on to explain that this first stage of Indianization consisted of "individuals or corporate enterprises, peaceful in nature, without a preconceived plan, rather than massive immigration" (Coedes 1968: 23). These early migrants and merchants were then followed by higher cast Brahmans and Buddhist monks who were more direct in their teaching of religion.

63.3 Islam

Indonesian Historian M.C. Ricklefs (1998: 12–13) notes that the "origins of Indonesian Islam are much disputed, probably fruitlessly." Like the two religions it supplanted, little is known about the spread of Islam, particularly in the early stages. Ricklefs surmises that:

> The Islamisation of Indonesia took place over a long period of time, during which Indonesia was a major crossroads of international trade, it is probable that Muslims from many parts of the Islamic world were present and played a part in the spread of Islam.... The means by which Islam spread has similarly been the subject of much conjecture and scholarly argument. There can be no doubt that trade and traders were of central importance, for it was by this means that communication was established between Indonesia and the Islamic World.

Through trade, Indonesians came into contact with Muslims and for whatever reason decided to convert. Also through trade foreign Muslims (Arabs, Indians, Chinese etc.) settled in Indonesia and married local woman who also chose to convert (Ricklefs 1993: 3). In describing the process of Islam spreading throughout Southeast Asia, noted historian Marshall Hodgson (1974: 546) suggests that "[e]very merchant was a missionary."

An Indonesian high school history text describes the spread of Islam as taking place through trade, *dakwah* (religious proselytizing), intermarriage, art and architecture. In reference to trade it states:

> Because of the influence of the monsoons, traders who came to Indonesia had to wait for a while for the reversal of the winds. While waiting for the winds to change, the traders would live in the settlements. They would visit with the locals and exchange ideas. From these discussions the locals would come to know of the common practices including the ways in which the merchants worshipped. Through this process the locals learned about Islam and eventually some of them converted (Indratno et al. 2007: 28–29).

One contemporary Muslim scholar from Indonesia describes the process this way: "Indonesia was not conquered by Muslim armies for Islam; rather, it was won by the piety and good examples of immigrant scholars, traders, and Sufi masters" (Shihab 2006).

63.4 Christianity

Historical accounts offer greater details about how trade helped (and sometimes hindered) the process of Christianization. Most of the accounts focus on missionaries, with Spaniard Saint Francis Xavier (co-founder of the Jesuit order) being one of the most noted. Missionary-minded merchants and colonial administrators also helped in the process.

At the colonial administrative level, conversions happened at the clan level as a means to gain power over rival Muslim clans (Aritonang and Steenbrink 2008: 32) and at the individual level through gifts and in hopes of gaining greater opportunity

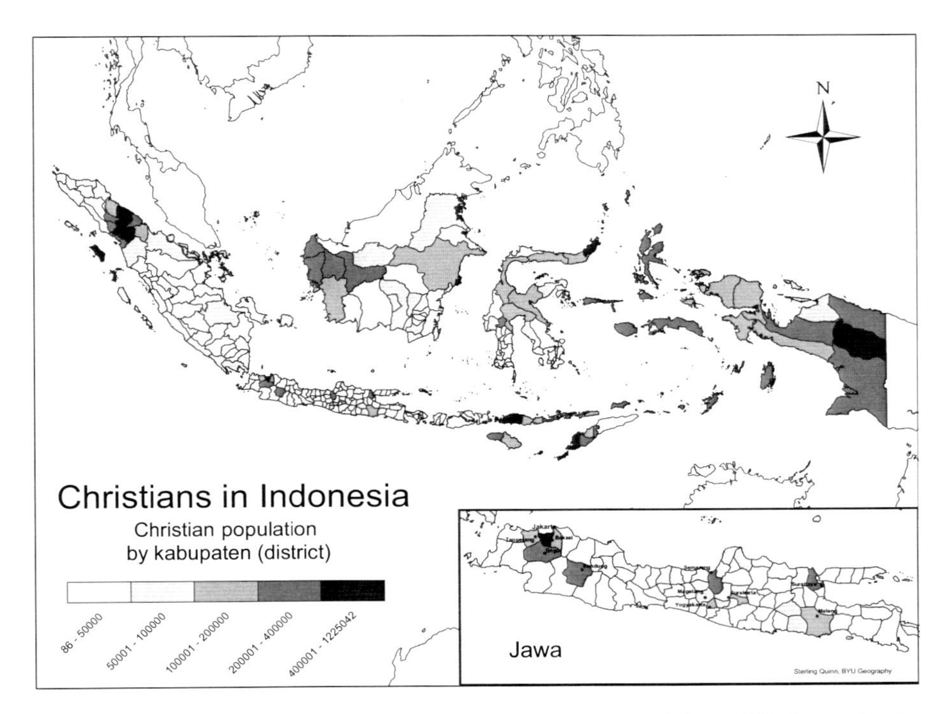

Fig. 63.2 Distribution of Christians in Indonesia (Map by Sterling Quinn, BYU Geography, for the author)

and access to the colonial economy (Aritonang and Steenbrink 2008: 107, 124). The "misconduct, tyranny, and corruption" of both Portuguese and Dutch colonial administrators also did much to stifle the spread of Christianity and in some instances resulted in many locals (re)turning to the greater stability and support of Islam (Aritonang and Steenbrink 2008: 29, 50).

The Portuguese had most of their success on the islands of eastern Indonesia where spices and sandalwood were plentiful and Islam was not as firmly planted. The lasting impact of these early conversions is readily seen in the contemporary distribution of Christians in Indonesia (Fig. 63.2). On the southeast islands of Solar, Timor and Flores, Portuguese merchants had plenty of time to interact with the natives while waiting for weeks or even months for a change in the monsoon winds to take them back to Malacca. On the sandalwood producing island of Solar, some of the visiting traders were described as "pious men who talked about their faith in Christ with the indigenous providers of wood" and thus helped to plant the seeds of Christianity (Aritonang and Steenbrink 2008: 74). Similar religious discussions occurred on other islands as traders awaited the monsoon winds. On Timor and neighboring islands, "a supposed chaplain to a merchant vessel, Fr. A. Taveira OP, is said to have baptized five thousand people" and on Flores a lay worker named João Soares converted around two hundred people (Aritonang and Steenbrink 2008: 74).

The Dutch also came to the East Indies to trade, but unlike the Portuguese and Spanish they had little initial interest in helping to facilitate the spread of Christianity. This was due in part to the fact that the East Indies was governed from 1602 to 1799 by the Dutch East Indies Company (VOC) which was singularly concerned with profitability and therefore only willing to allow missionary work to happen in areas where it would not challenge indigenous Muslims or threaten trade.

When conversions did happen, it was usually because it was good for trade. For example, in 1681 the VOC allied itself with local factions on the island of Roti and made it a supply base for surrounding islands as well as a source of slaves. In the eighteenth century the Rotinese began to take advantage of the VOC presence. As they interacted with the Dutch colonists "they slowly began adopting Christianity which offered higher social status, freedom from slavery and potential VOC favour." As more Rotinese accepted Christianity, the Dutch agreed to provide them with schooling, thus making the Rotinese "an educated elite in the region, which was to give them a leading regional role in the twentieth century" (Ricklefs 1993: 66).

The colonial policies of the Dutch government initially followed the model of the VOC and did not support Christianization—especially on agriculturally rich Java. Still during the nineteenth century, government economic and political policies unintentionally paved the way for the almost complete Christianization of Minhasa in the northeastern-most region of Sulawesi. The Spanish and then the Dutch were interested in this area because of its abundant rice production. Over time some coastal peoples adopted Christianity, while in the interior pagan beliefs prevailed. As the Dutch became increasingly involved in the region, they helped create a local elite that stifled any opportunities for social mobility. They discouraged local pagan ceremonies which made certain ways of gaining prestige irrelevant and they relocated villages which upset ritual sites, and they implemented compulsory coffee cultivation. According to Henley (1996: 52), one of the effects of these Dutch activities in northern Sulawesi was that many Minahasans began "experiencing a transformation in the conditions of their lives which was beyond their control as individuals and which called most of their collective social and religious assumptions into question." These Dutch induced transformations prepared the way for missionaries and mission schools to bring about the "remarkable spread of Christianity among the Minahasans during the nineteenth century."

Eventually calls in Holland for changes in colonial polices led to a more supportive role from the Dutch government. Missionary work even opened up on Java in the mid-1800s, prompted in part by a more secured Dutch presence (solidified by putting down the Diponegoro rebellion) and by the fact that an indigenous Christian community had already emerged in eastern Java independent of any Dutch directed missionary influence and under the influence of missionary minded Euro-Javanese Christians. In other instances, conversion came through direct contact with Dutch colonists. The first Christian in Indramayu on Java was a Chinese man who converted after reading a New Testament that was given to him by a Dutchman. He then became the founder and leader of a small Christian congregation (Aritonang and Steenbrink 2008: 654).

Dutch women acting as lay evangelists had quite an impact on spreading Christianity in Central Java. On the coffee plantation that her husband administered near Salitiga, Elizabeth LeJolla taught the Javanese workers about Christianity. Her efforts led to baptisms and the founding of a congregation of 50 people. In Banyumas, Johanna van Oostrom, who owned a batik industry, held church services with her female employees which resulted in nine of them being baptized. Johanna Philips, whose husband was involved in the indigo business in Purworejo, also invited workers to learn about Christianity. Her efforts led to the baptism of least 1,000 Javanese between 1860 and 1870. All of these women were aided in their efforts to teach and baptize by Dutch missionaries as well as local Christians. Unlike the official missionaries, these women did not feel bound to uphold traditional Dutch views of Christianity and they did not look down on the syncretic form of Christianity practiced by Javanese converts. These lay workers laid the foundation for official Dutch and German missionaries who would come to Java over the next few decades (Aritonang and Steenbrink 2008: 670–671).

Another example of missionary-minded colonists comes from the village of Ganjuran near Jogjakarta in Central Java where the pious Catholic Schmutzer family owned a sugar plantation. After study in the Netherlands, where they had become advocates of the ethical policy which called for better treatment of Indonesians, the two sons, Joseph and Julius, succeeded their father in running the plantation in 1912. They put their beliefs in to action with improved working hours, salary increases, vacations, and retirement and health benefits for their many workers. At the time of these changes all of the employees on the plantation were Muslim. Over time the two brothers introduced a health clinic (with the help of Julius' wife, who was a nurse) that evolved into a hospital and they took control of 12 primary schools. Teachers at these schools were trained at a Catholic run school and through their efforts many of their students converted to Catholicism. When the Schmutzer family retired from the plantation in 1934 there were already 1,350 Catholics in the area (Aritonang and Steenbrink 2008: 702–703).

There were also come pious Catholic administrators of plantations in the Malang area who sought to teach their workers about Catholicism. One such man was A.W.C. Blijdenstein who was an administrator on a Chinese owned rubber plantation from 1911 to 1933. He organized a Catholic congregation on the plantation that upon his departure numbered 700 Catholic Javanese. He then returned to the Netherlands where he trained to become a priest (Aritonang and Steenbrink 2008: 722).

In Surabaya a German–born watchmaker named Johannes Emde was encouraged by missionary Joseph Kam to help spread Christianity to both the native and Eurasian population in Surabaya. Johannes and his Javanese wife began holding religious services in their home. His wife invited many natives to the services and soon there was a small group of converts (Aritonang and Steenbrink 2008: 712).

63.5 Mormonism

One of the last religious sects to enter Indonesia was that of The Church of Jesus Christ of Latter-day Saints, more commonly known as the Mormon or LDS Church. The Mormon Church has an active proselytizing program throughout the world.

Since 1970 volunteer Mormon missionaries have served in Indonesia (young men for 2 years, young women for 18 months) and there now are about 7,000 Indonesian Latter-day Saints. Like earlier religions, the work of these formally trained and formally tasked missionaries was enabled and assisted by pious Mormon merchants who through various means also sought to spread their faith.[2] Interviews with some of these lay Mormon evangelists reveal interesting details and patterns which often mirror the accounts of earlier missionary minded Christian merchants. How these merchants helped to spread their faith might also help to inform our understanding of how the un-recorded efforts of Hindu, Buddhists and Muslims merchants transpired.

Perhaps the first Mormons to unofficially spread their beliefs in Indonesia were Maxine and Pete Grimm. They met during WWII in the Philippines. After the war they married and settled down in Manila where he ran a shipping company throughout Southeast Asia—including Indonesia. Much of his travel, often accompanied by Maxine and their two children, was aboard their private yacht the Lanikai (Fig. 63.3). As a devout Mormon (Pete converted in the late 1960s), Maxine was always a missionary. While never formerly called or set apart as a missionary her decades of travel to the many islands of Indonesia gave her a unique opportunity to spread her faith in a land not yet formerly opened or dedicated for Mormon missionary work. Throughout the 1950s and 1960s Maxine carried boxes of the Book of Mormon in English and

Fig. 63.3 Pete and Maxine Grimm at the helm of the Lanikai (Photo provided by Maxine Grimm)

[2] Latter-days Saints take quite literally the command of Jesus to take the gospel to all nations, kindreds, tongues and people. They also follow contemporary counsel from church leaders encouraging every member of the church to be a missionary as part of their everyday interactions with others.

Dutch as well as other church pamphlets on the Lanikai to hand out to whomever she might meet. Many of her efforts were directed at a captive audience—the channel pilots who would come on board to help the Lanikai navigate through the remnant of WWII water mines in Indonesia's channels and harbors. During their time at the wheel, Maxine said she would "indoctrinate" the Indonesian pilots with teachings about the church followed by a gift of a Book of Mormon (Grimm 2007).

In her decades of travels she helped spread the gospel to most every island. Her journal notes that while in Surabaya she taught two Indonesians and then in nearby Madura they had Sunday School on the boat and taught two more Indonesians. On the mostly Muslim island of Ternate she met some Pentecostals who were very interested in what she taught. When they asked her what would happen to them if they joined the Mormon Church she told them that they "would have the Truth and would be most joyful."

In route to visit a manganese mine in 1968, Maxine remembers travelling with two Indonesian Muslims named Ali and Abdulla. She wrote in her journal that one morning after breakfast she taught them a lesson about Joseph Smith and the Book of Mormon followed by presenting them with their own copies of the Joseph Smith pamphlet and the Book of Mormon.

On one of their early stops (early 1950s) in Jakarta, the Grimms met the enterprising Jan Walandouw—a Christian from Northern Sulawesi and a close friend of President Sukarno. Their friendship grew over the years as the Grimms channeled most of their business dealings through him. Walandouw reciprocated by opening doors and by introducing the Grimms to many of Indonesia's important citizens. Their friendship was not all politics and economics; it also had a strong religious component. Of all the people Maxine met in Indonesia, Walandouw was the most receptive to her religious distributions and discussions. He often expressed a desire to be the first Indonesian baptized into the Mormon Church.

Walandouw, for unknown reasons, never was baptized, but he and the Grimms worked together to encourage Mormon leaders to consider sending full-time missionaries to Indonesia. Maxine knew several members of the church hierarchy and so she let them know that she had a well-connected Indonesian friend who was willing to help get government permission for the Mormon Church to be officially recognized. Church leaders liked what they heard and set out to gain such permission—with the help of Walandouw (Grimm 2007). In October of 1969 the Mormon Church was officially granted permission to bring in missionaries. The first six missionaries entered the country in January 1970. Several months later when these young men received notice from the government that their visas were being revoked, Walandouw stepped in and through his high ranking connections was able secure new visas for the missionaries (Storer 2001).

Maxine Grimm may have been the first Latter-day Saint to preach the gospel in Indonesia, but she was not the only one. There were several expatriate Mormons who lived in Indonesia in the late 1950s and 1960s and befriended Indonesians, held church meetings, shared their beliefs and established useful government contacts. Most of these early Mormons in Indonesia came as employees of the U.S. government and as advisors to a young country. Instead of trading in commodities, they traded in expertise.

In 1959 George and Afton Hansen moved to Jogjakarta where George, a geology professor at Brigham Young University, was assigned by USAID to set up a geology department at prestigious Gadjah Mada University (Jones 1981). The Hansens hired a young man named Sutrisno who offered to work around their house in exchange for room and board and the opportunity to go to school (Hansen 1973). Sutrisno remembers that while the Hansens did not talk a lot about their church or offer to teach him about it, they did teach him much through their actions—including their keeping the Sabbath Day holy and their abstinence from alcohol and tobacco.

Following their return to BYU, the Hansens regularly corresponded with Sutrisno and kept him supplied with copies of church magazines. Meanwhile, Sutrisno sought work in Jakarta where his Hansen-enhanced good command of English helped him to gain work at the Ford Foundation. While at the Ford Foundation Sutrisno learned of a new consultant from Utah named Perry Polson. Knowing that the Hansens were from Utah, he set out to see if Polson perhaps knew the Hansens. He was surprised to find out that the Hansens did indeed know Polson (who was a BYU business education professor) and that they had requested that Polson look up Sutrisno. That meeting turned into a close friendship, strengthened in part through Sutrisno's assignment to teach Indonesian to Polson and his wife Gwen in their home.

One Sunday in the fall of 1968, the Polsons invited Sutrisno and his wife to their home for what turned out to be a 10:00 am worship service. The Sutrisnos joined in and were impressed when a 5-year -old Bradley Butler stood up to bear his testimony in which he expressed love for his parents, belief in Jesus Christ and happiness for being a child of God. Sutrisno thought how happy he would be to hear his own child do such a thing (Sutrisno 1990).

With that beginning, the Sutrisnos began to meet weekly with the small group of Jakarta Latter-day Saints. In addition to the Polsons, the group included several other expatriate families living in Jakarta. Dennis and Vernene Butler had arrived in Jakarta in January 1968 where he was first officer in the Canadian embassy. A few months later Ludy and Toontje VanderHoeven arrived in Jakarta where he worked as an auditor for USAID.

This small group of expatriate Mormons met for Sunday worship services in the Butler home. Sutrisno became a regular at these meetings and expressed interest in learning more, so Brothers Butler and VanderHoeven began to visit him in his home for regular lessons. On June 1, 1969, after more than a decade of close interactions with Latter-day Saints, Sutrisno was baptized a member of The Church of Jesus Christ of Latter-day Saints by Ludy VanderHoeven. He was the first Indonesian to join the Mormon Church and it all happened through the informal efforts of expatriate Mormons working in Indonesia (Butler 2011; VanderHoeven 2011).

Another merchant-missionary during the beginning years of the LDS Church in Indonesia was Frits Willem Tessers. He was born in 1929 in Makassar Sulawesi. During the tumultuous years leading up to independence, Tessers served in the Dutch military where he trained pilots. Following Indonesian independence, Frits fled the country first to the Netherlands (where he married) and then to the United States where the Tessers family settled in California. There at the invitation of their son's fifth grade teacher the family agreed to learn about the Mormon Church. The family liked what they heard and decided to be baptized. Shortly after joining the Church in 1968, Frits started an import/export business with Indonesia. He looked forward to his first trip back to Indonesia in over 20 years as a great missionary opportunity to share his new found faith. In preparation for the trip, Frits filled up a large leather suitcase with copies of the Book of Mormon and stacks of various pamphlets about the LDS Church with the intent to share them throughout his several months stay. When he arrived in Jakarta, a customs agent, upon inspection of his bags, was concerned that the large collection of English language materials might be subversive or illegal and so Frits was taken into detention. Soon a senior military officer wearing dark glasses entered the room and exclaimed to Tessers: "I'm the first pilot you taught to fly!" He then gave Frits a big hug. After renewing acquaintances, he asked Frits about all of the printed materials. Frits explained: "These are all materials that will help people to understand the LDS religion." The reply satisfied the officer who then offered military transport to Frits during his business travels. The only report of what Frits did with his many copies of the Book of Mormon was that with the approval of his military friend he was able to place them in hotels along the way (Tessers 2012).

Merchants of music who never set foot in Indonesia also had a lasting impact on the growth of the LDS Church in that land. The harmonizing Osmond Brothers, from one of the most famous Mormon families, catapulted to pop rock fame in the early 1970s. During this period of worldwide Osmondmania, the LDS Church began a period of increased emphasis on missionary activity. In 1974, Church President Spencer W. Kimball proclaimed that every young man in the church "should fill a mission," not by compulsion, but by choice (Kimball 1974). For those Osmond brothers in their late teens and early twenties—the normal time for voluntary 2-year missionary service—this increased emphasis on missionary service caused considerable concern about whether or not they should serve. On his web site, in answer to a question about his decision not to serve a Mormon mission, Donny Osmond (who turned 19 in 1976) states: "My parents and church leaders at the time all believed that I was able to do much more good if I remained in the public eye, so to speak, and lived the standards of our religion." He also explained: "I do feel I have had a rare and wonderful opportunity to share my beliefs to the world because of the media exposure that has surrounded my career in show business. My family and I have received countless reports of fans who have investigated the church because of the life style and harmony that exists within our family" (Osmond 2011).

Fans in far-off Indonesia were some of those who came into contact with the Mormon Church through the example of the Osmonds. One such fan was teenager Aischa Meyer, the daughter of an Indonesian father and a German mother. One Christmas while visiting Germany she saw the Osmond Brothers performing on television. She liked their songs and thought that they were descent people—something not too common at that time in the world of rock-and-roll. Back in Jakarta she discovered that some of her friends also liked the Osmonds and so they organized an Osmond Fan Club (Fig. 63.4).

Fig. 63.4 Aischa Meyer Tandiman in 1974 in the room in her home where the Osmond Fan Club met (Photo provided by Aischa Tandiman)

One of the early members of the club was already attending free English classes taught by Mormon missionaries at the Mormon meeting house. One day she asked Asicha if she would be interested in attending Mormon Church meetings. Aischa agreed. During the meetings she met two young missionaries and they arranged to come to Aischa's home for a religious discussion. Aischa's mom and younger brother soon started attending the weekly lessons. When asked if they would be interested in being baptized, Mrs. Meyer explained that she would consider it only after a trip back to Germany where she intended to see how the Mormon Church functioned there. She liked what she saw and so upon her return to Jakarta she and her two teenage children were baptized in 1975 into the Mormon Church (Tandiman 2001).

Lydia (not her real name), another teenage girl living in Jakarta, tells the story of stopping one day at the stall of a roadside magazine seller and noticing a wholesome looking family on the cover of a magazine. She read the accompanying article about the Osmond family and liked what she learned. At the end of the article it noted that there was a fan club in Jakarta and so she decided to check it out. Meetings were held on Saturdays at the home of Aischa. Lydia joined the club and eventually Aischa asked her new friend if she would be interested in attending the Mormon Church—the church to which the Osmond's belonged. Lydia agreed. There she too met some young "elders" and they inquired if Lydia would be interested in learning more. With her parents' permission, they taught Lydia for several months. Eventually, through study and prayer, weekly attendance at church meetings, and developing strong friendships with other young Latter-day Saints, Lydia expressed a desire to be baptized.

A third young woman named Steffi Hetarihon also first came in contact with the LDS Church through the Osmond fan club. Once baptized, these three young women introduced other family members and friends to the LDS Church—many of whom also chose to be baptized. These three also all became involved with formalized missionary work in Indonesia. Two served 18-month missions in Indonesia at age 21 and then again for 3 or 4 years with their husbands who were called to serve as president of the LDS Indonesia Jakarta mission. One moved to Utah where for many years she taught Indonesian at the Mission Training Center to departing Mormon missionaries. These three Osmond fans have all raised families in the church and sent their children on missions—some to Indonesia and some to other countries of the world (Subandriyo 2012; Tandiman 2001).

In January of 1990, Mike Swenson and Steve Smoot, two LDS businessmen from Salt Lake City, travelled to Indonesia in hopes of finding a distributor for their artemia products. Artemia is a small cyst that when hatched is used as food for baby fish and shrimp. Once there they looked at import records and found the name of an Indonesian businessman, Rusdi Lioe (Leo), who had imported a variety of other aquaculture feeds and products. They tracked him down and arranged to meet with him in a hotel lobby. Rusdi assumed that like other international businessmen he had worked with, these two men would also want to smoke and drink and be introduced to local women. He did not like having to help with such entertainment and was, therefore, delighted to find out that neither of them was interested in such vices – they did not smoke, drink or cheat on their wives. In addition, he found that they were "humble, honest, friendly and always willing to help." To Rusdi it looked as if there was a good possibility to do business (Lioe 2011). Mike and Steve felt the same way. Out of the many distributors they met with in Indonesia, they were most impressed with Rusdi's company and so they all agreed to do business.

Both of the Mormon men had served 2 year missions at age 19, but now they were in their 30s and more focused on family and career than on preaching the gospel. What preaching they did do in their travels was mostly by example. As they met and travelled together over the years, Mike never "pushed the gospel" on Rusdi, but when asked about his religious beliefs he was always willing to share. After 3 years of business dealings and a growing friendship, Rusdi travelled to Salt Lake City to meet with his American, Mormon business partners. While there, Michael invited him to attend an LDS Church service and Rusdi agreed. There he met nice people and "felt the spirit" (Swenson 2011). Additionally, he finally realized that his fun-loving business associates (now including Mike's brother Doug) were sincere in their religious beliefs and that their profession of faith was followed up by good works. They lived what they believed. Before his return to Jakarta, Mike presented Rusdi with a Book of Mormon in both English and Indonesian and then contacted LDS Mission headquarters in Jakarta with a request that full-time missionaries visit the Lioe family. That visit was never made and so a year later, Rusdi contacted the mission office on his own to ask for the missionaries to come and visit. An hour later Mission President Subandriyo, made a personal visit to Rusdi in his office where he apologized for his missionaries forgetting to come the first time. Subandriyo then made sure that this time the missionaries made the visit. They did and were invited to return weekly for the next 6 months as the family learned about Mormon beliefs.

Mike was not aware that the Lioe family was having serious weekly discussions with the missionaries. Then one day in September 1996 he received a call from Rusdi telling him that the three oldest members of the Lioe family (father, mother and oldest son) were going to be baptized (three younger daughters would be baptized once they turned 8 years old) (Leo 2011). Michael asked when and Rusdi replied in a couple of days. Excited to hear the news, Mike and Bruce Sanders (another business partner) "straightway booked a flight and flew over to Indonesia just to baptize [them]" (Lioe 2011). They landed on the morning of the baptism and went straight to the service. They left that evening to return to Salt Lake because of pressing business matters but did not want to miss the baptism of this great family (Swenson 2011). Bobby, the eldest of the Leo children, served a 2-year mission for the LDS Church in Singapore and Malaysia. Two of his younger sisters are now attending LDS owned Brigham Young University with a third at the University of Utah (Fig. 63.5).

Hal Jensen is an LDS businessman who first began working in Indonesia in 1968. While never residing full time in the country, he maintains a second home there. Over the past four decades he has been involved with a variety of business ventures and projects in Indonesia's oil, airline, infrastructure, telecommunication, and education sectors. Many of his projects have been in cooperation with the Indonesian government.

In the mid-1990s, Jensen was working on a project with the national sports ministry. One segment of the project involved a proposal for a scratch and win game included on tickets to sporting events. Jensen worried that such an activity would

Fig. 63.5 The Leo family (three daughters 3rd–5th from *right*, mother, son, father 7th–9th from *right*) in front of the Salt Lake LDS Temple with Steve (*far left*) and Doug Swenson (2nd from *right*), their wives and some of their children (Photo provided by Sheri Swenson)

not go over well with Indonesia's religious community and so he had the head of the sports ministry arrange for him to meet Abdurrahman Wahid aka Gus Dur, a nearly blind Muslim cleric who was the head of Nahdlatul Ulama (NU) – the country's largest Islamic organization, in hopes of obtaining religious approval for the scratch and win game.

In that meeting when Wahid offered Jensen some tea, Jensen gave his standard reply: "No thank you, I don't drink tea. I'm a Mormon." When Wahid heard that Jensen was Mormon he proceeded to tell Jensen all that he knew about the religion. For a Muslim from Indonesia his knowledge was extensive due to his very ecumenical view towards all religions. The Mormon and Muslim discussed religion for several hours and then Wahid invited Jensen back the next day for another lengthy visit. Wahid then insisted that Jensen come to visit him every time he was in Indonesia on business. From these initial meetings the two men developed a strong friendship based in part on their strong religious beliefs.

Following the fall of President Suharto, Wahid confided in Jensen of his desire to run for president. Jensen told him: "If you are going to be president, you're going to have to have your eyesight" and offered to fly Wahid to Salt Lake City's Moran Eye Center for an eye operation. Following the operation (which had only limited success) Wahid was able to meet with Gordon B. Hinckley, President of the LDS Church. At that time Hinckley gave Wahid a blessing in which he blessed Wahid with "vision to lead your nation." Following the blessing, Wahid told Pres. Hinckley: "when I am president we will invite your young men in white shirts to visit us." According to Hal Jensen this was not an invitation to have missionaries go teach Wahid, but rather an open invitation to have missionaries come into the country.

Wahid then returned to Indonesia where he jumped back into the ever surprising and often times troubled political landscape. After several months of political wrangling, Wahid emerged victorious as the first truly democratically elected president of the world's third largest democracy (Barton 2002).

Less than a month later, Wahid embarked on his first major international trip which included a stop in Salt Lake City for a follow up eye appointment (Barton 2002). During this visit he once again stayed in Hal Jensen's home—where he used part of his time to finalize the formation of his government. He was also able to meet for a second time with President Hinckley. During this visit Wahid offered a presidential invitation to President Hinckley to visit Indonesia as his official guest (Jensen 2003).

In January 2000 Gordon B. Hinckley became the first president of the Church to visit Indonesia. There he spoke at a gathering of 1,800 Latter-day Saints and he also met with President Wahid. President Hinckley had asked Mission President Subandriyo if he had any special requests for Wahid. Subandriyo, with a local missionary force of only a few dozen, requested foreign missionaries. The request was conveyed and surprisingly Wahid agreed. It took a while to negotiate the bureaucracy, but finally in early 2001, after a 20-year hiatus, a limited number of foreign Mormon missionaries were once again allowed to serve in Indonesia (Subandriyo 2002). One merchant turning down a cup of tea led to a significant boost in the missionary efforts of the Mormon Church in Indonesia.

63.6 Conclusion

These many examples from the diffusion of Christianity and, most specifically, Mormonism in Indonesia reveal interesting patterns and processes that when propelled back in time, might help to conceptualize how earlier merchants also helped in the spread of religions. For millennia, merchants have come to the Indonesian islands in search of wealth and employment. Some stayed for several months awaiting a reversal of the winds while others set up shop for years. More recently some have flow in for only a few days or weeks of business activity. Many of these merchants were religious and were willing to share their beliefs with others. Many felt compelled to do so because tenets of their faith encouraged lay members to evangelize. These efforts to spread the faith were most often non-formal, unplanned efforts of pious individuals who lived their religion and wanted to share it with others. Sometimes formal invitations were offered to study and learn, but in most instances it was the day-to-day practicing of one's religion that sparked an interest in others. While some of their contacts did chose to accept a new religion, there were many who did not. Once converted, the religion continued to spread through a process of contagious diffusion in which spouses, children, family and friends also became converted. The interactions of merchants with high ranking locals often helped provide the necessary connections and approvals for more formal missionary efforts.

Initially, most merchants were male and were unaccompanied by wives or female traders. More recently merchants were often married men and were accompanied by their wives—who in many cases were the main agents of religious conversion. The methods used included: intentional open dialogue, the good example of practicing the precepts of your religion, invitations to worship, the distribution of religious materials, and the hope of opportunities, perceived and real, for a better life provided through expanded opportunities facilitated by interaction with the merchant and acceptance of his/her religion.

The activities of merchants have diversified and changed over time. Originally it was spices and then other resources that brought the merchants. In more recent years those merchants have come to trade (both buying and selling) in a wider variety of commodities. They have also come offering expertise and training. In one instance merchants of music (the Osmonds) never set foot in Indonesia but modern technology and the media helped to disseminate their music and their family oriented lifestyle and beliefs to Indonesians.

It is highly unlikely that merchants of music from Hindu India or the Islamic Hadramawt helped in the spread of Hinduism and Islam to Indonesia, but it is very likely that, just like modern-day Mormons, the unrecorded efforts of pious merchants throughout the millennia have included living their religion, sharing their beliefs, distributing books, and making important friendships. Through these simple acts of religious devotion the religious landscape of Indonesia continues to evolve and change.

References

Aritonang, J., & Steenbrink, K. (Eds.). (2008). *A history of Christianity in Indonesia*. Leiden: Brill.

Barton, G. (2002). *Gus Dur: The authorized biography of Abdurrahman Wahid*. Sheffield: Equinox Publishing.

Butler, Dennis, & Vernene. Interview. March 10, 2011.

Coedes, G. (1968). *The Indianized states of southeast Asia*. Honolulu: The University Press of Hawaii.

Emmett, C. (2009). The siting of churches and mosques as an indicator of Christian-Muslim Relations. *Islam and Christian-Muslim Relations, 20*(4), 451–476.

Grimm, Maxine Tate. Interviews. February 28, 2007; October 26, 2007.

Hansen, A. H. (1973). *Under banyan trees in Indonesia*. (self-published).

Harsono, A. (2012, May 22). No model for Muslim democracy. *The New York Times*, A23.

Henley, D. (1996). *Nationalism and regionalism in a colonial context: Minahasa in the Dutch East Indies*. Leiden: KITLV Press.

Hodgson, M. (1974). *The venture of Islam, II*. Chicago: The University of Chicago Press.

Indratno, F., Sumardianta, J., Angkasa, I., & Purwanta, H. (2007). *Sejarah untuk SMA/MA Kelas XI IPA*. Jakarta: Grasindo.

Jensen, Hal. Interview. June 25, 2003.

Jones, G. N. (1981). *Spreading the gospel in Indonesia: A Jonah and a contagion*. Unpublished manuscript.

Kimball, S. W. (1974). Planning for a full and abundant life. *Ensign*. www.lds.org/ensign/1974/05/planning-for-a-full-and-abundant-life?lang=eng. Accessed 18 May 2012,

Leo, K. (2011). *Conversion story*. Unpublished manuscript.

Lioe, Rusdi Djamil. Interview. December 9, 2011.

Osmond, D. *Did you ever serve a mission?* www.donny.com/question/missions_and_missionaries-questions_about_jesus_christ-questions_and_answers/did-you-ever-serve-a-mission-and-were-you-born-in-the-church. Accessed 15 Nov 2011.

Park, C. C. (1994). *Sacred worlds: An introduction to geography and religion*. London: Routledge.

Proudfoot, I. (1998a). Historical foundations of Hinduism. In J. Fox (Ed.), *Religion and ritual* (pp. 42–43). Singapore: Archipelago Press.

Proudfoot, I. (1998b). Historical foundations of Buddhism. In J. Fox (Ed.), *Religion and ritual* (pp. 50–51). Singapore: Archipelago Press.

Ricklefs, M. C. (1993). *A history of modern Indonesia since c. 1300*. Stanford: Stanford University Press.

Ricklefs, M. C. (1998). Early history of Islam. In J. Fox (Ed.), *Religion and ritual* (pp. 12–13). Singapore: Archipelago Press.

Shihab, A. (2006, October 10). *Building bridges to harmony though understanding*. Forum address at Brigham Young University. http://speeches.byu.edu/reader/reader.php?id=11324. Accessed 27 May 2012.

Sopher, D. E. (1967). *Geography of religions*. Englewood Cliffs: Prentice-Hall.

Storer, Dale. Interview. April 6, 2001.

Subandriyo, Steffi Hetarihon. January 20, 2012 e-mail correspondence.

Subandriyo. Interview. June 8, 2002, Jakarta.

Sutrisno. (1990, April). Tanpa Saya, Gereja Dapat Terus Berkembang; Tanpa Gereja, Saya Tidak Dapat Berbahagia. *Terang OSZA* 34–37.

Swenson, Michael. Interview. December 9, 2011.

Tandiman, Aischa Meyer. Interview. May 2, 2001.

Tessers, Frits Rene. Interview. January 11, 2012.

VanderHoeven, Ludy, & Toontje. Interview. January 27, 2011.

Chapter 64
Tithes, Offerings and Sugar Beets: The Economic Logistics of the Church of Jesus Christ of Latter-Day Saints

J. Matthew Shumway

64.1 Introduction

The June 5, 2011 cover of *Newsweek Magazine* has a superimposed head of Mitt Romney on the body of what is obviously a Mormon missionary from the Tony award winning Broadway farce *The Book of Mormon*. The associated headline reads, "Mormons Rock!: They've conquered Broadway, talk radio, the U.S. Senate—and they may win the White House. Why Mitt Romney and six million Mormons have the secret to success" (Kirn 2011). The emergence of the Church of Jesus Christ of Latter-Day Saints (hereafter the Church) in the media stems from the growth of Church membership outside of the traditional Utah region and the increasing status of its members in politics, business, entertainment, and sports. While the increasing prominence of individual members attracts most of the media attention, it is the underlying growth of the Church that makes it significant.

Attempting to explain the current structure and organization of any entity, including one with a religious mission, requires the examination of three important components: (1) the nature of the entity; (2) its origin and history; and (3) its current orientation. This chapter will present information on the geographic representation of the underlying organizational structure that facilitates the growth and mission of the Church and discuss the Church's temporal affairs including financing of their temporal and spiritual operations through voluntary donations, investments and directly owned businesses.

J.M. Shumway (✉)
Department of Geography, Brigham Young University, Provo, UT 84602, USA
e-mail: jms7@byu.edu

S.D. Brunn (ed.), *The Changing World Religion Map*,
DOI 10.1007/978-94-017-9376-6_64

64.2 An Organized Religion: LDS Church Organizational Structure and Geography

64.2.1 Administrative Organization

Ecclesiastically the LDS Church is a hierarchy. The First Presidency, made up of the most senior apostle and his two councilors, constitutes the highest governing body of the Church (Fig. 64.1). Next in line is the Quorum of the Twelve Apostles, followed by various Quorums of the Seventy. The First Quorum of the

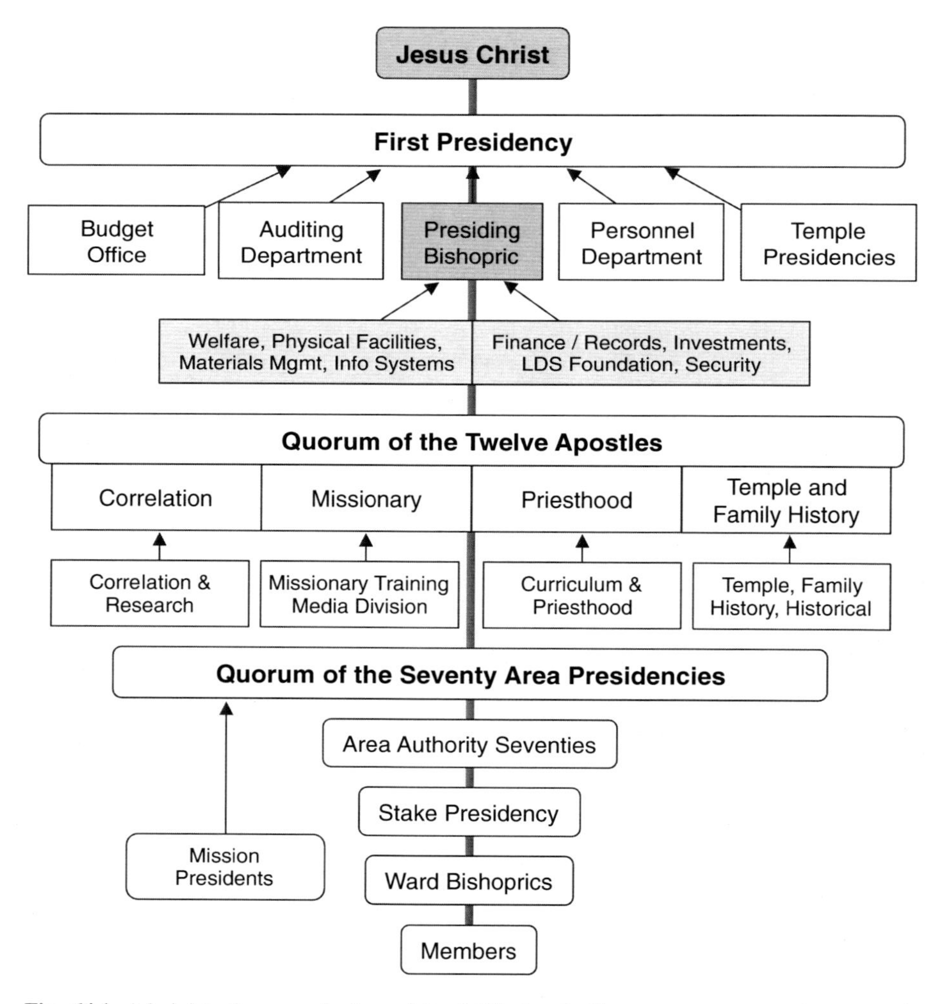

Fig. 64.1 Administrative organization of the LDS church (Source: www.mormonhaven.com/, used with permission)

Seventy has permanent membership and the other quorums consist of members serving approximately 5 years. Those holding these positions are commonly referred to as General Authorities, most of whom live and work at Church headquarters in Salt Lake City, Utah. Certain Seventies are assigned to live and work within a specific geographic region of the world. Additional ecclesiastical departments administer programs directed specifically towards women (Relief Society), youth ages 12–18 years (Young Men Young Women (YMYW) organization), and children ages 18 months – 11 years (Primary).

Local authorities are made up of a bishop, the leader of a local congregation called a ward, his councilors and the various local Relief Society, YMYW, and Primary leaders. A group of 5–12 wards make up what is called a Stake. "Stake" is taken from the Old Testament imagery of a tent (church) being held up by supporting stakes. Each stake has a Stake President with various councilors similar to other levels of the hierarchy. "Areas" are made up of a large number of stakes and are governed by Area Authorities. The number of Seventies and the number, size, composition, and geographic location of Area Authorities changes in accordance with areas of growth and decline in Church membership. For example, consolidation of Church units in Western Europe has taken place as growth in membership has either declined or remained steady. The number of congregations declined from 947 in 2000 to 846 in 2011 (Cumorah.com, 2012). On the other hand, some areas in Africa are experiencing rapid growth in membership. For example the number of members increased from approximately 9,000; 50,000; and 2,500 to 27,000; 100,000; and 10,000 in DR Congo, Nigeria, and Uganda respectively (Fig. 64.2).

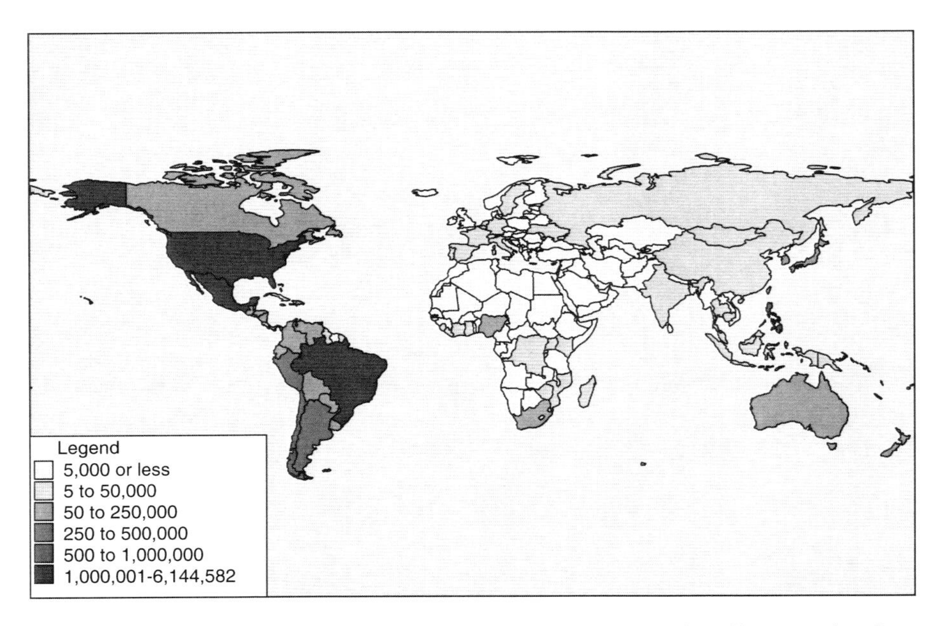

Fig. 64.2 LDS church membership, by country, 2011 (Map by J. Matthew Shumway, data from LDS Church Almanac, 2012)

Table 64.1 2011 LDS church statistics

Church units/organizations	Numbers
Total Church membership	14,441,346
Missions	340
Missionaries	55,410
Missionary training centers	15
Temples	136
Congregations	28,784
Universities & colleges	4
Countries receiving humanitarian aid (since 1985)	179
Welfare services missionaries (including humanitarian service missionaries)	9,251
Church materials languages	167

Data source: LDS Church

64.2.2 Geography of Church Members

As Table 64.1 indicates, Church membership at the end of 2011 was approximately 14.5 million organized into just under 29,000 congregations. Temples are arguably the best indicator of local membership strength, as the Church will not build a Temple without sufficient members and local leadership to patronize and staff it (discussed below). The highest concentrations of Church members and temples are in North, Central, and South America.

64.3 Church Growth and Organizational Transformation

The LDS Church was officially organized on April 6, 1830 in Palmyra, New York. It was not until 1947 that the Church counted one million members (117 years). In 1963 it reached the two million-member mark (16 years). The three million-member level was reached in 1971 – half that time (8 years). It took another 10 years to reach four million in 1980. Since 1980 the Church has added approximately three to four million members every decade (Fig. 64.3).

The location of Church membership has changed with its growth. In 1950 over 80 % of Church membership was located in the U.S. with about 80 % of those located in Utah and its surrounding states. In 2010, just over 40 % of Church membership is located in the U.S. with only about 50 % of members found in Utah and its surrounding states. However, the majority of members are still concentrated in the western part of the U.S. (Fig. 64.4). Table 64.2 shows the size and percent change in Church membership (1995–2010) as of 2010. If current patterns persist, we should see more members of the Church in Central and South America within the next 2–3 years.

Another indicator of the diffusion of the Church outside of Utah and the U.S. is seen in the growth of church meeting houses and temples. Meeting houses, which

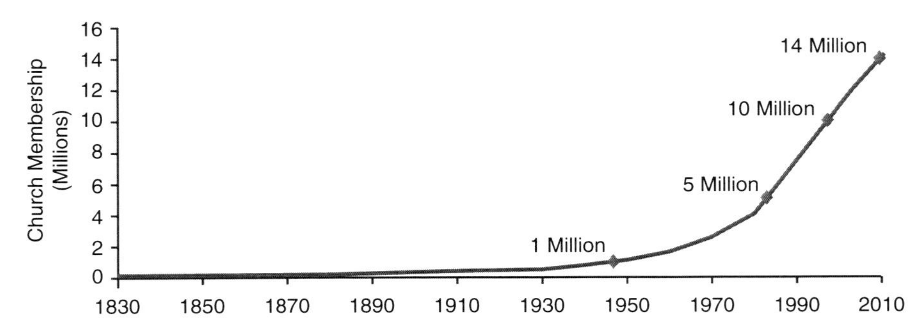

Fig. 64.3 Growth in church membership, 1830–2011 (Source: J. Matthew Shumway, using information on the Church of Jesus Christ of Latter Day Saints' website)

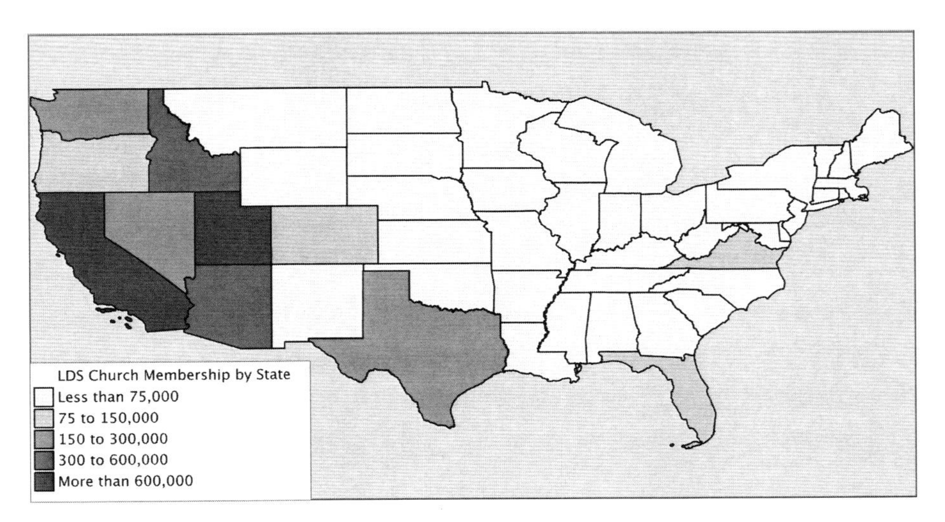

Fig. 64.4 LDS church membership, by state, 2011 (Map by J. Matthew Shumway, data from LDS Church Almanac, 2012)

Table 64.2 Geographic distribution of LDS church membership, by world region

Region	Current membership	Average annual percent change: 1995–2010 (%)	Number of missions	Number of temples
Africa	323,721	23	14	3
Asia and Pacific	1,467,794	21	53	20
Europe	493,418	19	56	13
Central and South America	5,600,686	14	126	42
U.S. and Canada	6,379,375	2	121	87

Data source: LDS Church

include a chapel, classrooms, offices, and often what is called a cultural hall (space for chapel overflow, cultural events, sporting and recreational activities, etc.), are used for weekly services and weekday/night activities. The Church finishes approximately 150 new meetinghouses a year, with the majority of these being built outside of North America (Uchtdorf 2011). Temples, viewed as sacred buildings in the LDS faith, are used by faithful members to make specific covenants with God and to participate in specific ordinances, including marriage. Temples are currently found throughout most world regions. The majority have been built in the last 20 years with approximately 55 % being built outside of the United States. Figure 64.5 shows the current spatial distribution of LDS temples across the globe with the highest concentrations in North, Central and South America. Figure 64.6 depicts the number of temples built by the Church over time.

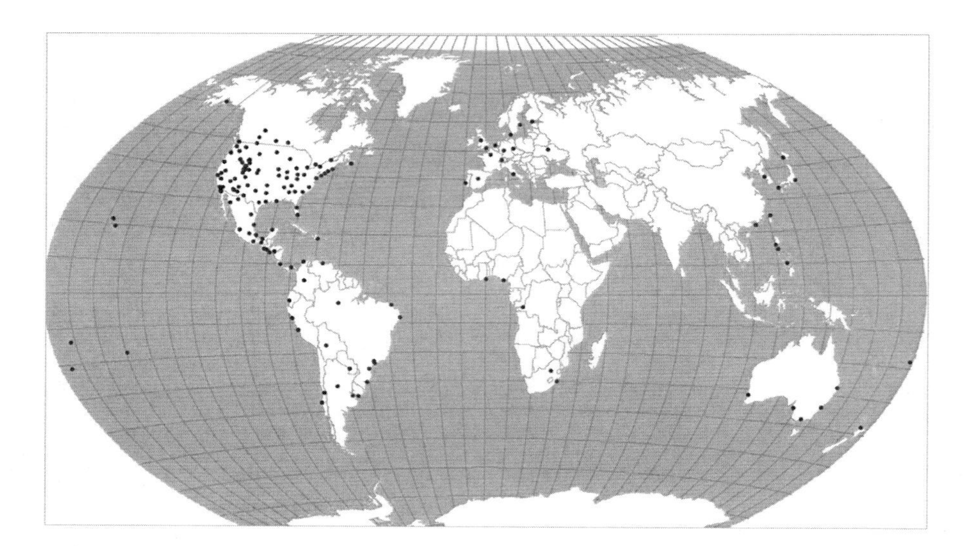

Fig. 64.5 Distribution of LDS temples, 2011 (Map by J. Matthew Shumway)

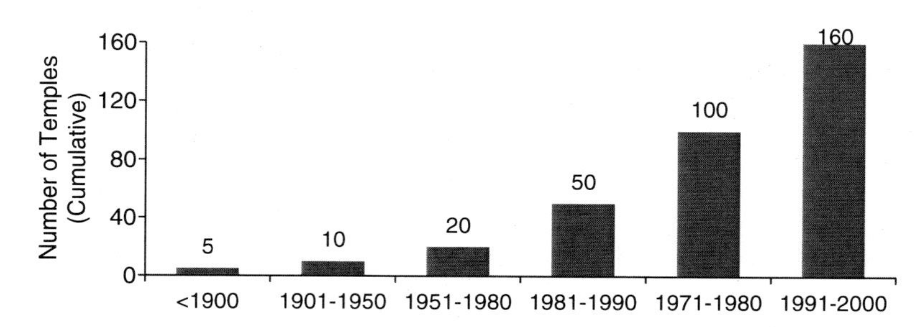

Fig. 64.6 Cumulative number of church temples, 1830–2011 (Source: J. Matthew Shumway, using information on the Church of Jesus Christ of Latter Day Saints' website)

64.4 The Changing Economic Foundations of the LDS Church

The economic history of the LDS Church can be disaggregated into three periods: the communal era, the tithing and fast-offering era; and the for-profit business era. These are overlapping arbitrary divisions as the transitions were/are neither abrupt nor all encompassing. Nevertheless, clear indicators of change are apparent and do mark some significant variations in approach.

The first era, starting with the formation of the Church but only fully realized in the early Utah period, was characterized by little to no separation between the Church and the local economy (Arrington 1958; Alexander 1991). This tight integration came about for two principle reasons, one spiritual and one practical. Spiritually, a theme found in the *Book of* Mormon (considered scripture by the Church) presents the effects of increased wealth and its concomitant increase in socioeconomic inequality as detrimental to the level of spirituality within the Church community (See *Book of Mormon* Alma 4:12, 3 Nephi 6:14). Therefore, the early Church sought to integrate its spiritual and temporal missions to minimize inequality between income and wealth and in so doing to relieve member suffering from temporal needs so that "things of the world" would not undermine the spiritual message and mission of the Church. This message included the practice of plural marriage. As a result, much of LDS society during this time was communal, directed by Church leaders, and designed to provide for the spiritual, social, and economic welfare of all members. The idea of social and economic equality, as embodied in the Church's *Law of Consecration and Stewardship*, was put into practice by asking members to live by a "United Order." Basically, the United Order was founded on collective and cooperative economic ownership and enterprise. All members' property and increase was given to the church, and in return the individuals/families had stewardship over these now Church-owned properties. Surplus was distributed among all the members. According to Arrington (1958), early Church practices developed as a consequence of the need to build a new social and economic order. Converts were strongly encouraged to leave their homes and "gather to Zion," where they would participate in the building of God's Kingdom here on earth, including living the United Order. The salvation of one's soul was as dependent on providing for the temporal needs of the community as it was on the spiritual message of Mormonism. Thus Arrington states:

> The establishment of God's Kingdom on earth ... required equal attention to the temporal and spiritual needs of man. Preaching and production work and worship, contemplation and cultivation–all were indispensible in the realization of the Kingdom. Indeed, the religious and economic aspects and problems of the individual and the group were viewed as incapable of disassociation .... Economics, and secular policy in general, were thus placed on par with–or incorporated in–religion. (Arrington 1958: 5)

In 1890 Church President Wilford Woodruff issued a "manifesto" ending the practice of plural marriage. The Church's communal view of social life and hostile attitudes towards outsiders along with the practice of plural marriage, versus the increasingly individual ethos and competition emerging in the U.S. prompted the U.S. government to enact laws designed to bring the Mormons in line with current social and economic norms. Foremost among these were the 1882 Edmunds and 1887 Edmunds-Tucker Acts which stripped the Church of its property, criminalized plural marriage and disenfranchised any members who practiced it (including most Church leadership), excluded all women from voting, took over the public school system in Utah, and abolished the Utah territorial militia (Alexander 1991). With the Church teetering on bankruptcy, adaptation to more secular social norms requiring the abandonment of the communal economy became the only way possible for continued survival as an ongoing religious organization (see Arrington 1958 for an in-depth discussion of this period). Ending plural marriage became necessary for the very survival of the Church organization and initiated the end of the communal period.

The second, more practical reason for the tight integration of the Church and economy during the communal era, was a means to provide important services that might not otherwise be available. The early Church was forced out of New York, Ohio, Missouri, and Illinois. Therefore the choice to settle in what was to become the state of Utah was based, in part, on its isolated location. This remoteness, together with the relatively small population, meant that the cost of many goods and services was prohibitive. The Church had to become self-sustaining. During the communal period, it was involved in many businesses not provided by private markets including news media, commercial agriculture such as sugar beets and sugar processing facilities, banking, health care, printing, retail cooperatives, and downtown accommodations (hotel and restaurants) for those visiting Salt Lake City. The Church was the only entity in the region that had enough capital on hand, or enough financial strength to borrow, and enough interest in serving the local market to engage in many of these businesses. It should be noted that at this time the Church viewed these businesses as an extension of the Church's mission to provide both spiritual and temporal salvation for its members. As a number of historians have pointed out, the Church did not see any difference between its temporal or worldly mission and its spiritual responsibilities (Arrington 1958; Alexander 1996; Quinn 1997).

Due to outside pressure from the U.S. government and the decreasing isolation with the arrival of the first telegraph line in 1861 and the transcontinental railroad in Utah in 1869, the Church became less directly involved in the creation and running of many of their businesses. They began to relinquish investments in and control of their unprofitable businesses, including major investments in sugar, rail, power generation, and manufacturing (Alexander 1996). The transition from a community based religious/economic/political entity where temporal and eternal salvation were one and the same, to a more simple religious organization with a focus on individual and not communal salvation was neither swift or easy and ended up taking decades. However, by the 1930s the transition was mostly complete (Alexander 1996). The economic entities that the Church continued to own and operate were those where the competition from private business was not as intense and/or in businesses that would help the Church deliver its spiritual message and defend the faith. Examples of the former included the Hotel Utah, and of the latter *The Deseret News* (and later media additions in radio and television). Perhaps the most important economic asset that the Church retained was agricultural land and land in and around Salt Lake City.

Ownership and control of land for agricultural production in rural areas and urban development in cities provided the foundation for the direction Church economics took in the next two eras. This transition from communal and community based economic organization to one of nascent assimilation into more modern economic relationships, characterizes the subsequent period of Church economic organization. Principle characteristics of this era included a complete abandonment of the United Order, an increased focus on financial commitments through tithing as the principle means of funding Church operations, and the development of fast-offerings and a new Church welfare system.

64.5 LDS Church Welfare: Principles and Organization

64.5.1 Principles and Policies

Transitioning from a policy of *Consecration and Stewardship*, to a lesser requirement of obedience to the *law of tithing* characterized the second period of the Church's economic organization. With the reality of governmental takeover accompanied by crippling debt (Alexander 1996), the Church had to find a way to fund its organization, its ongoing missionary efforts and to provide a spiritual purpose/identity that members could use to replace communal salvation. The answer was tithing for economic stability and a new welfare plan based on voluntary contributions by members associated with fasting.

Church president Lorenzo Snow (1898–1901) focused on members paying tithing. The "law of the tithe" was interpreted by the Church to be one-tenth of the increase or income of each member, and it was up to the member to decide whether or not their tithe was a full and honest tithe. It is important to point out that tithing has been, and still is, the principal means the Church uses to finance its operations and missions. Tithing funds are used primarily for building and maintaining meeting houses, temples, schools, and those facilities necessary for the continued growth and expansion of the Church and its mission to deliver its spiritual message. However, during this second era, tithing was not the only contribution required. Members were also asked to contribute resources (financial or in-kind) to the building and on-going maintenance of meetinghouses in their local areas, to any temples being built to which they were assigned, and to contribute to local congregational budgets necessary to run church auxiliaries. Although the Church no longer required a consecration of everything from members, they did require the members to shoulder a large proportion of the costs associated with the building and maintenance of local chapels, temples, and so forth.

64.5.2 *Fast Offerings*

The second most important financial component of the Church during this period was the increased emphasis on the payment of what is called "fast-offerings." The nature of the organization changed from one that would create spiritual and temporal self-sufficiency and salvation for the Church as a whole to one that would teach and help facilitate individual and familial self-sufficiency and salvation. In order to accomplish this mission the Church needed a strong, sound economic base outside of tithing and this is where fast-offerings came into the picture. Church members are encouraged to fast (go without food and water) one Sunday a month and donate the proceeds from those missed meals to the Church. These offerings are used to help the poor and needy. One Church leader stated:

> When we fast ... we feel hunger. And for a short time, we literally put ourselves in the position of the hungry and needy. As we do so, we have greater understanding of the deprivations they might feel. When we give to the bishop an offering to relieve the suffering of others, we not only do something sublime for others, but we do something wonderful for ourselves as well. (Wirthlin 2001)

Again, the Church is emphasizing the direct linkages between its spiritual mission and the temporal condition and needs of its members. This is a key element of the Church and, in fact, can be found in one book of their scriptures, the Doctrine and Covenants, where it states: "Wherefore, verily I say unto you that all things unto me are spiritual, and not at any time have I given unto you a law which was temporal…" (D&C 29:34). Quinn (1997: 136) quotes this scripture to argue that "Theologically, Mormonism has never accepted the "worldly" distinctions between secular versus religious, civil versus theocratic, mundane versus divine."

64.5.3 *A New Welfare Program*

While fasting and fast offerings were always part of the Church, it was not until 1936 that they received a renewed emphasis. Stung by the economic difficulties members were experiencing during the Great Depression, Church leaders created what is now known as the Church Welfare Program. The purpose of Church welfare is to teach self-reliance and provide goods to those who need them due to natural disasters, economic downturns and depressions or in times of family emergencies. Initially the production and processing of food, clothes, and other welfare items was mainly undertaken by stakes that had their own welfare farms or production facilities or at ward level by families who donated surpluses (Fig. 64.7). It was a decentralized system where local members responded to local needs through donations and by volunteering labor. While a decentralized approach was sufficient for identifying and satisfying local needs, there were areas of inefficiency. In locations where the Church was strong, wards and stakes over-produced many goods, but demand in outlying areas could not be met. In the 1980s, the Church began a consolidation and winnowing process in order to reap economies of scale and increase efficiencies

Fig. 64.7 LDS church dairy farm in Elberta, Utah (Photo by J. Matthew Shumway)

(Rudd 1995). Today most operations have a permanent staff of hired professionals supplemented with volunteers and members serving welfare missions (usually retired couples with some expertise in the product or process). The Church owns and operates welfare farms in the U.S., Canada, Mexico, South America (Chile, Argentina, Brazil), Europe, Australia, and Polynesia (Rudd 1995). Currently, over one million pounds (45 million kg) of grain, beans, meat, fruit, vegetables and other commodities are produced annually (LDS Church, Welfare Fact Sheet 2011).

64.5.4 Church Welfare Administrative Organization

All temporal affairs of the Church fall under the Presiding Bishopric Office (PBO), including Church Welfare (see Fig. 64.1). Church welfare is organized around three principal objectives: becoming self-reliant; caring for the poor and needy; and serving others. Relative to this era the most important elements are the creation of Deseret Industries, Church Employment Centers, and a network of Bishops' Storehouses.

64.5.5 Deseret Industries

Deseret Industries (DI) was originally organized during the Great Depression in the 1930s and modeled after Goodwill Industries (Rudd 1995). The basic premise of DI is the acceptance of donated goods, cleaning and reconditioning those goods, and them making them available for sale at DI stores. However, DI also provides

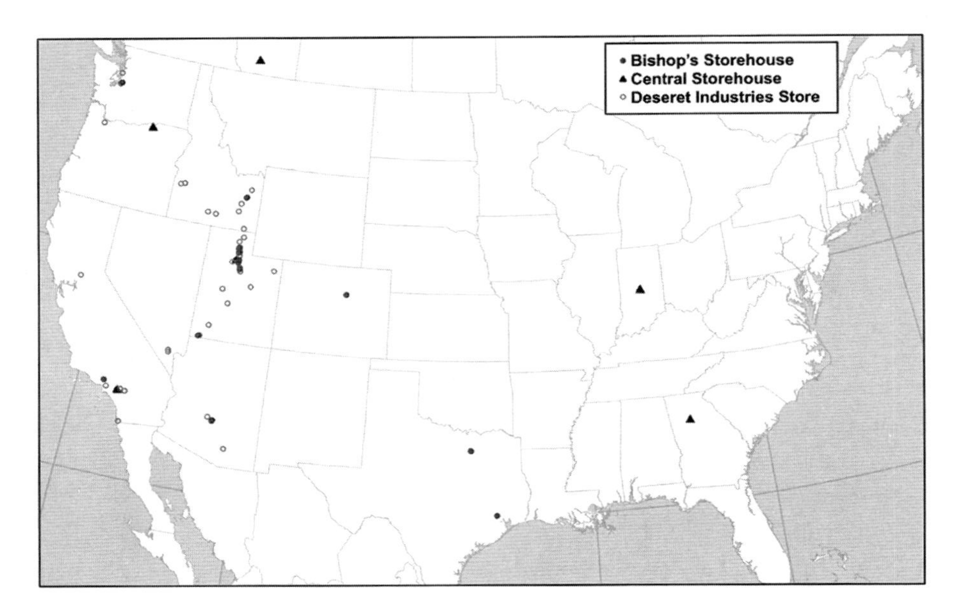

Fig. 64.8 Distribution of Deseret Industries Stores, Bishop's Storehouses, and Central Storehouses, 2011 (Map by J. Matthew Shumway)

on-the-job training in all aspects of its operations, career and technical training/education as part of the work experience – including teaching successful workplace behavior – and placing paid interns (paid by DI) with local businesses towards the goal of their eventual hire. Individuals must be recommended by either their Bishop if a Church member, or by local or state agencies if they are not. If there are openings available, then DI will generally accept most applicants. The on-the-job training is limited to a period of 18 months, which is usually sufficient for individuals to transition out of DI. Money earned through the stores is used to pay wages. As of 2011, there are 43 operating DI stores located primarily in the Western U.S. (Fig. 64.8). While DI stores and their associated Employment Services divisions are found primarily in the Western U.S., other LDS Employment Service Offices (LDSES) can be found throughout the world (Fig. 64.9). LDSES are designed to identify individual needs, assess skills, education, and knowledge, train and assist with obtaining employment, and provide an "office" address. According to Church compiled statistics, in 2011 there were just over 147,000 employment and training placements made by DI and LDSES (LDS Church, Welfare Fact Sheet 2011).

64.5.6 Bishops' Storehouse

Central to the operation of Church welfare is the Bishops' Storehouse. The Bishops' Storehouse is the physical manifestation of the donated fast offerings set apart to help the poor and the needy. Currently there are 142 Bishops' Storehouses located

Fig. 64.9 LDS Employment Resource Centers (Map by J. Matthew Shumway)

throughout the world, 33 of which are now located outside of North America. Storehouses are stocked with items found in most grocery stores with some variation in size and diversity of goods depending on locality. Service and/or welfare missionaries run each storehouse with volunteer help from Church members including those who have received Church welfare. Storehouse items come from three sources: those items that are produced by the Church, items bought from local retailers, and donated items. The Church is involved in the production and distribution of approximately 80 % of the goods in each storehouse. Products come from Church owned and operated farms and facilities that produce a variety of crops/products such as fruit, honey, dry beans, dairy, hay, sugar beets, raisins, peanuts, grains, pork, turkeys, and beef. These goods are then canned, cleaned, packaged or otherwise processed in Church owned facilities and shipped by the Church's Deseret Transportation fleet to six Bishops' central storehouses. These are located in Hermiston, Oregon; Mira Loma, California; Lethbridge, Alberta, Canada; Indianapolis, Indiana; Atlanta, Georgia; and the sixth and main storehouse located in Salt Lake City, Utah (Table 64.3 presents statistics on the storehouse in Salt Lake City).

If the Church does not produce the needed commodities, funds from fast offerings are used to purchase those necessary items from local markets. Donations are also accepted to help stock the storehouses. The central Bishops' Storehouses also house the Church's disaster relief materials – including the Church's emergency short-wave radio system, and between 10 and 20 "spearhead units" (tractor-trailers) that contain everything necessary to support 1,000 people for 48 h (Rudd 1995).

As would be expected, the geographic distribution of the Bishops' Storehouses essentially mirrors the geographic distribution of Church members (see Fig. 64.8). In areas with production and/or processing facilities, local congregations are given "welfare assignments" to provide service for help in the handling of storehouse

Table 64.3 Characteristics of the main Bishops' Storehouse in Salt Lake City, Utah

Characteristics	Statistics
Miles driven by Deseret Transportation drivers	3,500,000
Square feet at office and storehouse	535,966
Pallet capacity	65,000
Types of food and commodities	143
Employees	92
Trucks in and out daily	50
Loading docks	44
Full-time truck drivers	43
Acres at storehouse property	36
Regions that receive supplies from storehouse	5

Data source: LDS Church

commodities. For example, local stakes and wards where canneries are found (Utah (3), Idaho (2), Denver, Colorado; Mesa, Arizona; Houston, Texas; Indianapolis, Indiana; Atlanta, Georgia, and Cleveland, Ohio) will be asked to go to the cannery and spend at least 4 h helping to can commodities for delivery to central storehouses (Fig. 64.10). While the majority of the goods produced by the Church are for the Bishops' Storehouse system, members may patronize the canneries and home storage facilities to process materials for their own families. Outside of the U.S. the Church uses local fast offerings and funds sent from Church headquarters to support local welfare needs. Because of the lack of adequate Church infrastructure outside of the U.S., the Church simply provides funds in local currencies to purchase what local members need. If the Church is responding to a disaster or another type of humanitarian project, it usually works with local charities and international NGOs to deliver Church supplied commodities. For example, after the Indonesia earthquake in 2006, the Church sent $1.6 million worth of emergency supplies to devastated areas with Islamic Relief Worldwide providing the necessary transportation (Los Angeles Daily News, 2006, N3). Emergency relief for the Haiti earthquake consisted of two planes loaded with over 80,000 lb each of food and emergency resources such as tents, tarps, water filtration bottles and medical supplies donated by the Church. Airline Ambassadors and Food for the Poor provided the transportation (LDS Church 2010).

64.6 A Global Transformation

During the first two-thirds of the twentieth century the Church continued to struggle financially. Quinn (1997) confirms that the Church had as many yearly financial deficits as not up through the early to mid-1960s, and argues that most of the financial problems, such as running yearly deficits and thus having to borrow money, stemmed

Fig. 64.10 Bishop's Storehouse, Welfare Square, Salt Lake City, Utah (Photo by J. Matthew Shumway)

from poor financial choices made by Church leaders. This changed when members with more business experience were appointed to key leadership positions. By the mid-1970s the Church was on a more secure financial foundation, and in the mid to late 1980s the Church was financially secure enough to eliminate all required contributions by members except for tithing and fast offerings. Church headquarters presently pays for all programs and physical facilities in the Church. Voluntary contributions for fast offerings, temples, humanitarian relief, and other programs are encouraged and accepted, but not required.

Currently the Church is more financially secure than any time in its history due to the focus on eliminating debt, a conservative investment strategy, and the shift in involvement of Church leadership away from business operations to concentrating on ecclesiastical issues (Quinn 1997). This, along with the consolidation and centralized direction of all Church welfare farms, processing facilities, and so forth, are two of the indicators that the Church has entered a new economic phase. There are three principal elements of this new economic era: the aforementioned shift of how the Church pays for its operations; an increasingly global and humanitarian outlook; and a growth in Church owned for-profit businesses.

64.6.1 A Global and Humanitarian Focus

The primary reason for decreased Church leadership involvement in its financial affairs was the rapid growth of the Church from the 1970s forward, particularly growth outside of the U.S. in South and Central America, the Pacific Islands & Asia and now in Africa South of the Sahara. The majority of converts to the Church in these areas are substantially less well off financially than members in the U.S., which in turn has influenced Church policy. Most members in less industrialized countries do not have the financial capability to contribute to the same extent as those in more advanced economies. According to Quinn (1997) tithing funds originating in areas outside of the U.S. and Canada were and are not sufficient to support Church growth in those areas. Tithing and other funds collected outside of North America stay in the local areas and are supplemented by funds from Church headquarters. In other words, there has been and continues to be a net flow of money from Church headquarters in Salt Lake City, Utah to other areas of the world in order to build meetinghouses, temples, supplement local missionary programs, and to provide welfare services in these areas. According to one Church leader, there is one new church building completed every other workday somewhere in the world, all of which are paid for out of tithing funds (Uchtdorf 2011). Such international growth provides increased knowledge of and opportunities to participate in more ecumenical humanitarian and relief efforts outside of member oriented welfare programs.

One of the criticisms of the Church during the welfare period is that although the welfare system supported Church members, there appeared to be a lack of concern for those outside of the faith. This situation was understandable given the Church's lack of financial resources through most of its history and its limited ability to deal with its own growth (Quinn 1997). This has clearly changed – especially with regards to humanitarian projects and disaster relief. Dieter Uchtdorf, a member of the Church's First Presidency, stated:

> It is important to recognize that the growth of the Church is not merely about numbers, our mission is to bring souls unto Christ. … Divine leadership principles are based on the commandment that ye love one another, and that is irrespective of religion. By reemphasizing this commandment, the Savior has made feeding His sheep one of our ongoing responsibilities, which cannot be dismissed.

This clarifies the increased emphasis in the Church on humanitarian services to people not of the LDS faith. Geographically, this arm of Church welfare starts at the

Humanitarian Center located in Salt Lake City, Utah. Here all of the excess items donated to the Church are sorted, cleaned, repaired, and made available for distribution either to central Bishops' Storehouses, to other humanitarian or relief organizations, or to commercial companies that sell second-hand goods throughout the world. Where the Church has an established presence and the necessary infrastructure, they will respond directly through LDS Humanitarian Services and LDS Charities. Where they do not, they work with relief agencies, governments, and NGOs to identify specific needs, and transport and deliver disaster relief or humanitarian supplies, or provide money to local Church leaders and the other organizations to purchase the necessary goods and supplies.

Administratively Church welfare and humanitarian relief are under the direction of the Presiding Bishopric's Office. The PBO utilizes retired couples called to serve humanitarian missions to identify needs, support local Church leaders, coordinate with other relief organizations, and help to distribute relief and humanitarian supplies. Proselyting missionaries will also help in times of disaster and in the distribution of supplies when needed. Each area of the world has an Area Authority who must approve all requests for humanitarian help within their area.

Since the Church's humanitarian program was started in 1985, the Church has provided more than $1.4 billion in total assistance to individuals in 179 countries. Most of these goods originate at the Humanitarian Center, which prepares emergency relief supplies for shipment worldwide. On average, the Humanitarian Center ships about 12 million pounds of clothing, 1 million hygiene kits, and 1 million pounds of medical supplies to relieve suffering in more than 100 countries per year (LDS Church, Welfare Fact Sheet 2011). Humanitarian services also attempt to address some serious and more entrenched human needs. For example, Church humanitarian services are currently engaged in five global projects including neonatal resuscitation training, clean water projects, wheelchair distribution, vision treatment, and measles vaccinations. Church members often donate to the humanitarian fund at their own discretion, but when serious disasters occur the Church will often ask members for specific donations. The Church also receives donations from people and organizations around the world, all of which go directly to relief and humanitarian efforts (the Church covers all of the overhead).

64.6.2 For Prophet and Profit?

Arguably the most controversial aspect of current Church economic organization and practice is the growth of Church owned and operated for-profit businesses. In the 1970s historian Marvin S. Hill argued that the Church, as a body, views itself as surrounded by a hostile and threatening world, and given its early history, with good reason (Hill 1989). He went on to suggest that this view has had a profound effect on not only how the Church and its individual members view themselves, but in Church policies designed to protect itself against outside threats. Specific policies have adapted over time due to changing circumstances, but the overriding motivation for all of them, including the current practice of a large and diverse set of Church owned for-profit businesses, is to lesson Church dependence on the outside world (Quinn 1997).

In the late 1990s, TIME Magazine, in its article on the LDS Church entitled "Kingdom Come," reported that,

> The top beef ranch in the world is not the King Ranch in Texas. It is the Deseret Cattle & Citrus Ranch outside Orlando, Fla. It covers 771,000 acres (312,000 hectares); its value as real estate alone is estimated at $858 million. It is owned entirely by the Mormons. The largest producer of nuts in America, AgReserves, Inc., in Salt Lake City, is Mormon-owned. So are the Bonneville International Corp., the country's 14th largest radio chain, and the Beneficial Life Insurance Co., with assets of $1.6 billion.

As TIME rightly points out, the LDS Church is different because of how it invests its surplus:

> most of its money is not in bonds or stock in other peoples' companies but is invested directly in church-owned, for-profit concerns, the largest of which are in agribusiness, media, insurance, travel and real estate. (Van Biema et al. 1997)

However, as the Church has repeatedly pointed out, it sees its mission in both temporal and spiritual terms. The acquisition of wealth is pursued for the purpose of extending the gospel of Jesus Christ and improving the lives of both members and non-members of the Church. In addressing issues of Church owned business and its overall level of wealth, Gordon B. Hinckley, a high ranking Church leader gave the following statement:

> The Church does have substantial assets, for which we are grateful. These assets are primarily in buildings in more than eighty nations. They are in ward and stake meeting facilities. They are in schools and seminaries, colleges and institutes. They are in welfare projects. They are in mission homes and missionary training centers. They are in temples, of which we have substantially more than we have ever had in the past, and they are in genealogical facilities. But it should be recognized that all of these are money-consuming assets and not money-producing assets. They are expensive to build and maintain. They do not produce financial wealth, but they do help to produce and strengthen Latter-day Saints. They are only a means to an end. They are physical facilities to accommodate the programs of the Church in our great responsibility to teach the gospel to the world, to build faith and activity among the living membership, and to carry forward the compelling mandate of the Lord concerning the redemption of the dead. We have a few income-producing business properties, but the return from these would keep the Church going only for a very short time. Tithing is the Lord's law of finance. There is no other financial law like it. It is a principle given with a promise spoken by the Lord Himself for the blessing of His children. When all is said and done, the only real wealth of the Church is the faith of its people. (Hinckley 1985: 50)

Commercial businesses owned by the Church are controlled and coordinated by the Deseret Management Company (DMC). DMC is responsible for all of the finances of each of the separate businesses including profits/losses, auditing, producing financial statements, paying taxes, and reviewing business operations and plans. Companies founded under the DMC umbrella include Beneficial Life Insurance Company, Bonneville International Corporation, Deseret Book Company, Deseret News Publishing Company, Deseret Digital Media, KSL Broadcast Division,

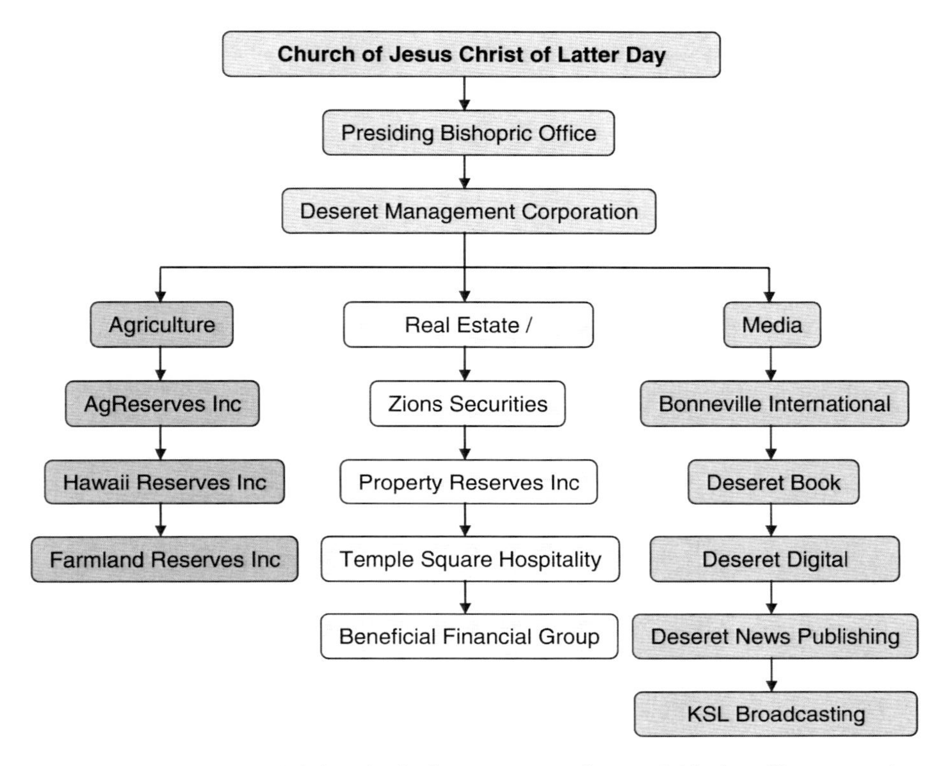

Fig. 64.11 LDS church administrative business structure (Source: J. Matthew Shumway, using information on the Church of Jesus Christ of Latter Day Saints' website)

Temple Square Hospitality, and Zions Securities Corporation (develops and manages Church property), AgReserves Inc., (manages agricultural properties in Utah, Washington, California, Montana, Wyoming, Kansas, Nebraska, Florida, and in Canada, Chile, Brazil, Argentina, and Australia), Laie Resorts, Inc., and the Polynesian Cultural Center in Hawaii. Figure 64.11 shows business under DMC.

Probably the most ambitious and highly criticized property development is the newly (2012) opened City Creek Center (managed by City Creek Reserve, Inc.) in downtown Salt Lake City. The City Creek project is a cooperative mixed-use residential and retail urban development project between the LDS Church, Salt Lake City and private businesses. According to the Downtown Alliance (entities involved in developing Salt Lake City), City Creek Center is one of the largest mixed-use developments in the U.S. It includes high-end retail, 700 residential units and 5000 underground parking stalls (Downtown Alliance 2012). The total cost of all projects combined is estimated at $5 billion with City Creek Center costing $1–1.5 billion – all provided by the Church's profits from its business, not from tithing or other donated funds from members (LDS Church 2006). While the Church has and will likely continue to be criticized for such a great expenditure on what is seen as business, the underlying purpose of City Creek and other Church owned and maintained property in Salt Lake City is similar to many policy decisions made by the Church over the years – to protect itself from outside forces. In other words, the Church wants to have as much control over its temporal destiny as possible and one way it can do this is through ownership and control of property around its worldwide headquarters.

Salt Lake City is the headquarters of the LDS Church and a part of its historical core identity. It is not in the best interest of the Church to allow the deterioration of downtown Salt Lake City, so it invests in property in and around downtown to ensure that the city remains viable (LDS Church 2006). In the late 1960s and early 1970s the Church recognized the importance of having a vibrant downtown area in and around Temple Square and Church headquarters and became more involved in Salt Lake City real estate. In the late 1960s the Church chose to lease to Salt Lake County, for one dollar per year, the property on which Symphony Hall and the Salt Palace Convention Center are now located (Brady 1992). In addition, income from Church business operations permits participation in local community causes without using the tithing of members from around the world. Church businesses provide charitable contributions to many community organizations in and around Salt Lake City for similar reasons. Through the LDS Foundation, the Church contributes to local hospitals, the Salvation Army, Saint Vincent De Paul Center for the homeless, the Salt Lake Homeless Shelter, the Utah Symphony, Ballet West, the United Way, and related cultural organizations.

64.7 An Organized Religion

The Church of Jesus Christ of Latter-day Saints is the physical and temporal manifestation of a spiritually conceived and based set of religious doctrines, values, beliefs, and practices. The Church can be thought of as the instrument used to sustain, maintain, share, and perpetuate its religious/spiritual beliefs. Without the organization, the ability to fulfill what it views as its divinely inspired message is difficult if not impossible to achieve. Nevertheless, like all increasingly complex organizations, once created with all of the rules – both spoken and unspoken – regulations and obligations, the organization itself can, and often does, subvert the purpose for which it was initially organized. Empie (1966) calls this "the dilemma of organized religion." As he says:

> The logistics of outreach and communication in the twentieth century involve planning, the training of specialized personnel, the use of printed and audio-visual media, financing, and the maintaining of administrative direction to such degrees that without organization and the creation of institutions, religious groups would face a hopeless task. Only through such instruments can they function effectively amid the power structures, which surround them. There is always the danger that organized religious groups, through power structures of their own, will lapse into the faults common to all institutionalized forms of society-indeed, this happens all too frequently. But the risk must be taken unless religious groups are willing to abandon a dynamic role and settle for futile navel-gazing and breast-beating. (Empie 1966: 364)

In order for organized religion to be successful it must have both a compelling message and the organizational structure necessary to sustain and perpetuate whatever that spiritual message may be. Obviously this involves money. While the good news is certainly free to all who wish to hear it, the capabilities associated with delivering the message, along with all of the other charitable and humanitarian acts that the Church engages in, are not free. All organized religions/churches must have some form of financial support in order to carry out their mission. Before nineteenth century America, funding typically came either directly or indirectly from the state. Olds (1994) stated that historically religion was viewed as a service industry with (mostly) positive externalities. Because these externalities accrue to members of society who may not be associated with a particular religion, religion was often viewed as a public good and supported by the state through various forms of taxation. This changed in the U.S. starting in the late 1700s as states began to disestablish their state-supported religions. In other words, religion was privatized. This privatization meant that churches could no longer rely on state supported tax funding and had to find new ways to finance their operations. This has been and still is an ongoing and important issue for all churches that are not supported by a state.

In order for a religious message to have resonance with as many people as possible, for the religion associated with the message to be a dynamic and ongoing concern, and for the religion to add adherents, an organization must be created, exist outside of state control and financed so that it can continue to exist. The LDS Church exists and is growing not simply because of its spiritual message, but because it has organized itself in a way that facilitates its continued expansion. Gottlieb and Wiley (1984: 65) make this same point:

> In Mormon terms the LDS church is not a commercial rather than a religious institution, but the LDS church is commercial because it is religious. Likewise, Mormonism's aims are not temporal rather than spiritual, but its aims are temporal because they are spiritual...these are money managers, but unlike any other kind of money managers ... the wealth and power, in the end, come down to the essentials: The church is in the business of expanding the church . . . a temporal structure whose major goal is spiritual—the building of the Kingdom of God on earth in preparation for the millennial reign of Jesus Christ.

With most of the growth of the Church occurring outside of the United States the Church was faced with the dilemma of providing the necessary infrastructure to support that growth (for example, meetinghouses, temples) from an increasingly poorer per capita membership and increasing their humanitarian and disaster relief efforts – something expected from a global religious body. Doing both is expensive and a drain on Church resources. However, they solved the dilemma through the use of tithing funds for Church building and maintenance; fast-offering funds, Church welfare farms, and donations to support the Church welfare program; and profits from Church-owned businesses to supply funds for humanitarian efforts, disaster relief, and stipends for general authorities. The current economic structure of the Church is an extension of the three principal threads running through its history: first, spread the gospel of Jesus Christ; second, to explicitly link together temporal and spiritual salvation; and third, to protect itself, as much as possible, from outside threats. Although exactly how the Church is organized both administratively and geographically changes over time, and will likely change in the future, the underlying patterns are more easily interpreted if those three threads are kept in mind.

References

Alexander, T. G. (1991). *Great Basin kingdom revisited: Contemporary perspectives*. Logan: Utah State University Press.

Alexander, T. G. (1996). *Mormonism in transition: A history of the Latter-day Saints, 1890–1930*. Urbana: University of Illinois Press.

Arrington, L. J. (1958). *Great Basin kingdom: An economic history of the Latter-day Saints, 1830–1900*. Cambridge, MA: Harvard University Press.

Brady, R. H. (1992). Church participation in business. In D. H. Ludlow (Ed.), *Encyclopedia of Mormonism* (p. 242). New York: Macmillan.

Church of Jesus Christ of Latter-day Saints. (2006). *News of the Church: City Creek*. https://www.lds.org/ensign/2006/12/news-of-the-church?lang=eng. Accessed 3 May 2012.

Church of Jesus Christ of Latter-day Saints. (2010). *Church sends additional aid to Haiti earthquake victims*. http://www.mormonnewsroom.org/article/church-sends-additional-aid-to-haiti-earthquake-victims. Accessed 24 May 2012.

Church of Jesus Christ of Latter-day Saints. (2011). *Welfare Services Fact Sheet*. http://www.lds.org/bc/content/shared/content/english/pdf/welfare/2011-welfare-services-fact-sheet.pdf. Accessed 1 May 2012.

Deseret News. (2012). *2012 Deseret News Church Almanac*. Salt Lake City: Deseret News.

Downtown Alliance. (2012). *City Creek. Downtown rising*. http://www.downtownrising.com/. Accessed 8 May 2012.

Empie, P. C. (1966). Can organized religion be unethical? *The Annals of the American Academy of Political and Social Science, 363*(1), 70–78.

Gottlieb, R., & Wiley, P. (1984). *America's saints: The rise of Mormon power*. New York: Harcourt Brace Jovanovich.

Hill, M. S. (1989). *Quest for refuge: The Mormon flight from American pluralism*. Salt Lake City: Signature Books.

Hinckley, G. B. (1985, November). Questions and answers. *Ensign*, pp. 49–52. http://www.lds.org/ensign/1985/11/questions-and-answers?lang=eng. Accessed 1 May 2012.

Kirn, W. (2011). Mormons Rock! *Newsweek*. http://www.thedailybeast.com/newsweek/2011/06/05/mormons-rock.print.html. Accessed 24 Apr 2012.

Los Angles Daily News. (2006, May 31). Relief effort for Java quake lagging. p. N3.

Olds, K. (1994). Privatizing the church: Disestablishment in Connecticut and Massachusetts. *Journal of Political Economy, 102*(2), 277–297.

Quinn, M. D. (1997). *The Mormon hierarchy: Extensions of power*. Salt Lake City: Signature Books.

Rudd, G. L. (1995). *Pure religion: The story of church welfare since 1930*. Salt Lake City: The Church of Jesus Christ of Latter-day Saints.

Uchtdorf, D. (2011, November). Providing in the Lord's way. *Ensign*, pp. 53–56.

Van Biema, D., Gwynne, S. C., & Ostling, R. N. (1997). Kingdom come. *Time Magazine*. http://www.time.com/time/magazine/0,9263,7601970804,00.html#ixzz1ub44hXzd. Accessed 4 May 2012.

Wirthlin, J. B. (2001, April). The law of the fast. *Ensign*, pp. 134–138.

Chapter 65
Entrepreneurial Spirituality and Community Outreach in African American Churches

James H. Johnson Jr. and Lori Carter-Edwards

65.1 Introduction

In the current era of political gridlock in Washington, DC and the defunding of entitlement programs at both the national and state levels of government, faith-based institutions will have to play a greater role in addressing pressing societal problems in the years ahead. To do so, church leaders will have to become far more creative and entrepreneurial in their outreach ministries to parishioners and communities.

In this chapter, we introduce the concept of *entrepreneurial spirituality* as an avenue for churches and other faith-based institutions to address unmet social and economic needs in our communities. We begin by situating this concept within the extant research literature on the link between spirituality and entrepreneurship and illustrating how it is being used today by the leaders of two mega-Black churches to foster and facilitate local community economic development. Next, we shift our attention to the issue of racial health disparities in America and present a concrete example of a church-based social entrepreneurial venture that is designed to reduce falls–related injuries and deaths among the black elderly. We conclude with a discussion of both the challenges and opportunities to scale entrepreneurial spirituality approaches to poverty alleviation, job creation, and community development in America.

J.H. Johnson Jr. (✉)
Kenan-Flagler Business School and Urban Investment Strategies Center,
University of North Carolina, Chapel Hill, NC 27599, USA
e-mail: johnsonj@kenan-flagler.unc.edu

L. Carter-Edwards
Gillings School of Global Public Health, Public Health Leadership Program,
University of North Carolina, Chapel Hill, NC 27599, USA
e-mail: lori_carter-edwards@unc.edu

S.D. Brunn (ed.), *The Changing World Religion Map*,
DOI 10.1007/978-94-017-9376-6_65

65.2 Background and Context

Considerable research exists on the link between entrepreneurship and spirituality. Much of it focuses on understanding how individuals integrate mind, body, and soul into their work (Kauanui et al. 2008, 2010; Kosmin et al. 2011; Fernando 2007). Several tenets run through this research.

One tenet is based on understanding how spirituality influences an entrepreneur's decision to initiate and maintain a business. No cohesive explanation has emerged from this research. A second tenet is based on understanding the entrepreneur and how spirituality impacts individual decision-making within firms. This body of research has identified a continuum of ways in which entrepreneurs incorporate spirituality into their decision-making. The continuum ranges from those who integrate business in their personal life by doing something that they love (the "make me whole" group) to those who keep business and their personal life separate (the "strictly business" group). Between these two extreme tenets are individuals who reportedly (a) use personal expression and self-motivation to express spirituality at work (the "soul seekers"), (b) perceive work and spiritual life as separate but also seek both internal and external motivators for success (the "conflicting goals" group), or (c) connect entrepreneurship on a limited basis and define success in both monetary and non-monetary terms (the "mostly business" group).

There is a third tenet in research on the connection between entrepreneurship and spirituality. It focuses on some of the highly visible, large scale examples of entrepreneurial spirituality, where Black clergy have successfully integrated entrepreneurship and spirituality to build high-impact outreach ministries. Two specific examples of faith-based entrepreneurial spirituality are highlighted here: the efforts of Bishop T.D. Jakes and the efforts of Kirbyjon Caldwell.

Bishop T.D. Jakes, considered one of the top ten religious leaders today, is the senior pastor of The Potter's House, a 30,000-member non-denominational ministry in Dallas, Texas. This megachurch is a place where multiple denominations – Baptists, Methodists, and Pentecostals – intersect (Pappu 2006). Jakes' journey into ministry was strongly influenced by the death of his father when he was 16, which created a need for a more fulfilling spiritual life, and by a set of financial challenges early in his adult life, which forced him to recognize the need to creatively diversify his income. Before launching his spiritual career, Jakes worked as a processor at a chemical plant, a position he acquired immediately after successfully completing his GED.

Jakes' early preaching on the issue of women's empowerment spurned a best-selling book, W*oman, Thou Art Loosed* (Jakes 2004), which helped catapult his financial success. Through his entrepreneurial and spiritual experiences, and his enormous ministry and outreach, The Potter's House sends volunteers to serve the elderly, conducts drug and alcohol counseling in disadvantaged communities, and works with victims of domestic abuse, just to name a few. Internationally, his church has brought water, medicine, and ministry to communities in Africa (PBS 2000). He is the author of over 20 books, fueling the significant financial success that simultaneously has increased his personal wealth and benefited the larger community. Bishop Jakes' experience reflects the influence of entrepreneurship on spirituality, where the "soul seeking" nature of his decision-making eventually transformed into a "mostly business" nature.

Kirbyjon Caldwell is the senior pastor of the 14,000-member Windsor Village, in Houston, Texas, the largest United Methodist Church in the U.S. He was raised in a middle-class neighborhood in Texas. After receiving a Master's in Business Administration from the University of Pennsylvania's Wharton School of Business, he worked as an investment banker in New York City. Although he was a successful Wall Street executive, Caldwell realized he wanted to go into the ministry and become a pastor. Thus, he abruptly quit his job on Wall Street and moved to Houston where he started Windsor Village with a small group of parishioners. Leveraging the skills acquired at Wharton and the experience and connections gained in corporate America, Caldwell has been able to successfully combine biblical and entrepreneurial principles to address the needs of the community. Along the way, Windsor Village's membership base grew rapidly and the church is recognized as a major change agent in the city of Houston.

Caldwell's best-selling book, *The Gospel of Good Success*, states that true success in work and life comes from combining spirituality with action (Caldwell 1999). In another book, *Entrepreneurial Faith: Launching Bold Initiatives to Expand God's Kingdom*, Caldwell et al. (2004) discuss the importance of using one's faith to innovatively and collaboratively invest in meeting the needs of the church. These principles are reflected in the outreach and community development ministries of Windsor Village.

Windsor Village's 42-acre all-purpose center houses a number of non-profit organizations, including a tutoring program, a program for abused children, and a matched mentoring program. To accommodate the growth of the congregation and to foster and facilitate Black community redevelopment and revitalization in Houston, Windsor Village renovated a former department store and created a "Power Center," which includes a school and medical clinic, a Women, Infant, and Children (WIC) nutrition program, AIDS outreach program, satellite classrooms for a local community college, and even a bank branch. Windsor Village also operates an independent living facility, for those ages 62 years and older, located next to a YMCA.

Through uniquely different paths, these two pastors demonstrate large scale success of integrating entrepreneurship and spirituality. They both continue to write and preach about their motivations to serve others by developing and implementing spiritually-led and entrepreneurially-oriented strategic plans. Key elements from these and other successful church-based entrepreneurial initiatives can be used to devise faith-based training programs in entrepreneurial spirituality.

Most Black clergy are not leaders of mega churches, and many have not had the training and/or do not have ready access to the appropriate resources to begin. But developing such a training program, as described below, will teach Black clergy how to forge the requisite strategic alliances with other key stakeholders in the communities they serve to generate the resources required to address unmet needs in their communities.

Below we describe a concrete example of a church-based social entrepreneurial venture designed to prevent fall-related injuries and deaths among the Black elderly. We also present the broad outlines of an entrepreneurial education and training program for Black clergy.

65.3 A Critical Area of Need

Despite the recently enacted Affordable Care Act, there is perhaps no area in which entrepreneurial spirituality strategies and approaches are more urgently needed today than in addressing the health status and challenges of one of the nation's most vulnerable populations, the Black elderly. One of the nation's most rapidly growing populations, research confirms that Black seniors have a disproportionately high prevalence and severity of many chronic diseases compared to other ethnic groups (Howard et al. 2007; Lapane and Resnik 2005; Hall et al. 2003; Hsu et al. 2003; Jackson 1988).

65.4 Demography and Health Status of the Black Elderly

The U.S. population is simultaneously growing older and becoming more racially diverse (Johnson and Kasarda 2011; Johnson et al. 1997; Johnson and Parnell 2013). Between 2000 and 2010, the senior population (15.1 %) grew more rapidly than the general population (9.7 %). In 2010, there were 40.3 million people who were age 65 years and older and 5.5 million who were 85 years or older (Werner 2011). Between now and 2050 there will be a 1.5-fold increase in the U.S. population, with a projected 2.4-fold increase in persons over 65 and 4.5-fold increase in the over 85 population (Census Bureau, May 14, 2009). This anticipated rapid growth of the U.S. senior population is commonly referred to as the "graying" of America (Johnson and Kasarda 2011; Johnson and Parnell 2013).

In conjunction with aging is the "browning" of America—an increasing racial and ethnic diversity of the population driven in part by heightened immigration from abroad and partly by below replacement level fertility among the non-Hispanic white population (Johnson et al. 1997; Johnson and Kasarda 2011). As a consequence of these demographic dynamics, the Black share of the U.S. elderly population is expected to steadily increase over the next four decades, reaching 12 % by 2050 – up from 9 % in 2010 (Vincent and Velkoff 2010).

Over the past quarter century, as the U.S. population has both aged and become more diverse, Blacks have experienced significant increases in educational attainment and income. They are also living longer today than 25 years ago. However, compared to the elderly population at large, disparities in both socioeconomic and health status persist (Appold et al. 2012).

In 2008, the Black elderly median household income ($35,025) was almost $10,000 less than the median for all elderly households ($44,188). With regard to poverty, the Black elderly poverty rate (20 %) was more than twice the poverty rate for all elderly (9.7 %). These disparities reflect a host of social ills in American society (housing and workplace discrimination, environmental injustice, etc.) which, not only significantly impact the incidence and prevalence of many diseases but, also create inequities in health awareness as well as in access to and utilization of health care services.

Compared to other race/ethnic groups, the Black elderly are at greater risk of disease, disease complications, and mortality, all of which significantly influence the cost and modes of care delivery (Somers 1978). More than 50 % of black elderly in America are in poor health. They have higher rates of multiple chronic illnesses, especially hypertension, obesity, and diabetes (Manton and Gu 2001; Jackson 1988). Untreated, or treated inconsistently, these chronic conditions can progress and lead to other health complications, including stroke, blindness, loss of an extremity, impaired mobility, kidney failure, heart disease, cancer, and dementia (Harper 1990; Jackson 1988; Baker 1988). Decreased self-image, self-worth, feelings of helplessness, hopelessness, and frustration often accompany these conditions. All of these disorders and medical conditions are typically associated with persistent poverty, long-term high-risk lifestyle factors, and inadequate coping or problem-solving skills (Jackson 1988).

For Black elderly, the high prevalence of chronic diseases is due in part to their lower engagement in healthy lifestyles (for example, low fat diets and exercise), limited access to care, lack of regular physician services, and the high cost of prescription medications. These problems are more common for the Black elderly who live in economically marginalized and socially isolated urban and rural communities where access to healthy foods and safe recreational environments are almost non-existent (Appold et al. 2012). It is particularly within these types of communities that we believe the Black church can play a greater role in health promotion and disease prevention (Petchers and Milligan 1998).

65.5 The Black Church: A Nucleus for Elderly Health Promotion

Rooted in the second Great Awakening (1790–1840s) (Foster and Dunnavant 2004), today's Black churches are the oldest organizations within most Black communities (Braithwaite et al. 2000; Banks 1999–2001; Battle 2006; Lincoln and Mamiya 1990) and represent one of the viable environmental settings for health promotion (Wimberly 2001; Resnicow et al. 2000). In a recent survey, approximately half of the Black respondents reported that they had attended church in the past week (Barna Group 2007). Further, research suggests that churches are often the first source of support for health promotion and disease prevention in low income and minority communities (Goldmon and Roberson 2004; Olson et al. 1988; Carter-Edwards et al. 2002; Maddox 2000; Mullen 1990). And, in nearly all Black communities, churches encourage and promote holistic health, often seen as places of comfort, guidance, and inspiration, offering a variety of resources that can be beneficial for health promotion efforts (Goldmon and Roberson 2004).

Given the relatively high rate of Black church attendance, especially among the Black elderly, and the emerging role of the Black church in promoting health, Black churches should be considered an integral part of the health promotion and care supply chain. In a 2008 report, *Retooling for an Aging America:* ***Building the Health Care Workforce***, the Institute of Medicine concluded that the "geriatric competence of virtually all members of the health care workforce needs to be improved through significant enhancements in the educational curricula and training programs" (IOM 2008). Since they not only set the tone but also are instrumental in the successful implementation of health programs, Black clergy should be among those receiving training. They require training on how to meet the individual needs of their elderly congregants through health counseling and on how to maximize positive health impact within as well as beyond their church walls.

Black churches can play a pivotal role in eliminating health disparities and promoting healthy lifestyles among the Black elderly. Studies suggest that church-based healthy lifestyle interventions are becoming increasingly popular and may be moderately effective (Physical Activity and Nutrition Branch 2010; Wilcox et al. 2007). However, two issues challenge the viability and sustainability of faith-based health and wellness programs.

First, evidence suggests that pastor or clergical support and involvement are critical determinants of the success of these interventions. But it is not clear how many pastors address physical wellness or how often this issue is addressed in church health programs and activities. Black clergy typically receive formal training to minister and manage the spiritual needs and routine life stressors that their parishioners face. But most ministers do not receive formal training in physical wellness and health promotion.

Second, the current economic recession—the worst since the Great Depression—has adversely affected the financial health of many churches. Parishioners' struggles with job loss, housing foreclosure, and other recession-related crises, have led to a decrease in church contributions and compromised endowments (where they exist). As such, financial support for a wide range of ministries is not as abundant today as in the past. In fact, church budgets can be in unstable flux, thereby further constrained (for example, adverse weather events that can and have forced churches to cancel Sunday services).

Despite these potential challenges, great benefits can accrue to the Black elderly from well-designed, faith-based health and wellness programs (Physical Activity and Nutrition Branch 2010). According to prior research, seven key elements are required to establish successful church-based community health programs: (1) partnerships; (2) positive health values; (3) availability of services; (4) access to church facilities; (5) community-focused interventions; (6) health behavior change; and (7) supportive social relationships (Petersen et al. 2002). To successfully establish and maintain these critical elements, particularly in today's resource-constrained environment, requires *lateral capacity building*, which involves investments in the training of church leaders and in facilitating strategic alliances with other key community stakeholders who can play a pivotal role in improving health promotion and disease prevention among the Black elderly (Fullan 2008; Christiansen 2009). Clergy whose churches, or groups of churches, already possess these attributes or successfully practice these types of efforts are engaging in what we will call "entrepreneurial spirituality."

65.6 A Faith-Based Entrepreneurial Venture

The clergy leadership of a Durham, NC-based Baptist church is drawing on the fundamental principles of entrepreneurial spirituality to launch a program called SAFE—Secure Audits for the Elderly—which is designed to reduce the incidence of fall-related injuries and deaths among seniors in its congregation and in the local communities surrounding the church. Research indicates that more than one-third of Americans over the age of 65 experiences falls—most often in their homes—each year. In 2005, nearly 16,000 seniors died from injuries sustained in such falls. An estimated 1.8 million seniors were treated in hospital emergency departments for non-fatal falls in 2005. Nearly a quarter were ultimately admitted to the hospital as patients (Stevens et al. 2006).

These falls impose a heavy financial burden on family and caretakers as well as hospitals, insurance companies, the government, and the victims, especially the uninsured and indigent (CDC 2013; Centers for Medicare and Medicaid Services 2009). Studies reveal that fatal falls cost the nation an estimated $179 million and non-fatal falls an estimated $19 billion in 2000 (Stevens et al. 2006). These direct, medical costs do not even reflect the enormous emotional, psychological, and physical health-related cost these catastrophic events impose on family and caregivers (CDC 2013).

If viable interventions are not developed, the problem will worsen and the cost will escalate in the years ahead. Over the next 20 years, seniors will be the most rapidly growing segment of our population. Constituting 12.5 % of the U.S. population today, the population 65 or older will comprise 20 % of the nation's population by 2030 (Vincent and Velkoff 2010). In North Carolina, the senior population is projected to reach 2.2 million by 2030, up from 12.1 % (1.1 million) in 2007. By 2020 the annual direct and indirect cost of fall-related injuries nationally is expected to reach $54.9 billion (in 2007 dollars) (CDC 2013).

Recognizing the severity of the problem, the Durham-based Baptist church leadership proposes to engage key stakeholders in the university community, government, and the private sector in mutually beneficial strategic alliances aimed at attacking the fall prevention problem (Fig. 65.1). More specifically, it will work with: (a) the Center for Aging and Health at the University of North Carolina at Chapel Hill to design a program to train and certify lay health professionals to conduct fall prevention education campaigns and senior home audits; (b) UNC-Chapel Hill's Kenan-Flagler Business School to handle strategy and marketing, program logistics, and evaluation; (c) the Employment Security Commission,

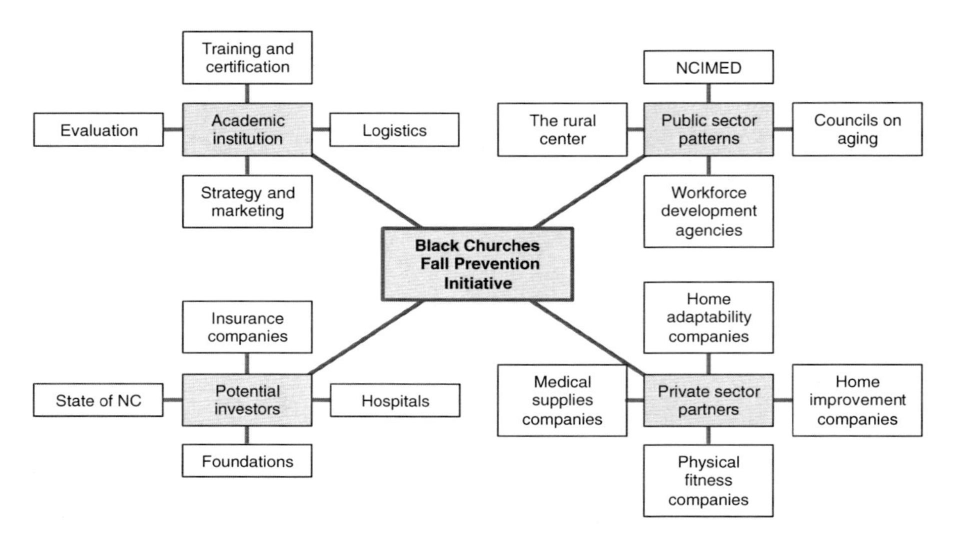

Fig. 65.1 Black churches fall prevention initiative strategic alliance partnerships (Source: James H. Johnson, Jr. and Lori Carter-Edwards)

other workforce development agencies, and The North Carolina Rural Center to recruit unemployed North Carolinians for slots in the lay health professional training program; and (d) North Carolina-based home construction professional associations and the North Carolina Institute for Minority Economic Development to establish standards as well as recruit and certify small construction contractors for SAFE senior home adaptability renovations.

The church also will work with the relevant government agencies, where appropriate, to leverage the Low Income Housing Tax Credit, the New Markets Tax Credit, the Renewable Energy Tax Credit, and the Historic Rehabilitation Tax Credit in the senior home adaptability process. To sustain SAFE operations, the church leadership will aggressively seek angel and equity investments from insurances companies, hospitals, home improvement and medical supply companies, and other businesses (for example,, appliance manufacturers and home security firms) that stand to benefit from its efforts to reduce falls among the state's elderly and to renovate their homes according to SAFE standards.

The church's SAFE program will contribute to longer term independent living among seniors by decreasing the incidence of falls, physical injuries, and injury-related deaths. It will reduce costs associated with fall-related hospitalizations and medical treatments as well as stress on family/caregivers. And it will create jobs and contribute to the growth and expansion of small businesses in the state (Johnson 2002).

The church will pilot test SAFE by recruiting the senior members of its 6,000 member congregation and enrolling them in the program. Upon successful completion of the pilot test, church leadership will devise a roll-out strategy for implementing the program statewide.

A statewide, faith-based, Black elderly-targeted health education and training program must be designed to achieve five objectives:

1. Highlight the current and projected health issues impacting the Black elderly.
2. Increase awareness of substantial benefits of Black-elderly targeted health promotion and disease prevention programs.
3. Introduce the fundamental principle of entrepreneurial spirituality and explain how churches can apply these principles to address social, economic, and health issues of the Black elderly.
4. Outline the specific steps in the entrepreneurial venture creation process required to develop sustainable Black elderly-focused health promotion and disease prevention programs.
5. Create opportunities for Black clergy to develop a specific plan for their congregation or community.

Building on the strategies employed in entrepreneurial training programs for other nonprofit sector leaders (Orton et al. 2007; Baker et al. 2006; Johnson et al. 2006; Mims 2006; Scotten and Absher 2006; Umble et al. 2006; Walls 2006), the proposed statewide Black clergy health promotion training program was jointly developed by a team of academic health professionals, who are experienced in working with both Black clergy and the Black elderly, and a team of academic business professionals, who have expertise in social and civic entrepreneurship.

The proposed program is comprised of six core components. The first, which sets the context for the training program, focuses on both the current and projected primary health issues impacting the Black elderly. Through their pastoral experience, most clergy are aware of the health challenges faced by older adults in their congregation and community. However, they may not be aware of the broader impact and implications of these challenges. The goal here is to provide Black clergy with an in-depth perspective on the nature and magnitude of the challenges, with the goal of helping them: (a) improve, and perhaps broaden the types and levels of service they currently provide; and (b) more effectively plan ministries for future generations of elderly congregants.

The remaining five core components of the proposed training program are designed to develop and/or fine-tune the entrepreneurial and financial literacy skills of Black clergy and facilitate their ability to better manage money, people and information. The overarching emphasis is on "coopetition" (cooperative competition) or competitive collaboration—the ability to build strategic alliances within the community and between fellow clergy to address elders' health issues and concerns. Research reveals that such collaborative approaches are far more advantageous than individual competitive approaches (Brandenburger and Nalebuff 1996).

Black clergy program participants will acquire the requisite skills in entrepreneurial coopetition and organizational management through the following course offerings:

1. *Civic entrepreneurship* – insight on entrepreneurial approaches and strategies for generating new revenue within Black communities, integrating clergy's experience and spiritual calling to improve the social condition of Black elderly;

2. *Finance and accounting for non-financial managers* – effective management of scarce and continually dwindling resources from congregation members, other community sources, and state and federal government;
3. *Human resource management and communications* – management in an ever-increasing workforce comprised of the elderly, with bi-directional exchange of skills, experience, and technology between elderly and youth;
4. *Information management* – leveraging the power and influence of the Internet and World Wide Web for comparative advantage; and
5. *Strategic management* – strategic thinking about market-based solutions to pressing public health problems impacting Black elderly.

At UNC-Chapel Hill, we currently offer these courses through our Management Academy for Public Health, a joint venture between the Gillings School of Global Public Health and the Kenan-Flagler Business School (www.maph.unc.edu). Over the past decade, we have trained over 1,000 state and local public health officials how to generate alternative streams of revenue to sustain their operations and programmatic initiatives (Orton et al. 2009).

Given the busy schedules of Black clergy and many competing demands, instruction will be offered in three distinct formats: (a) elective, semester long courses for full time divinity school students; (b) divinity school-based continuing education course for ordained clergy offered in a 3–4 day boot camp format; and (c) continuing education certification in entrepreneurial spirituality offered during annual ministerial or national denomination conferences. Regardless of the instructional format, learners will have access to an executive coach, participate in group brainstorming exercises to foster strategic alliance building and skills identification, and engage in a venture development exercise designed to address elements of a current health or health care issue impacting Black elderly in their community.

65.7 Summary and Discussion

The disproportionate burden of health problems and associated costs among a growing Black elderly population indicate a need for innovative, sustainable solutions for health promotion. The Black church is the cornerstone of many Black communities, and Black church leadership is the key to any health program implementation within congregations. As such, Black clergy are in a prime position to play a critical role in the identification and development of health solutions. However, given limited resources within many Black communities and the defunding of entitlement programs at both the federal and state levels of government, collaborative ventures built on strategic alliances will be necessary to promote sustainable solutions. Gil Rendle, in his book on leading change in the congregation, states that, “a posture of non-change in an environment of great change is not faithful leadership” (Rendle 1998: 17). Thus, educating and training Black clergy to actively build capacity to serve the elderly through entrepreneurship and social networks is not only prudent but critical in times when resources will continue to be limited. *Entrepreneurial spirituality* can provide a practical platform by which clergy can actively work together to address health issues plaguing the elderly, as well as the social, political, and economic factors associated with these outcomes.

One of the strengths of this approach is that it allows clergy, in this case, Black clergy, to capitalize on their pre-existing groups and social support systems (Yancey et al. 1999), which are critical for program sustainability. Clergy have demonstrated the capacity to forge such strategic alliances through their concerted efforts to mobilize resources and provide services to communities devastated by natural disasters such as Hurricane Katrina and the Haiti earthquakes. The same entrepreneurial spirit can be leveraged to effectuate dramatic change in the health status of the elderly.

Large churches such as those led by T.D. Jakes and Kirbyjon Caldwell are clear examples of how the principles of entrepreneurial spirituality can be leveraged to improve the quality of life in economically distressed communities. Through their existing relations, conglomerations of small churches with limited resources, if properly mobilized, can achieve similar outcomes. Education and training in entrepreneurial spirituality can help them develop strategies to maximize their utility for the common good of improving health in general and elder health in particular.

Another strength of this approach is that it promotes organizational skills-building at multiple levels – for the clergy, elderly, youth, general church body, private and public health sector, and academic institutions. So, instead of developing skills for individuals, this approach develops collaborative work attributes, where engagement with others to address a common goal is a higher premium for success than individual attributes. Thus, as implied in Genesis 48:4, it is the "community of peoples" that brings about the increase and ultimately sustainability through generations, rather than any individual.

A limitation to education and training using this approach is that clergy are historically independent, making collaborative ventures a challenge. There is an inherent mindset, contrary to the message in Genesis 48:4, which can inhibit or even stifle collaborative growth. Clergy working together within a community may vehemently disagree with the aims of the collaborative venture enough to thwart progression of partnerships at multiple levels. However, in a time of limited resources, this can be costly, not only financially, but also spiritually. Social innovation and civic responsibility can certainly tap individual initiative, but success depends on clergy's ability to work *together* for a common good.

Another limitation of this strategy is that some clergy may be uncomfortable with the notion of integrating entrepreneurship and spirituality. They may believe that business, competition, and financial success violate the Church's teachings of modesty and humility. However, in addressing the daily needs of their aging congregants, clergy are constantly attending to practical needs and employing a combination of practical and spiritual resources. Clergy are called upon to help aging congregants with a host of practical concerns, such as balancing a household budget or paying taxes, accessing healthy foods where supply is scant or non-existent, increasing mobility, or juggling limited funds among multiple competing needs.

Hence, embracing a combined model of entrepreneurship and spirituality may be appropriate in accepting inevitable, revolutionary change in health and health care.

To implement an education and training program for Black clergy in entrepreneurial spirituality that targets the health needs of Black elderly, the following first steps are recommended:

Introductory and cross-linked coursework. Several universities, according to the Good MBA Guide, already offer MBA/Masters of Divinity dual degree programs. Emory, Campbell, Howard, and Eastern are reportedly the topic universities (http://goodmbaguide.com/mdiv-mba-dual-degree/). Yale University (http://divinity.yale.edu/admissions/joint-degree-religion-management) and Samford University (http://almanac.logos.com/Beeson_Divinity_School,_Samford_University) also offer joint degree programs in business and divinity. For these and other universities which aspire to develop such dual degree programs, an introductory course on entrepreneurial spirituality can be created once business professionals and clergy with health experience are identified. .

This course would be based on the principles of coopetition and collaborative relationship development. Accordingly, one of the clergy's first assignments should be to formally assess the resources and potential partnerships within their communities. The course can be cross-referenced in business and divinity schools. Alternatively, if developing a separate course on the topic is not feasible, efforts can be made to cross-reference business classes within the divinity school for divinity students to take, either as an elective, or as a required course by the divinity school.

Online and regional satellite course offerings. Given geographic, economic, and time constraints, offering courses online and in remote regional locations may help maximize clergy participation. Clergy who are younger, more educated, and of higher socioeconomic levels may be more likely to use the Internet and thus have access to online courses for continuing education credit or certification. Online courses and group learning, perhaps paired with subsequent face-to-face training, may be quite attractive to some clergy. For those who are less apt to use the Internet and prefer face-to-face training, but are remote from institutions of higher education (for example, clergy in rural churches), satellite offerings of entrepreneurial spirituality coursework might maximize their participation. This may be done through course offerings at community colleges within a given area.

Engaging the elderly during the education and training. In order for any program to be sustainable among a group of people, they need to be engaged in the decision-making process. The same holds true for the elderly, in particular Black elderly, whose voices often go unheard. Representation by elderly impacted by the health condition for which an intervention will be designed, during the training and during the actual development and execution of an initiative, is very important (for example, elderly males recuperating at home following hip replacement might be part of planning an intervention on falls prevention). This completes the partnership link in that the persons utilizing the services provide both content and context regarding the health issue and its impact. In the development phase, consumers' voices are critical from a cost and time perspective. In the implementation phase, their feedback can be useful in making appropriate modifications at multiple levels of the partnership. Clergy are important to this process because they serve as the liaison between the elderly and the external partners working together to meet the elders' needs while building an infrastructure for health promotion change.

References

Appold, S. J., Johnson, J. H., Jr., & Kasarda, J. D. (2012). *The economic impact of racial health disparities in North Carolina*. Chapel Hill: Frank Hawkins Institute of Private Enterprise, UNC-Chapel Hill.

Baker, F. (1988). Dementing illness and black Americans. In J. S. Jackson (Ed.), *The black American elderly: Research on physical and psychosocial health* (pp. 215–236). New York: Springer.

Baker, E. L., Jr., Fox, C. E., Hassmiller, S. B., Sabol, B. J., & Stokes, C. C. (2006). Creating the Management Academy for Public Health: relationships are primary. *Journal of Public Health and Management Practice, 12*(5), 426–429.

Banks, W. L. (1999–2001). *The black church in the U.S.* Haverford: Infinity Publishing.com.

Battle, M. (2006). *The black church in America: African American Christian spirituality*. Malden: Blackwell Publishing.

Braithwaite, R. L., Taylor, S. E., & Austin, J. N. (2000). The black faith community and public health. In *Building health coalitions in the Black community*. Thousand Oaks, CA: Sage Publications.

Brandenburger, A. M., & Nalebuff, B. J. (1996). *Coopetition: A revolutionary mindset that combines competition and cooperation*. New York: Doubleday.

Caldwell, K. J. (1999). *The gospel of good success: A six-step program to spiritual, emotional, and financial success*. New York: Simon & Schuster.

Caldwell, K. J., Kallestad, W., & Sorenson, P. (2004). *Entrepreneurial faith: Launching bold initiatives to launch God's kingdom*. Colorado Springs: WaterBrook Press.

Carter-Edwards, L., Fisher, J. T., Vaughn, B. J., & Svetkey, L. P. (2002). Church rosters: Is this a viable mechanism for recruiting African Americans for a community-based survey? *Ethnicity and Health, 7*(1), 41–55.

CDC. (2013). *Costs of falls among older adults*. Available at http://www.cdc.gov/homeandrecreationalsafety/falls/fallcost.html. Accessed Apr 5.

Centers for Medicare and Medicaid Services, Office of the Actuary, National Health Statistics Group. (2009). *2007 National health care expenditures data*. Available at http://www.cms.gov/Research-Statistics-Data-and-Systems/Statistics-Trends-and-Reports/NationalHealthExpendData/Downloads/tables.pdf.

Christiansen, C. M. (2009). The White House Office on social innovation: A new paradigm for solving social problems. *Huffington Post*. Retrieved from http://www.huffingtonpost.com/clayton-m-christensen/the-white-house-office-on_b_223759.html

Fernando, M. (2007). *Spiritual leadership in the entrepreneurial business: A multifaith study*. Northampton: Edward Elgar.

Foster, D. A., & Dunnavant, A. L. (2004). *The encyclopedia of the Stone-Campbell Movement: Christian Church (Disciples of Christ), Christian Churches/Churches of Christ, Churches of Christ*. Grand Rapids: Eerdmans Publishing.

Fullan, M. (2008). *The six secrets of change: What the best leaders do to help their organizations survive and thrive*. San Francisco: Jossey-Bass.

Goldmon, M. V., & Roberson, J. T., Jr. (2004). Churches, academic institutions, and public health: Partnerships to eliminate health disparities. *North Carolina Medical Journal, 65*(6), 368–372.

Hall, W. D., Clark, L. T., Wenger, N. K., Wright, J. T., Jr., Kumanyika, S. K., Watson, K., Horton, E. W., Flack, J. M., Ferdinand, K. C., Gavin, J. R., 3rd, Reed, J. W., Saunders, E., & O'Neal, W., Jr. (2003). African-American Lipid and Cardiovascular Council. The metabolic syndrome in African Americans: A review. *Ethnicity and Disease, 13*(4), 414–428.

Harper, M. S. (Ed.). (1990). *Minority aging: Essential curricula content for selected health problems and health professions* (DHHS Publication No. HRS P-DV-90-4). Washington, DC: U.S. Government Printing Office.

Howard, D. L., Hakeem, F. B., Njue, C., Carey, T., & Jallah, Y. (2007). Racially disproportionate admission rates for ambulatory care sensitive conditions in North Carolina. *Public Health Reports, 122*(3), 362–372.

Hsu, C. Y., Lin, F., Vittinghoff, E., & Shlipak, M. G. (2003). Racial differences in the progression from chronic renal insufficiency to end-stage renal disease in the United States. *Journal of the American Society of Nephrology, 14*(11), 2902–2907.

Institute of Medicine (IOM), Committee on the Future Health Care Workforce for Older Americans, Board on Health Care Services, Institute of Medicine. (2008). *Retooling for an aging America: Building the health care workforce*. Washington, DC: National Academies Press.

Jackson, J. S. (Ed.). (1988). *The black American elderly: Research on physical and psychological health*. New York: Springer.

Jakes, T. D. (2004). *Woman thou art loosed*. New York: The Berkeley Publishing Group.

Johnson, J. H., Jr. (2002). A conceptual model for enhancing community competitiveness in the new economy. *Urban Affairs Review, 37*(6), 763–779.

Johnson, J. H., Jr., & Kasarda, J. D. (2011). *Six disruptive demographic trends: What Census 2010 will reveal*. Chapel Hill: Frank Hawkins Kenan Institute of Private Enterprise, Kenan Flagler Business School, University of North Carolina at Chapel Hill.

Johnson, J. H., Jr., & Parnell, A. (2013). *Aging in place in the Carolinas: Demographic highlights and programmatic challenges and opportunities*. Chapel Hill: Frank Hawkins Kenan Institute of Private Enterprise, UNC-Chapel Hill, January.

Johnson, J. H., Jr., Farrell, W. C., Jr., & Guinn, C. (1997). Immigration reform and browning of America: Tensions, conflicts and community instability. *International Migration Review, 31*, 1029–1069.

Johnson, J. H., Jr., Sabol, B. J., & Baker, E. L., Jr. (2006). The crucible of public health practice: major trends shaping the design of the Management Academy for Public Health. *Journal of Public Health and Management Practice, 12*(5), 419–425.

Kauanui S. K., Thomas K. D., Sherman C. L., Waters G. R., & Gliea, M. (2008). *Entrepreneurship and spirituality: An exploration using grounded theory*. International Council of Small Businesses. Retrieved from http://www.icsb.org/assets/entrepreneurship_and_spirituality.pdf

Kauanui, S. K., Thomas, K. D., Sherman, C. L., Waters, G. R., & Gilea, M. (2010). *Entrepreneurship and spirituality: An exploration using grounded theory*. Retrieved from http://www.icsb.org/assets/entrepreneurship_and_spirituality.pdf

Kosmin, B. A., Mayer, E., & Keysar, A. (2011). *American religious identification survey*. Hartford: Trinity College.

Lapane, K. L., & Resnik, L. (2005). Obesity in nursing homes: An escalating problem. *Journal of the American Geriatric Society, 53*(8), 1386–1391.

Lincoln, C. E., & Mamiya, L. H. (1990). *The Black church in the African American experience*. Durham, NC: Duke University Press.

Maddox, M. (2000). Spiritual wellness in older women. *Journal of Christian Nursing, 17*(1), 27–29.

Manton, K. G., & Gu, X. (2001). Changes in the prevalence of chronic disability in the United States black and nonblack population above age 65 from 1982 to 1999. *Proceedings of the National Academy of Sciences, 98*(11), 6354–6359.

Mims, S. (2006). A sustainable behavioral health program integrated with public health primary care. *Journal of Public Health and Management Practice, 12*(5), 456–461.

Mullen, K. (1990). Religion and health: A review of the literature. *International Journal of Sociology and Social Policy, 10*(1), 85–96.

Olson, L. M., Reis, J., Murphy, L., & Gehm, J. H. (1988). The religious community as a partner in health care. *Journal of Community Health, 13*(4), 249–257.
Orton, S., Umble, K., Zelt, S., Porter, J., & Johnson, J. H., Jr. (2007). Management academy for public health: Creating entrepreneurial managers. *American Journal of Public Health, 97*(4), 601–605.
Orton, S. N., Menkens, A. J., & Santos, P. (2009). *Public health business planning: A practical guide*. Sudbury: Jones and Bartlett Publishers.
Pappu, S. (2006). The Preacher. *Atlantic Magazine*. Retrieved from http://www.theatlantic.com/magazine/archive/2006/03/the-preacher/4606/
PBS Religion and Ethics Newsweekly. (2000, January 14). Bishop T.D. Jakes, Episode No. 320.
Petchers, M. K., & Milligan, S. E. (1998). Access to health care in a black urban elderly population. *Gerontologist, 28*(2), 213–217.
Petersen, J., Atwood, J. R., & Yates, B. (2002). Key elements for church-based health promotion programs: Outcome-based literature review. *Public Health Nursing, 19*(6), 401–411.
Physical Activity and Nutrition Branch. (2010). *African-American churches eating smart and moving more: A planning and resource guide*. Raleigh: NC DHHS, NC Division of Public Health. Available at http://www.eatsmartmovemorenc.com/AfricanAmericanChurches/Texts/ESMM_AACguide.pdf
Rendle, G. R. (1998). *Leading change in the congregation: Spiritual and organizational tools for leaders*. Herndon: Alban Institute, Inc.
Resnicow, K., Wallace, D. C., Jackson, A., Digirolamo, A., Odom, E., Wang, T., Dudley, W. N., Davis, M., Mitchell, D., & Baranowski, T. (2000). Dietary change through African American churches: Baseline results and program description of the eat for life trial. *Journal of Cancer Education, 15*(3), 156–163.
Scotten, E. S., & Absher, A. C. (2006). Creating community-based access to primary healthcare for the uninsured through strategic alliances and restructuring local health department programs. *Journal of Public Health and Management Practice, 12*(5), 446–451.
Somers, A. R. (1978). The high cost of health care for the elderly: Diagnosis, prognosis, and some suggestions for therapy. *Journal of Health Politics, Policy and Law, 3*(2), 163–180.
Stevens, J. A., Corso, P. S., Finkelstein, E. A., & Miller, T. R. (2006). The costs of fatal and non-fatal falls among older adults. *Injury Prevention, 12*(5), 290–295.
The Barna Group. (2007). *Church attendance*. Cited 2008 January 3. Retrieved from http://www.barna.org/FlexPage.aspx?Page=Topic&TopicID=10
Umble, K. E., Orton, S., Rosen, B., & Ottoson, J. (2006). Evaluating the impact of the Management Academy for Public Health: developing entrepreneurial managers and organizations. *Journal of Public Health and Management Practice, 12*(5), 436–445.
Vincent, G. K., & Velkoff, V. A. (2010). *The next four decades: The older population in the United States: 2010 to 2050. Population estimates and projections*. Current Population Reports P25-1138. Washington, DC: U.S. Census Bureau.
Walls, B. E. (2006). The Management Academy for Public Health: Transforming the business of healthcare. *Journal of Public Health and Management Practice, 12*(5), 480–481.
Werner, C. A. (2011). *The older population 2010*. 2010 Census Briefs C2010 BRR-09. Washington, DC: U.S. Census Bureau.
Wilcox, S., Laken, M., Bopp, M., Gethers, O., Huang, P., McClorin, L., Parrott, A. W., Swinton, R., & Yancey, A. (2007). Increasing physical activity among church members: community-based participatory research. *American Journal of Preventive Medicine, 32*(2), 131–138.
Wimberly, A. E. S. (2001). The role of black faith communities in fostering health. In R. L. Braithwaite & S. E. Taylor (Eds.), *Health issues in the black community* (pp. 129–150). San Francisco: Jossey-Bass.
Yancey, A., Miles, O., & Jordan, A. (1999). Organizational characteristics facilitating initiation and institutionalization of physical activity programs in a multiethnic urban community. *Journal of Health Education, 30*(2), 44–51.

Chapter 66
Environmental Governance, Property Rights and Judeo-Christian Tradition

Kathleen Braden

66.1 Framing the Question and Considering the Role of Geography

Although discussion of how the Judeo-Christian tradition relates to an ethic for human-nature relations dates back thousands of years, writers in recent decades have at times implied that this is a new avenue of inquiry in environmental governance. The birth of "ecotheology" has variously been attributed to Lynn White's famous *Science* essay of 1967, Pope Paul VI's address to the Stockholm United Nations Conference on the Environment, or even American Lutheran, Joseph Sittler's "A theology for earth" (White 1967; Paul VI 1972; Sittler 1954).

But how much do Judeo-Christian traditions in the secularized, western world really have an impact on political attitudes of people who claim the Christian religion? This question can be set within the framework of ideas on cultural change, particularly political attitudes as a subset of culture. I apply this question here to the realm of environmental governance, with a focus on attitudes toward private property rights and government regulations to protect the environment in the United States. The Pew Forum on Religion & Public Life provides survey data that 78.4 % of American adults identify with a Christian religious tradition, with 51.3 % Protestant including 26.3 % evangelical denominations. In terms of social and political attitudes, survey results confirmed:

> the close link between Americans' religious affiliation, beliefs and practices, on the one hand, and their social and political attitudes, on the other. Indeed, the survey demonstrates that the social and political fault lines in American society run through, as well as alongside, religious traditions. The relationship between politics and religion in the United States is particularly strong with respect to political ideology and views on social issues such as

K. Braden (✉)
Department of Political Science and Geography, Seattle Pacific University,
Seattle, WA 98119, USA
e-mail: kbraden@spu.edu

S.D. Brunn (ed.), *The Changing World Religion Map*,
DOI 10.1007/978-94-017-9376-6_66

> abortion and homosexuality, with the more religiously committed adherents across several religious traditions expressing more conservative political views. On other issues included in the survey, such as environmental protection, foreign affairs, and the proper size and role of government, differences based on religion tend to be smaller. (Pew 2007)

The results on environmental protection attitudes were confirmed in a 2010 survey, with Pew reporters noting:

> There is only a modest religious element in attitudes about environmental protection. Solid majorities of all major religious traditions favor stronger laws and regulation, including 73 % of white evangelical Protestants, 79 % of black Protestants, 85 % of Catholics and 84 % of the unaffiliated. Religion has far less influence on opinions about environmental policy than other factors do. Just 6 % say that their religious beliefs have had the biggest influence on what they think about tougher environmental rules. Education and what people hear or read in the media are the strongest drivers of opinions about environmental regulations; roughly three-in-ten cite their education (29 %), and 26 % mention the media as having the most influence on their thinking about this issue.

However, among Christian denominations, white evangelicals, sometimes called the "Christian Right," reported the largest opposition (22 %) to tougher environmental regulations (Pew 2010).

James Davison Hunter of the University of Virginia in his book, *To Change the World: The Irony, Tragedy, & Possibility of Christianity in the Late Modern World*, examined how disillusioned the Christian Right has become with the path American society has taken, resulting in an angry response of perceived marginalization and a hope for recapturing "traditional values related to neoclassical capitalism and conservative social norms" (Hunter, 201: chapter II-3). But the Christian Left also embeds its theological interpretation in political action for what it perceives as a more just society, the interpreted heart of the message of Jesus Christ. Hunter looks at these groups, along with NeoAnabaptists, in his book, all making utilization in various ways of political culture, all fractured in disagreement about Biblical tradition and political action.

Hunter has also questioned whether Christians are very good at effecting changes in popular culture. He argues that Christian groups have often misunderstood the way culture evolves and therefore may be better at *assuming* their religion will impact culture than actually achieving culture shifts. "America was never, in any theologically serious way, a Christian nation, nor the West a Christian civilization" (Hunter 2010: 280).

Correspondingly, geographer James Proctor has examined religion as a source of moral voice in the United States (Proctor 2006). His emphasis on religion as trust in authority reveals a "fractured American religious landscape" with strong regional differences mirroring the famous red and blue zones of presidential politics. An important moment for purposes of this essay is Proctor's conclusion that, while geographers need to pursue this inquiry further, institutional religion often resembles non-religious regimes. As we consider the dominant Judeo-Christian religious claim of Americans, we may ask whether core faith formation truly has an impact on attitudes toward environmental governance. Geographer Janel Curry engaged this question in her examination of how various Protestant eschatologies and biblical interpretations affected worldviews in the environmental arena, echoing Proctor's call for further research in this area: "Most studies on the relationship between religion and environmental attitudes deal with vague notions like belief in God and church affiliation. Such studies are insufficient. Though I attempt to get beyond such vague notions in this study, I offer limited evidence of how these religious beliefs affect practice, an area where more research is needed" (Curry-Roper 1990: 167).

Thus, the role of religion in society has long been investigated by geographers. In fact, the relationship between human society and the physical environment is a major defining theme in the discipline. Many geographers have brought these two themes together, understanding how much underlying religious ethics or sacred practices and texts often relate to human views of nature that *may* inform environmental governance.

Recently two reflective summaries on these geographic themes have been provided in the *Annals of the Association of American Geographers*: (1) Karl Zimmerer constructed an intellectual history of nature-society geographies from 1911 until 2010 and identified core clusters of ideas in the field during various periods. He notes that environmental policy studies have been a major focus in the more recent periods (Zimmerer 2010: 1084); (2) A rich forum of essays on geography and religion was published likewise in the *Annals* in 2006, with contributions by James Proctor, Michael Ferber, Julian Holloway, and Anne Buttimer, who concluded her essay by stating that: "Geography, in its broadest sense, as lived reality and as academic subject, affords a fertile ground for serious research on religion and environment" (Buttimer 2006: 198).

Within the theme of human-environment interaction, I examine below a basic element governing human use of nature, viz., the notion of property rights. While Diana Liverman correctly noted: "Geography has much to offer a world in which environmental change is widespread and where new actors, scales, and metrics are transforming environmental decisions," she also pointed out that "markets in environmental services are becoming the dominant approach to managing and protecting the environment in the twenty-first century" (Liverman 2004: 734, 735).

Within this mix of markets, environmental regulations, and value systems, geographers stand at the confluence of scholarly inquiry on human-nature relations and the role of religion in cultural change. They join with other social scientists, theologians, and legal experts to examine whether the Christian faith truly has an impact on environmental governance practice. I argue here that in the case of United States, debates on property rights, represent a fragmented and contested relationship.

66.2 Select Empirical Investigations on Religions Convictions and Environmental Politics

In addition to James Proctor's research on the American religious landscape and moral authority and the Pew Forum surveys cited above, scholars in other fields have also tried to establish through empirical data a link between religion and environmental politics. Three such articles provide a good snapshot of the complexities in that relationship:

Research reported in the *American Journal of Political Science* in 1995 attempted through national survey data in the United States to establish a relationship between the religious views of various Protestant groups and their attitudes towards the environment. The authors hypothesized that evangelicals would be least supportive of environmentalism. While the outcome ultimately confirmed that supposition, the authors also noted "Not surprisingly, environmentalism is strongly associated with political-identity variables" (Guth et al. 1995: 373). Also although conservative eschatologies were indeed suggestive of a negative attitude toward the environment, results suggested that the relationship might be reversed in terms of initial cause: "Not surprisingly, political-identity variables emerge as strong predictors of environmental sympathies" (Guth et al. 1995: 376). The authors further concluded from the results of their data that religious tradition/social group may have more powerful impacts on attitude than the actual religious belief.

Calling upon political scientists to take religious perspectives seriously in exploring public attitudes toward the environment, the authors pointed out the questions left unanswered in their study about the role of organizational elites in influencing religious socialization (and therefore environmental attitude), particularly among conservative evangelicals.

Similarly complicated results in the European context were shown by scholars at Göteborg University investigating opinions about genetically-modified crops and air pollution as indicators of attitude toward environment. Biel and Nilsson (2005) noted the role of various political and other attitudinal elements as much as religion based on empirical investigation:

> this proposition does not lend itself to a proper test. In particular, do difficulties arise since values and beliefs are confounded with other determining factors such as political ideology and wealth? Perhaps one has to do with piecemeal evidence. (Biel and Nilsson 2005: 190)

The authors used the term "situational cues" (I would suggest akin to the definition of Guth et al. on social groups and the role of elites), but could not report that their study revealed that religious values were activated by the presentation of the genetically-modified crop issue. Empirical verification of Lynn White's contention, therefore, remained elusive for Biel and Nilsson.

The third article of interest here examined social aspects of Christian affiliation in the United States and environmental positions. Political scientists investigating data on congregational membership of Christians reinforced the argument that social forces in churches, rather than core doctrinal expression, may shape environmental attitudes (Djupe and Hunt 2009). The role of socialization, predispositions toward existing political attitudes, and the way filtered information reaches congregation members may be the most significant items creating environmental outlook. However, as with the 1995 study on conservative eschatologies, Djupe and Hunt found that Biblical literalism was a good predictor of environmental attitude (Djupe and Hunt 2009: 678). The authors concluded:

> Thus, our account casts doubt on explanations in the literature that rely heavily on religious beliefs. In our view, it is not enough to know whether individuals hold dominion or stewardship beliefs. Instead, we need to know why they have come to hold those beliefs because the

various routes to holding an opinion dictate different persuasion strategies. Are beliefs truly intermediaries between social communication and opinion change or are members' beliefs reformulated to correspond to shifts in their opinions due to social communication? (Djupe and Hunt 2009: 682)

I noted above the contention by James Davison Hunter that cultural change is impelled largely by elites in a top-down manner; thus, the notion revealed by these analyses that religion as a social-milieu construct may be a more powerful correlate with environmental attitudes pairs well with looking at the role of the clergy and Christian leaders. But how are we to understand the ways Judeo-Christian traditions inform the worldview of these elites? I turn now to the example of evangelical Christians who are voices (though not always from mainstream academic affiliations or positions) in advocacy non-governmental organizations dealing with property rights and environmental governance.

66.3 Property Rights, Evangelicals and Environmental Politics

While potential regulation regarding human impact on climate change is the main focus today in the debate over religious values and environmental policy among evangelical Protestants (Curry 2008), I take up here the question of property rights because it represents core values of a land ethic that scholars such as White *assumed* must have a biblical basis for present day Christians. In fact, I contend that the underlying notions of property rights impel most of the environmental governance debates in the United States, from climate change to water regimes to endangered species regulation.

66.3.1 Two Examples of Takings Controversies in the United States

"Takings" is a word with negative connotations that raises the hackles especially of landowners in the United States West. More properly expressed, "takings" can refer to implementation of environmental laws designed to protect social goods or resources and the term in a broader sense refers to the "Takings Clause" of the United States Constitution's Fifth Amendment, that is, private property shall not be taken for public use without just compensation. The government can exercise the right of eminent domain to *take* land away from private ownership, but this is a different action than restricting the *use* rights of a property owner.

The 1973 federal Endangered Species Act has resulted in many legal questions about rights of private property holders. Habitat protection to preserve species designated as endangered may result in restrictions on usage of land or delivery of water to land holders. Particularly scrutinized and contested is the role of the U.S. government under the Constitution Takings clause in terms of providing just compensation against the common good of conserving threatened wildlife (Thompson 1996–1997; Scaccia 2010). Critics of the original regulation argue that it is a disincentive for private landowners (geographic space where the majority of endangered species is located) because it in essence penalizes them for upholding the social good. Compensation provisions proposed in Congress in 2005 were argued by environmentalists to constitute an entitlement to landowners (Olivetti and Worsham 2003). Re-formulated policy on takings and the Endangered Species Act is still not settled.

Perhaps even more contested is the issue of water takings. The landscape of the West, the high percentage of land owned by the federal government, historic circumstances, and even culture may all be factors that place water takings at the front lines in the environmental governance-property rights battles. In addition, water resource management is a particular subset of environmental governance that presents special complications wherever withdrawals occur due to the fleeting resource characteristics of water. Issues that complicate the matter further include common property resource management, territoriality and boundaries, or damage to water beyond its re-usability (Matthews 2003).

The increasing popularity of rainwater harvesting makes this type of water use a good case study of the role of property rights and underlying ethical principles. Both graywater reuse and rainwater harvesting represent capturing water or rain respectively on a small scale before it becomes runoff. Sometimes the phrase is suggested that rainwater otherwise would be "wasted" (Bretsen 2011: 159) or, as noted in a publication of Colorado State University extension, "putting it to beneficial use" by capturing it, implying the water is not useful otherwise (Waskom and Kallenberger 2012).

The Colorado Division of Water Resources is obligated to protect all vested water rights and administer allocation of the resource to users and deals with issues found elsewhere in some western states of the prior appropriation doctrine. This "first in time, first in right" principle makes sorting out whose property rights are being diminished a thorny problem for regulators. In 2009, Colorado passed laws SB 09–080 to allow rooftop precipitation collection systems and HB 09–1129 to allow pilot projects for larger collection systems. The advent of these laws raised questions about whether "senior" users are being deprived of property rights under the prior appropriation tradition in the state. Furthermore, questions have arisen as to whether senior users are truly damaged and due compensation.

James Proctor found in his research that there was more of a negative correlation between theocracy and ecology in the American South than the West, but argued for more region-specific research (Proctor 2006: 193). In fact, urban areas of the West and the influence of the coast may mask the Christian culture of western rural farmers and ranchers in the data and the strong presence of the Mormon faith further complicates the scene at various scales. Brehm and Eisenhauer, in their 2006 empirical investigation of Mormon culture and environmental attitudes, found there was a relationship among those surveyed of what would be interpreted as conservative political views on the environment. But the authors correctly also pointed out that *regional* culture, such as the Sagebrush Rebellion attitude of the rural West, may also have a strong influence on political attitudes and factoring out the role of religion may then be complicated (Brehm and Eisenhauer 2006: 39). We know that politically conservative voices argue that environmental objectives would be best served by a private market for water or endangered species laws that do not provide *disincentives* for landowners who find such species on their land. But are these so-called "wise use" or market-based environmental approaches in the secular arena consistent with Judeo-Christian ethics for ownership and human relations with the earth? And how might evangelical advocacy groups make use of theological argument to prop up political stances?

66.3.2 Contested Biblical Interpretations

The debate over biblical calls to dominion and/or stewardship in human relationship to the earth has dominated the literature among scholars of Christian ecotheology. Along with Catholic scholars and activists (see De Cosse 1996, for example on the property rights debate in Catholic circles), evangelical Protestants have particularly engaged in this contested narrative. The creation in 1993 of the Evangelical Environmental Network and the publication by EEN *This Land is your Land: This Land is God's Land: Takings Legislation versus the Judeo-Christian Land Ethic* defied conservative politics and neoliberal economics that informed many evangelical leaders. When the authors wrote, "no one has an inherent 'right' to do anything with property unless God, as its ultimate owner, permits it" (Alexander et al. 1999: 17), the libertarian Acton Institute responded. Peter J. Hill in the Institute's *Religion & Liberty* publication argued that "the EEN document was fundamentally flawed because an orderly regime for private property rights allows for individual choice and the best outcome for both owner and neighbor" (Hill 1999: 6–7).

Likewise, Christians who espoused the "wise use" movement argued that a proper understanding of Genesis chapters leads to a view for human dominion over nature and natural resources. The Cornwall Alliance for the Stewardship of Creation was formed in the year 2000 (earlier called the Interfaith Stewardship Alliance) as a counterbalance to EEN, putting forth its argument that economic freedom leads to the best environmental stewardship and that environmentalism has become a religion and is inherently "totalitarian". The December 19, 2012 newsletter, for example, featured an article "Good news for environmentalists-and other sinners" (Beisner 2012).

How can evangelicals, beginning with similar regard for a biblically-based worldview, arrive at such different political attitudes on environmental governance? I summarize the opposing arguments on governance and property rights in Table 66.1. The sacred texts of Judaic and Christian traditions do indeed have much to offer about land ethics, accumulation of property and wealth, justice, and treatment of nature and of neighbor. The commands for stewardship and/or dominion

Table 66.1 Contested views on environmental governance by two evangelical organizations

Key concept	Acton institute	EEN
Stewardship	Stewardship of earth best accomplished via market and property rights	Stewardship and dominion of earth should reflect humanity's servanthood; landowner role should not be higher than steward role
Common good	A few people (property owners) should not bear burden for common good	Humans not allowed to diminish fruitfulness of land; good of the individual should not trump common good
Human labor	Person owns fruits of his/her labor	The poor have a right to portion of harvest
Human sin	Coercive power does not negate power of sin	Humans called to follow Christ and not harm neighbors or nature
Human place in natural order	Humans have privileged position as sole ones made in God's image	Christ's servant heart is model for human attitude
Role of governments	Governments also subject to sin and therefore should be restricted	Laws set up by governments define property rights to begin with
Love of neighbor	Love of neighbor is expressed via individual choices, not coercion by authority	Love of neighbor is one of most vitally important commandments

Source: Kathleen Braden

derive from the opening chapters of Genesis in the Hebrew Bible. However, the search for Judaic roots in ecotheological land ethics and property rights regime sometimes risks jumping straight into the "love thy neighbor" rule of the New Testament and by-passing the justice arguments of the Hebrew Bible. Geographer Jeanne Kay has a perhaps difficult proposition for modern environmentalists looking for an ethic that will lead to gentle treatment of nature. She rightly pointed out that God has another purpose in mind for nature with the ancient Hebrews because it becomes "a barometer of society's contractual relationship with God" (Kay 1989: 217). There are many constraints on how the ancient Hebrews were to treat and take care of the land, but ultimately, nature can be punishing when disobedience occurs. Yes, human beings have dominion over nature, but the model of Divine Justice takes precedence. In the Hebrew Bible, nature is not forgiving when humans are disobedient to the covenant with their Creator.

Likewise, Gerald Blidstein of Ben-Gurion University has examined the ethics of the Talmud with respect to property rights. Here again, a just order is the ultimate aim. He cited Deuteronomy 22:1–3, which admonishes the people of God to restore lost property to its rightful owner. The ethics of the ancient Hebrew Bible are embedded in arguments not about some altruism of collective property rights and sharing the riches of the earth, but about a logical and proper order of ownership, restitution, and compensation (Blidstein 2009).

Even the notion of a sabbatical year when the earth rests, often used by modern environmentalists arguing for a minimalist approach to land use, is centered on a reaffirmation of God's ownership rights to the land before those of humans. In short, one might argue that the Hebrew Bible provides guidance less for an environmental governance ethic and more a reminder of humanity's relationship with a sovereign supreme being.

66.3.3 Is There a Role for Christian Precepts in Environmental Law?

John Copeland Nagle has noted that "A Christian approach to environmental law must be based not only on a Christian view of creation, but a Christian view of law" (Nagle 2001: 442). Development of an environmental governance regime cannot escape the realm of property rights and the story of these rights within the Christian faith is a multi-century history of disagreement, ambiguities, and inconsistencies. Dominion and stewardship discussions sit on the question of *use*. One of the early passionate disagreements over ownership and use erupted in the Middle Ages with the Franciscan Order's vow of poverty. Does holding property lead to sin? Did appropriation of resources only occur when sin entered the world after the Fall from grace? Did Jesus and the Apostles actually *own* anything they used?

These were not esoteric questions for the medieval church, but struck at the heart of church power and wealth, often involving ownership of vast tracts of land estates. The relationship between individual property rights and the needs of the community was explored by scholars such as St. Thomas Aquinas, St. Bonaventure, and Marsilius of Padua, a fourteenth century scholar whose work presaged later treatises on government, natural law, and ownership rights. Much of the debate over the moral question of property ownership centered on the issue of using up (that is, extinguishing the object) and therefore has relevance to later, modern debates on environmental policy (Lee 2009). Recent ecotheological disagreements over dominion and stewardship have skimmed lightly over what happens once the principle is established under Judeo-Christian precepts, but does dominion imply "using up to extinction"? Does stewardship imply "owning lightly"?

Perhaps a key moment in this long history of debate erupted with the advent of John Locke and other scholars associated with Classical Liberalism and the Enlightenment. Given the enormous influence Locke had on the western world, particularly notions of property rights (and therefore land use and environmental governance) in the United States, we might ask whether Locke had any theological convictions that formulated his views on property (Menashi 2012). Locke was baptized and raised in the Anglican Church, graduated from Christ's Church at Oxford, and later espoused a brand of liberal Protestantism. However, many scholars have argued that he came to entertain deep skepticism of Christian ethics as a basis for civil law. While acknowledging that God provided all that humans needed on the earth, Locke wrote that individuals had the right to appropriate this commonly-held stock of resources according to the outcome of their labor. Governments are created to help the preservation of property rights: "Legitimizing the desire for material gain – against religious teachings that condemn it – redirects human energies and transforms social attitudes. In doing so, Locke's theory of property entails significant challenges to the biblical tradition" (Menashi 2012: 227).

But this contention would not sit well with people of the Acton Institute or Cornwall Alliance who profess a strong, evangelical faith *and* like property rights just fine. One of the criticisms of a neoliberal argument portrayed by an organization claiming foundations in biblical principles, such as the Acton Institute, is the notion that property is an inherent concept derived from natural law. Blumm and Ruhl, in responding to conservative economic stances, wrote: "to libertarians … property is an idealized abstraction: a pre-political, natural law concept that cannot be altered by democratic processes like regulation. This largely static view of property is mistaken; property is actually a product of the state and subject to different interpretations by the state's legislature and its courts over time" (Blumm and Ruhl 2010: 811). Compare this statement to a similar, but Biblically-based interpretation by authors of the EEN publication, *This Land Is Your Land*: "property ownership in Scripture consists only of whatever rights God has chosen in His wisdom to allow the owners-His stewards-to exercise" (Alexander et al. 1999: 12). The authors in fact contend that the very *idea* of property is created by God's law *and* human law.

But will the government understand correctly the collective good and come up with a system that allows for the best environmental governance of endangered species or a resource like water? Peter Hill, writing for the Acton Institute, counters that the government can no better guarantee good outcomes for the environment than can the market because both are human institutions and therefore subject to sin and corruption: "Implicit in the analysis of *This Land Is Your Land* is the assumption that whatever takes place in the private sector reflects human greed and fallibility but that actions in the government sector are in the public interest and not marred by sin … Is it not far more scriptural to assume that human sinfulness is pervasive and can affect all human endeavors and humanly designed institutions?" (Hill 1999: 6)

Therefore, when we look at modern governance issues such as endangered species regulation or rainwater harvesting and the passion with which some Americans hold to the foundational principles of private property rights, we find little agreement among activists and scholars who hold common ethics from the Judeo-Christian Bible. But the path of the disagreement is unclear: do differences of theological understanding and emphases drive the actors to different environmental attitudes or, conversely, do environmental governance politics trump theologies and merely inspire searches for justification in the sacred texts?

66.4 Questioning Assumptions and Confusing Principles with Policy

Does the Christian faith equal anti-environmental attitude? Empirical inquiries suggest that Lynn White was *partially* right in attributing environmental politics to Christianity, but it seems that there is no unanimity among Christians, even within the realm of Protestant evangelism, admittedly a fractured branch of the religion.

I am suspect of the notion that modern, secular humanism has killed the relationship between Christian piety and an ethic for relationship with the earth. If we return to White's original contention, he accuses Christianity through its valuing of progress as a notion, ultimately a value that emerges with a technology that is ruthless toward nature. Ironically, Locke and White both express skepticism about intentions of the sacred Judeo-Christian text. In praising St. Francis and an "alternative Christian view," White likes the humility of this saint and his attitude toward nature. But it may have been not only a perceived pantheism in Francis that threatened the church establishment, but also the danger to church property rights posed by the saint's poverty. The battle over religious interpretation and an ethic for how humans treat the earth is an ancient one because it is a struggle of power and authority.

I contend that both environmentalists and property rights activists who claim Biblical priorities idealistically hope to find justification for political attitudes in the sacred text. John Copeland Nagle, who was cited above with his argument for a Christian view of the law, still cannot respond affirmatively when he poses the question whether obedience to Christian teaching make some more likely to obey the Endangered Species Act as landowners: "In other words, not enough people feel a religious or ethical obligation to manage their land in a way that preserves endangered species" (Nagle 2001: 445). Nagle goes on to note that he might expect a Christian landowner to heed the legislation due to religious convictions, but that the laws and the Constitution's Takings Clause were not written with the notion that the earth ultimately belongs to God and that some Christians will endorse the idea of species protection, but others will not, just as outside the Christian community (Nagle 2001: 449).

Perhaps the most blunt skepticism about looking for environmental governance outcomes via the path of Christianity in the United States is found in William Stuntz's (Harvard Law School) review of the book *Christian Perspectives on Legal Thought*. In the review, he gave a pointed warning about confusing the purpose of faith: "Why should anyone think about law in Christian terms? Perhaps the answer is, no one should. That is surely the conclusion most American law professors would reach. Religion is not a topic of much conversation in the law school world; what little discussion there is tends to treat serious religious commitment as a disease – call it the germ theory of religion – perhaps especially if the religion is Christianity. If that is a correct view of Christianity, the law should stay as far away from it as possible, so as not to catch the virus. One might frame the impulse in terms more favorable to religion and say that the threat of infection runs the other way, that law and Christian faith belong in separate spheres in order to protect the latter from the former. Either way, the conclusion is the same: in America's legal conversation, Christianity is an unwelcome guest" (Stuntz 2003: 1707).

An evangelical Christian himself, Stuntz tragically died in 2011 from cancer at the age of 52. He had written in his review: "the Christian Bible is a *story*, a narrative. Law plays a part in the narrative, but it is plainly not the main actor; its point is to focus attention away from itself and toward the One who *is* the main actor" (Stuntz 2003: 1746). In agreement with Janel Curry's argument that the Judeo-Christian story is one about *relationships*, he writes, "For Christianity is not, above all, about the business of making better rules. Relationships – especially the one between us and our Creator – matter infinitely more" (Stuntz 2003: 1749).

James Davison Hunter ultimately ended his book on the role of Christianity in the late modern world by suggesting that faithful presence to society, giving way on the battlefield of political ideas, is perhaps the best course for the Christian faith in America: "In public discourse, the challenge is not to stifle robust debate, but rather to make sure it is real debate. The first obligation for Christians is to listen carefully to opponents and if they are not willing to do so, then Christians should simply be silent." (Hunter 2010: 266)

Geographers as social scientists are well positioned to explore the question of *whether* religious values inform the part of cultural change we call environmental governance and evaluate judgments about the *value* of such a process. As scholars with the long view of how humans relate to the earth, we are well positioned to ask about the deep complexities of this relationship to the sacred realm of human life. Anne Buttimer summarized the confluence of geography, religion, and environment: "Geography, in its broadest sense, as lived reality and as academic subject, affords a fertile ground for serious research on religion and environment" (Buttimer 2006: 198).

Acknowledgements The author acknowledges support from the Council for Christian Colleges and Universities grant program on Free Market Economics, Dr. Caleb Henry, Seattle Pacific University, principle investigator.

References

Alexander, A., Clark, F., Krueger, F., & LeQuire, S. (1999). *Takings legislation versus the Judeo-Christian land ethic*. Wynnewood: Evangelical Environmental Network.

Beisner, C. (2012, December 19). Good news for environmentalists – and other sinners. *Cornwall Alliance*. Newsletter. Retrieved January 29, 2013, from http://www.cornwallalliance.org/newsletter/issue/newsletter-december-19-2012/

Biel, A., & Nilsson, A. (2005). Religious values and environmental concerns: Harmony and detachment. *Social Science Quarterly, 86*, 178–191.

Blidstein, G. (2009). Talmudic ethics and contemporary problematics. *Review of Rabbinic Judaism, 12*, 204–217.

Blumm, M., & Ruhl, J. (2010). Background principles, takings, and libertarian property: A reply to Professor Huffman. *Ecology Law Quarterly, 37*, 805–841.

Brehm, J., & Eisenhauer, B. (2006). Environmental concern in the Mormon culture region. *Society and Natural Resources, 19*, 393–410.

Bretsen, S. (2011). Rainwater harvesting under Colorado's prior appropriation doctrine: Property rights and takings. *Fordham Environmental Law Review, 22*, 159–231.

Buttimer, A. (2006). Afterword: Reflections on geography, religion, and belief systems. *Annals of the Association of American Geographers, 96*, 197–202.

Curry, J. (2008). Christians and climate change: A social framework of analysis. *Perspectives on Science and Christian Faith, 60*, 156–164.

Curry-Roper, J. (1990). Contemporary Christian eschatologies and their relation to environmental stewardship. *The Professional Geographer, 42*, 157–169.

De Cosse, D. E. (1996). Beyond law and economics: Theological ethics and the regulatory takings debate. *Boston College Environmental Affairs Law Review, 23*, 829–849.

Djupe, P., & Hunt, P. (2009). Beyond the Lynn White thesis: Congregational effects on environmental concern. *Journal for the Scientific Study of Religion, 48*, 670–686.

Guth, J., Green, J., Kellstedt, L., & Smidt, C. (1995). Faith and the environment: Religious beliefs and attitudes on environmental policy. *American Journal of Political Science, 39*, 364–382.
Hill, P. (1999). Takings and the Judeo-Christian land ethic: A response. *Religion & Liberty, 9*, 5–7.
Hunter, J. D. (2010). *To change the world: The irony, tragedy, & possibility of Christianity in the late modern world*. New York: Oxford University Press.
Interfaith Council for Environmental Stewardship and Acton Institute for the Study of Religion and Liberty. (2000). *Environmental stewardship in the Judeo-Christian tradition*. Retrieved March 21, 2012, from http://www.acton.org/public-policy/environmental-stewardship/theology-e/environmental-stewardship-judeo-christian-traditi
Kay, J. (1989). Human dominion over nature in the Hebrew Bible. *Annals of the Association of American Geographers, 79*, 214–232.
Lee, A. (2009). Roman law and human liberty: Marsilius of Padua on property rights. *Journal of the History of Ideas, 70*, 23–44.
Liverman, D. (2004). Who governs, at what scale, and at what price? Geography, environmental governance, and the commodification of nature. *Annals of the Association of American Geographers, 94*, 734–738.
Matthews, O. (2003). Simplifying western water rights to facilitate water marketing. *Water Resources Update (Universities Council on Water Resources), 126*, 40–44.
Menashi, S. (2012). Cain as his brother's keeper: Property rights and Christian doctrine in Locke's *Two Treatises of Government. Seton Hall Law Review, 42*, 185–273.
Nagle, J. C. (2001). Christianity and environmental law. In M. McConnell, R. Cochran, & A. Carmella (Eds.), *Christian perspectives on legal thought* (pp. 435–452). New Haven: Yale University Press.
Olivetti, A., & Worsham, J. (2003). *This land is your land, this land is my land: The property rights movement and regulatory takings*. New York: LFB Scholarly Publishing LLC.
Paul VI, P. (1972). *A hospitable earth for future generations*. United Nations Conference on the Human Environment, Stockholm. Retrieved February 1, 2013, from http://conservation.catholic.org/pope_paul_vi.htm
Pew Forum on Religion & Public Life. (2007). *U.S. religious landscape survey*. Retrieved February 6, 2013, from http://religions.pewforum.org/reports
Pew Forum on Religion & Public Life. (2010). Few say religion shapes environment, immigration view. Retrieved February 9, 2013, from http://www.pewforum.org/politics-and-elections/few-say-religion-shapes-immigration-environment-views.aspx#2
Proctor, J. (2006). Religion as trust in authority: Theocracy and ecology in the United States. *Annals of the Association of American Geographers, 96*, 188–196.
Scaccia, B. (2010). "Taking" a different tack on just compensation claims arising out of the Endangered Species Act. *Ecology Law Quarterly, 37*, 655–682.
Sittler, J. A. (1954). A theology for earth. *The Christian Scholar, 37*(3), 367–374.
Stuntz, W. (2003). Review of M. McConnell, R. Cochran, A. Carmella (2001). *Christian perspectives on legal thought*. New Haven: Yale University Press in *Harvard Law Review*, (116), 1707–1749.
Thompson, B. (1996–1997). The Endangered Species Act: a case study in takings and incentives. *Stanford Law Review*, 49, 305–380.
Waskom, R., & Kallenberger, J. (2012). *Graywater reuse and rainwater harvesting*. Factsheet 6.702. Colorado State University Extension. 8/3/2012. Retrieved September 30, 2012, from http://www.ext.colostate.edu/pubs/natres/06702.html
White, L. (1967). The historical roots of our ecologic crisis. *Science, 155*(3767), 1203–1207.
Zimmerer, K. (2010). Retrospective on nature-society geography: Tracing trajectories (1911–2010) and reflecting on translations. *Annals of the Association of American Geographers, 100*, 1076–1094.

Chapter 67
The Camel and the Eye of the Needle: Religion, Moral Exchange and Social Impacts

Lucas F. Johnston and Robert H. Wall

67.1 Introduction

Jesus of Nazareth reportedly said (to paraphrase) that the chance of a rich man getting to heaven was about the same as the chance that a camel could fit through the eye of the needle—in other words, highly improbable.[1] But as historian Donald Worster put it, "Apparently, in America, however, our camels were smaller and our needles larger" (1993: 14). Religious production is typically most intense and diverse at the peripheries, the "ecotones" between cultural, ethnic or religious groups. North America, a space once defined by its expanding frontiers (Turner 1893 [1956]: 3) and a variety of cultural backgrounds and modes of exchange, provided a unique breeding ground for innovative religious communities and practices. These novel religious forms were forged in the fires of capitalism as it spread across the continent, consuming resources and promoting cultural homogenization.[2] But in many cases religious groups which were born in these diverse and highly productive peripheries have, through command of financial and thus political power, moved into the mainstream.

[1] See Matthew 19:24 in the Christian Bible: "it is easier for a camel to go through the eye of a needle than for someone who is rich to enter the kingdom of God." See also Mark 10:24–25, Luke 18:24–25.

[2] Examples are legion, but one illustration from North America is the widespread removal of American Indian children from their homes and their involuntary enrollment in schools where they were forbidden to speak their languages, or participate in traditional spiritual practices.

L.F. Johnston (✉)
Department of Religion and Environmental Studies,
Wake Forest University, Winston-Salem, NC 27109, USA
e-mail: johnstlf@wfu.edu

R.H. Wall
Counsel, Spilman Thomas & Battle, PLLC 110 Oakwood Drive Suite 500,
Winston-Salem, NC 27103, USA
e-mail: rwall@spilmanlaw.com

S.D. Brunn (ed.), *The Changing World Religion Map*,
DOI 10.1007/978-94-017-9376-6_67

Taking as qualitative data analyses of the ways in which exchange relations and their theological justifications impacted various efforts to map, classify, and conquer (Chidester 2004), this analysis focuses on the nexus of economic activity, specifically investments, colonialist expansion, and contested spaces. First, discussion of the historical relationship between capital investment and a capitalist-democratic society will be illustrative. The intention here is not to provide an exhaustive history, but rather a representative analysis of theological and social developments related to money and investment, and how they in turn were shaped by heterotopic geographies wherever they unfolded. Second, we will review some specific examples of how religious individuals and organizations have highlighted the supposed authenticity of their religious identity for personal or community gain. It will become clear that currency, property and investments are not doing *only* God's work. Rather, their accrual and exchange have long provided material expressions of the authenticating discourses deployed by various religious authorities.

The decimation of native populations in North America, the Australian continent, and beyond is sufficient reminder that relationships between money, political power, and social efficacy always also have significant moral import. In some cases religious aims were achieved through economic means, and economic need sometimes shaped theological interpretations. The so-called gospel of prosperity is perhaps the quintessential example of a theology which reflects the materialistic tenor of its cultural context. The focus here is on such rich terrain, where theological justifications and explanations relate to exchange relations and investment placements, which in North America at least, have taken a multitude of guises.

Ultimately religions are ecological patterns of psychological and biological expressions (Sperber 1985). For instance, televangelism—the use of television as a medium for growing a ministry—was peculiar to the global north, where conspicuous consumption was considered by some to be a virtue. Such a narrative presupposes a class society, specifically one in which the upper classes are set apart by their command and procurement of capital.

Take for example the Methodist-Pentecostal-Evangelical pastor Oral Roberts, who in January of 1987 declared on television that if he did not raise four and a half million dollars by the end of March, God would "call him home."[3] Presumably, Roberts' God did not intend to send him back to Ada, Oklahoma, but rather to his eternal heavenly abode. When the deadline passed, Roberts emerged from his immense "prayer tower" on the university campus which bore his name, and

[3]This identification acknowledges that he was raised in a nominally Methodist home, and gravitated toward Pentecostal and charismatic forms of practice early in his public career. Broadly speaking evangelicalism is a movement not limited to a specific denomination. It is characterized primarily by belief in the Bible as the authoritative source for understanding how to live on earth, and the belief that the "good news" of salvation must be actively proclaimed (for a more detailed definition and description of evangelicalism see Marsden 1991, especially pp. 1–3 for definitions).

announced that he had raised the funds required to delay his heavenly journey.[4] But his pilgrimage toward the prosperity gospel began much earlier, at the age of 29, when the then poor, itinerant preacher picked up his Bible, and it fell open to the third letter of John. Verse two caught his eye: "I wish above all things that thou mayest prosper and be in health, even as thy soul prospereth."[5] Roberts concluded that this was a sign that God intended him to be rich. So the next day Roberts purchased a Buick, and afterward perceived that God spoke to him and told him to heal people (Reed 2009). Taking full advantage of radio and television to promote his novel interpretation of the gospel, his ministry began to grow, reaching its height in the 1980s. Ultimately, both Roberts and his son Richard were implicated in lawsuits over their lavish and fraudulent use of funds. Neither was convicted of any wrongdoing, though Oral was forced to sell some of his vacation properties and luxury automobiles to pay debts (Lobdell 2009; Reed 2009).

Roberts pioneered both televangelism and the popularization of the Gospel of Prosperity (Schneider 2009). Yet with regard to the latter, Roberts has been overshadowed by pastor personalities that dwarf his ample income, and who surpass Roberts in the savvy with which they have invested the spoils from spreading God's word for politically beneficial outcomes. For instance Kenneth Copeland, one of these prosperity preachers, in conversation with self-styled historian David Barton, has argued on television that only certain forms of taxation are appropriate. The existing tax system (under U.S. President Obama) was labeled "extortion," while taxation for the purpose of national defense (that is, military spending) was viewed as biblically mandated (KCM 2012). Copeland announced that:

> They [liberals/socialists] didn't say a lot against Christians until the last fifty years, [when] we began to preach prosperity. [Then,] here they came! Because we began to prosper, and it's the ministries that are doing the work prospering.[6]

The fruits of such supposedly sacred labors can be better grasped when the material causes and consequences of such movements are analyzed. The success of the presently dominant mode of capitalism owes much to its amenability to religious anthropologies and sentiments.

---

[4] Roberts founded Oral Roberts University in 1963 (see www.oru.edu, accessed 23 July 2012).

[5] This version of the text is from the King James Version (KJV) of the Christian Bible, considered authoritative by most Pentecostals and many evangelicals. The translation of the KJV, however, is considered to be highly problematic by many biblical scholars. Interestingly, not all translations of the verse include the term "prosperity." The New Revised Standard translation, for instance, reads: "I pray that all may go well with you and that you may be in good health, just as it is well with your soul." For more on the problematic King James translation, see Ehrman (2007).

[6] See http://www.kcm.org/media/webcast/kenneth-copeland-and-david-barto/120830-gods-way-to-increase. The program originally aired 30 August 2012. It should be noted that there was no recognition that today's Republicans and Democrats (the two currently dominant political parties in the United States) have different platforms than they have historically. "Liberalism," in this episode, was equated with "socialism," "communism," and godlessness, and even referred to as "pagan," which of course illustrates a rather obvious muddling and misuse of these terms.

67.2 A Case in Point: The Spirit of Capitalism and the Gospel of Prosperity

An introductory religious studies text by scholar of religion and multiculturalism Malory Nye suggests that what he refers to as the "Faith and Prosperity movement" ultimately emerged from Pentecostal roots (Nye [2009] (2007)). In some Pentecostal communities, evidence of divine favor supposedly manifests in publicly visible material practices. For instance, speaking in tongues, or in a small number of Pentecostal churches, practices such as dancing with venomous snakes without injury, demonstrate that an individual is "under the anointment of God."[7] In charismatic churches, born from Pentecostal movements, the visible signs of divine favor or grace were imagined to be related to financial gain. Ultimately these Pentecostal and charismatic forms of Christianity have even deeper roots which reach back to the emergence of evangelical and Pietist movements in Europe in the 1700s. Interestingly, although evangelical, Pietist and Pentecostal movements were typically viewed—at least in the cultural milieus from which they sprang—as counter-hegemonic groups, they would become, in the form of people like Oral Roberts, decidedly mainstream.

Although the celebration of financial wealth as a manifestation of divine grace came to its most obvious fruition among twentieth century televangelists, the stage was set for capital accumulation and investment at least as early as the Protestant Reformation in Europe. Protestant movements began in part because of a widespread perception that the Catholic Church was economically exploiting its constituents, amassing significant wealth often displayed in ornately decorated churches, and demanding payment for salvation (that is, the selling of "indulgences," which ensured forgiveness of sins). Thus, the first Protestant movements, such as the one spearheaded by John Hus in the late fourteenth century, or Martin Luther's later protestations, frowned upon ostentatious public displays of wealth. Hussites (as followers of Hus were known) and later Anabaptists from central Europe retained a community-minded ethos, but rejected the rigid hierarchy and lavishness of the Catholic Church. Others, such as those who followed more closely on Luther's heels, endorsed a more radically individualist (as opposed to community-based) ethos. To divest priests of their power over citizens, some of these early Protestants emphasized the soteriological importance of grace over works (reversing the Catholic Church's position), and focused on a personal relationship with God (rather than one mediated by a priest).[8] The personal and affective relationship with the divine was the theological seed that, when exposed to the vast and largely rural

[7] This is a quote from a young preacher in a Pentecostal snake-handling church in Eastern Tennessee, which was featured on CNN (see www.cnn.com/video/?hpt=hp_c3#/video/bestoftv/2012/07/05/dnt-snakehandling-church.wmc, accessed 5 July 2012).

[8] The notion that grace is more important than works was related to the idea of predestination, discussed below. If God is indeed omnipotent and knows before an individual's birth the fate of her soul, then one cannot simply earn her way to heaven by doing good deeds. Only divine grace, so the logic went, could be responsible for saving a person's soul.

populations of the expanding United States, would flower into the highly emotional religious movements associated with the First and Second Great Awakenings in North America, including Pentecostalism.

It was this latter religious form associated with individualism which the sociologist Max Weber analyzed in what might be termed a sociology of rationalization (Weber [1930] 2005).[9] Weber noted the above historical trends, and highlighted how one particularly strong mode of Christianity—Calvinism—as well as those manifestations of Christianity which were particularly important in the development of the United States (for example, Puritanism), provided the widespread psychological conditions for capital accumulation. Calvinism and Puritanism both emphasized the notion of an omnipotent, omniscient deity who knew, prior to each individual's birth, whether that soul was destined for heaven. Though God might know who was destined for heaven, such theologies fueled a pervasive anxiety among Christians about the individual soul's ultimate abode. This *anomie*, Weber believed, combined with a capitalist ethos which promoted rationalization, modernization, and individualization, and which resulted in the cultural homogenization of Europe and North America.[10] From a theological perspective, the only way to tell if one was "favored" was to see the fruits of godliness manifested in personal success or well-being. From a capitalist perspective, the measure of success was the same—personal wealth and well-being. This was what Weber referred to as an "elective affinity" between the Protestant ethos and the spirit of capitalism (Weber 2002).[11] Importantly, however, while wealth was considered evidence of divine favor, displaying this wealth was frowned upon. As Weber put it:

> asceticism looked upon the pursuit of wealth as an end in itself as highly reprehensible; but the attainment of it as a fruit of labour in a calling was a sign of God's blessing. And even more important: the religious valuation of restless, continuous, systematic work in a worldly calling, as the highest means to asceticism, and at the same time the surest and most evident proof of rebirth and genuine faith, must have been the most powerful conceivable lever for the expansion of that attitude toward life…called the spirit of capitalism. ([1930] 2005: 116)

Investment and re-investment of the profits from business ventures—the process of capital accumulation—demonstrated God's grace *and* provided an avenue by which capital could be collected without overtly celebrating what was imagined as an ultimately fallen material world. The emergence of modern capitalism was thus driven by, or at least buttressed by, theological developments.

---

[9] Weber believed that the result of the Protestant ethic was evidence of the loss of a sense of enchantment in the profane world and of the rise of modernism.

[10] The term anomie describes the dissolution of social bonds and resultant feelings of existential crisis. The term is often attributed to the French sociologist Emile Durkheim, who popularized the term (though he did not coin it). Although Weber's work has explanatory power, it is clear that the continent was certainly not homogenous. Those who wielded increasing economic power through the colonial period, though, did share some theological and practical commitments.

[11] The translation by Talcott Parsons, cited below, translates the term "wahlverwandtschaften" as "correlations" rather than "elective affinity," which appears in the translation from Baehr and Wells. Thanks to Bernie Zaleha, who pointed out to us this difference in translations, and who buttressed our interpretation of Weber's work throughout.

As a further illustration, consider the Moravians, who ultimately trace their theological genealogy back to John Hus. The Moravians funded their extensive missionary work across the globe by exploiting exchange networks in North America, the West Indies and Europe. In the late 1750s, the Moravians created the Commercial Society, which combined private and church capital, and effectively "structurally and legally connected religious work and transatlantic commerce" (Engel 2007: 120). This group provides a provocative illustration of Weber's thesis that religious ideologies could play the role of a switchman (Weber [1948] (1958)), displaying elective affinities with economic ideologies, and actualizing particular historical outcomes and foreclosing others. It also challenges, however, Weber's assumption that rationalization, bureaucratization, and individualization resulted in pervasive cultural homogeneity. For the Moravians founded their important colonial-era settlements in Europe and the United States on distinctly communitarian and collectivist values, in some ways radically opposed to the individualistic ethos of other mainstream Christian groups. But they also participated as much, if not more fully, in the transatlantic economy than many of their contemporaries. They thus contributed to the spirit of capitalism without embracing all of the features of Weber's Protestant ethic. The story, then, is more nuanced than Weber imagined, though it is clear that despite different values and motivations, there were many constituencies (including but not limited to mainstream Protestant groups) who shaped the ways in which capitalism and capital accumulation proceeded in the modern period.

While Weber's work is helpful in illuminating the relationships between ideologies and social developments, it is clear that these variables are also subject to geographic and demographic pressures that were largely outside his purview. The European Renaissance, indirectly and perhaps ironically spurred by several attempts to capture an important sacred space (Jerusalem), generated an increasingly wide-reaching series of exchange routes, and fueled European cultural elites' appetites for Asian goods. Political and religious leaders shepherded the search for luxury goods and cheap resources to feed the growing political economy of Europe.

Even earlier demographic changes provided a foundation for the later spread and success of Christian colonialism. For instance, by the turn of the Common Era both Europe and Asia had developed densely populated urban areas, and in Europe burgeoning populations often lived in close proximity to domesticated animals. Such dense population centers facilitated the rapid mutation and transmission of pathogens (many of which were first transmitted from non-human to human animals), which significantly impacted Europeans. Their most visible and widespread effects, however, were among the indigenous cultures that were the victims of colonization. As soldiers, priests and bureaucrats were unleashed across the oceans (Axtell 1992: 72) in large vessels drawn from budding European industrial areas, they mapped, named and then transformed the landscapes they encountered. The first zoos and botanical gardens appeared in Europe to showcase the creatures (human and non-human) gathered from far-away lands (Chidester 2004). Bureaucrats laid claim to lands by cataloguing and re-naming the world, a process which reflected the "encyclopedia" mentality of Renaissance Europe. To illustrate, Queen Isabella of Spain supposedly asked the royal historiographer Elio Antonio de Nebrija about the purpose of the first published grammar of the Castilian language in the vernacular, to which he reportedly replied: "Your Majesty, language has always been the companion of Empire" (Sale 1991: 18). Indeed, the geographic expansion of European cultures during the age of colonization provided not only metals and riches, but also important food sources, trade goods, raw materials and medicines (Sale 1991: 46), changing the world's religious landscape in significant ways.

Thus, the capitalist mode of production (including the global expansion of capitalist markets, and significant increases in manufacturing efficiency, consumption, and wealth disparity) was constrained by geographic and biological features of habitats, and in turn shaped the theology and practice of religious individuals and groups. For instance, the unique communitarian ethos of the Moravian church weakened and eventually dissolved as their settlements transformed from utopian frontier communities into centers of commerce and industry. The group grew into a relatively mainstream denomination that in some ways is virtually indistinguishable theologically from other mainstream Christian groups. It is clear, then, that the affinity between the individualistic Protestant ethos and a form of capitalism which maximized rational pursuit of self-interest and promoted bureaucratization sometimes eroded the counter-hegemonic energies of evangelical and charismatic groups, bending them toward the political and social center. Likewise, the gospel of prosperity, birthed in movements that were primarily rural and non-hierarchical, spread through populations like a contagious cultural virus, and moved into the mainstream as capitalistic individualism became its own global religion (Loy 2000).

The relationship between religion, money, and investments, then, is complex and multi-faceted, as Weber himself insisted. Religious ideologies were often related to, or grew from, specific social and economic relations. And in many cases, such ideologies were used to authenticate these social and economic relations. As the religion scholar David Chidester has argued, ideologies perform religious work to the extent that they channel economic relations in particular directions, focus community desire, and bind communities of people together with reference to sacred values (2005: 8). The jargon of authenticity, cloaked in religious concepts, metaphors and justifications, has always been the handmaiden of cultural changes which shift power from center to periphery, or draw power to the center. Authenticating discourses have been deployed by those in power to solidify and expand their exploitation of resources (human and natural), as well as exercised by marginalized constituencies to challenge the values at play in economic, social and imaginary worlds. So when Oral Roberts (and others in the late twentieth and early twenty-first centuries) drew on religious concepts to make normative claims about why their deities wanted them to be rich, they were drawing on a long history of religious justifications for exchange relations which have oscillated between preserving the status quo, and challenging it.

When anyone from Terry Jones (of "World Koran Burning Day" fame) to Fred Phelps (founder of www.godhatesfags.com) to Rick Santorum (U.S. Senator from Pennsylvania and former U.S. Presidential candidate) can all claim the legitimacy of their own distinctive interpretation of Christianity, it is important to analyze the financial and social impacts of such claims to authenticity. To provide further texture to the varied relationships between religious individuals and groups and their financial management, some specific examples will be illustrative. In some instances religious groups have invested human and financial resources in social activism. At other times, religious justifications (both explicit and implicit) have been utilized to undermine or restrict the rights of particular constituencies. This raises questions about the relationship between religious groups and individuals' investments and the ongoing relationships between missionary work and economic expansion. What follows is a description of some contemporary examples of these multifarious relationships.

67.3 A Moral Market? Religion-Resembling Economic Discourse and Responsible Investing

Economic markets and investment strategies have always been used to exercise power over others, and the exercise of such power is often related to specific sacred places. But there have been a variety of religious or religion-resembling responses to the coupling of economic well-being and religion. Relating the market and morality, some economists have suggested that the sort of valuation endorsed by prevailing economic and political powers (using Gross Domestic Product, for example, as a measure of social success) is misguided. In other cases, religious organizations have helped to pioneer socially responsible investing (SRI), a rapidly growing area of interest and an increasingly potent force for social change.

Examples of the former arguments include a response by the editors of the well-known journal *The Ecologist* to Donnella Meadows et al.'s classic *Limits to Growth* (1972). Their response, titled *Blueprint for Survival*, aimed to promote what they referred to as "a new philosophy of life" which might bring on "the dawn of a new age in which Man (*sic*) will learn to live with the rest of nature rather than against it" (Goldsmith et al. 1972: vi). The authors approvingly cite a lecture given by Bishop of Kingston in which the Bishop[12] provides a new set of commandments:

> … there must be a fusion between our religion and the rest of our culture, since there is no valid distinction between the laws of God and Nature, and Man must live by them no less than any other creature. Such a belief must be central to the philosophy of the stable society, and must permeate all our thinking. (Meadows 1972:165)

In this interesting passage, the journal editors used religion in a broad sense to refer to core values and deep beliefs, and to claim that appropriate investments reflect such values, rather than aiming for mere accumulation of wealth. Former World Bank economist Herman Daly (1996: 218) has argued that:

> we need a new central organizing principle—a fundamental ethic that will guide our actions in a way more in harmony with both basic religious insight and the scientifically verifiable limits of the natural world. This ethic is suggested by the terms "sustainability," "sufficiency," "equity," "efficiency."

[12] The reader is introduced to the bishop quite suddenly, with no real explanation of who he is, or why he is important to the authors.

Like the Bishop, Daly provided an "11th commandment": "Thou shalt not allow unlimited inequality in the distribution of private property" (1996: 206).[13]

It is not unusual for scientists and economists to use highly affective and often religious language to describe alternative economic priorities. In some cases religious individuals and groups have also used their social visibility and financial muscle to invest in companies and corporations that are aligned with their values. The Quakers were perhaps the first to engage in SRI in the 1700s, when they channeled their resources into investment opportunities that were perceived to be nonviolent. They retain their interest in SRI today, and in fact, now most of the so-called "world religions" have targeted equity or mutual funds which seek to maximize these alternative, life-affirming values.[14] Some Catholic and conservative Protestant Christian SRI funds, for instance, avoid investments which directly or indirectly support organizations engaged in family planning or termination of pregnancy. Muslim investment groups may avoid supporting corporations engaged in the manufacture of alcohol, pork, or tobacco.

Do such faith-based investment sacrifice financial returns by limiting their stock options? There is some debate here, but some scholars suggest that faith-based investments generally perform better than the market average, and in some cases better than combined SRI funds (see, for example, Lyn and Zychowicz 2010). Interestingly, the stated values of such faith-based investment firms seem to be directed primarily at investment prohibitions, rather than providing positive guidance about what businesses should be supported.

67.4 Render unto Caesar

Unique to the United States, one important area in which religion intersects with finance and investment is in the realm of taxation, namely, the status of religious organizations under federal tax laws. Since the ratification of the Bill of Rights and the specific prohibition against establishment of a state sanctioned belief systems or state impediments to the free exercise of religion, religious organizations in the United States have existed in a realm in which oversight by taxing and regulatory authorities has been in flux—a sort of tax purgatory, where neither the higher (governmental) power nor the adherent are bound by a clear set of guidelines.

Historically, religious organizations have been governed by their own set of rules within a broad framework of federal and state laws and regulating authorities. Disputes in the U.S. between governing authorities and religious organizations predate the passage of the Sixteenth Amendment in 1913, out of which arose the

[13] Daly grounds his argument on what he calls a biblical basis, which provides a particularly cutting commentary on the ideas of equity and economics.

[14] For examples, see Friends Fiduciary (Quaker), www.friendsfiduciary.org/socially-responsible-investing/; Amana Mutual Funds Trust (Islamic), www.amanafunds.com/; the U.S. Conference of Catholic Bishops' SRI guidelines, www.usccb.org/about/financial-reporting/socially-responsible-investment-guidelines.cfm; Guidestone Financial (Baptist), www.guidestone.org/; Everence (Mennonite), www.everence.com/, accessed 24 July 2012.

income tax. But the number of disputes between taxing authorities and religious organizations has increased dramatically since that time, particularly in the last 40 years.

As churches and religious organizations have evolved into stewards of material wealth, taxing authorities have reacted with oversight, while ever-aware of the limitations set forth by the First Amendment. Religious organizations are not explicitly required to seek tax-exempt status under federal law—such a requirement would be a potential violation of the Establishment Clause. In the event that adherents of a religious organization seek to receive the tax benefit of tithes, offerings and donations to a particular religious organization, however, said organization must establish that it has a religious purpose pursuant to federal law. Presently, federal tax law defines neither "church" nor "religious organization," lest it violate the broad brush of First Amendment protections. Under its regulatory authority, however, the Department of the Treasury attempts to more narrowly define "Church" through questions on its form for exemption regarding matters of "religious creed," "hierarchy," "members," and "ecclesiastical government" (see Internal Revenue Service Form 1023, Schedule A).

Religious organizations have consistently sought status as "churches" for federal tax purposes to take full advantage of the tax exemption afforded churches for unrelated business taxable income. The Internal Revenue Code does not define the term "church;" however, the term is defined in Treasury Regulation § 1.511-2(a)(3)(ii), pertaining to organizations exempt from taxes on unrelated business income. The regulation provides in part:

> (ii) The term 'church' includes … a religious organization if such … organization
>
> ((a)) is an integral part of a church, and
>
> ((b)) is engaged in carrying out the functions of a church, whether as a civil law corporation or otherwise.
>
> In determining whether a religious … organization is an integral part of a church, consideration will be given to the degree to which it is connected with, and controlled by, such church. A religious … organization shall be considered to be … carrying out the functions of a church if its duties include the ministration of sacerdotal functions and the conduct of religious worship … [which is to be determined based upon] the tenets and practices of a particular religious body constituting a church.

Original jurisdiction for cases involving a determination or revocation of tax-exempt status of religious organizations lies with the Court of Federal Claims and not typically with the federal district court in which the organization is based. Located in Washington, D.C., the Court of Federal Claims is unattainable for many religious organizations simply due to its centralized location. Interestingly, this Court has noted as recently as 2009 "that it is, and believes it should be, uncomfortable with the criteria used by the IRS for determination of the church status of religious organizations." Foundation of Human Understanding v. U.S., 88 Fed. Cl. 203 (2009). In dicta unrelated to the ruling of the case, the Court strongly suggested that the government failed to address significant First Amendment problems as they relate to taxation of churches and the property of churches. In declining to address the Constitutional issues related to the definition of "church," the Court went on to state:

The criteria used by the IRS to determine church status for tax purposes… bear a striking similarity to the topics of the questions contained in the 1906 Census of Religious Bodies. This resemblance strongly suggests that defendant's criteria are time-conditioned and reflect institutional characteristics that no longer capture the variety of American religions and religious institutions in the twenty-first century. The regime appears to favor some forms of religious expression over others in a manner in which, if not inconsistent with the letter of the Constitution, the court finds troubling when considered in light of the constitutional protections of the Establishment and Free Exercise Clauses.

While the language of the Court is not conclusive, it certainly provides some insight into why the government and the Internal Revenue Service may be hesitant to seek further determinative and binding rulings pertaining to the government taxing authority over religious organizations.

As the Court noted, what counts as a religious organization is, despite the Constitutional limitations, undoubtedly restricted to a Judeo-Christian worldview. In the Islamic tradition, for example, mosques and Islamic Centers maintain no membership rolls to substantiate that an adherent is a member or a participant in activities of the center. Similarly, most Buddhist sects lack what would properly be deemed to be an ecclesiastical government.

With the advantages that come with recognition of tax-exemption under federal guidelines, so too come the restrictions. Within the restrictions of federal tax law exists the regulatory framework that shapes many of the investments of modern religious organizations. Religious organizations, like all "charitable" organizations, avoid paying taxes on items of income that are related to the religious purpose of the organization. Like other similarly situated tax-exempt organizations, religious organizations are subject to a tax on unrelated trade or business income. The major question raised by these restrictions is what may qualify as being "related to" the religious purpose of the organization? Additionally, the taxing authorities have to closely examine whether the activities in question are in direct competition for revenue with other non-religious, for-profit ventures. Under most circumstances, these questions are answered either through the legal system or by the Internal Revenue Service. For instance, an organization that utilizes a commercial broadcasting license to broadcast religious television programs for substantially all of its broadcast time is subject to the exemption. Any revenue derived from programs broadcast for purely commercial purposes, however, is unrelated business and is thus subject to unrelated business income tax (see Revenue Ruling 78–385, 1978–2 CB 174). An organization that provides religious-based travel, which also includes social and recreational activities, however, is not operated exclusively for exempt religious purposes (see Revenue Ruling 77–366, 1977–2 CB 192).

Generally, investment income, in its broadest scope, is expressly excluded from taxation when those investments are owned by religious entities. Dividends, interest, annuities and other similar investments are exempt from taxation. In certain circumstances, including debt financing and utilization of corporations under the control of the religious organization, such income may be included as taxable by the religious organization. Likewise, royalties, rents, gains from the sale of property, income from the lending of securities, and income from the lapse of options is generally excluded from taxation. On a much smaller scale, sales of donated merchandise and commercial sale of farm products produced by a farm on which labor was derived from members who had taken a vow of poverty were related to the religious purpose of the organization (see St. Joseph Farm 85 TC 9, 1985).

In addition to legal issues of unrelated business taxable income, legal issues dominating the intersection of churches, investments and taxes primarily arise as they relate to state and local tax issues, property ownership, and private inurement. Very rarely do these topics rise to the level of a Constitutional dispute worth of certiorari by the United States Supreme Court; however, the modern Court has opined on tax issues related to federal unemployment taxes levied against religious organizations due to employment. St. Martin Evangelical Lutheran Church v. South Dakota, 451 U.S. 772, 780–85 (1981). The Court has, however, declined to render opinions on many of the issues related to taxation and investments simply because very few cases involving religious organizations continue through the legal process to such a result. That is, either the government or the organization is hesitant or can afford to press the matter to that conclusion.

All 50 states have some sort of property tax exemption for religious organizations and churches. State laws generally provide generous property tax exemptions for personal property and real property used for religious purposes; however, several states narrowly tailor property tax exemptions to only provide relief for designated "houses of worship." These narrowly construed rules have given rise to a number of disputes between state revenue authorities and religious organizations as to the religious purpose of the property at issue. *Holy Spirit Association v. Tax Commission*, 55 N.Y.2d 512, 518 (1982). The Supreme Court has, however, ruled that government administration of the property tax exemption does not violate the Establishment Clause because although it involves the government in religious institutions, it does not involve "judicial inquiry into dogma and belief" *Walz v. Commissioner*, 397 US 664 (1970).

Most recently, states have seen challenges in which groups litigating against property tax exemptions for religious organizations have argued that state property tax exemptions are constitutionally prohibited under the Establishment Clause. Specifically, a current ongoing legal dispute has been allowed to proceed against the California Department of Revenue and the Internal Revenue Service challenging parsonage exemptions under 26 U.S.C. § 107 and the corresponding California revenue statute. *Freedom From Religion Foundation, Inc. v. Geithner*, 715F. Supp. 2d 1051 (DC Ca 2010).

Churches, much as all charitable organizations, are specifically forbidden to benefit private individuals through financial gain or benefit. As discussed above, with the rise of certain evangelical movements related to God ordained material blessing on those in whom He finds favor, an increasing area in which churches and the tax system have clashed pertains to the benefit of individuals involved in religious organizations, and who derive significant benefit therefrom. Recent cases have involved the Internal Revenue Service criminally prosecuting a couple who served as pastors in a large evangelical church in Charlotte, North Carolina for underreporting taxable income by accepting what they deemed to be gifts from congregants, and not including them on their income tax returns. *US v. Jinwright*, 683 F3d 471 (CA 4 2012). In yet other cases, the Internal Revenue Service has been hampered in its examinations of these types of transactions by procedural rules and restrictions on investigating churches that have been put in place by Congress. *US v. Living Word Christian Center*, 102 AFTR 2d 2008–7220 (DC Mn 2008).

Thus far, Internal Revenue Service investigations into the financial benefit of individuals associated with various religious movements have not produced a significant shift in the legal positions of many of the most prevalent evangelical movements. Recent activity by congressional committees, however, has brought to light potential changes in the legal ramifications of such religious beliefs. In February 2009, Senator Charles Grassley, a Republican from Iowa, and ranking member of the Senate Finance Committee, launched a Senate investigation of six of the highest paid media savvy evangelists. Senator Grassley openly investigated Paula White, Joyce Meyer, Creflo Dollar, Eddie Long, Kenneth Copeland, and Benny Hinn. To date, no conclusive findings have been released by the committee. Interestingly, at the time of the inquiry, three of the individuals under investigation—Creflo Dollar, Benny Hinn, and Kenneth Copeland—served in some capacity, in leadership roles at Oral Roberts University (CBS News 2012).

Another possible legislative challenge emerged during the 2012 election cycle when several U.S. pastors declared October 7 to be "Pulpit Freedom Sunday" (Fox Insider 2012). Organized by a group called the Alliance Defending Freedom, which is itself a 501 (c)(3) tax-exempt ministry, the goal of the event is to encourage pastors to discuss the policy platforms of the two presidential candidates (Barack Obama and Mitt Romney), and then make a specific recommendation to their congregants regarding their vote (Alliance 2012). Event organizers have specifically stated that their goal is to challenge a 1954 amendment to the IRS tax code which prohibits tax exempt organizations from participating in a political campaign on behalf of a specific candidate. Their claim is that the tax code violates both their right to free speech and violates the Free Exercise Clause of the First Amendment of the US Constitution.[15] The argument is that if a particular religious group perceives that they have a religious obligation to participate in politics, then it is discriminatory to forbid that group to do so. In a sense, such argumentation pits the Free Exercise Clause against the Establishment Clause. If it is ultimately determined that the tax system discriminates against political speech in a religious setting, the entire tax code related to religious organizations would be delegitimated. While the complexity of these rules and tax codes have provided cause for great concern among both elected officials and religious organizations seeking guidance for rule compliance, the emergence of the evangelical movements of the last half of the twentieth century and the first decades of the twenty-first century and the events noted above indicate that religious organizations have grown more aggressive in their pursuit of the accumulation of wealth at the expense of taxing authorities. Interestingly, however, the federal government and the Internal Revenue Service have specifically restricted themselves in the further examination of the religious activities of religious organizations as it relates to the business interests of the organization (see 26 C.F.R. §301.7605-1(i)(4)). So although Jesus suggested that his followers ought to "render

[15] The First Amendment to the U.S. Constitution reads: "Congress shall make no law respecting an establishment of religion [the Establishment Clause], or prohibiting the free exercise thereof [the Free Exercise Clause])

unto Caesar those things that are Caesar's, and unto God the things which are God's" (Matthew 22:21), in the twentieth and twenty-first centuries, some Christians have been more inclined to find that money and investments are "God's things."

67.5 Missionary Investment and the Colonial Impulse

Missionary work has been one of the most obvious and large expenditures from religious organizations historically. Missionizing was both a financial investment, and in some sense also an investment in the future of the church. Most often, from the age of European colonization through the nineteenth century, missionaries were accompanied by military units and bureaucrats. These colonizers operated in two ways (sometimes concurrently, sometimes sequentially). In some cases, the aim was to exploit natural resources which had been held under traditional or indigenous land tenure. Religious conversion was often either a catalyst for this transfer of resources, or a product of it. For instance, in some cases high level actors in traditional societies were converted (sometimes voluntarily, sometimes not), and with their significant sway over land management decisions, they ceded power to colonizers. In other cases, traditional cultures' inability to halt rapid destruction of natural resources and the spread of disease prompted them to search for religious explanations for their plight. Prophets often appeared who adopted and adapted the Christian message and its salvific concepts as a salve for their cultural disruption (see, for example, Wright 2009). In still other instances saving "savage" souls was the explicit motivation for colonial endeavors, even if the most immediate gains were manifest in financial prosperity of the colonizers.

Colonial administrators in Bali, for instance, replaced remarkably complex water management systems that had developed over thousands of years with bureaucratic, Westernized modes of management. The result was ecologically disastrous, but the negative effects went unnoticed until the 1970s (Lansing 1991). The anthropologist Michael Taussig noted the ways in which traditional devil-focused religious practices in Columbia morphed as indigenous communities were exposed to capitalist modes of production. For the natives, capitalism masked the violent and dehumanizing lifeways which fueled its expansion, and was thus the most obvious way in which the devil manifested in the real world (1980). Robin Wright, an anthropologist who lived and worked in Brazil for decades, noted that religious non-governmental organizations (NGOs) disrupted traditional social relations, resulting in an increase in both evangelism, prophetic zeal, and assault sorcery (2009). While the NGOs intended to provide avenues for indigenous populations to engage in the global market, in this case by selling their traditional basketry, they contributed to the erosion of social bonds and the rise of interpersonal and intergroup conflict.

Such cultural disruptions have often proceeded under the guise of co-called "sustainable development" projects, intended to raise the standard of living of particular constituencies by including these marginalized characters in the global economy. As often as not, however, the result has been a transformation that fixes socioeconomic relations into a self-perpetuating rationalized, efficiency-maximizing state—something

Max Weber predicted would necessarily follow from a truly global form of capitalism. Drawing on Richard Baxter, an expert in Puritan ethics, Weber noted that

> in Baxter's view the care for external goods should only lie on the shoulders of the 'saint like a light cloak, which can be thrown aside at any moment.' But fate decreed that the cloak should become an iron cage. (2005 [1930]: 123)

This iron cage, the cultural inertia of capital accumulation through investment in a system that maximizes individualization and bureaucratic development, is where we find Oral Roberts and his ilk, utilizing the levers of economic and social power to increase their possessions in the name of God.

67.6 Conclusion

The developments noted above illustrate the ways in which the global distribution of religious beliefs and practices has shifted significantly over time. In addition to shifting religious allegiances, the physical map of the world has changed, as capitalism, "discovery" and colonization, and more recently, corporatism (what has been called the "McDonaldization" of the globe) have been exported from North America, and resources (both human and non-human) have been more efficiently exploited. What remains to be seen is how these developments will play out. Economic globalization has also resulted in the exportation of peculiarly western ideals, such as multiculturalism, religious tolerance, and democracy. This raises and important question: do concepts related to democratic participation and life-affirming values (however defined), packaged for the developing world in terms of "freedom and democracy" or "global ethics," act as evangelists for a global faith that merely perpetuates the central values of Western colonialist cultures (Chua 2003)? At the very least, they seem to reinvigorate the emergence of strong religious groups who respond to cultural pluralism and tolerance with heightened rhetoric, intensified proselytization, and occasionally violent reactions (Jurgensmeyer 2008). The future is uncertain, but past and current impacts are clear—the globe has been religiously and physically re-shaped by capital accumulation, investment, and expansion.

References

Alliance. (2012). www.alliancedefendingfreedom.org/. Accessed 26 Sept 2012.

Axtell, J. (1992). *Beyond 1492: Encounters in colonial North America*. New York: Oxford University Press.

CBS News. (2012). *Senate Panel Probes 6 Top Televangelists*. www.cbsnews.com/8301-500690_162-3456977.html. Accessed 16 Sept 2012.

Chidester, D. (2004). Classify and conquer: Friedrich Max Muller, indigenous religious traditions, and imperial comparative religion. In J. K. Olupona (Ed.), *Beyond primitivism: Indigenous religious traditions and modernity* (pp. 71–88). New York: Routledge.

Chidester, D. (2005). *Authentic fakes*. Berkeley: University of California Press.

Chua, A. (2003). *World on fire: How exporting free market democracy breeds ethnic hatred and global instability*. New York: Anchor Books.
Daly, H. E. (1996). *Beyond growth: The economics of sustainable development*. Boston: Beacon.
Ehrman, B. (2007). *Misquoting Jesus: The story behind who changed the Bible and why*. New York: HarperOne.
Engel, K. C. (2007). Commerce that the Lord could sanctify and bless: Moravian participation in transatlantic trade, 1740–1760. In M. Gillespe & R. Beachy (Eds.), *Pious pursuits: German Moravians in the Atlantic world* (pp. 113–126). New York: Berghahn Books.
Fox Insider. (2012). *Pastors to challenge IRS by preaching politics on 'Pulpit Freedom Sunday'*. http://foxnewsinsider.com/2012/09/25/pastors-to-challenge-irs-by-preachingpolitics-on-pulpit-freedom-sunday/. Accessed 26 Sept 2012.
Goldsmith, E., Allen, R., Allaby, M., Daroll, J., & Lawrence, S. (1972). *Blueprint for survival*. Boston: Houghton Mifflin.
Jurgensmeyer, M. (2008). *Global rebellion*. Berkeley: University of California Press.
KCM. (2012). *God's way to increase*. http://www.kcm.org/media/webcast/kenneth-copeland-and-david-barto/120830-gods-way-to-increase.
Lansing, S. (1991). *Priests and programmers: Technologies of power in the engineered landscape of Bali*. Princeton: Princeton University Press.
Lobdell, W. (2009, December 16). Oral Roberts dies at 91; Televangelist was pioneering preacher of the "prosperity gospel." *LA Times*. Available at www.latimes.com/news/obituaries/la-me-oral-roberts16-2009dec16,0,3407978.story. Accessed 28 Aug 2012.
Loy, D. R. (2000). The religion of the market. In H. Coward & D. C. Maguire (Eds.), *Visions of a new earth: Religious perspectives on population, consumption, and ecology* (pp. 15–28). Albany: State University of New York Press.
Lyn, E., & Zychowicz, E. J. (2010). The impact of faith-based screens on investment performance. *The Journal of Investing, 19*(3), 136–143.
Marsden, G. M. (1991). *Understanding fundamentalism and evangelicalism*. Grand Rapids: Wm. B. Eerdmans.
Meadows, D. (1972). *The limits to growth: A report for the club of Rome's project on the predicament of mankind*. New York: Universe Books.
Nye, M. [2009] (2007). *Religion: The basics*. New York: Routledge.
Reed, C. (2009, December 15). Oral Roberts Obituary. *Guardian*. Available at www.guardian.co.uk/world/2009/dec/15/oral-roberts-obituary. Accessed 28 Aug 2012.
Sale, K. (1991). *The conquest of paradise: Christopher Columbus and the Columbian legacy*. New York: Plume.
Schneider, K. (2009, December 15). Oral Roberts, Fiery Preacher, Dies at 91. *New York Times*. Available at www.nytimes.com/2009/12/16/us/16roberts.html?pagewanted=all. Accessed 28 Aug 2012.
Sperber, D. (1985). Anthropology and psychology: Towards an epidemiology of representations. *Man, 20*, 72–89.
Taussig, M. (1980). *The devil and commodity fetishism in South America*. Chapel Hill: University of North Carolina Press.
Turner, F. J. [1893] (1956). *The significance of the frontier in American History*. Ithaca: Cornell University Press.
Weber, M. [1948] (1958). The social psychology of the world religions. In H. H. Gerth & C. W. Mills (Eds.), *From Max Weber: Essays in sociology* (pp. 267–301). New York: Oxford University Press.
Weber, M. [1930] (2002). *The Protestant ethic and the "sprit" of capitalism and other writings*. New York: Penguin Books.
Weber, M. [1930] (2005). *The Protestant ethic and the spirit of capitalism* (trans: Talcott Parsons). London: Routledge.
Worster, D. (1993). *The wealth of nature: Environmental history and the ecological imagination*. New York: Oxford University Press.
Wright, R. (2009). The art of being *Crente*: The Baniwa protestant ethic and the spirit of sustainable development. *Identities: Global Studies in Culture and Power, 16*, 202–226.

Chapter 68
Law and Religion: The Peculiarities of the Italian Model—Emerging Issues and Controversies

Maria Cristina Ivaldi

68.1 Introduction

In this chapter, the perspective used to study religion and its changes is the law. As is well-known, Italy is a civil law country. There is a specific branch of Italian law, called "Diritto ecclesiastico," devoted to investigating the treatment of the religious factor, especially in terms of religious freedom and the relationship between the Church and the State.[1]

After a brief analysis concerning constitutional provisions for religion and the principle of secularism, as defined by the Constitutional Court since the late 1980s, the essay examines the issues and controversies among those not yet resolved in the Italian order. The questions have a background in the deep changes occurring in the religious landscape that are caused by the increasing secularization of society, which has significantly eroded the number of individuals identifying themselves as Catholics, the historical religion of the majority of citizens and the advent of new religious movements not traditionally present in Italy. Circumstances that arose as a result of the meaningful shift of believers to other religious faiths, in particular Jehovah's Witnesses, but even more by the new demography caused by the waves of immigrants who have hit the Country in recent decades, including a significant portion of whom are Muslim.[2]

We focus on topics that appear more moderate and more critical, including the multidimensional nature of religion and the transformation of the legal framework in the country that are caused primarily by Italy's accession to supranational bodies

[1] On this topic, see, among others, Casuscelli (2012), Finocchiaro (2012), Vitali and Chizzoniti (2012), Musselli and Tozzi (2007) and, for the European context, see Doe (2011).

[2] For changes in religious affiliation, see Davie (1994) and Hervieu-Léger (1993).

M.C. Ivaldi (✉)
Dipartimento di Scienze Politiche "Jean Monnet", Seconda Università degli Studi di Napoli, Viale Ellittico, 31, Caserta 81100, Italy
e-mail: mariacristina.ivaldi@unina2.it

S.D. Brunn (ed.), *The Changing World Religion Map*,
DOI 10.1007/978-94-017-9376-6_68

such as the COE, OSCE and EU. Undoubtedly, the presence of religious symbols in the public sphere (and especially in Italy the case of the crucifix) and questions concerning places of worship (which today are significantly if not exclusively about Islam) seem to be of interest even outside the Country.

68.2 Constitutional Framework

The Italian Constitution, in force since 1948, contains several provisions, directly or indirectly, concerning religion that, taken together, outline a system in which the religious factor is not confined to the private sphere as in other European Countries.

First, the Constitution recognizes and guarantees "the inviolable rights of the person, both as an individual and in the social groups where human personality is expressed" (article 2, Const.), in a context of equality before the law "without distinction of sex, race, language, religion, political opinion, personal and social conditions" (article 3, paragraph 1, Const.).[3]

More specifically, article 19 of the Constitution is dedicated to religious freedom, under which "anyone is entitled to freely profess their religious belief in any form, individually or with others, and to promote them and celebrate rites in public or in private, provided they are not offensive to public morality."

Here, the Constitution speaks only of religious freedom and not explicitly of "freedom of thought, conscience and religion," according to the denomination accepted in some international instruments of protection of fundamental rights (for example, article 18, ICCPR) or supranational (article 9, ECHR and, now, article 10, Charter of Fundamental Rights of the European Union) in which Italy participates.

Nevertheless, the Italian Constitution contains a specific provision for the right to freely express thoughts (articles 21), and article 19 Const. also protects freedom of atheism, as explicitly endorsed by the Constitutional Court's judgment, no. 117/1979, concerning the oath in criminal trials.[4]

In this perspective, we must always remember the central role in protecting the religious factor, for a long time ignored, of article 20 Const. under which "No special limitation or tax burden may be imposed on the establishment, legal capacity of any organization on the ground of its religious nature or its religious or confessional aims."

Conversely, as pertains to the relationship between the State and religious groups, fundamental Italian law has two separate provisions, the first of which is reserved exclusively for the Catholic Church. We refer to article 7 Const., which declares that "The State and the Catholic Church are independent and sovereign, each within its

[3] In this chapter, the English Presidency of the Republic website is used. Retrieved: January 18, 2013, from http://www.quirinale.it/qrnw/statico/costituzione/pdf/costituzione_inglese_01.pdf

[4] The Constitutional Court's pronouncements can be read at http://www.cortecostituzionale.it/actionPronuncia.do, where the database search form is available.

own sphere. Their relations are regulated by the Lateran Pacts. Amendments to such Pacts which are accepted by both parties shall not require the procedure of constitutional amendments."[5]

The provision includes the particular historical background relating to the well-known events concerning Italian unification (the 150th anniversary was celebrated in 2011), specifically the resolution of the so-called Roman Question as well as the continuing recognition of the opportunity to settle bilaterally the alleged *rex mistae* (that is the subjects that are of interest to the State and for the Catholic Church) also passed the totalitarian fascist era in which the same Pacts were undersigned.

The following article 8 Const., instead, is divided into three paragraphs. The first concerns the entities identified by the term "religious denominations" (*confessioni religiose*),[6] which is not recognized with formal equality before the law but only with equality in freedom. Instead, the two following paragraphs pertain to denominations other than Catholicism. The paragraphs consider "the right to self-organization according to their own statutes, provided that these statutes do not conflict with Italian law" and the possibility of regulating by law "their relations with the State," based on "agreements with their respective representatives."

Although it is not appropriate to mention the many constitutional rules relevant in different ways to the study of religion, we must remember that reforming Title V of the Constitution, provided by law no. 3/2001, has involved a significant transformation in the allocation of legislative and administrative powers among State and other local authorities (Regions, Provinces, Municipalities and Metropolitan Cities), with remarkable consequences for regulating the religious factor.

68.3 The Principle of Secularism as a Product of Constitutional Jurisprudence

As anticipated, from all these provisions the Constitutional Court has deduced secularism (*laicità*) is the supreme principle of legal order. This is according to a particular meaning, from the well-known judgment no. 203/1989 concerning the problem of compulsory religious education in public schools as governed by the agreement between the Holy See and the Italian Republic modifying the Lateran Concordat, signed on February 18, 1984, and enforced by law no. 121/1985.[7]

---

[5] The Lateran Pacts were signed between the Kingdom of Italy and the Holy See on February 11, 1929, and executed with law no. 810/1929. These include, in particular, the Treaty and the Concordat.

[6] Regarding the uncertain clarification of this expression, see the Constitutional Court's judgments no. 195/1993 and 346/2002. For the doctrine, see Colaianni (2000: 363).

[7] For this and the other Italian laws quoted, see www.normattiva.it which is the Italian legal portal.

We should start by specifying that the category of Supreme Principles has been developed to refer only to the relationship between the State and Catholic canon order. It allows for a constitutional review of the norms proceeding from the Concordat, otherwise precluded by virtue of the particular coverage provided by article 7 Const., which is the norm for legal sources (judgment no. 30/1971 as well as 16/1982 and 18/1982). This category deals with principles that cannot be subverted or changed in their essential content, even by law for constitutional amendments or other constitutional laws (judgment no. 1146/1988).

In accordance with judgment no. 203/1989 and the subsequent case law,[8] we summarize two principal readings. The first and more general identifies in articles 3 and 19, together with articles 7, 8 and 20 Const., the "values" contributing "to structur[ing] the supreme principle of State secularism, which is one of the profiles of the form of [the] State outlined by the Constitution of the Republic." The second, more specific and functional, concerns the unambiguous wording of the content, stating that the principle "entails non-indifference of the State for religions but guarantee of the State for the safeguard of religious freedom, in a regime of confessional and cultural pluralism." In other words, State actions have to be equidistant and impartial when referring to all religious denominations.

The constitutional judgment supports a positive viewpoint of secularism that meanwhile prohibits any appreciation of worth regarding the religious factor and the use of a sociological or numerical standard to legitimize situations of privilege (Casuscelli 1999: 440). But that reading does not forbid actions aimed at sustaining religion it focused on as fulfilling a worthy interest.

In any case, the same principle of secularism conveys tensions as revealed by looking at the initiatives, on the opposite side, that were submitted during the just-ended Legislative session (16th) and designed to amend the Constitution. Thus, together with the direct proposal to revise articles 7, 8, 19 and 20 Const. to strengthen the secularism of the Republic[9] there coexist bills that are designed to include a reference to the Christian heritage, even though in formal deference of secularism,[10] in this echoing the diatribes already disclosed during the drafting of the unratified Treaty establishing a Constitution for Europe in 2004 (Ivaldi 2008: 49). This reference is not unknown in the same Italian regional order (Ivaldi 2011: 328).

Nevertheless, among the issues that mostly seem to call this principle into question, we must emphasize, above all, those pertaining to the presence of religion in the public sphere. The presence, albeit with different accents, affects several other European States and questions the doctrine and the need to rethink the division between public and private space (Foblets 2012: 1).

---

[8] The Constitutional Court, subsequently, has intervened several times, evoking secularism. However, it is first and foremost about blasphemy and vilification, which have registered overriding references to secularism. For a global analysis, see Ivaldi (2004: 235).

[9] Constitutional bill 29 April 2009, C 241/2009. For these and the other projects of the parliamentary array listed below, refer to these websites http://www.camera.it and http://www.senato.it

[10] Constitutional bills S 320/2008, C 1483/2008, C 2374/2009 and C 2457/2009.

68.3.1 The Longstanding Question of the Presence of Religious Symbols in Public Space: The Case of the Crucifix

In Italy, this topic assumes a very particular feature because the specific meaning of secularism as upheld by Italian law is observed wholly different, for example, from that the French where it is expressly recognized at the constitutional level (article 1, Const. 1958), and which imports the radical observance of the principle of separation – already enshrined in the famous law 9 December 1905. It provides for significant outcomes even for religious freedom such as the existence of a sharp and severe distinction between the private sphere and the public sphere. In France, in this direction alone, major and recent legislative actions have led to restrictions on the use of religious symbols by individuals in certain fields.[11] If the ban on displaying religious symbols by students in schools, expressly Islamic headscarves, appears debatable, a somewhat different evaluation seems to motivate the most recent law on the *burqa*, which instead is based, *inter alia*, on shared protection choices for public safety.

Instead, in Italy symbols worn by believers do not appear to engender particular criticism, except the case of the so-called integral veil (that is, *burqa* and *niqab*) and similar other religious symbols, which is part of the ongoing need to avoid religious dress codes that prevent the identification of individuals.[12]

The core argument, in reverse, concerns the presence of religious symbols in public institutions, especially schools (as places for teaching and when used as polling stations) or in courthouses.

Limiting our analysis to public schools, we emphasize that the presence of the crucifix is provided by a bylaw dating back to the 1920s[13] that describes this symbol (together with benches, desk, etc.) among the furniture present in each classroom.

Confining ourselves to the circumstances that led Italy to present itself before the Court of Strasburg (that is, to the renowned Lautsi case), the case stemmed from a complaint brought before the Regional Administrative Tribunal (TAR Veneto) in 2002 by a woman of Finnish origin, on her own and on behalf of her two children. She observed that the crucifix displayed in public school classrooms constituted an incompatible infringement of her sons' freedom of belief as well as of their right to obtain an education in accordance with the family's religious and philosophical convictions.

The lawsuit crossed different levels of judgment, including meaningful suspension to allow examination by the Constitutional Court of the constitutional validity raised

---

[11] Namely, laws no. 2004–228 and 2010–1192 can be viewed at www.legifrance.gouv.fr

[12] Article 5, law no. 152/1975 "Disposizioni in materia di ordine pubblico," as modified by laws no 533/1977 and 155/2005.

[13] That is, articles 118, royal decree no. 965/1924 and 119, royal decree no. 1297/1928. These norms are examples, along with others, of the re-establishment of Catholicism as the State Church in the Italian fascist era that culminated with the signing of the Lateran Pacts in 1929.

against the dispositions providing the crucifix exposition. See articles 2, 3, 7, 8 and 19 Const., that is, arguing breach of the principle of secularism. The argument was declared inadmissible, as the proposal regarding the bylaw fall outside the Court's constitutional review which was limited to acts having the force of law (judgment no. 389/2004).

Consequently, the competent Tribunal, resuming the trial, rejected the complaint in judgment no. 1110/2005. This measure was confirmed by the Council of State (the Supreme Administrative Court) decision no. 556/2006.[14] These judgments are not fully persuasive and reveal a certain uncertainty in defining secularism. The vagueness was also caused by the enduring silence of the Legislature. In fact, both looked at the crucifix substantially (neither as furniture nor as an object of worship), but as a sign of the historical and cultural heritage of the Country, and of the identity of the Italian people. According to this measure the crucifix is a symbol suitable for expressing in the best way fundamental Italian values and, therefore, does not contrast, but even confirms the principle of secularism that characterized the Republican State.

The present case proves interesting because it exemplifies the multilevel protection of fundamental rights endorsed by Italy, that is, full participation in the system of the European Convention for the Protection of Human Rights and submission to the jurisdiction of the European Court in Strasbourg.

The Court resorted to by the applicant with the Chamber's judgment (Second Section) November 2, 2009 (app. no. 308147/069) declared unanimously that the right to education (article 2, protocol no. 1) was violated along with the freedom of thought, conscience and religion (article 9 ECHR), and ordered Italy to pay non-pecuniary damages.[15] Specifically, the Court considers that "the compulsory display of a symbol of a particular faith in the exercise of public authority in relation to specific situations subject to governmental supervision, particularly in classrooms, restricts the right of parents to educate their children in conformity with their convictions and the right of schoolchildren to believe or not believe." The Court continued, affirming that it "is of the opinion that the practice infringes those rights because the restrictions are incompatible with the State's duty to respect neutrality in the exercise of public authority, particularly in the field of education" (para. 57).

The opening of the Second Section, which aims to define the state entity in terms of neutrality and impartiality, corresponds to the judgment handed down March 18, 2011, which originated from the referral to the Grand Chamber brought under article 43 ECHR. It overturned the earlier decision. During the exercise of continuously balancing work between the effectiveness and uniqueness requirements of the Convention and differentiated constitutional identity and by invoking the doctrine of margin of appreciation, the better placement of national courts and the absence of

[14] The administrative judges' acts can be read at http://www.giustizia-amministrativa.it

[15] That judgment actually cannot lead to the conviction of positive behavior or, that is to say, to the removal of crucifixes from classrooms. All European Court judgments are downloadable from the search HUDOC case law at http://www.echr.coe.int

uniform standards, starting from the recognized subsidiarity of the conventional protection mechanism, are taken into account. The margin of appreciation embraces a rather broad notion, given the acknowledged lack of European consensus on the issue of the presence of religious symbols in the public space (para. 70). Involvement of different stakeholders and non-governmental organizations as well as various governments, mainly representing Orthodox Christianity, is relevant. The judgment, widely noted for the debate that arose, must be critically evaluated from several perspectives. In fact, in addition to marking a setback to the evolving trend that has marked conventional case law, the judgment shows the difficulty of achieving stable and shared solutions on issues, such as religious symbols, which seems to assume, in a different point of view, a growing importance in the European context.[16]

68.4 Overview of Enduring Questions in Achieving a Constitutional Plan

If the constitutional framework, as a whole, seems to guarantee suitable protection of religion and its public outward expression, individual or collective, delays persist in fully implementing the framework.

One issue concerns the status of religious denominations other than Catholicism. Indeed, two problems are linked. Not all groups access agreements that bilaterally regulate relationships with the State. Undeniably, by virtue of governmental discretion, several groups fail to access even the negotiating phase. For example, until now and despite attempts by some organizations,[17] no agreements with Islam were pledged. The Government has refused to proceed by taking refuge behind the impossibility of finding unified representation within the Muslim Community. Not considering the different treatment of various Christian denominations, which, however, the State has endorsed, separately, in several texts.[18] However, agreements already signed at the governmental level such as the one with Jehovah's Witnesses, the first in 2000 and another in 2007, have failed to see the process, which ends in a parliamentary law, completed.[19]

The legal context is complicated by the circumstance that denominations without agreement are still placed under outdated legislation, law no. 1159/1929

[16] This decision extensively annotated is also in English. For an overview of the Lautsi case, in light of the jurisprudential trend of this Court, see Ventura (2011: 293).

[17] That is, "Associazione Musulmani Italiani," "Unione delle Comunità ed Organizzazioni Islamiche in Italia" and "Comunità Islamica in Italia."

[18] For example, see the agreements with the Waldesians, Pentecostals, Adventists et al. For further details, please refer to the Italian government website's section, dedicated to religious denominations http://www.governo.it/presidenza/usri/confessioni/intese_indice.html#2

[19] Unlike two non-Christian denominations or those traditionally present in the Country (as was instead for Judaism whose accord goes up again to 1987), or Buddhism and Hinduism which saw their agreement concluded in 2007 and implemented in law before the end of the 16th Legislature.

and implementing royal decree no. 289/1930. But having been purged from the most macroscopic illegality by the Constitutional Court (judgment no. 59/1958) that suffers from an authoritarian approach. The use of non-constitution-oriented terminology, as it happens when using the term "cults allowed," is characteristic. Moreover, we must consider the impossibility to achieve the enactment of a general law on religious freedom that replaces these rules, despite the great number of bills drawn up by the Government or Parliament (Senate and Chamber of Deputies) since the beginning of the 1990s.[20]

These suggestions describe, at the same time, issues that are still disregarded as well as emphasize how the tensions in question concern only two denominations, Jehovah's Witnesses and Islam, which for different reasons are widely practiced within the Country. The absence of actionable agreements concluded with these cults is also reflected in practical problems and questions of liberty for which it is difficult to provide a satisfactory answer, as is clear, if we look at the issue of places of worship.

68.4.1 The Unresolved Issue of Muslim Places of Worship

The theme concerning places of worship is paradigmatic, by virtue of unavoidable links with freedom of religion (article 19 Const.), which is constitutionally oriented in light of the principle of equality (article 3 Const.).

First, we have to distinguish topics concerning the different legislative[21] and administrative[22] prerogatives of the competent territorial authorities (affecting, mostly, the identification of areas to be allocated for that purpose, the supply of contributions, the construction rules, etc.). These are closely related and concern the status of which single buildings can benefit,[23] which are all connected to the exercise of constitutionally recognized rights.[24]

This connection is well presented in the Constitutional Court, which in one of the first actions, judgment no. 59/1958, ruled that royal decree no. 289/1930 was partially unconstitutional due to infringement of article 19 Const., insofar as it provided the authorization requirement for opening temples and chapels, even if this was

[20] For close examination, see Tozzi et al. (2010).

[21] Under the cited reform of Title V, legislative powers in land use planning are allocated concurrently to the State and the Regions. Conversely, to the residual exclusive jurisdiction of Regions lies in the building matter. In any case, these prerogatives must be exercised "in compliance with the Constitution and with the constraints deriving from EU legislation and international obligations" (article 117, Const.).

[22] Administrative functions, at first, expressly conferred pursuant to article 118 Const., to municipalities and based on the principles of subsidiarity, differentiation and adequacy.

[23] It is a profile that falls outside this discussion, as well as those issues concerning the possible qualification of a place of worship as a cultural asset in compliance with the requirements.

[24] For doctrinal insights, see Persano (2008).

simply a means for an autonomous profession of religious faith. It also allowed for the right to hold religious ceremonies or carry out other acts of worship in these buildings, on the condition that the meeting is chaired or authorized by a minister, duly approved by the Interior Ministry.

The judgment deals with various themes including which one of the possible risks lies in the fact that the local level does not guarantee concrete compliance with the principle of equality of individuals and of equal freedom of different denominations, disregarding the principle of secularism (Floris 2010: 17) as sometimes happens in contiguous territories.

Incidentally, important rulings by constitutional judges such as judgments no. 195/1993 and 346/2002 have been handed down concerning the constitutional validity of certain regional norms of planning regulations concerning places of worship proposed by Jehovah's Witnesses, which have led to the unlawfulness of provisions excluding denominations without agreement from government grants for this purpose.

Recently, the Council of State, the Fourth Section, with decision no. 8298/2010, emphasized that is "is [the] duty of local authorities to ensure that are given to all religious denominations can freely exercise their activities, including identifying suitable areas to accommodate the faithful." The Court also affirmed that the right to the free exercise of worship cannot exempt individuals "from the observance of planning regulations that, in the essential content, aims explicitly to balance the different possible land uses," referring implicitly to the incorrect stratagem to change the intended use of a property held for other purposes, by various Muslim communities throughout the Country.[25]

As for the *jus condendum,* we should note the unsuitability, even unconstitutionality (Marchei 2010: 107), of the provisions referred to in various bills introduced during the 16th Legislature. We refer, for example, to C 552/2008, C 1246/2008 and S 1042/2008.[26]

Draft C 1246, using the non-constitution-oriented terminology "cults allowed," provides a submission to a referendum for people concerned about a local authorization measure concerning the construction, renovation or change of use of a building to be destined for worship (article 2, para. 1).

Conversely, a more favorable evaluation seems to be offered by bill C 2186/2009, which, based on the recognized freedom of religion and of the objectification of the constitutional principle of secularism, states that "measures of implement or detail of whichever authority [...] cannot produce anyway discriminatory effects, even indirect, to the detriment of a denomination or of her members" (article 2, para. 4).

Partly equivalent reflections deserve the norms aimed at this purpose, contained in different proposals for fulfilling the right of religious freedom (Mazzola 2010: 192).

[25] Council of State, Fourth Section and decision no. 4915/2010, more explicitly, had previously described the facts alleged as exclusively urban, not detecting "any profile relating to the freedom of worship, which may find other more suitable opportunities for expression."

[26] For the texts, see the websites http://www.camera.it and http://www.senato.it

In addition, satisfactory solutions are not offered by the "Committee for Italian Islam" found with advisory functions at the Ministry of the Interior on February 11, 2010, in place of the "Council for Italian Islam" established five years before, not least because in it are not represented large portions of the national Islamic panorama, essentially the most problematic, as those imputable to the Union of Islamic Communities and Organizations in Italy.

However, the same opinion entitled "Places of Islamic worship" released on January 27, 2011, does not appear to be a harbinger of developments fully in accordance with the constitutional dictates[27] especially insofar as is required of Muslims a "greater availability aimed at ensuring full transparency and willingness to effective integration in the context of settlement." Increased availability invoked by bill C 3242/2010 focused on establishing a national register of mosques and imams.

It remains to be seen whether the new legislature (the 17th) will record significant progress in this field, where the religious element in different contexts often assumes an identity-making value. The circumstances duly reported by the authoritative doctrine emphasize the frequency with which local administrators "appear inclined to forms of 'autarchy' in the religious sphere, in defense of the 'identity' of the communities who elect them," based on the incorrect premise to consider themselves "the only regulators of 'religious market' in their territories, detached from the rules of higher-level law (national, international and EU)" (Casuscelli 2009: 12).

68.5 Provisional Remarks: The Italian Model in Light of the Growing Multilevel Constitutionalism and the Judicial Competition

It is difficult to predict future developments in the topics considered such as religious symbols and places of worship and other outstanding issues related to religion. As shown in the introduction, the description would not be complete without additional lines about the relevance of the religious factor in the most complex supranational body in which Italy participates, that is, the EU in light of the new primary law in force since December 1, 2009 (Lisbon Treaty).[28]

Although, in compliance with the principle of subsidiarity, the *status* of churches and non-confessional organizations (article 17, para. 1 and 2 of the Treaty on the Functioning of European Union, hereafter TFUE) are beyond the EU competence (that is, they fall within the prerogative of national states). Thus it is equally impor-

[27] Opinion retrieved: January 19, 2011, from http://www1.interno.it/mininterno/export/sites/default/it/assets/files/20/0457_Luoghi_di_culto_islamici_-_Parere_del_Comitato_per_lxIslam_Italiano.pdf.

[28] Each EU document can be viewed at http://eur-lex.europa.eu

tant that in the "dialogue open, transparent and regular" under article 17, para. 3, the TFEU emphasizes the role of these entities in the European context.[29]

However, what seems to assume a greater importance, because it may interfere with national legislation, is undoubtedly the principle of non-discrimination (article 19 TFEU), pursuant to which European institutions can take appropriate action to combat discrimination based on religion. This is not a merely theoretical statement, devoid of practical consequences; rather it is based on this provision, the EU may issue directives that Member States are required to implement.[30]

However, the most interesting suggestion seems to be offered by the Court of Justice case law.[31] We refer mainly to the recent Grand Chamber judgment released on September 5, 2012,[32] in joined cases C-71/11 and C-99/11 Germany v Y and Z. In addition to constituting an important precedent, this judgment represents the first decision in which the Court of Justice ruled specifically, using as the interpretation parameters the EU norms concerning the right to freedom of religion, as laid down in article 10 of the European Charter of Fundamental Rights, which is at the same time narrowing the content. This judgment did not reserve anything more than a simple and straightforward formal warning to the Conventional system *sub species* article 9 ECHR. It avoided mentioning the articulated and long-standing Strasbourg case law.

Although detailed analysis of this case (Ivaldi 2012: 7) is deferred to more suitable places, it is nevertheless desirable to recall the statements made, pursuant to which, regarding freedom of religion in all faculties inherent in the external and public manifestation in which religious affiliations are included. The leading role in protecting the fundamental rights that the EU wants to ascribe to itself in the international arena is fully materialized.

In so doing, the Court of Justice scores a point in its favor in the "cultural competition" between supranational courts among them and the national courts (albeit within the competences ascribed by law), which should be based on authority and the ability to offer the best effective protection, based on the most suitable reference text (Ruggeri 2012: 21–23).

In substance, the picture as described above, that is, a national or supranational context, reveals a trend of the *judicialization of political-religious issues* well described in doctrine (Hirschl 2008: 191). It is an evolving phenomenon that in addition to reinforcing the centrality of case law, with an intra-systemic circulation of the models of guardianship, confirms the possibility of looking at the European supranational boundary as a particular example of judicial globalization.[33]

---

[29] For the EU point of view, see the activities of the Bureau of European Policy Advisers (BEPA) http://ec.europa.eu/bepa

[30] For example, see Council Directive 2000/78/EC, especially para. 23.

[31] Decisions by EU justice bodies before 17 June 1997 can be viewed on the European legal website mentioned above while those made after that date are available on http://www.curia.europa.eu

[32] The complaint originated from the denial of asylum and protection are governed by Directive 2004/83/EC was opposed by German authorities for two Pakistani citizens, despite the risk of persecution to which they were subject, in their home Country because of their membership in the Ahmadiyya movement for the reform of Islam.

[33] This expression was used, for the first time by the American doctrine (Slaughter 2000: 1103).

References

Casuscelli, G. (1999). Uguaglianza e fattore religioso. In *Digesto delle discipline pubblicistiche*, Vol. 15 (pp. 428–449). Torino: UTET.

Casuscelli, G. (2009). *Il diritto alla moschea, lo Statuto lombardo e le politiche comunali: le incognite del federalismo* (pp. 1–14). Retrieved January 19, 2012, from http://www.statoechiese.it/images/stories/2009.9/edit7m.09.pdf

Casuscelli, G. (Ed.). (2012). *Nozioni di diritto ecclesiastico* (4th ed.). Torino: Giappichelli.

Colaianni, N. (2000). Confessioni religiose. In *Enciclopedia del diritto*, Vol. 4 Agg. (pp. 363–380). Milano: Giuffrè.

Davie, G. (1994). *Religion in Britain since 1945. Believing without belonging*. Oxford: Blackwell.

Doe, N. (2011). *Law and religion in Europe. A comparative introduction*. Oxford: Oxford University Press.

Finocchiaro, F. (2012). *Diritto ecclesiastico* (4th ed.). Bologna: Zanichelli.

Floris, P. (2010). *Laicità e collaborazione a livello locale. Gli equilibri fra fonti centrali e periferiche nella disciplina del fenomeno religioso* (pp. 1–25). Retrieved January 19, 2012, from http://www.statoechiese.it/images/stories/2010.2/floris_laicit.pdf

Foblets, M. C. (2012). Religion and rethinking the public-private divide: Introduction. In S. Ferrari & S. Pastorelli (Eds.), *Religion in public spaces. A European perspective* (pp. 1–21). Farnham/Burlington: Ashgate.

Hervieu-Léger, D. (1993). *La Religion pour mémoire*. Paris: Cerf.

Hirschl, R. (2008). The Judicialization of politics. In K. Whittington, D. Kelemen, & G. A. Caldeira (Eds.), *Oxford handbook of law and politics* (pp. 119–141). Oxford: Oxford University Press.

Ivaldi, M. C. (2004). *La tutela penale in materia religiosa nella giurisprudenza*. Milano: Giuffrè.

Ivaldi, M. C. (2008). *Diritto e religione nell'Unione europea*. Roma: Edizioni Nuova Cultura.

Ivaldi, M. C. (2011). Sussidiarietà, diritto e fattore religioso. In M. Sirimarco & M. C. Ivaldi (Eds.), *Casa Borgo Stato. Intorno alla sussidiarietà* (pp. 261–339). Roma: Edizioni Nuova Cultura.

Ivaldi, M. C. (2012). Il fattore religioso nella giurisprudenza della Corte di giustizia. Prime note sulla sentenza 5 settembre 2012, cause riunite C-71/11 e C-99/11. In M. C. Ivaldi (Ed.), *Scritti di diritto ecclesiastico* (pp. 6–43). Roma: Edizioni Nuova Cultura.

Marchei, M. (2010). Gli edifici dei "culti ammessi": una proposta di legge coacervo di incostituzionalità. *Quaderni di diritto e politica ecclesiastica*, 1, (pp. 107–128).

Mazzola, R. (2010). La questione dei luoghi di culto alla luce delle proposte di legge in materia di libertà religiosa. Profili problematici. In V. Tozzi, G. Macrì, & M. Parisi (Eds.), *Proposta di riflessione per l'emanazione di una legge generale sulle libertà religiose* (pp. 192–208). Torino: Giappichelli.

Musselli, L., & Tozzi, V. (2007). *Manuale di diritto ecclesiastico. La disciplina giuridica del fenomeno religioso* (4th ed.). Roma: Laterza.

Persano, D. (Ed.). (2008). *Edifici di culto tra Stato e confessioni religiose*. Milano: Giuffrè.

Ruggeri, D. (2012). Prospettiva prescrittiva e prospettiva descrittiva nello studio dei rapporti tra Corte costituzionale e Corte EDU (Oscillazioni e aporie di una costruzione giurisprudenziale e modi del suo possibile rifacimento, al servizio dei diritti fondamentali) (pp. 1–24). Retrieved January 19, 2012, from http://www.associazionedeicostituzionalisti.it/sites/default/files/rivista/articoli/allegati/Ruggeri_6.pdf

Slaughter, A. M. (2000). Judicial globalization. *Virgin Journal of International Law*, 40, (pp. 1103–1124).

Tozzi, V., Macrì, G., & Parisi, M. (Eds.). (2010). *Proposta per l'emanazione di una legge generale sulle libertà religiose*. Torino: Giappichelli.

Ventura, M. (2011). Conclusioni. La virtù della giurisprudenza europea sui conflitti religiosi. In R. Mazzola (Ed.), *Diritto e religione in Europa. Rapporto sulla giurisprudenza della Corte europea dei diritti dell'uomo in materia di libertà religiosa* (pp. 293–362). Torino: Giappichelli.

Vitali, E. G., & Chizzoniti, A. G. (Eds.). (2012). *Diritto ecclesiastico. Manuale breve* (7th ed.). Milano: Giuffrè.

Printed by Books on Demand, Germany